Integrated Graphic and Computer Modelling

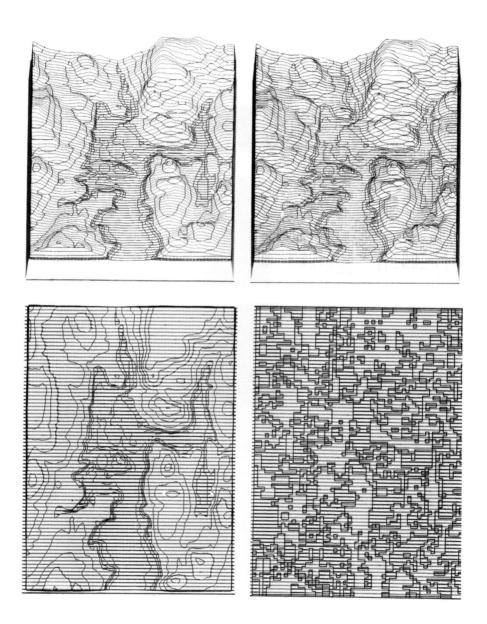

Frontispiece: *OBLIX: Honey hill dam location, planning studies: texture mapping and scan conversion Harvard University 1969*

Adrian Thomas

Integrated Graphic
and Computer Modelling

 Springer

Adrian Thomas, B.Arch, PhD, MCD
University of Sussex
UK

ISBN: 978-1-84800-178-7   e-ISBN: 978-1-84800-179-4
DOI: 10.1007/978-1-84800-179-4

British Library Cataloguing in Publication Data
A catalogue record for this book is available from the British Library

Library of Congress Control Number: 2008927486

Printed on acid-free paper

9 8 7 6 5 4 3 2 1

Springer Science+Business Media
springer.com

Dedicated to

Elizabeth, Jessica and Christopher

Preface

Background and Motivation

When setting out to write this book the initial target was to produce a relatively straightforward introduction to basic computer graphics algorithms. As work has progressed the goal posts have moved! This is partly because there were already many excellent books covering this aspect of the subject, but partly because there seemed to be other important approaches to the subject that needed to be addressed beyond merely how to automate the production of drawings and pictures.

The evolution of information processing technology and its deployment is still under way, and the role of computer graphics within it is being extended and modified year by year. Though it remains necessary to present the technical side of the subject in order to understand the constraints that limit its use on one hand and establish the potential it has on the other: it now seems important to present graphics as one among a series of modelling techniques that can be integrated together using computers, in a way that makes a far better use of the new technology than merely automating what was previously done.

The subject of computer graphics draws on results from a variety of disciplines: the properties of analogue and digital electronics, and even device physics that allow computing and display devices to work; the computer science topics that cover the computer languages, operating systems, data-structures and algorithms needed to program these systems; the perception studies that range from the experience of the artist to the scientific experiments employed in perception psychology that underpin the way interactive and animated displays can be designed: are all important but they only make up part of the subject. There are also all the application areas that create their own specialist demands and contributions to the subject: cartography, engineering and architectural design, scientific visualisation, medical imaging, robot control, as well as a full range of educational and entertainment applications. This spread of relevant topics by itself poses a presentation problem, let alone providing the background needed to make intelligent or creative use of material they offer.

The contribution made by these disciplines to the subject of computer graphics can be grouped under various headings. A useful classification scheme is that given in Figure 1. This presents three axes, the vertical one being concerned with representing information. This covers modelling types and their structure and properties: mathematics, both computing and natural languages, analogue systems and graphic and scale modelling systems. The applications of these representational schemes to the real world, based on experiments and empirical observation, essentially cover scientific research and its topics can be laid out along the B axis. Finally axis C represents activities that apply the results from A and B to the design and production of new goods and services.

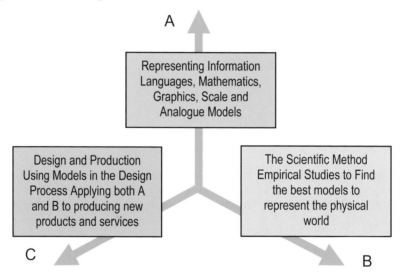

Figure 1 A modelling classification space

Clearly the various disciplines listed above occupy different places in this classification space. Although the treatment of the subject of this book would seem to lie along the A axis, the decision to present it from a programmer's perspective, moves the approach towards the C axis. Designing and producing programs to support graphic and other modelling systems will need to draw on results from both A and B, but will be a study of how algorithms can be implemented in a working system as much as a study of the mathematical properties that make them possible.

The approach adopted in this book is therefore to treat the subject as a design and modelling task. The design process can be presented in a variety of ways. Computer science is rediscovering and renaming processes well known in other older design disciplines. In particular a graphics-modelling scheme with an associated language called the Universal Modelling Language UML, is being developed to help design large software systems. The problem this addresses is the difficulty seeing structure in thousands of lines of complex programming code: more to the point managing the choices and changes that are integral to their design and development work. A simple schematic for the design process is given in Figure 2. An important new term for an old idea is that of "refactoring", which covers the tasks involved in design integration and optimisation

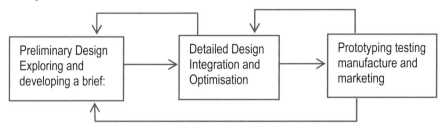

Figure 2 The design process

   The starting idea was to present the design and implementation of a basic two and three-dimensional graphic modelling system. This would allow related topics to be examined only in the depth needed to make design choices, but this still presented difficulties. If existing graphic libraries are examined it is clear that in order to provide a fully working system there is a lot of code which is there to provide error-checking, diagnostic messages, recovery routes for example if *"undo"* commands are provide – all very necessary in interactive environments but obscuring the structure of the key modelling operations. Consequently a compromise approach has evolved. The first two stages, in Figure 2, are explored for important display and modelling tasks but some of the work is only presented as "work-in-progress". This allows key issues that affect the design and programming choices to be highlighted and examined that might otherwise be overlooked, but it means that some of the code and examples are at different levels of development and in some cases all that is done is to show that links between the more detailed studies are possible.

Setting a Design Brief

The first step in any design study is to explore the context for the new product. This includes the environment that determines the service the new system or product is expected to provide, but also the technical constraints, arising from materials, expected costs, and the speed or size of available components, which may affect the performance that can be provided.
   Traditional graphics is divided into many camps, depending on the medium employed. Line drawing, engraving, watercolour painting, oil colours, fresco painting all require considerable skill to master and therefore tend to separate out as independent areas of expertise. There are further subdivisions depending on the preferred subject matter: high art, portraiture, cartooning and many more. Research is being carried out to allow graphic display devices to simulate the colour and texture effects produced by traditional media such as watercolour or oil colour paints. Not only does the use of computer graphics offer a unifying force to this area, but it also links in sculptural art forms and three-dimensional shape modelling. Underpinning computer graphics are various forms of spatial models based on the mathematical

representations of geometric-relationships. These modelling schemes also support numerically controlled machine tools; laser lithography and related techniques, and robot controlled manufacturing cells, and potentially provide a unifying link between two dimensional graphic models and fully three dimensional models. In order to cover the full range of possibilities for new systems it seems necessary to place the use of graphics within this larger evolving framework of compatible computer-based facilities.

The Context: Computer Based Information Processing Systems

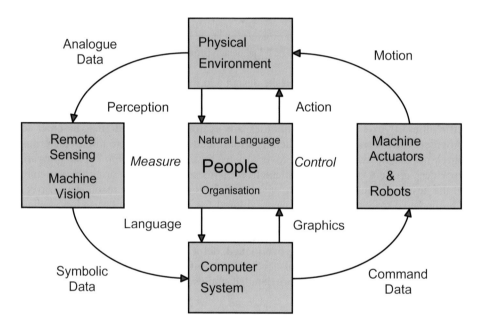

Figure 3 The place of graphics in future computer based systems

As a change in basic technology, automating information processing systems fundamentally affects the cost, speed and size of many processes and products. This in turn rearranges many existing patterns of use and in turn employment, removing some ways of earning a living but creating many new ways to replace those that are lost. To provide a context in which to understand and predict the consequences of these developments, it seems necessary to study the flows of information that support the essential activities in the system as well as the activities that are being modified by the changes. This involves identifying those tasks traditionally carried out manually that are now more effectively automated, those where human capabilities are still essential, and perhaps most important of all, in a way, the form of the communication interface between the two, needed to maintain an operational and efficient working system. The diagram in Figure 3 gives an abstract framework for this kind of information flow analysis.

Computer systems, which generate more work than they remove, for the same result, clearly are not worth developing. A variation on this observation was a criticism of the San Francisco Bay Area planning simulation model: that it would take more people to collect the data to feed the model than the model in action would service. This can be contrasted with many practical and successful remote-sensing systems where data capture is automated.

One explanation for the desire of governments to issue electronically read identity cards to people is that the automatic logging of transactions such as occurs in supermarkets provides the kind of human activity data that would allow economic planning simulation models to become more practical. The fear of Big Brother is clearly a sensitive political issue, here. A balanced flow of information, or the data that carries it, has to be maintained without bottlenecks if efficient but practical working computer-based systems are to evolve successfully.

Another framework that is necessary to consider when working with information processing applications is the level at which information is represented: the nature of the data. The input of image data in Figure 3 to the computer system is at a completely different level to the language data input by human beings. As they stand they are incompatible without human intervention. The development of machine vision algorithms is necessary if this divide is to be crossed automatically. Figure 4 outlines the layers in information processing systems that have evolved or are evolving as the technology expands and matures and diffuses into everyday usage.

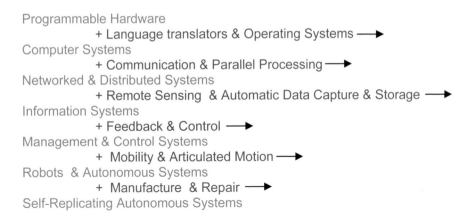

Figure 4 Computer Based Information System Hierarchy

Finally the third contextual framework, which needs to be explored, is the communication between people and machines within computer-based information systems. In early systems the mismatch between the speed of central processing units and data entry led to the creation of multi-tasking and then time-sharing systems. For human beings the eyes provide a very fast input for information. This allows dangerous situations to be recognised and avoided, it allows body language to be interpreted, and signalling using hands and arms to be responded to in real time.

More sophisticated physical forms of communication that require the use of the eyes: such as drawings, diagrams, maps and pictures, all take time to construct and are not generally the basis for real time interactive communication. In contrast hearing has supported the development of language forms of communication. Speech and conversation clearly support real time interaction.

Eyes: Vision Based ◄—— Symbolic Interface ——► Hearing Based: Ears

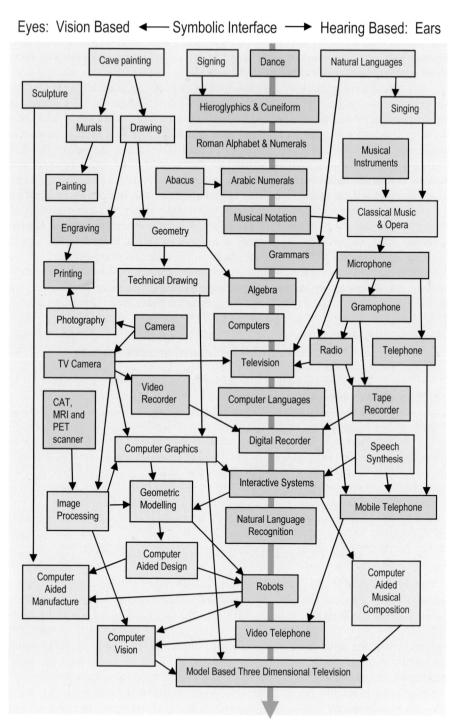

Figure 5 Evolving communications: Vision v Hearing: Graphics v Speech

Speech is ephemeral. People need to remember what has been said for its content to have any lifetime. In contrast graphics and text are physical artefacts that last for thousands of years, and as long as there are people who know how to interpret them, provide a long term memory or storage for the information they contain, particularly from generation to generation as the basis for education, maintaining cultural systems, and providing a form of memory for the whole community.

The evolution of the relationship between these two modes of communication based on hearing and seeing can be laid out against a rough time line in the way shown in Figure 5.

The ability to create animated images in real time from symbolic language inputs and the automated translation of spoken to text forms of language-input introduces a new "threshold of practical usage" for graphic modes of communication which is only now beginning to be explored. Similarly the development of machine vision systems that can create higher-level models of the environment and bridge to language levels of communication again offer a new "threshold of practical usage" for systems such as TV broadcasting, particularly as they become digitally based. The transfer of storage from graphics and text forms to computer memory systems itself offers new possibilities since this data is physical and can have a long life, but can be turned from passive storage forms into active usable forms without necessarily involving a human agent to interpret it.

These new options cannot be developed by considering computer graphics in isolation. All the related modelling schemes for representing the real world and any virtual worlds under discussion need to be brought together. This is again too large a brief for one book. Consequently the idea is to present graphics algorithms as the primary theme but explore it as a thread through a wider domain.

As an analogy consider prospective developments to be an archaeological site. It is possible to cut "trenches" across the site to get a broad view of what is hidden and it is also possible to dig "pits" more carefully examining material in greater detail where key finds appear. The main story line from chapter to chapter can be considered to be a linking "trench", while the separate studies in each chapter correspond to the more detailed studies in the "pits"! This reflects the hope that an understanding of the overall potential can be provided, but acknowledges it leaves "unexplored" regions that could still contain important material and even contain "finds" capable of changing the interpretation of the whole "site". This approach also allows the specialist information needed to interpret the material in the "pits" to be presented only in the outline needed to give the bigger picture or support the main theme, but will demand further reading from other sources if more detail is required!

Specific Objectives

As an introduction to this book it is necessary to identify who might be interested in its contents. The initial objective was to present *"the essentials of computer graphics"*, in other words to enable people to create graphical products using a computer. The problem with this statement is that there are many levels at which this task can be approached in a modern computer system.

If a computer system is set up with a graphics user interface already in place, then learning to use it can be considered to be very close to learning *"the essentials of graphics"*. This is because such a computer system can provide a working context similar to that in which artists, designers and draughtsmen have traditionally worked. Physically moving a "mouse" can be used to push a "pointer" around a display surface leaving marks of various kinds. Different "brushes" can be associated with the mouse, which allow areas to be painted, textured or shaded, as the artist or designer chooses. Similarly text and symbols can be created, selected and placed as desired anywhere on the display surface.

Consequently the *"essentials of computer graphics"*, has been taken to mean the task of making a computer system produce displays without an interactive display system being in place between the user and the computing system. In this exercise the initial targets of the presentation are constructing the components of a system, which can provide the display environment required by an interactive graphics user interface.

What is the essential ingredient of this approach? It is that all images or drawings have to be constructed by issuing computer language commands. For an artist this amounts to producing a painting or drawing "indirectly", issuing verbal instructions to a second party while blindfold. This restricts the process to those actions that can be defined in unambiguous language statements. A very difficult task if all the sophisticated internalised knowledge and experience of the artist is to be accessed. In fact it is virtually impossible which is why a "computer-aided" environment employing an interactive graphics-user-interface (GUI) must be provided for many application that need to use feedback from this kind of human expertise.

At the risk of oversimplifying the situation, there seem to be two communities of readers for a book outlining the essentials of a computer graphics system in this way. Those who already can produce graphics manually, or within a graphics painting or drawing environment, but have not either needed to, nor had the opportunity to learn to express their actions in an explicit computer language form. And those who can program computers and therefore are accustomed to working in this way, but who have not yet explored the use of a computer language to generate pictures.

The aim is to satisfy both these groups, because once this has been done then a next stage can be moved into: *"using an interactive graphics environment"* to build customised application systems: a subject of interest to both communities. This can then be extended to address the more general theme, which can be presented as *"integrated computer and graphic modelling"*. In this context graphics becomes a primary medium for communication between users and the computing system. The emphasis is providing support for content generation in specialist application work.

For example making a system to help in the post-production work on films, but not the post-production work itself. Where the interest becomes "content generation" for a particular application this becomes the subject matter for a different book.

In conclusion the technical foundation of this subject can be summarised as the use of algebra to represent geometrical relationships in order to use computer language statements to create pictures. The overall subject of graphics as a modelling medium opens up a wide range of topics that are too extensive to cover even in a pair of books. Consequently only the main ideas are presented that are required for the graphics and spatial modelling tasks that are examined in the text.

For some of the mathematical topics a full treatment will have to be found in specialist texts. Similarly a full treatment of the programming language Java will need to be found from other sources. Java code is provided wherever possible because in general-outline, an algorithm can appear simple, but the real programming difficulties lie in the special cases and the details of implementing the algorithm as a robust working program.

What is interesting is the range of mathematics that is needed to set up useful graphic and three dimensional models within the context of computer programming is small compared with the range of mathematical results that exist, that are potentially applicable to this area of work. The starting point could hardly be simpler: write a program to generate a list of properties such as colour values for each cell in an array of pixels and allow the display hardware to present this list as an image.

Selected Java programs to support the text will be added to the website at www.springer.com/978-1-84800-178-7.

Acknowledgments

The work in these two books is the result of an extended research effort that started in an undergraduate study in Liverpool University, School of Architecture, called Models in Design Procedures, supervised by Dr Morcos. The scope of this work was extended in the Department of Civic Design on being introduction to the use of computer based urban simulation models to predict the impact of a variety of planning actions, from road layouts to locating housing, shops and other essential amenities.

A Kennedy Memorial Scholarship to Harvard and MIT allowed this study of computer applications to architecture and planning work to be continued. The initial focus of the work under the guidance of Howard Fisher and Alan Schmidt was the application of computer cartography to urban and regional planning problems. Carl Steinitz, extended this experience in a course which involved constructing an Urban Data Base for the Metropolitan area of Boston, and then using a variety of modelling schemes some of them computer based, predicting the future growth and development of the region. The frontispiece showing studies using Honey Hill data for a dam location project shows studies of hidden-line removal, texture mapping and scan conversion completed in 1969 in the Laboratory for Computer Graphics and Spatial Analysis, in Harvard University.

Howard Slavin a fellow student convinced me that programming computers to carry out useful tasks could be done by correctly implementing the Sutherland line-clipping algorithm as a "joint" homework for the PDP 1 computer. Subsequently Thomas Waugh worked with me to produce several basic computer programs for graphic and spatial operations. Perhaps the most interesting being a pre-processor to a redistricting program for the Democratic Party, to avoid gerrymandering. The important developments were experiments using the OBLIX program to develop an early hidden-line removal algorithm for curved surfaces 1969, and with Tom Waugh the development of a full Geographic Information System called GIMMS (Geographic Information Management and Mapping System) 1970.

Work was continued in Edinburgh University on a Ph. D. subject titled "Spatial Models in Computer Based Information Systems". My supervisor for this work was Professor Coppock. Collaboration on the GIMMS system continued with Tom Waugh, to develop the Thiessen polygon and Delaunay tessellation algorithms for the mapping package and the network overlay program for sieve mapping. Work then progressed to three-dimensional modelling interacting with the Architectural Research unit. Initially this extended the networking approach used in the cartographic system to model object surfaces. This led to a hardware design to display volume models based on Boolean expressions, again related to the overlay labelling process developed for GIMMS. John Oldfield and John Gray provided tutorials on logic design that supported this work. John Downie by using a hybrid computer simulation helped the development of these ideas, in particular the perspective transformation of plane surfaces to the unit display cube, which made the subsequent digital designs possible.

I owe both John Downie and Ian Morrison in the Geography department in Edinburgh University a major vote of thanks for help and moral support when the work hit dark days!

Support for the work on the display processor was taken up by Professor Heath in Heriot Watt University and a Science Research Council grant was obtained to develop "Aspect01: A sequential parallel electronic colour terminal". Again tutorial instruction from Bernard Howard on advance logic design was important in taking this work to a successful conclusion. Research colleagues: Patrick O'Callaghan, provided an introduction to programming in C, and John Mclean provided much needed practical electronics advice that allowed the prototype systolic processor eventually to work! Jim Braid and his colleagues in Ferranti Edinburgh were also involved in developing a corresponding prototype hardware system with a view to developing an integrated circuit to provide a fast processor for real time avionic displays.

An application for a further research grant to develop a model based machine vision system to link to the real-time display system unfortunately did not get support. The concept of model based TV was regarded as fanciful in 1978!

The work was moved to Durham University in 1980, where a further research grant from the Science and Engineering Research Council in 1983 provided support to continue the exploration of the VLSI circuit design. Tector Ltd a small simulator company was able to apply the basic display algorithm to introduce a target aircraft into its pilot training simulator

In 1986 the project was moved to Sussex University where related work on simulators had been started. A research group to study "Model Based Animation and Machine Vision" was set up and carried out a series of studies related to the model based TV project. Acknowledgement for the work of Ph.D. students in this group go to A Cavusoglu, H. Sarnel, H. Sue, G. Jones, U. Cevic, N. Papadoupolos, D. Joyce, A. Lim, S. Zhang, B. Rey and C. Morris. In particular Hoylen Sue produced the texture-mapped duck in the title of the preface. An EPSRC, DTI Faraday project was carried out in collaboration with John Patterson from Glasgow University capturing 3D shape from image, equal brightness contours. The objective was identifying shape in the presence of specular reflection. Professor J Herschfeld from the Mathematics department provided a basic geometric construction that gave a way to factor a specular reflection field into two or three overlaid Lambertian reflection fields.

Finally thanks must go to Professor John Vince from Bournemouth University, who provided the initial push and then the continued positive support needed to get this book into production. The project has taken a long time as work has had to be carried out as a spare time activity. The tolerance of Springer's Editors waiting for the text also needs to be acknowledged with thanks.

Contents

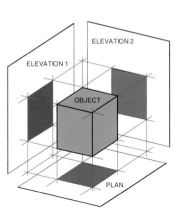

1

Models: Language, Graphic, Analogue, Scale, Mathematical & Computer Models

Introduction

Interactive graphics with computer languages have between them become the primary forms of communication between the computer system and its users for a large and growing number of application systems.

The developing emphasis on computer based modelling systems employing this new picture-making or image capturing technology, is reflected by the adoption of system design frameworks with the title of *information system*, rather than the older, more restrictive sub-system titles: computer graphics, image processing, machine vision, image databases and the adoption of the overall subject-area title of *informatics* to cover all the topics relevant to information capture, representation and processing.

This widening of the working context for system analysis and design, allows "digital convergence" to be explored and fully exploited: the unifying effect that adopting digital systems can have, and in fact needs to have to obtain the greatest advantage from using computers. It also gives a better starting point for developing the new or modified products that the changes in information technology are making possible.

Models

Modelling or representing and describing aspects of the real world, is the underlying theme of this book. It is also an essential supporting activity for most design work. At first sight models are distinct from description, though both are used to represent

A. Thomas, *Integrated Graphic and Computer Modelling*,
DOI: 10.1007/978-1-84800-179-4_1, © Springer-Verlag London Limited 2008

"the real world". The difference lies in the way that relationships in a physical model are constrained by the modelling medium. It is possible to describe in words many relationships that are impossible to realise within the real world. The reason for building physical models is usually to determine the consequence of interacting physical constraints. A very simple example would be to define a triangle with sides of length 2, 3 and 9 units, attempting to build it physically quickly shows it is not possible! Models can be single purpose in the sense that one resulting relationship is sought as a consequence of setting up initial model conditions, or multi-purpose where for example, the interaction of many relationships, which have to be co-ordinated in a design study, needs to be investigated.

There are several well-established classes of model: the simplest being the prototype. This is the object itself, but created as a trial example before say a mass production run, to ensure problems posed by a new product are reduced to a minimum. Prototyping is expensive for large objects, inappropriate for one-off objects such as buildings, and often not very useful for very small objects. A solution to cost and size for many such modelling tasks is to change the scale of models.

Scaled Models: Geometrical and Physical Models

Scale models are of two types: geometric models where spatial relationships are the primary issue, and physical models where not only scaled spatial relationships are important but also the physical properties of the materials used to build the models. The distinction between the two can be highlighted by the difficulty of scaling the density of water to estimate the behaviour of a half scale model of a boat. Similar problems accompany the use of wind tunnel models for variously sized test objects.

These difficulties are not insuperable, but their solution depends on empirically establishing the relationships that hold between the physical properties of the system, which is being modelled. What can be done is based on identifying "dimensionless products" of these properties in the full size system, and then ensuring the values of these products are maintained in the scaled model. In other words by multiplying together the measured values of properties which in combination give a dimensionless value, a pure number, rather than values such as velocity which would be measured in distance per unit time interval, it is possible to obtain usable results from scaled physical systems.

Geometrically scaled models have been used for design purposes since early on in history. If the historical record is to be believed, many of the large-scale engineering works of the ancient world were conceived and tested-out by using models of this type. Although closely related: the use of drawings to support the same design work was probably a later development. The rigorous study of the geometric rules needed to construct two-dimensional models, (as drawings in a sandpit), is generally attributed to Greek mathematicians, though other ancient civilisations: the Egyptians, Sumerians, Chinese, and the people who lived in the Indus valley appear to have had a working knowledge of many of them.

Drawing and Two Dimensional Scale Models

There is a series of basic geometrical relationships that are necessary to construct any line drawing or to program matching computer graphic algorithms. These basic relationships need to be reviewed, before the more complex operations, which depend on them, can be developed. The simplest geometrical relationships are those between straight lines and angles. Where two lines cross, opposite angles are equal, and angles on the same side of a straight line add up to 180°.

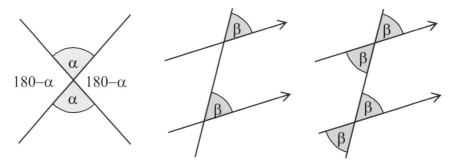

Figure 1.1 Parallel lines alternate and complementary angles

Lines that intersect a transverse line at the same angle are parallel. This gives the angular relationships shown in Figure 1.1. This allows the sum of the angles of a triangle to be calculated in the way shown in Figure 1.2.

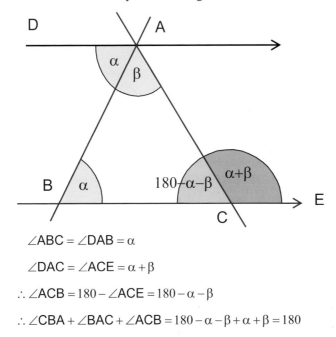

$$\angle ABC = \angle DAB = \alpha$$

$$\angle DAC = \angle ACE = \alpha + \beta$$

$$\therefore \angle ACB = 180 - \angle ACE = 180 - \alpha - \beta$$

$$\therefore \angle CBA + \angle BAC + \angle ACB = 180 - \alpha - \beta + \alpha + \beta = 180$$

Figure 1.2 Angles in a triangle

Congruent Triangles

Demonstrating the conditions that show two triangles have an exactly matching shape, in other words are congruent, is often used as a bridge to establish other relationships required in more complex geometric reasoning. There are several standard tests for congruence.

If two triangles have sides of the same length then they are congruent.

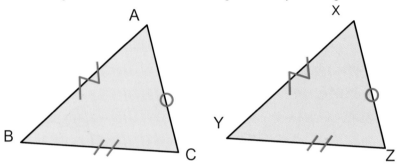

If two triangles have two sides of the same length and the same angle between them then the triangles are congruent.

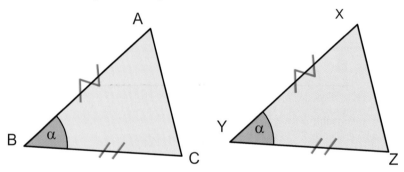

If two triangles have two angles the same and a corresponding side of the same length then the triangles are congruent.

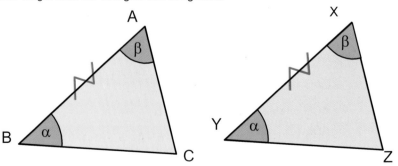

Figure 1.3 Congruent triangles

Pythagoras

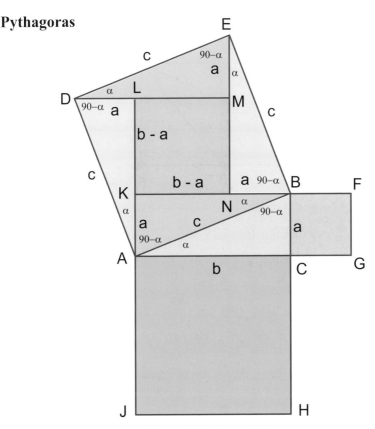

Figure 1.4 Pythagoras' theorem

The next relationship relating the sides of a triangle that is important is provided by Pythagoras' theorem for a right angle triangle. In Figure 1.4 the triangle ABC is a right angled triangle with sides of length BC = a, AC = b and AB = c. On each side of this triangle squares BADE, CBFG and ACHJ are constructed. The objective is to show that the area of BADE is equal to the sum of the areas of CBFG and ACHJ. If the sides of the squares BF and AJ are extended in the way shown, and lines DM and EN constructed parallel to these sides to generate the triangles ADL, DEM, EBN and BAK and the rectangle NKLM, then the following relationships hold. The triangles ADL, DEM, EBN and BAK are all congruent, they each have a side of length c and all their corresponding angles are equal. This makes the rectangle a square with sides a-b. The area of the square ABED is therefore the sum of the four triangles and this square NKLM:

$$ABDE = ADL + DEM + EBN + BAK + NKLM \ = \ c^2$$

$$c^2 \quad = \ 4 \times ADL + NKLM \ = \ 4.\left(\tfrac{1}{2}.a.b\right) + (a-b)^2 \ = \ 2.ab + a^2 + b^2 - 2.ab$$

$$c^2 \quad = \ a^2 + b^2$$

The area of a triangle can be calculated from the construction given in Figure 1.5.

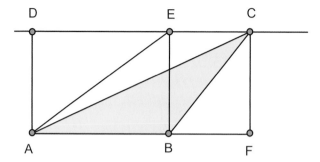

Figure 1.5 Area of a triangle: half base times perpendicular height

Area $\triangle AEB = 1/2$ Area $ADEB = 1/2 \times EB \times AB$ AED conguent to AEB

Area $\triangle ACF = 1/2$ Area $ADCF = 1/2 \times EB \times AF$ ACD conguent to ACF

Area $\triangle BCF = 1/2$ Area $BECF = 1/2 \times EB \times BF$ BCE conguent to BCF

Area $\triangle ACB = $ Area $\triangle ACF - $ Area $\triangle BCF = 1/2 \times EB \times AF - 1/2 \times EB \times BF$

Area $\triangle ACB = 1/2 \times EB \times (AF - BF) = 1/2 \times EB \times AB$

Once two triangles have been shown to be congruent then each pair of corresponding sides and angles are equal. However two triangles can have the same shape without necessarily being the same size.

The area of a triangle can be used to demonstrate an important relationship between the angles and the sides of a triangle in the following way.

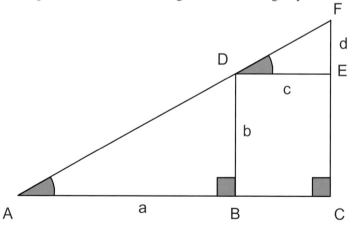

Figure 1.6 Angle ratios

The area of triangle AFC can be calculated in two ways:

Area $\triangle AFC = 1/2.(a+c).(b+d) = 1/2(a.b + a.d + b.c + d.c)$

Area $\triangle AFC =$ Area $\triangle ADB +$ Area $\triangle DFE +$ Area $\square$ DECB

Area $\triangle ADB +$ Area $\triangle DFE +$ Area $\square$ DECB $= 1/2.a.b + 1/2.c.d + b.c$

$\therefore$ $a.b + a.d + b.c + d.c = a.b + c.d + 2.b.c$

$\therefore$ $a.d = b.c$

$\therefore$ $\dfrac{b}{a} = \dfrac{d}{c}$ these two ratios relate to the equal angles $\angle$FDE and $\angle$DAB

also $\dfrac{c}{a} = \dfrac{d}{b}$ because $\triangle$ADB is the same shape as $\triangle$DFE

Triangles ADB and DFE are not congruent because they can be any size, but they have the same shape because their corresponding angles are the same. Such triangles are called *similar* triangles. In this case the two triangles are similar but they are also right angled triangles. This means the sides of the triangles AD and DF can be calculated using Pythagoras and their ratio can be calculated in the following way:

$\dfrac{AD}{DF} = \dfrac{\sqrt{a^2 + b^2}}{\sqrt{c^2 + d^2}}$ However $a = \dfrac{b.c}{d}$

$\therefore \dfrac{AD}{DF} = \dfrac{\sqrt{\dfrac{b^2.c^2}{d^2} + b^2}}{\sqrt{c^2 + d^2}} = \dfrac{\sqrt{\dfrac{b^2.c^2 + b^2.d^2}{d^2}}}{\sqrt{c^2 + d^2}} = \dfrac{\sqrt{\dfrac{b^2.(c^2 + d^2)}{d^2}}}{\sqrt{c^2 + d^2}}$

$\therefore \dfrac{AD}{DF} = \dfrac{b}{d} = \dfrac{\sqrt{a^2 + b^2}}{\sqrt{c^2 + d^2}}$

$\therefore \dfrac{b}{\sqrt{a^2 + b^2}} = \dfrac{d}{\sqrt{c^2 + d^2}}$ which again are ratios relating to equal angles

A similar substituti on of $d = \dfrac{b.c}{a}$ into $\dfrac{\sqrt{a^2 + b^2}}{\sqrt{c^2 + d^2}}$ gives the ratios :

$\dfrac{a}{\sqrt{a^2 + b^2}} = \dfrac{c}{\sqrt{c^2 + d^2}}$

The way these ratios of the sides of similar, right angled triangles give the same value for the same angle whatever the size of the triangle allows these ratios to be used as a measure of the angle.

The simplest condition to establish this relationship of *similarity* is that two triangles have identical corresponding angles. This leads to a more general set of relationships for all similar triangles between their sides and angles, that is easiest to demonstrate using the trigonometrical functions which are the ratios of the sides of right angled triangles shown in Figure 1.7.

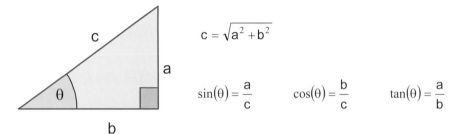

$$c = \sqrt{a^2 + b^2}$$

$$\sin(\theta) = \frac{a}{c} \qquad \cos(\theta) = \frac{b}{c} \qquad \tan(\theta) = \frac{a}{b}$$

Figure 1.7 Trigonometric functions

The trigonometric functions of double angles provide key relationships with many applications.

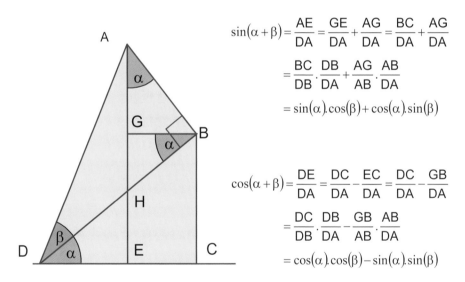

$$\sin(\alpha + \beta) = \frac{AE}{DA} = \frac{GE}{DA} + \frac{AG}{DA} = \frac{BC}{DA} + \frac{AG}{DA}$$

$$= \frac{BC}{DB} \cdot \frac{DB}{DA} + \frac{AG}{AB} \cdot \frac{AB}{DA}$$

$$= \sin(\alpha).\cos(\beta) + \cos(\alpha).\sin(\beta)$$

$$\cos(\alpha + \beta) = \frac{DE}{DA} = \frac{DC}{DA} - \frac{EC}{DA} = \frac{DC}{DA} - \frac{GB}{DA}$$

$$= \frac{DC}{DB} \cdot \frac{DB}{DA} - \frac{GB}{AB} \cdot \frac{AB}{DA}$$

$$= \cos(\alpha).\cos(\beta) - \sin(\alpha).\sin(\beta)$$

Figure 1.8 Double angle functions

The geometrical construction given in Figure 1.8 sets up a right angled triangle ABD. By drawing vertical lines through points A and B and a horizontal line through B, angles BDC and DBG both equal α because GB is parallel to DC. Angle GBA is 90-α which makes angle GAB also equal α. This construction allows the formulae for the sine and cosine of the sum of two angles α+β to be determined in the way shown in Figure 1.8.

In the triangle ABC in Figure 1.9 if a perpendicular is dropped from vertex A to the base BC at the point D, then the double angle formula allows the relationship between the length of the sides of the triangle and the angles at its vertices to be set up in the following way.

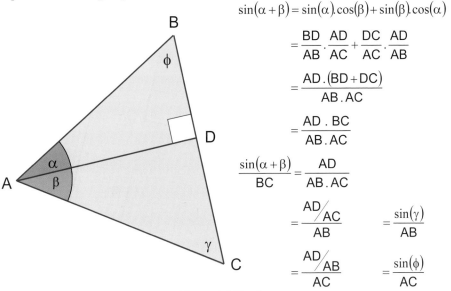

$$\sin(\alpha+\beta)=\sin(\alpha).\cos(\beta)+\sin(\beta).\cos(\alpha)$$

$$=\frac{BD}{AB}.\frac{AD}{AC}+\frac{DC}{AC}.\frac{AD}{AB}$$

$$=\frac{AD.(BD+DC)}{AB.AC}$$

$$=\frac{AD.BC}{AB.AC}$$

$$\frac{\sin(\alpha+\beta)}{BC}=\frac{AD}{AB.AC}$$

$$=\frac{AD/AC}{AB}\qquad=\frac{\sin(\gamma)}{AB}$$

$$=\frac{AD/AB}{AC}\qquad=\frac{\sin(\phi)}{AC}$$

Figure 1.9 Sine law

If two triangles are similar and therefore have a matching set of angles, the ratio between the length of the corresponding sides from the two triangles will be the same.

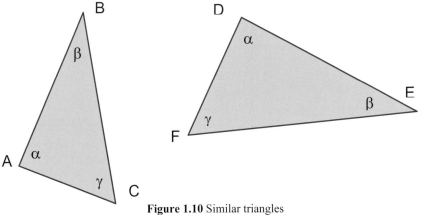

Figure 1.10 Similar triangles

$$\frac{\sin(\alpha)}{BC}=\frac{\sin(\beta)}{AC}=\frac{\sin(\gamma)}{AB} \qquad \frac{\sin(\alpha)}{FE}=\frac{\sin(\beta)}{FD}=\frac{\sin(\gamma)}{DE}$$

Hence for similar triangles : $\quad\dfrac{FE}{BC}=\dfrac{FD}{AC}=\dfrac{DE}{AB}$

The double angle formula can also be used to extend Pythagoras to cover triangles that are not right angle triangles in the following way:

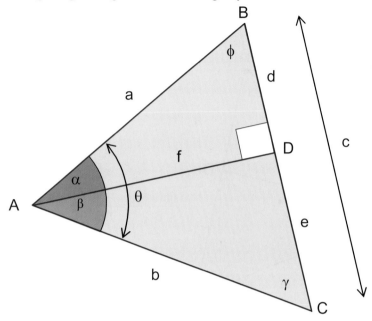

Figure 1.11 Cosine law

In triangle ABD $\qquad\qquad a^2 = f^2 + d^2$

In triangle ADC $\qquad\qquad b^2 = f^2 + e^2$

Adding these two results together

$$a^2 + b^2 = d^2 + e^2 + 2.f^2$$

$$c^2 = (d+e)^2 = d^2 + e^2 + 2.d.e$$

$$a^2 + b^2 = c^2 + 2.f^2 - 2.d.e = c^2 + 2.\left(f^2 - d.e\right)$$

However

$$\cos(\alpha + \beta) = \cos(\alpha).\cos(\beta) - \sin(\alpha).\sin(\beta) = \frac{f}{a}.\frac{f}{b} - \frac{d}{a}.\frac{e}{b} = \frac{1}{a.b}.\left(f^2 - d.e\right)$$

$$\therefore\quad a^2 + b^2 = c^2 + 2.a.b.\cos(\alpha + \beta)$$

$$\therefore\quad a^2 + b^2 = c^2 + 2.a.b.\cos(\theta)$$

Many of these triangle properties can be extended to polygons by partitioning the polygons into triangles in the following way:

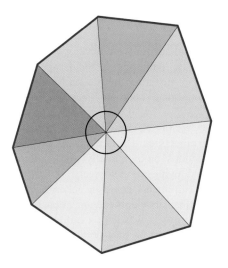

Figure 1.12 Polygon properties

Sum of a polygon's angles:

Add the angles of the triangles formed by linking each polygon vertex to a point inside the polygon then subtract the sum of the angles round this point:

$$\text{Sum} = n.180° - 360°$$

$$= n.\pi - 2.\pi \quad \text{radians}$$

Polygon area:

Add the areas of the set of triangles formed by partitioning the polygon into non-overlapping triangles.

Similar triangles can be used to give the same number of equally spaced subdivisions in lines of different length, using a ruler. If a triangle is formed by drawing a line at an angle through one end of the target line, in the way shown in Figure 1.13, then the number of equal length subdivisions can be marked off along the new line using a ruler. If the ends of these two lines are then linked to complete the triangle then a set of parallel lines constructed through the marks on the measuring line will intersect the original line to give the same number of equally spaced points.

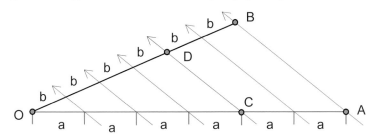

Figure 1.13 Geometrically scaling a measuring line

All the triangles formed by the lines parallel to the base of the triangle OAB are similar, as their corresponding angles are all the same. This generalises to give the relationships:

$$\frac{OD}{OB} = \frac{OC}{OA} = \frac{DC}{BA} \;\rightarrow\; \frac{OD+DB}{OD} = \frac{OC+CA}{OC} \;\rightarrow\; 1+\frac{DB}{OD} = 1+\frac{CA}{OC} \;\rightarrow\; \frac{DB}{CA} = \frac{OD}{OC}$$

Projections in three dimensions can be carried out in a similar way using parallel planes through lines and points, transferring relationships from one location to another in a variety of useful ways.

Three Dimensional Graphic Models

Geometric techniques for drawing three-dimensional scenes were developed in Italy during the Renaissance to give accurate realistic paintings of landscapes and buildings. However, it was a French mathematician, just after the French revolution, Gaspard Monge who first established the geometrical approach that led to the technical drawing techniques in use today.

Figure 1.14 illustrates a manual, graphic technique based on "descriptive geometry" which represents three-dimensional objects as a set of drawings using orthographic projections: in other words projections onto mutually perpendicular plane surfaces to give *plan*, *section* and *elevation* drawings of the objects.

The advantage of this approach was that measurements could be taken directly from the drawings. The French army used this development of "descriptive geometry" to layout building installations, and the way it was kept a military secret for quarter of a century is some indication of its effectiveness over previous methods. The import of these techniques to Britain, by Isimbard Brunell's father must have contributed greatly to his son's prodigious output of engineering projects during the 19th century.

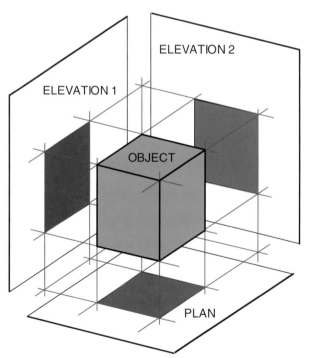

Figure 1.14 Descriptive-geometry: orthographic projections

If a point is recorded using perpendicular projections to three plane drawing surfaces that are mutually orthogonal, then the position of the point is uniquely defined, in the same way as giving it a Cartesian co-ordinate. This property and the use of parallel projection techniques gave a flexible and easy-to-use system for generating a model of an object in the form of cross-referenced drawings. A parallel projection of lines was generated by setting up a set of planes that were all perpendicular to a common plane (usually the new drawing) through the lines. A property of this system was that these planes through parallel lines were themselves parallel, and so would cut other transverse planes in new sets of parallel lines. This made it possible to manipulate the projected drawings of objects with very few geometric rules and to construct them for objects in any required orientation relative to the orthogonal drawing planes.

In Figure 1.15 the process of generating the projected drawings of a rotated cube, is illustrated. The first step is to rotate the plan view about the chosen axis. This will not affect the vertical heights of the cube's vertices, shown in the existing elevation-view: only their horizontal positions. These can be re-established, by projecting the new vertex positions from the cube's changed plan-view back to the elevation drawing, then matching and intersecting horizontal projection lines for corresponding vertices from the original elevation drawing. This will create a new elevation drawing for the rotated cube. This new representation can then itself be rotated about the other axis perpendicular to the elevation plane by a similar sequence of operations. The two operations together will allow any orientation of the object to be constructed. This process is simple to implement. Rotating or translating a copy of an object's projection, using tracing paper, into a new position, can be used to freely rearrange its orientation in any elevation plane. The corresponding projections in other elevation-drawings can then be matched-up using parallel lines. The tee-square and setsquare, compass and drawing-scale being the only tools, along with sheets of tracing paper, needed to put together complete descriptions of complex engineering shapes.

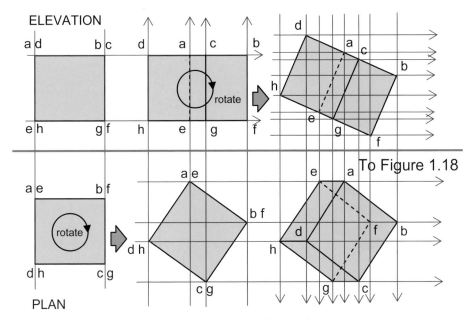

Figure 1.15 Rotation using orthographic projections

This use of drawings as a modelling process for engineering and design work has to be distinguished from the development of perspective-drawing which evolved during the early Renaissance in Italy, as a technical aspect of producing realism in paintings. It is possible, employing the rules of descriptive geometry, to generate perspective drawings from plan section and elevation drawings using the constructions shown in Figure 1.16. The perspective projection results from intersecting the rays of light from points in a scene that enter the eye of an observer, with a transparent viewing screen. Rays from this "*perspective*" projection onto this screen will consequently be indistinguishable to the eye, from those coming from the original scene.

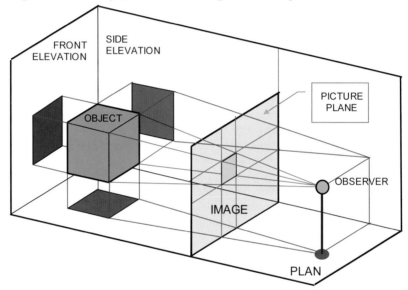

Figure 1.16 Perspective projection from orthographic projections

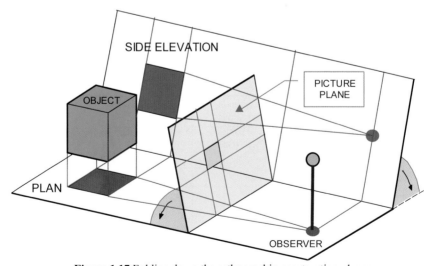

Figure 1.17 Folding down the orthographic construction planes

It is possible to use perpendicular projections of these viewing rays onto the orthogonal planes of appropriately chosen plan, sections and elevation drawings, to construct the position at which these viewing rays will pass through the picture plane or viewing screen, in the way shown in Figure 1.16. Figure 1.17 shows the way in which the orthographic projections of the elevations can be folded down, along with the picture plane, onto the plan to give one drawing. By carefully placing these drawings as overlays on one sheet of paper it is possible to move from one to the other, transferring scaled size-information in a procedure which allows the perspective drawing of the object to be constructed in a relatively easy way.

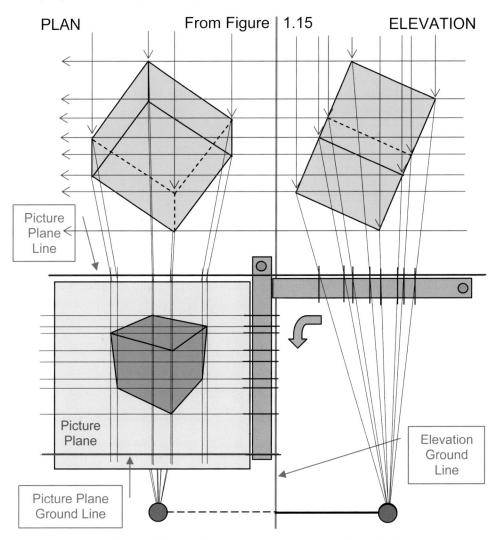

Figure 1.18 Perspective projection from orthographic projections

In Figure 1.18 the picture plane has been folded down and slid towards the observer. The ground line indicates the position it would have cut the plan, marked by the "picture-plane" line on both plan and elevation drawings. This arrangement means the projections of viewing rays onto the plan can be continued downwards by parallel lines on the picture plane drawing, from the points where they intersect the "picture-plane" line on the plan drawing. The equivalent intersections with the picture plane for the projections of these rays on the side elevation (folded sideways) have to be transferred by a ruler (in magenta) to the side of the picture plane drawing, measured up from the ground line, or by using the construction shown in Figure 1.19.

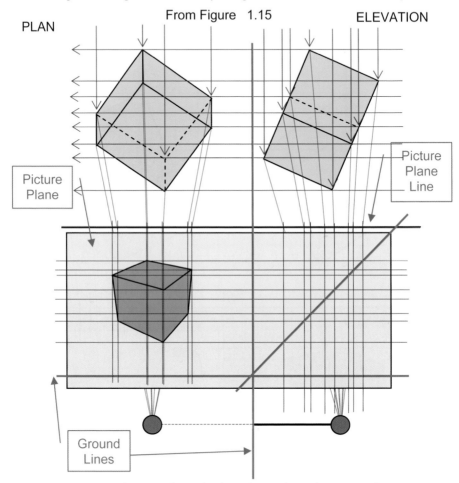

Figure 1.19 Perspective projection using an alternative construction

This approach gives a set of simple rules, which are easy to use. The only draw back to this approach, in operation, is that very high accuracy is needed in projecting points from one drawing to another, and errors can be cumulative. For this reason, and to allow perspectives to be set up directly from measurement information, an alternative approach to their construction has evolved. This involves the use of vanishing points and measuring points. The image of converging railway lines meeting at a point on the horizon is the usual example of the geometric property, on which this approach depends. It can be illustrated and explained using the various relationships, which can exist between three planes in three-dimensional space, shown in Figure 1.20.

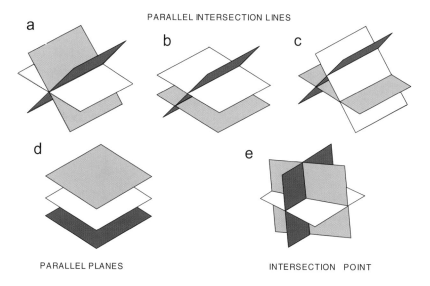

Figure 1.20 Interactions of three planes

Three planes can form a "pencil" of planes shown in 1.20a, where all the planes intersect in a common line. If two planes are parallel then the third will intersect them in two parallel lines, as shown in Figure 1.20b. Alternatively where a plane intersects two other planes which are not parallel in two parallel lines, the two original planes will themselves intersect in a third line parallel to the first two lines, as shown in Figure 1.20c. In contrast three planes which are parallel will never intersect shown in Figure 1.20d, and planes which intersect in non-parallel lines will intersect in a single point, as shown in 1.20e.

Where the rays of light from the end points of two lines which are parallel to the picture plane are projected to the eye, they define two planes as shown in Figure 1.21. The relationship between these planes and the picture plane and their intersection lines corresponds to Figure 1.20c. This means that the lines projected on the picture plane are themselves parallel. If the original pair of parallel lines is now rotated about the observer's position the result will be that shown in Figure 1.22. The relationship between the planes, generated by the rays from the end points of these parallel lines to the eye, and the picture plane will be that shown in Figure 1.20e. In other words the three planes will intersect in a single point. Since the images of the original lines will fall on the intersection lines between their projection planes and the picture plane, it can be seen that the image lines will converge and if they were extended they would like the railway lines in Figure 1.23 meet at a single point, their vanishing point.

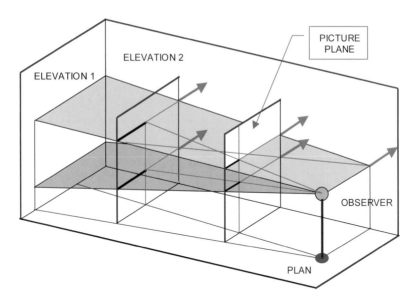

Figure 1.21 Lines parallel to the picture plane project as parallel lines

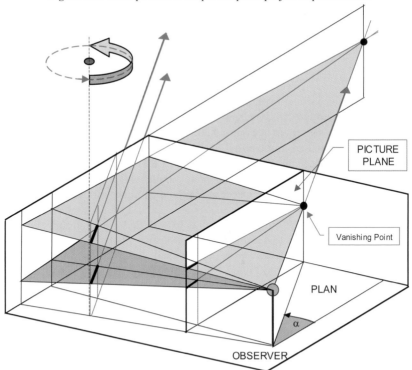

Figure 1.22 Projection of a rotated set of parallel lines to give vanishing points

This construction allows vanishing points to be set up for each set of parallel lines in a scene. If a construction technique based on drawing boxes is employed then again a relatively easy-to-follow drawing procedure evolves. The vanishing points can be constructed by projecting the three lines through the eye, parallel to the box's edges, onto the plan or ground plane, and onto the elevation planes, and then onto the picture plane as before. All subsequent lines parallel to the box's sides can then be set up to pass through one of the three related vanishing points.

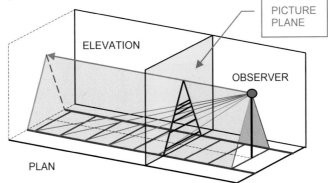

Figure 1.23 Railway lines perpendicular to the picture plane

Vanishing points make it easy, centred on the eye, to accurately project lines in a scene onto the picture plane. And where the projection of an edge onto the picture plane surface is constructed in this way, it allows its scaled size to be measured onto the drawing as though onto the picture plane directly. This mechanism provides a neat way of constructing correctly dimensioned elements and relationships in a perspective picture-space, in the way illustrated in Figure 1.24.

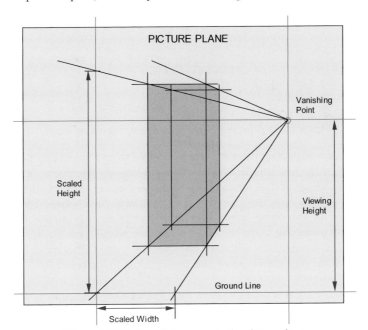

Figure 1.24 Scaling distances onto the picture plane.

An extension of the same idea is shown in Figure 1.25 where by setting up a vanishing-point for the diagonals of a square checkerboard, aligned with the picture plane, depth measurements can be scaled onto the picture plane and then projected correctly back "into" the drawing. In this example these direct measurements are made along the ground line. The use of square tiles in Figure 1.25 to transfer distances along the ground line to distances in depth along lines perpendicular to the display screen is a special case. Vanishing points for measurement lines can be set up for any set of parallel lines, which converge to the same vanishing point in the way shown in Figure 1.27.

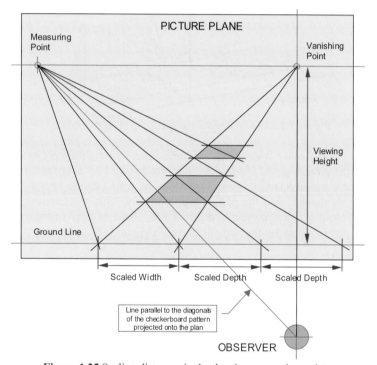

Figure 1.25 Scaling distances in depth using measuring points

The secondary vanishing point is usually referred to as a measuring point. In the case of lines in planes parallel to the ground plane the technique is to project distances from the ground line onto oblique lines in the way shown in the plan view in Figure 1.26 using isosceles triangles. The vanishing point for the base of these triangles gives the measuring point for oblique lines in the way shown. This can be constructed by projecting a line through the viewing position parallel to the base of these isosceles triangles to intersect the picture plane. This position can be constructed on the plan drawing and then projected up to the horizontal line through the vanishing point.

The measuring point is the same distance from the vanishing point as the viewing position, and traditionally has been constructed using a compass centred on the vanishing point: striking two arcs of the same radius, the first through the viewing position and the second through the measuring point on the plan view.

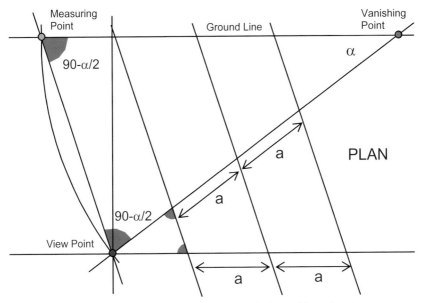

Figure 1.26 Scaling distances in depth along oblique lines

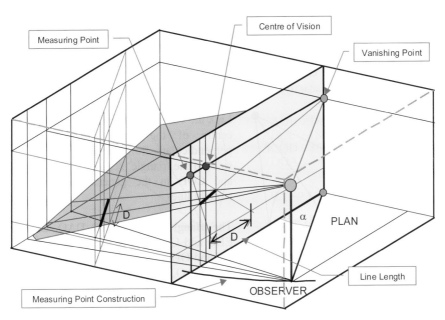

Figure 1.27 Scaling line lengths using measuring points

The compass is an important tool for the draughtsman: it allows equal length lines to be marked off along straight lines, and to be transferred to different places in a drawing. It also allows circles to be drawn. The geometric relationships between

for film and cartoon work. This is usually done, by setting up a grid framework within which, more complex details can be drawn in, accurately enough to provide a working model for many applications. In essence this provides a three dimensional measuring grid or framework projected onto the picture plane.

Where this framework is made up from lines parallel or perpendicular to the display plane the result is one point perspective (a single vanishing point and a single measuring point). Where the framework is rotated about a vertical axis through the viewing position in the way shown in Figure 1.22, the result is two-point perspective (two vanishing points and two measuring points). A useful relationship for setting up this scheme for sketching is shown in Figure 1.32. The two vanishing points and the viewing point on the plan view must lie on a semi-circle. This arrangement ensures the direction of the two line sets are at right angles to each other: the angle subtended by the diameter of a circle at any point on its circumference being 90°.

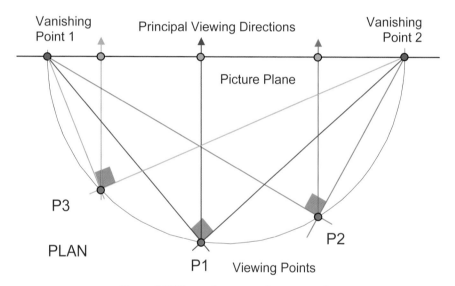

Figure 1.32 Two point perspective construction

Circles and spheres are also useful for visualising and setting up the relationships between vanishing points and viewing points for a three dimensional grid when the framework is rotated about a horizontal axis through the viewing point. The same approach is applicable: lines parallel to the grid lines can be drawn through the viewing position, and where they hit the picture plane defines their vanishing points. This gives three-point perspective (three vanishing points and three measuring points). The lines linking the vanishing points to the viewing position form a tetrahedron with its triangular base on the picture plane. Each of the other three faces of the tetrahedron still have to obey the relationship shown in Figure 1.32 as these faces are right angled triangles. The circumcircles for each of these triangles constructed in the planes of the faces all lie in the surface of the same sphere in the way illustrated in Figure 1.33. If the faces of the tetrahedron are reflected across their base edge to create a complete rectangular box in the way shown in Figure 1.33, then symmetry shows that a single sphere passes through the viewing position and the three vanishing points.

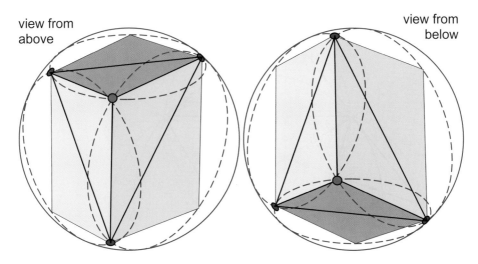

Figure 1.33 Three point perspective: three vanishing points and a viewing point

The circle round the triangle faces will be the circular edges of slices cut through this sphere. This property makes it clear that once the vanishing points have been set up on a picture plane the position of the principal viewing position (the perpendicular projection of the viewing position on the picture plane), is also fixed by the relationships shown in Figure 1.34.

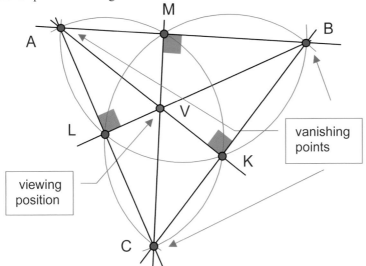

Figure 1.34 Three-point perspective: defining the viewing position

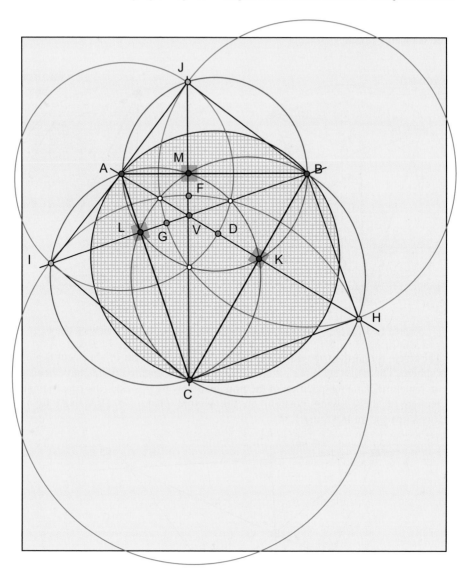

Figure 1.35 Constructing vanishing points and measuring points

Figure 1.35 shows the construction lines in the picture plane used to establish the relationships between the viewing-position the vanishing points and the measuring points. A, B, C are the vanishing points and D, G, F are their corresponding measuring points. The lengths AD, BG and CF are the same lengths as the corresponding sides of the viewing tetrahedron AI, BJ and CH respectively. This gives the isosceles triangles needed to define the measuring points. This construction also shows that the corresponding sides of the triangles CHB, AJB, AIC, the faces of the viewing tetrahedron, match, constructed (in the way shown in Figure 1.32) using the blue circles which have the sides of the main triangle ABC as diameters.

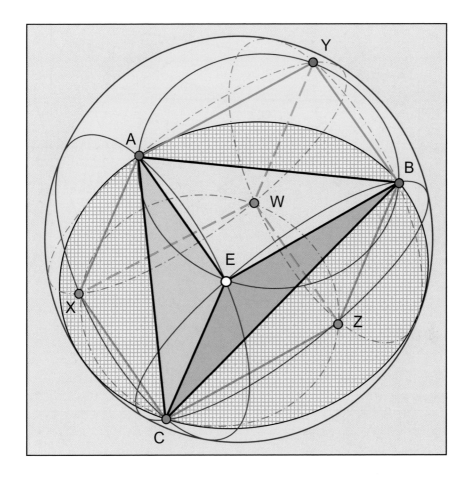

Figure 1.36 Constructing the viewing pyramid

This match can be seen in Figure 1.35, looking at the circles drawn with red lines. Figure 1.36 shows how these triangles can be folded upwards to meet giving the right-angled viewing pyramid with its apex at the eye E. Figure 1.36 shows the viewing pyramid embedded in a sphere. The Circumcircle of triangle ABC shown crosshatched in both Figures 1.35 and 1.36 is the intersection of the picture plane with this sphere. The facets of the viewing pyramid lie in the planes cutting this sphere shown as blue circles in Figure 1.36. These circles correspond to the blue circles in Figure 1.35, which have the edges of the triangle ABC as diameters. The eight vertices of the orthogonal box AYBECXWZ lie on the surface of the sphere.

These geometric constructions used in combination allow shadows to be constructed, as well as reflections and other illumination effects. Drawing rectangular objects is easy, curved objects are much more complex but can be done by using a sequence of profile sections. Systematic ray tracing and projection can even construct reflections in glass spheres, though manually it is a tedious and time-consuming task!

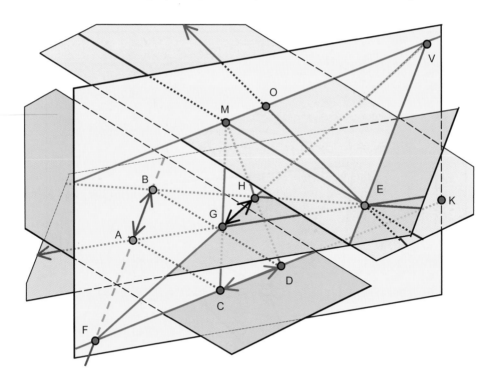

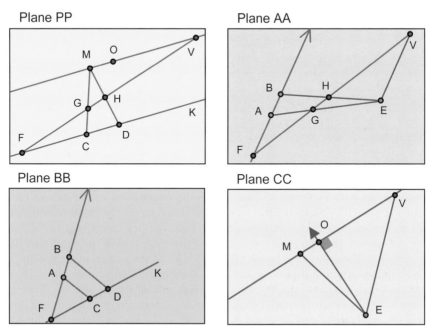

Figure 1.37 Constructing measuring points for oblique lines

In the construction for three-point perspective given in Figures 1.35 and 1.36 the measuring points lie on the lines joining the related vanishing point and the principal viewing point -- the perpendicular projection of the position of the eye onto the display surface of the picture plane.

Just as in Figure 1.27 showing the construction for a two-point perspective, measurements cannot be made along these lines. Figure 1.37 gives the relationships that illustrate how the measuring points can be used in the general case. The picture plane is shown as PP. The eye's position is labelled E and the principal viewing point O. The plane CC is the blue plane through O, the vanishing point V and E. The vanishing point V is where the projection of a line through E parallel to the line AB in the scene space, hits the picture plane. If the line AB is projected until it hits the picture-plane at F it gives a new plane labelled AA, coloured pink, containing V, F and AB. Finally a fourth plane BB, coloured green, can be set up through the line FK parallel to OV, and the line BAF. M is placed on line OV so that MV is the same length as VE, to give the necessary isosceles triangle MVE to transfer measurements into the scene space. The similar triangles in these planes establish the following relationships.

In the picture plane PP : MV is parallel to FK

therefore $\dfrac{MV}{FC} = \dfrac{VG}{FG}$ Similar triangles MVG and FGC

and $\dfrac{MV}{FD} = \dfrac{VH}{FH}$ Similar triangles MVH and FHD

in plane AA : AB is parallel to EV

therefore $\dfrac{VG}{FG} = \dfrac{VE}{FA}$ Similar triangles VEG and FGA

and $\dfrac{VH}{FH} = \dfrac{VE}{FB}$ Similar triangles VEH and FHB

but $VE = MV$ by construction $\triangle MVE$ isosceles

therefore $FC = FA$ and $FD = FB$

therefore $\triangle FAC$ and $\triangle FBD$ are isosceles

therefore $AB = CD$

Consequently measurements made along line FK projected back onto the measuring point M will give the correctly scaled length for the perspective projection of the line AB as GH on the picture plane. In Figure 1.38, measuring points and measuring lines are used to draw a cube using the construction given in Figure 1.35. An edge of the cube through the vertex S is projected to hit the picture plane at T, and the intersections of planes parallel to the cube's facets are constructed as lines through T parallel to the lines KA, LB and MC to give the measuring lines in the picture plane.

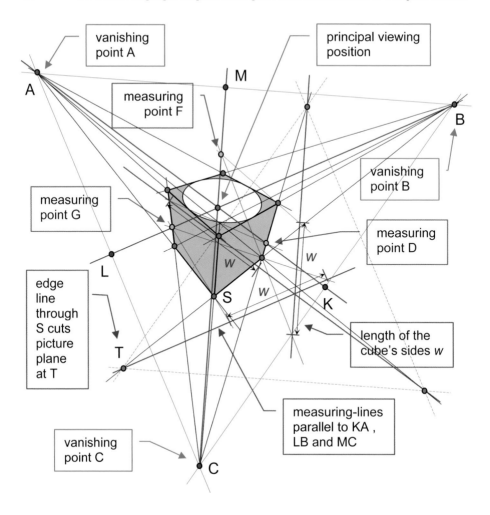

vanishing
point A

principal viewing
position

M

measuring
point F

vanishing
point B

measuring
point G

measuring
point D

edge
line
through
S cuts
picture
plane
at T

length of the
cube's sides w

measuring-lines
parallel to KA ,
LB and MC

vanishing
point C

Figure 1.38 Constructing a cube using three point perspective

Though the drawing in Figure 1.38 looks complicated, in practice, construction lines are drawn in and erased as they are needed giving a clear and simple process that allows perspective drawings of complex scenes to be constructed relatively easily.

Analogue Models

The analogue model is one where one physical system which maintains the same relationships among a set of properties as those among a target-set of properties in a second physical system, is used as a surrogate to investigate the behaviour of the target system. As a physical system once it is set, it has the advantage of maintaining complex multiple relationships in parallel in real time. For this reason the analogue and physical model can often provide single step parallel modelling operations, rather than the sequential or incremental model building provided, for example, by a language model where the required relationships can only be presented one at a time. Setting up the physical model essentially establishes all the required relationships at the same time. A simple example is sorting a set of values. If a rod with an appropriately scaled length models each value, lining up all the ends of a bundle of rods on a flat surface allows the largest to be located in one step. If the largest rod is removed in sequence the result is an ordered list of values. The sorting operation to find the largest is a single step operation once the system has been set up. The consequences of changing one relationship on the others in the system, is found in one step. This distinction is important when considering the behaviour of computer and mathematical models. The distinction between this one-shot evaluation using a physical model and the sequential step-by-step evaluation of an operation such as finding the largest in a set of numbers, using a computer program, is discussed in the next chapter on basic Java programming. In a slightly different but related way quantum-computing promises to establish all the required relationships between input values as a single operation. The difficult task in both cases is getting the resulting information from the models.

Geometric Graphical Models

A geometric diagram such as a graph can be thought of as an analogue model. Drawing out the graph of two functions allows their common values to be determined by the positions where the two function lines cross.

Language Models

The fastest rough and ready "model" to generate, is probably a verbal description of an object. However, to define an object accurately it is necessary to specify all the relationships that are necessary to determine its structure in a non-ambiguous manner. One way in which this can be done is to describe the building of a prototype, scale or graphic model of the object, in steps that will lead to its correct construction.

There are many natural languages and they can be used to express the same information in completely different ways. Where they are represented in text format the alphabets may be different. Even when they are not they will be made up from different sequences of characters. These different representations have many structural similarities. It is possible within the rules of each language to establish the strings of characters that form correct words and which in turn obey the grammar rules that allow a person who knows the language to understand what language statements mean.

Mathematical Models

A mathematical model can be regarded an extension of this kind of language model, but it is based on more exacting rules than those that govern natural language descriptions, to ensure that its statements remain correct and reliable. Basic relationships between entities in the model are defined, and from these only correct deductions, or consequential relationships, are then permitted.

Algebraic models represent relationships symbolically, where the symbols can be read with meaning, but they can also have their symbolic structure rearranged following formal rules. The resulting symbolic rearrangement can still be meaningfully read to represent a consequentially true relationship derived from the first symbolic arrangement. It is this ability to rearrange the structure of a symbolic model and still maintain a true statement that allows computer models to be constructed.

Computer Models

Computer models are constructed from algebraic symbolic data that can be manipulated formally to give new results, where the manipulation is handled by automated data processing machinery, which allows the process to be handled fast enough to execute usefully complex operations.

Computer languages allow computer models to be built and used for many information-processing tasks. They can be regarded as a way to extend and apply algebraic symbolic representation, and also as simplified versions of natural languages. They consist of sequences of characters grouped together to form words, and they also have to conform to sets of grammatical rules. Like natural languages there are many computer languages: all capable-of-expressing the same information in different ways.

It is possible to classify programming language facilities into different types. The commonest is probably the imperative language statement essentially a command expressing operations on data objects. Then there is the functional programming language approach, where statements have a more mathematical flavour, based on nested function calls. Logic languages employ statements defining true and false relationships, while concurrent languages allow parallel processing schemes to be defined so unconnected operations can be carried out independently of each other, even where they are being executed in a machine system that can only carry out one operation at once.

To a greater or lesser extent, most common computer languages, support each of these ways of expressing information processing tasks, more or less efficiently depending on the application. In this book the computer modelling techniques are illustrated using Java. This language provides facilities using each of these language constructs. It is also an object-oriented language built round the unifying idea that everything can be considered as an object or an operation on an object. This provides a convenient starting point for computer-based model building, where the object becomes any entity represented by a language, mathematical or computer model.

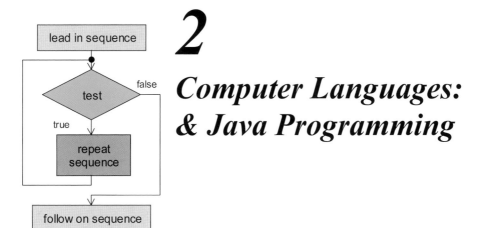

2

Computer Languages: & Java Programming

Introduction

It is assumed that the reader has a working knowledge of some computer language. However, key aspects of programming are summarizes in this chapter using the Java language to support the illustrations developed later in the text that demonstrate how graphic algorithms can be implemented in a practical way.

Libraries accessed through high level programming languages provide most graphics facilities now in common use. The demonstrations and illustrations in this section are all prepared using Java's standard window display libraries. Some of the topics, which will be explored in later chapters, will be presented using specially constructed high-level graphic and geometric-modelling language facilities that are extensions not supported by standard high-level languages. In order to outline how these extensions might be implemented it is necessary to provide an introduction to computer language processing, and it is convenient, to overlap this task with a summary of the Java language facilities used through out this book.

Later chapters set out to present the way in which a high-level computer language like Java is processed in stages to control the hardware of a computer system in the desired way. This is done using a series of simplified system-simulations. These are written in Java, using Java graphics user interface facilities to visualize the operation of the computer system and to illustrate the way that the various language levels translate from one to the other. The lowest level is micro coding and machine language programming, the next level up is an assembly language translation system, and finally at the top of the language system hierarchy is Mini-JC the kernel of a high level Java or C like programming language. Access to graphic facilities can be made at each of these language levels. Although display processors need to be examined briefly at the hardware level to understand their fundamental capabilities, programming them, can be done at all language levels above their machine code.

A. Thomas, *Integrated Graphic and Computer Modelling*,
DOI: 10.1007/978-1-84800-179-4_2, © Springer-Verlag London Limited 2008

Structured Programming Constructs

The core of an imperative-language system is based on four kinds of construction.

- Simple command statements and sequences.
- Conditional statements and sequences.
- Repeat statements and sequences.
- Sequences of statements in a hierarchical block structure.

In Java these occur in the following ways:

Names

In order to issue commands in any language it is necessary to identify the objects to which the commands will be applied. In natural language these references are names or nouns. In computer languages there are two kinds of references to simple objects. The first are called literal references and in a sense they are the objects they represent. Examples are numbers such as *2.304*, which though they are character strings, directly represent the particular numbers they encode. In order to treat this sequence of characters merely as a sequence of characters it is necessary to enclose them within quotation marks. This identifies them as a character string literal: a *String*, *"2.304"*. Character strings must be represented using double quotes: *"67"*, *"234 items"*, or *"Fred"*, single characters using single quotes: *'k' or 'G'*. Names are character strings (but not in quotes) that start with an alphabetic character and optionally continue with further alphabetic or numerical characters. They must be treated like algebraic variables in that they are a name that can represent any value or object. A numeric variable has to be given a value before an expression containing it can be evaluated. Variables can be rearranged in algebraic expressions in "valid" ways, independently of the values they represent. Strings and characters can be reordered to implement such text manipulation within larger language statements. Naming allows general commands to be expressed that can be applied to many different particular values or objects that a name could represent.

Simple Command Statements

Many simple commands are in reality sub-program names, for example:

```
IO.writeString("Hello World");
```

The command *IO.writeString( )* is a call to a sub-program elsewhere in the system, which takes the data *"Hello World"* and writes it out to the display screen. Other simple commands duplicate, rename or generate objects using assignment statements.

```
number1 = 3.78;
number2 = IO.readInteger( );
number3 = number2;
number4 = number3*number1;
```

These assignment statements are commands, in the first case to associate a variable name *number1* with the real value, 3.78: In the second to call the sub-program *IO.readInteger( )* to get a number from the computer keyboard and store it as a variable called *number2*, in the third as a command to transfer the value stored in *number2* to another variable called *number3,* and finally in the fourth the values of two variables are combined in an arithmetic expression creating a new value which is then assigned to the variable *number4*. The assignment statement can be interpreted in two ways in Java. In the first, shown below, it is a copy command. In other cases it can be thought of as a renaming or multiple naming-command.

In this example if **a** and **b** are simple variable names referring to data representing numbers or characters, or truth values, then the assignment operation is one of duplication. The simplest way of thinking about the operation is that **a** and **b** represent boxes, and the assignment takes a copy of what is in box **b** and places it in box **a**.

If on the other hand, **a** and **b** are the names of a more complex object, then the assignment $a = b;$ means that the object named **b** can also be referred to or called by the name **a**. The simplest way of visualising this case is as follows.

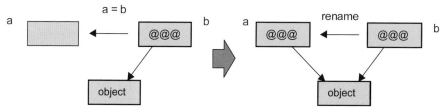

Where the data object referred to by a name is small, the box associated with the name is used to hold the data. Where the data object referred to by a name is too large for this or has a variety of possible sizes, then the name-box holds an indirect reference to the object shown in the diagram above by @@@. This means the assignment can be executed by exactly the same operation as that used in the simple case: copying the reference @@@ from one box to the other, but to get a duplicate or new object the statement has to be written $a = b.copy( );$

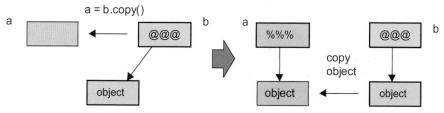

Clearly, further operations using the new names in each of these two cases must be handled differently. In the first case there are two objects, which can be acted on independently using their corresponding names, in the second there is one object and either name will select it. If the first name is used to change the object, then, when the second name is used to access the object that it refers to, following the change, it will address the modified object; not the unchanged one it originally referred to.

Declaring Variables and Initialising Objects

In carrying out an assignment unless the operation is policed by the system, any "type" of object could be associated with a name. If the name was intended to represent a number and by accident a truth-value is assigned to it, then clearly unlooked for results would be produced! Most programming languages type-check variables and restrict assignments to permitted associations. In order to do this a necessary part of the program writing is to specify the type of data a variable can hold. This is done in declaration statements.

```
int  variable1,  variable2,  variable3;      // integer variables
double   variable4 = 3.2,   variable5 = 6.1;  // double length floating point variables
static TextWindow   IO   = null;              // a null reference to a TextWindow object
```

It is generally a good idea to initialise variables with start up values like the assignments shown in the second example, even if these are going to be changed later. Where a variable refers to a more complex object, then this initialisation may be deferred in the way shown for the *TextWindow* object. Where it is not then an object has to be constructed by the system. This is achieved by issuing the **new** command. A useful example of this is initialising a text window to support input and output to the program, which can be done in the following way:

```
static TextWindow   IO   = new TextWindow (20, 30, 500, 60 );
```

The name of the object type: *TextWindow* precedes the variable name: *IO*, in the same way used to declaring simple objects. However, the **new** command is required to initialise a new *TextWindow*, by a call to the sub-program that sets up an object of this type, before its reference can be assigned to the variable name *IO*. In Java the procedure for building more complex objects is called a *"constructor"*, and it has the same name as the type of object it generates. The qualification of this declaration, by the keyword *static* makes the variable a *class* variable. This allows it to be used directly in simple programs.

A Simple Java Program

The classic example of a simple program is one that outputs the greeting:

Hello World

```
public class Program1 {
    static TextWindow IO = new TextWindow(20,170,500,200 );
    public static void main(String[] args){
        IO.writeString(" Hello World ");
    }
}
```

Notice that the output command *writeString("hello world")* is coupled to the object name *IO* by a full stop character. This is because the *TextWindow* object: *IO* carries out the operation of displaying the message, and the sub-program that does the work, is part of this more complex object. This program when it is run generates the following display on the screen

The command *writeString()* is a call to a sub-program, elsewhere in the system, which is part of the definition of the text window object, and which takes the data *"Hello World"* and writes it out to the display screen. It is an example of several commands needed to enter and display data. Essential, if useful programs are to be written.

The facilities in Java to handle the input and output of data are very flexible to cover a variety of different modes of interaction between the user and the system, and consequently are fairly complex. In order to simplify most of the examples of programs presented in this book, a reduced set of commands solely for entering and displaying text information in a text window is provided.

```
class TextWindow extends JFrame{

    TextWindow(int col, int row,int width, int height){}    // Constructor

    public String readTextString(){}
    public String readString(){}
    public char readCharacter(){}
    public void readSpaces(){}
    public String readLine(){}
    public byte readByteInteger(){}
    public short readShortInteger(){}
    public int readInteger(){}
    public long readLongInteger(){}
    public float readReal(){}
    public double readLongReal(){}

    public void writeString(String str){}
    public void writeLine(){}
    public void writeCharacter( char ch){}
    public  void writeByteInteger(byte number, int align){}
    public  void writeShortInteger(short number, int align){}
    public  void writeInteger(int number, int align){}
    public  void writeLongInteger(long number, int align){}
    public  void writeReal(float number, int align,int frac){}
    public  void writeLongReal(double number, int align,int frac){}
    public void newLine(){}

    public void quit()
}
```

This *TextWindow* object *IO*, once it is set up, by calling its constructor in the way illustrated in the "*Hello World*" program, supports the list of commands or methods given above. Since a subprogram with a particular name can only return one type of data object, this list of methods introduces most of the basic types of data that the system user handles as text. The first eleven methods return numbers of different types, which can be assigned to variables of matching types in the following way:

```
char  character  = IO. readCharacter( );      // single characters
String  string1  = IO.readTextString( );      // character strings
String  string2  = IO.readString( );          // character strings
String  string3  = IO.readLine( );            // character strings
byte number1     = IO.readByteInteger( );     // 8 bit integers
short  number2   = IO.readShortInteger( );    // 16 bit integers
int      number3 = IO.readInteger( );         // 32 bit integers
long number4     = IO.readLongInteger( );     // 64 bit integers
float number5    = IO.readReal( );            // 32 bit floating point values
double number6   = IO.readLongReal( );        // 64 bit floating point values
```

These newly defined variables *character, string1, string2, string3, and number1.. number6,* will receive input from the keyboard. Their contents can in turn be output to the display in the *TextWindow* by the matching commands:

```
IO.writeCharacter( character);
IO.writeString( string1);
IO.writeString( string2);
IO.writeString( string3);
IO.writeByteInteger( number1, 5);
IO.writeShortInteger( number2, 5);
IO.writeInteger( number3, 5);
IO.writeLongInteger( number4, 10){}
IO.writeReal( number5, 10, 5);
IO.writeLongReal( number6, 10, 5);
```

Input to the system is best thought of as a flow of characters, in a single-file stream from the keyboard, in the order in which they are typed into the system. This will include character codes for formatting the text display like the carriage return, which finishes one line and starts the next in the display. The *IO.readLine()* command returns all the remaining text in the current line of input up to and including the carriage return. The *IO.newLine()* command does the same thing but does not return the String of characters making up the rest of the line. Its useful function is to clear input stream characters such as the carriage return that might interfere with subsequent read commands. In a related way the *IO.writeLine()* command places a carriage return character into the output steam of characters being sent to the text window. The *IO.readTextString* command and *IO.readString* command returns the next sequence of characters, upto the next "space" character code. The *IO.readTextString()* command removes any leading space characters before looking for input. The sequence *IO.readSpaces(); IO.readString();* being equivalent to *IO.readTextString().*

Formulae, Expressions and Equations

A natural extension to the assignment of simple variables is made, where the assignment transfers the result of evaluating an expression into a new variable box. In this case the assignment can be viewed as a way of generating a new variable value.

$$c = (f - 32.0) / 9.0 * 5.0;$$

This provides the simplest route into writing useful programs. Many scientific relationships and results are recorded mathematically in the form of formulae and equations. These usually translate into assignment statements and expressions in a conveniently directly way.

A simple program to convert a Fahrenheit temperature value into its corresponding Celsius value can be written in the following way.

```
public class Program2{
    static TextWindow IO = new TextWindow(20,170,500,200 );
    public static void main(String[] args){
        IO.writeString("Please enter a Farenheit value: ");
        double f = IO.readLongReal(); IO.newLine();
        double c = (f-32.0)/9.0*5.0;
        IO.writeString("The Celsius value is: ");
        IO.writeLongReal(c,6,3);
        IO.writeLine();
        IO.writeString("calculation complete \n");
    }
}
```

```
Text Window                                          _ |□| ×|
Please enter a Farenheit value: 68
The Celsius value is: 20.000
calculation complete
```

Statement Sequences: Blocks and Subprograms

A small program like this is made up from a list of simple commands. However the flow of control from one statement to the next, often needs to be rearranged to follow more complicated routes through different statement-sequence blocks, and facilities have to be provided to move from block to block in a program in a controlled manner. One of these block-structuring approaches has already been mentioned, and consists of giving a name to a commonly used block of code as a sub-program, and accessing it by issuing its name as a single command. The *readInteger()* command given above refers to such a block of code which carries out the reading operation demanded by the program, obtaining the number from the keyboard as it is typed into the computer system.

Conditional Statements

The simplest command that requires a statement sequence to be divided into blocks is the conditional command, which is illustrated by the two examples given below. In each of these cases a special pair of symbols < and > is used as brackets to indicate a section of code that has not been expressed using correctly structured commands. These brackets are called meta-symbols because they are not part of the language, but denote either an *approximation name* or an *abstract name* for an operation, which will be replaced in the final completed program by an equivalent sequence of correct commands. They are useful to denote "pseudo" code, in other words, statements, which are a rough approximation of what the final program is intended to do, but need to be distinguished from finished code. This helps to develop a program in organised steps, as it is being set up and designed. The meta symbols < >, in a similar way but more formally, are also used to define the grammatical components of larger language structures. An example would be a sentence defined as:

<sentence> := <subject> <verb> <object>

In order to get a correct sentence the elements in < > brackets have to be replaced by the real words that make up, or are deemed appropriate as, the subject, the verb or the object of the final sentence.

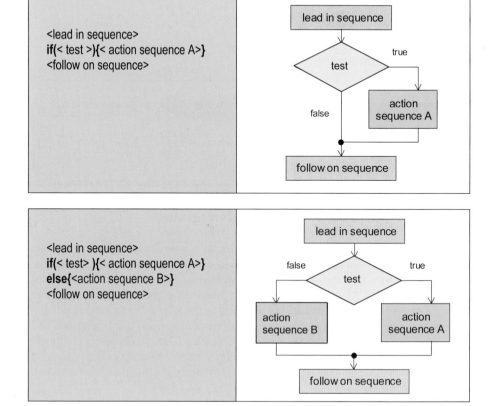

The conditional consists of a block of code, which must be processed only if a certain condition is met. A test statement specifies the condition, and the condition is met when this test is found to be true. Its associated operation can then be executed.

An extension of the same idea is given in the second example. Here there are two alternative sequences of code. The first is executed when the condition is met in other words when the test is true, the second is executed when the test fails and gives a false result. It is convenient to visualise the flow of control defined by these statements in the way shown above in the diagrams on the right.

```java
public class Program3{
    static TextWindow IO = new TextWindow(20,170,500,200 );

    public static void main(String [ ] args){
        IO.writeString("Please enter two numbers: ");
        double a = IO.readLongReal();
        double b = IO.readLongReal();

        IO.writeString("ascending order \n");
        if(a<b){
            IO.writeLongReal(a,10,3);
            IO.writeLongReal(b,10,3);
            IO.writeLine();
        }else{
            IO.writeLongReal(b,10,3);
            IO.writeLongReal(a,10,3);
            IO.writeLine();
        }
        IO.writeString("descending order \n");
        if(a>b){
            IO.writeLongReal(a,10,3);
            IO.writeLongReal(b,10,3);
            IO.writeLine();
        }else{
            IO.writeLongReal(b,10,3);
            IO.writeLongReal(a,10,3);
            IO.writeLine();
        }
    }
}
```

```
Text Window                        _ □ ×
Please enter two numbers: 23 89
ascending order
   23.000   89.000
descending order
   89.000   23.000
```

A program to order two numbers for output can be written using a single conditional test in the way shown above. However it is possible to place conditional statements within the statement sequences already controlled by other conditional statements.

This gives a nested structure, which can be used to reorder more than two numbers and output them as an ordered list, in a single step. This is loosely comparable to the analogue use of measured rods to order a collection of values in a single step, described in chapter one as a "one-shot" operation.

```
public static void main(String[] args){
    IO.writeString("Please enter three values: ");
    int a = IO.readInteger(); int b = IO.readInteger(); int c = IO.readInteger();

    if(a<b)
        if(a<c)
            if (b<c) { IO.writeInteger(a,9);IO.writeInteger (b,9); IO.writeInteger (c,9);}
            else   { IO.writeInteger(a,9); IO.writeInteger (c,9); IO.writeInteger (b,9);}
        else if (b<c) { IO.writeString("Impossible case :except ring order ");}
            else   { IO.writeInteger(c,9); IO.writeInteger (a,9); IO.writeInteger (b,9);}
    else if(a<c)
            if (b<c) { IO.writeInteger(b,9); IO.writeInteger (a,9); IO.writeInteger (c,9);}
            else   { IO.writeString("Impossible case :except  ring order ");}
        else if (b<c) { IO.writeInteger(b,9); IO.writeInteger (c,9); IO.writeInteger (a,9);}
            else   { IO.writeInteger(c,9); IO.writeInteger (b,9); IO.writeInteger (a,9);}
}
```

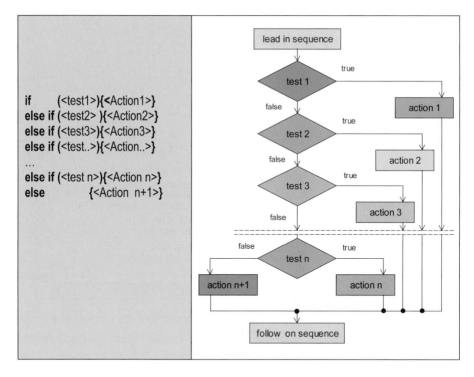

A special case of this nested arrangement is provided by the example shown above where the tests are applied sequentially in a list. In this example a list of conditional tests is processed and whenever a test is successful the dependent code sequence is executed after which, control leaves the conditional statement and passes to the next statement. To order a sequence of numbers using this construction would require the following arrangement.

```
public static void main(String [ ] args){
    IO.writeString("Please enter three values: ");
    int a = IO.readInteger(); int b = IO.readInteger(); int c = IO.readInteger();

    if    ((a<b)&&(b<c))  { IO.writeInteger(a,9); IO.writeInteger (b,9); IO.writeInteger (c,9);}
    else if((a<c)&&(c<b)) { IO.writeInteger(a,9); IO.writeInteger (c,9); IO.writeInteger (b,9);}
    else if((c<a)&&(a<b)) { IO.writeInteger(c,9); IO.writeInteger (a,9); IO.writeInteger (b,9);}
    else if((b<a)&&(a<c)) { IO.writeInteger(b,9); IO.writeInteger (a,9); IO.writeInteger (c,9);}
    else if((b<c)&&(c<a)) { IO.writeInteger(b,9); IO.writeInteger (c,9); IO.writeInteger (a,9);}
    else if(((c<b)&&(b<a)) { IO.writeInteger(c,9); IO.writeInteger (b,9); IO.writeInteger (a,9);}
    else IO.writeString("Impossible case :except  ring order ");
}
```

In this example the tests are clear but more complex. Each test is an expression, which combines the truth-values of two simple binary relationship tests such as (a<c) to give a final single truth-value, the result of evaluating the overall test expression.

These truth-values are called Boolean values, which can be represented in the program by literal representations in the same way that numerical values can. In this case they are one of two values represented by the words *true* and *false*. Boolean expressions are formed by combining Boolean variables using the operators *"not"*, *"and"* and *"or"*. These are represented in Java by the characters: !, &&, ||, respectively. The following operator, *"truth-tables"* define their actions.

input		output
A	▶	$!A$
true	▶	false
false	▶	true

input	input		output		output
A	B	▶	$A\&\&B$	▶	$A\|\|B$
true	true	▶	true	▶	true
true	false	▶	false	▶	true
false	true	▶	false	▶	true
false	false	▶	false	▶	false

Expressions can be constructed using operator precedence rules (*not* > *and* > *or*) and brackets in much the same way that arithmetic algebraic expressions are built up and the resulting values can be assigned to Boolean variables.

```
boolean itIsRaining = true;
boolean result = !((a < b) && (b < c)&& itIsRaining || (a == b))||(x != y)
```

The *<switch statement>* provides an alternative statement to the nested conditional that allows a similar kind of multiple-choice. In this case there is not a sequence of tests for true or false, but a function that generates an integer value. This number is used to select the label case-number that it matches, and then the code associated with the case is executed. Notice in this case that the *break* statement is necessary to pass control on to the next command. Where it is missing control passes to the next case in sequence, below. In the example shown above, if the selection function gives an integer value 7 then actions 3 and actions 4 are executed. If the selection function gives 8 or 9 then only action 4 is executed.

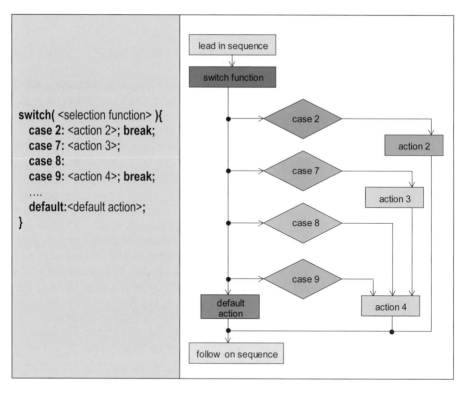

```
switch( <selection function> ){
    case 2: <action 2>; break;
    case 7: <action 3>;
    case 8:
    case 9: <action 4>; break;
    ....
    default:<default action>;
}
```

The application of this statement to sorting three numbers can be carried out in the following way.

```
public static void main(String[] args){
    IO.writeString("Please enter three values: ");
    int a = IO.readInteger(); int b = IO.readInteger(); int c = IO.readInteger();

    int j = 0;
    if(a<b)j=j+1;
    if(a<c)j=j+2;
    if(b<c)j=j+4;

    switch(j){
        case 0: IO.writeInteger(c,9); IO.writeInteger (b,9); IO.writeInteger (a,9);break;
        case 1: IO.writeInteger(c,9); IO.writeInteger (a,9); IO.writeInteger (b,9);break;
        case 3: IO.writeInteger(a,9); IO.writeInteger (c,9); IO.writeInteger (b,9);break;
        case 4: IO.writeInteger(b,9); IO.writeInteger (c,9); IO.writeInteger (a,9);break;
        case 6: IO.writeInteger(b,9); IO.writeInteger (a,9); IO.writeInteger (c,9);break;
        case 7: IO.writeInteger(a,9); IO.writeInteger (b,9); IO.writeInteger (c,9);break;
        default:IO.writeString("Impossible case :except  ring order ");break;
    }
    IO.writeLine();
}
```

Decision Tables

These three examples lead to a useful programming construct, which is helpful in designing programs that handle complex relationships. This is the decision table. When a relationship test is evaluated or when a *boolean* variable is defined, combinations of their truth values can be used to select different actions. When these conditions and actions are complex then it is useful to set out all the possible combination of *boolean* variable values to ensure that each outcome, from all possible inputs, is allocated to an appropriate action. In essence this is giving all input combinations of values an output action rather than a value, which is done in the case of the truth table definition of *boolean* operator functions. Laying out these relationships in a table allows the relationships between the tests and the actions they require to be systematically examined and reduced to their simplest form, it also saves mistakes resulting from unaccounted cases being overlooked.

The ordering of three numbers depends on the primitive operation of comparing pairs of numbers. There are six relationship pairs generated by three variables {a, b, c} which are (a, b), (a, c), (b, c) and (b, a), (c, a), (c, b). If these are all tested using the < test, then this gives six *boolean* values and six possible tests of the form *if(a<b) A1 else A2, if (b<a) A3 else A4* etc.. However the results of these tests are not all independent. If *(a<b)* is true then *(b<a)* will be false. These tests are not opposites and cannot therefore be treated by one test *if(a<b) then A1(or A4) else A2 (or A3).* The problem is the case where *(a==b)*. Where *(a==b)* both *(a<b)* and *(b<a)* will be false. When the actions A1 and A2 are considered for ordering the two values *a* and *b*, it is clear that where *(a == b)* either outputting *(a, b)* or *(b, a)* gives the same result. Analysing the actions in relationship to the tested conditions in this case allows a single test to be used to replace two. A decision table for this problem can be set up as follows:

(a<b)	true				false			
(a<c)	true		false		true		false	
(b<c)	true	false	true	false	true	false	true	false
Action	A7	A6	A5	A4	A3	A2	A1	A0
Output	a, b, c	a, c, b	error	c, a, b	b, a, c	error	b, c, a	c, b, a

In each of these examples the program is designed to select and then output the sorted numbers in one step as a complete ordered list of numbers. There is a limit to the number of elements that can be treated in this "one-shot" way, with three numbers there are only 3! in other words 6 possible output lists. When the number of elements in the unordered list is raised to 4, then the number of output orders goes up to 24. With seven numbers, this number would expand to 7! in other words 5040 output statements would be needed in the program to carry out the task in the same way. Laying out the decision table would show that 7 numbers would require 7*6/2 binary relationship tests. As a decision table this would set up 2^{21}, in other words over 2 million potential actions. Since only 5040 of these are not error messages, this approach clearly has strict practical limits.

The sorting program can be greatly simplified if the sorted output is built up step by step rather than being generated in a single step as a one-shot operation. If each step removes the current largest element in the input list, only **n** steps will be required to create the output list in the modified operation to order **n** values. This sequential process can be greatly simplified as a program if it can be expressed as an operation, which can be applied repeatedly, to the same set of data.

Repeat Statements

The next commands that modify the way sequences of code are processed are the repeat commands. There are three main forms commonly used. The first is the *for*-loop. This contains three fields in a control section followed by a sequence of dependent statements within { } brackets. The first field sets up initial conditions, usually a counting variable. The second field sets up a finishing condition usually a relationship test applied to the counting variable and finally the third field defines the changes that must be executed at the end of each repetition cycle. The latter is usually the increment or decrement of the counting variable.

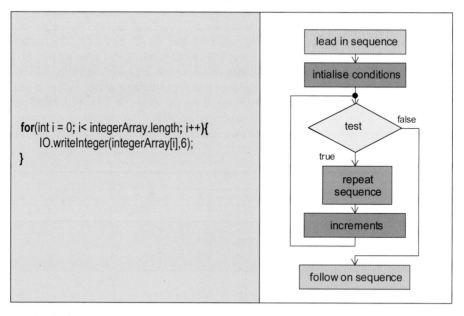

```
for(int i = 0; i< integerArray.length; i++){
    IO.writeInteger(integerArray[i],6);
}
```

The *for*-loop is often associated with actions on arrays. An array is a collection of objects where each object can be accessed by giving the array name followed by the index of the object in the array, in square brackets. If the counter in a *for*-loop is used to index elements in the array then each object in it can be visited in order. A program to write out all the integers in an array of integer numbers can be written in the way shown above. The other two repeat commands are more primitive in that they merely control the repetition of a block of code by a terminating test. The two forms of this command apply this test at the beginning and at the end of the repetition cycle respectively. The following diagrams illustrate the flow of control set up by these statements using the equivalent programs to the one above for writing out the contents of an array.

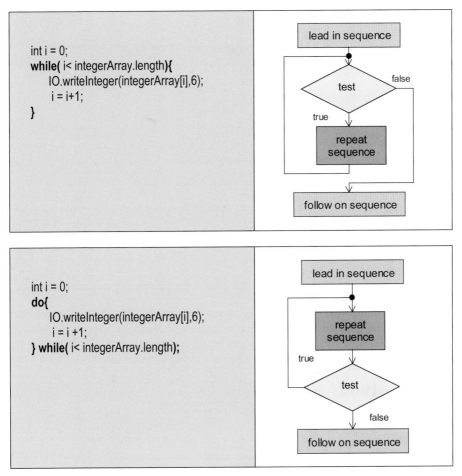

The difference between these two examples is that the first can cope with an array with no contents, while the second needs the array to have at least one entry if an error is not to be generated by the system attempting to access a non-existent element.

Using the repeat command allows a set of numbers held in an array to be ordered by a simple repetitive swapping operation. If the largest element is selected each cycle through an array and the value is stored at the beginning of the array, then after **n** passes for an array of **n** elements the array will end up being sorted into descending order.

```
public static void main(String[] args){
    int[] integerArray = new int[7];
    IO.writeString("Please enter seven values: ");

    for(int i=0;i<7;i++) integerArray[i] = IO.readInteger();
    for(int j=0;j<7;j++){
        for(int i=j+1;i<7;i++){
            if(integerArray[j]<integerArray[i]){
                int temp=integerArray[j];
                integerArray[j]=integerArray[i];
                integerArray[i]=temp;
            }
        }
    }
    for(int i=0;i<7; i++)IO.writeInteger(integerArray[i],6);
    IO.writeLine( );
    for(int i=6;i>=0;i--)IO.writeInteger(integerArray[i],6);
}
```

```
Text Window                                                    _ □ ×
Please enter seven values: 2 45 56 7 22 99 16
 99  56  45  22  16   7   2
  2   7  16  22  45  56  99
```

The application of the second form of repeat command allows the program to be set up to handle variable amounts of input data. A program to convert Fahrenheit temperature measures to Celsius values can be written in the following way, where the system asks at the end of each calculation if another is required.

```
public class Program 4{
    static TextWindow IO = new TextWindow(20,170,500,200 );

    public static void main(String[] args){
        String str = "";
        do{ IO.writeString("Please enter a Fahrenheit value: ");
            double f = IO.readLongReal();IO.newLine();
            double c = (f-32.0)/9.0*5.0;
            IO.writeString("The Celsius value is: ");
            IO.writeLongReal(c,6,3); IO.writeLine();
            IO.writeString("Do you wish to continue? y/n: ");
            str = IO.readString();IO.newLine();
        }while(str.equals("y"));
        IO.writeString("calculation complete \n");
    }
}
```

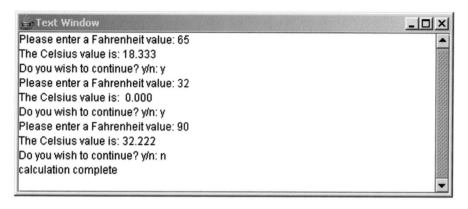

The sequences of commands in a repeat loop have to be designed carefully, to be applicable again and again as the program cycles. An example of a construction, which can be applied repetitively to a sequence of data, can be seen in the case of the formulae used to define the sum, the average, and the variance of a list of values.

$$\text{Sum:} \qquad S = \sum_{i=1}^{i=n} x_i$$

$$\text{Average:} \qquad \bar{x} = \sum_{i=1}^{i=n} \frac{x_i}{n} = \frac{1}{n} \cdot \sum_{i=1}^{i=n} x_i$$

$$\text{Variance:} \qquad \sigma^2 = \sum_{i=1}^{i=n} \frac{\left(\bar{x} - x_i\right)^2}{n}$$

$$\sum_{i=1}^{i=n} \frac{\left(\bar{x} - x_i\right)^2}{n} = \sum_{i=1}^{i=n} \frac{\left(\bar{x}^2 - 2.\bar{x}.x_i + x_i^2\right)}{n}$$

$$= \frac{1}{n} \sum_{i=1}^{i=n} \bar{x}^2 - 2.\bar{x}. \sum_{i=1}^{i=n} \frac{x_i}{n} + \sum_{i=1}^{i=n} \frac{x_i^2}{n}$$

$$= \bar{x}^2 - 2.\bar{x}.\bar{x} + \sum_{i=1}^{i=n} \frac{x_i^2}{n}$$

$$= \sum_{i=1}^{i=n} \frac{x_i^2}{n} - \bar{x}^2$$

The sum is simple to calculate within a *"for"* loop or a *"while"* loop. The average also can be calculated in a single repeat loop, in one of two ways depending on whether the length of the list is known at the beginning of the repeat command or only when the list has been completely processed. In the first case if the length of the list is known then each element can be divided by the list length and then added to the total. In the second case a count has to be kept of each new element added to the sum of list elements, when the list is complete the answer is the sum divided by the number of elements.

The variance in contrast, appears to require two loops the first to calculate the average, the second to calculate the variance. Rearranging the formula, algebraically in the way shown above allows the variance to be calculated in one loop. By calculating the sum of the squared elements, $(x_i.x_i)$ and the average $\bar{x}$ within the loop, the final result can be obtained by squaring the average and subtracting it from the average of the sum of the squares. This is one example from a variety of different "recurrence relationships" designed to use repeat commands to provide compact and efficient program code.

Another example of this process occurs with the choice of names for variables. A program to generate the Fibonacci series can be set up by defining the nth element as the sum of the $n\text{-}1^{th}$ and the $n\text{-}2^{th}$ elements in the series.

$$0, 1, 1, 2, 3, 5, 8, 13, 21, 34, \ldots$$

```
nextElement = lastElement + lastButOneElement;
```

Once this statement has been executed the *lastElement* and the *lastButOneElement* no longer hold the values appropriate to their names. In order to complete a sequence of statements that can be repeated, it is necessary to redefine these variables in the following way:

```
nextElement = lastElement + lastButOneElement;
output(nextElement);
lastButOneElement = lastElement;
lastElement = nextElement;
```

Exactly the same redefinition is needed to draw a curve based on plotting a string of line segments.

```
plotLine(leadingPoint, laggingPoint, colour);
laggingPoint = leadingPoint;
leadingPoint = calculateNextPoint();
```

A more complex repeat pattern can be set up to implement a "finite state machine". This is a program where the current state determines the action, and the next state is determined by the combination of new inputs and the current state. A diagram of this kind of mechanism is a very useful tool for programming a variety of problems. It is often easier to visualise the interactions needed to make a program function correctly if bubbles labelled by the state names are drawn out to represent the states, and state transitions are shown by arrows, from one bubble to another, with each arrow associated with the inputs that cause the transition.

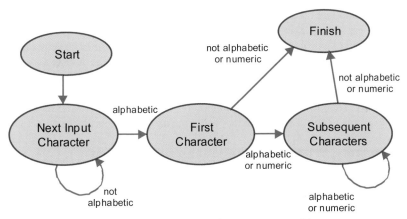

Figure 2.1 State transition diagram for a name recognition program

```
public static void main(String[] args){
    String str = ""; char ch = '\0';
    boolean notFinished = true;
    int start = 0, nextCharacter = 1;
    int firstCharacter= 2, subsequentCharacters = 3, finish=4;
    int state = start;
    while(notFinished){
        switch(state){
        case 0:state = nextCharacter; break;
        case 1:ch = IO.readCharacter();
                if(((ch>='a')&&(ch<='z'))||((ch>='A')&&(ch<='Z'))) state = firstCharacter;
                break;
        case 2:str = str+ch;
                ch = IO.readCharacter();
                if(((ch<'a')||(ch>'z'))&&((ch<'A')||(ch>'Z')) &&((ch<'0')||(ch>'9')))state = finish;
                else state = subsequentCharacters;
                break;
        case 3:str = str + ch;
                ch = IO.readCharacter();
                if(((ch<'a')||(ch>'z'))&&((ch<'A')||(ch>'Z'))&&((ch<'0')||(ch>'9'))) state = finish;
                break;
        case 4:notFinished= false; break;
        }
    }
    IO.writeString(str + "\n");
}
```

A switch statement, acting on a state variable, contained within a repeat loop, can be used to implement this kind of process in the way shown in Figure 2.1 to recognise a name within a sequence of characters. Though this structure is very powerful and will cope with many programming tasks, its limitation is its fixed number of state variables. Many programming tasks require more memory than this in order to build up a varying number of partial results, before the overall task can be completed, the amount of memory depending on the nature of the input received: the classical example of this kind of task is evaluating an expression.

Sub-programs, Procedures, Functions and Methods

One way of getting this extra memory as it is needed is to use recursive procedures – sub-programs, which call themselves. The terms: sub-program, subroutine, procedure, function, and method; are with small variations interchangeable and depend on the computer language being used. In object oriented languages the preferred term referring to a sub program is the term "method". In Pascal and Modula-2 the terms procedure and function-procedure are used to distinguish sub-programs which returned no value, and sub-programs which returned a value, like *value = sin(alpha);*. In Java and C, because statements are considered to have values, (allowing strings of assignments such as a = b = c = d; to be written) any subprogram can be written as a function. However, methods are allowed to return a void value, and therefore can behave as procedures that cannot be used in assignment statements!

```
Public class Program5{
    static TextWindow IO = new TextWindow(20,170,500,200 );
    static double mul (double x,double y)   {return x*y;}
    static double div  (double x,double y)   {return x/y;}
    static double sub (double x,double y)   {return x-y;}
    static double add (double x,double y)   {return x+y;}
    public static void main(String[] args){
        double a=2,b=3,c=4,d=10,e=1,f=4,g=2;
        double expression1 = a*b+c*(d-f)/(e+g);
        double expression2 = add(mul(a,b),mul(c,div(sub(d,f),add(e,g))));
        IO.writeString("exrpession 1: "+ expression1+"\n");
        IO.writeString("exrpession 2: "+ expression2+"\n");
    }
}
```

The declaration of functions in a simple program to evaluate an arithmetic expression using function calls is shown in Program 5. As before it is necessary to qualify the definition of each method by the keyword *static*. In this example the same expression is presented in two different ways. In the first it is written in the conventional form using arithmetic operators and brackets. In the second each operator is replaced by a function call. Each function method executes its corresponding arithmetic operation in a standard way on the two values passed to it and returns the result. This example illustrates the way methods are defined and the way they are called. In the function definition the type of the return value has to precede the function name, and the values passed to the function have to be given working names in order for the function code to be written, and each has to have its type defined. When the function is "called" the real parameter names matching the dummy, working names in the function definition have to be placed in the method's argument list, in the correct order to match the dummy arguments in the method definition. This matching process allows the function to be applied to any variables that the calling statement specifies.

When a method is called, control is passed to the sub-program code. The parameter values in the calling statement are copied to the dummy variables in the sub-program code. When the method's computation is complete its resulting value is passed back, and treated, as a value associated with the function name in the calling statement, as if the calling name were a simple variable.

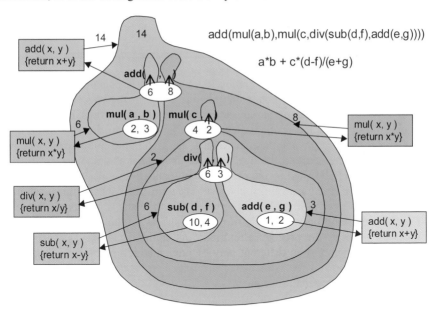

Figure 2.2 Parameter matching for sub routine calls

When a more complex object is passed as an argument to a sub-program, its indirect reference will be copied to a local variable in the method. This means, in this case, the object is not duplicated. For simple objects such as integers the value is passed and therefore it is duplicated. Arrays are more complex objects and they have their references passed to methods rather than the whole array being duplicated. Consequently an array with one element will act as a convenient example of a pass by reference. An example, where this treatment makes a difference is the subprogram to swap the contents of two variables.

```
Public class Program6{
    static TextWindow IO = new TextWindow(20,170,500,200 );
    static void swap1(double x, double y){ double temp = x;  x = y; y = temp; }
    static void swap2(double [] x, double[] y){ double temp = x[0];  x[0] = y[0]; y[0] = temp;}

    public static void main(String[] args){
        IO.writeString("please enter two different numbers: ");
        double a = IO.readLongReal();  double b = IO.readLongReal();
        double[] c = new double[]{a}; double[] d = new double[]{b};
        swap1(a,b);
        IO.writeString("swap1: " + a + " "+ b +"\n");
        swap2(c,d);
        IO.writeString("swap2: " + c[0]+ " " + d[0] +"\n");
    }
}
```

```
Text Window                                                    _ □ ×
please enter two different numbers: 56 34
swap1: 56.0  34.0
swap2: 34.0  56.0
```

In program 6, swap1 shows a swapping function which exchanges the contents of the local parameter variables, but which has no effect in the space of the calling program. In contrast swap2, by passing the references to two arrays, by exchanging their contents, provides the result back to the calling program, because it uses the same references to the arrays in the function that are used in the calling program.

When a subroutine is called, storage space for its internal and parameter variables, (local variables) are allocated to the program automatically by the language system. This extra memory space is arranged in a stack data-structure, where the last element added to the stack, is the first element returned from the stack. If a procedure calls itself then it will build up a sequence of memory spaces on the stack, which will be taken off the stack as the procedure returns to its calling statement. This recursive procedure calling supports a different way of implementing a repetitive operation. For example writing out the contents of an array can be done either forwards or backwards, by the following procedures:

```
static void forwardOrder(int i, double[] array){
    if(i<array.length)IO.writeLongReal(array[i],6,2);else return;
    forwardOrder(i+1, array);
}
static void backwardOrder(int i, double[] array){
    if(i<array.length)backwardOrder(i+1, array);else return;
    IO.writeLongReal(array[i],6,2);
}
```

In these examples each call to the routine will set up a local variable *i*, which will be placed in the stack starting with 0. A new value for *i* will be generated, and incrementally increased, by each routine call, until it is equal to the length of the array, when the procedure will return through all its intermediate calls back to its start, releasing memory space for *i* as it goes. By placing a write statement before the recursive call the local values of *i* will be used as they increase, by placing the write statement after the recursive call the local value of *i* will be used as they decrease during the return path of the calling sequence. It is essential that some way of stopping such a chain of recursive calls be built into recursive procedures, in this case testing to see if the end of the array has been reached stops the sequence of calls.

Two standard data structures can be handled in an elegant way using this approach. The first is the simple list, either as an array in the way shown in the example, or as a dynamic linked list data structure. The second is the tree data structure where elements are hierarchically linked to two or more lower level elements. The expression used in Program 5 can be represented as the operator tree shown diagrammatically in Figure 2.3. This will be explored more fully in a later chapter.

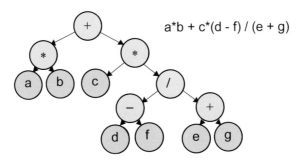

a*b + c*(d - f) / (e + g)

Figure 2.3 An arithmetic expression operator tree

Recursive programs can handle more complex data structures than simple finite state machines, mainly because of the way they can manage the growth and release of data stored on a stack.

Associated with the tree structure is a programming strategy called "divide and conquer". Where this approach is applicable it usually provides a more efficient algorithm, than alternatives. Consider the sorting problem. To order a list of values that can appear in any order, it is necessary to remove the largest value *n* times from a list of *n* elements, and each selection will take *n* comparisons. Overall this requires n^2 comparison operations. If the original list is divided into two halves, and this is done recursively until there is only one element in each list, then the return operation can be one of merging lists in order. At any level this will consist of systematically merging lists already ordered lower down the recursive chain. Program 7 shows an example of a merge-sorting algorithm of this type. Figure 2.4 shows that the number of comparisons in this approach is reduced to the order of *n.log(n)* for a list of *n* elements.

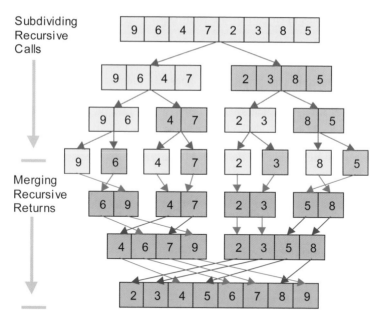

Figure 2.4 Divide and conquer mergeSort procedure

```
public class Program7{
    static TextWindow IO = new TextWindow(20,170,500,200 );  static int num= 9;
    public static void main( String[ ]  args ){
        double[][] a = new double[2][num];  int level=0, left=0, right =num;
        IO.writeString("please enter "+num+" different numbers: ");
        for(int i=0;i<num;i++) a[0][i]= IO.readLongReal();
        mergeSort(level,left,right,a);
        for(int i=0;i<num;i++) IO.writeLongReal(a[0][i],6,2);
    }
    static void mergeSort( int level, int left, int right, double[ ][ ] a){
        int nextLevel = (level+1)%2;
        if((right - left)==1){ a[1][left] = a[0][left];  return;}
        int middle = (left+right)/2;
        mergeSort( nextLevel, left, middle, a);   mergeSort(nextLevel, middle, right, a);
        merge(nextLevel, left, middle, right, a);
    }
    static void merge( int level, int left, int middle, int right, double[ ][ ]a ){
        int r = (level+1)%2, s = level, t, i =left ,j = middle, k = left;
        while(((i<middle)||(j<right))){ t = 0;
            if((i>=middle)|| ((j<right) &&(a[s][j]<=a[s][i]))) {a[r][k++]= a[s][j];t=t+1;}
            if((j>=right) || ((i<middle)&&(a[s][j]>=a[s][i]))) {a[r][k++]= a[s][i];t=t+2;}
            switch( t ) {case 1: j++; break;  case 3: j++; case 2: i++; }
        }
    }
}
```

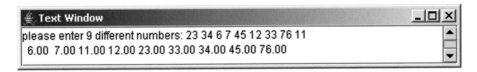

please enter 9 different numbers: 23 34 6 7 45 12 33 76 11
  6.00 7.00 11.00 12.00 23.00 33.00 34.00 45.00 76.00

The merging operation is a linear sequential operation analogous to the action of closing a zip fastener, in the way shown schematically in Figure 2.5!

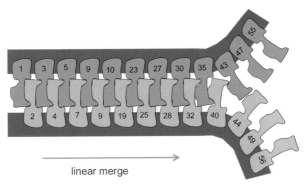

linear merge

Figure 2.5 Linear merging

Types, Classes and Objects

Java is an object-oriented language. This means that it provides more than the basic structured programming constructs discussed above. The object-oriented approach extends the way data types are handled, including more complex data structures and algorithms within a common, unified conceptual framework.

A data-type appears to be a relatively simple idea when viewed at the text level. The symbols used to represent numbers and words immediately indicate that their interpretation must be handled in a different way. They are different data types. At the machine level in the computer system all the information is in the same form: strings of binary digits. In this setting the data gives no indication of its type. Its type is only determined by the context in which it can be used correctly. A bit string is a number when it is processed as a number, but it could equally well be processed as a character string if it were passed to the input-stream of a printer. A data type is defined by the permitted operations that can be carried out on data of that type. The structure of the data and the valid operations on it are inextricably entwined.

Data Structures and Algorithms

The way data structures and algorithms have to be considered together can be illustrated by the following example. If an eight-bit data value is used to represent whole numbers then it can be used to represent positive integers in the range 0 -- 255, but the information defining the use that can be made of these bit patterns has to be held outside the pattern itself. If the number is negative this will change its type and hence for example, the way it can be added to a positive number. It is quite possible to use one bit from the eight bits to determine which of these two types is present, though the number range held in the remaining 7 bits of data would be limited to half the original, only giving: 0 to 127: and the program to process the two data formats would still have to be different.

If the bit patterns, in positive binary numeric order are placed round a circle from 0 to 255, then the value range from –128 to +127 can represented by a shift round the circle: shown by the red and blue labels in Figure 2.6. If 0 to 255 is represented by 00000000 to 11111111 (blue) then –128 to 0 to +127 can be represented by the sequence 10000000 to 00000000 to 01111111 (red). This allows the addition of positive and negative numbers to be the same operation merely giving a result offset along the original positive number line. To turn a positive number to a negative number in this representation, or vice versa, merely requires the number's bit pattern to be mapped horizontally across the circle, shown by the green arrows in Figure 2.6

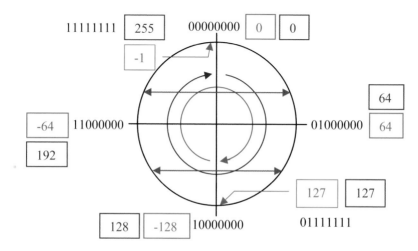

Figure 2.6 Two integer types: mapped onto the same bit patterns

This mapping gives the two's complement representation for negative numbers. 00000001 as a positive 1 becomes 11111111 as negative 1, and 127 with bit pattern 01111111 becomes 10000001 as –127. If the binary values in each bit position are inverted (0⇨1 and 1⇨0), and a binary 1 added to the new overall 8-bit binary number, the result is a conversion from a positive to a negative number, and the reverse process converts values the other way. In this case the number range for positive or negative numbers is still halved but the permitted operations become the same for both.

Element Names, Object Names and Collection Names

Simple variable names are ways of accessing memory locations, which contain changeable bit patterns. The type associated with the name determines the correct operations that can be carried out on these bit patterns. Object names in contrast are also ways of accessing memory locations that hold bit patterns, but these bit patterns are the address of objects, which can consist of many words of data. The first is a direct reference to the value the second is the reference to a reference in other words an indirect reference. The two names directly reference different types of data, the first a variable value the second an address or a pointer. Simple variable names are not sufficient to allow useful programs to be written. Names for more complex objects are necessary and this can be demonstrated by attempting to sort a collection of numerical values stored as individually named, simple variables for example *{a, b, c, d, e, f }*. If the *"repeated selection of the largest"* algorithm is used then the only way it can be coded is as a sequence of commands of the form:

```
int a= 5, b=8, c= -9, d=23, e=14, f=3, t=0;
int m= Integer.MIN_VALUE;
if(m<a){ t= a; a=m; m=t;}
if(m<b){ t= b; b=m; m=t;}
if(m<c){ t= c; c=m; m=t;}
if(m<d){ t= d; d=m; m=t;}
if(m<e){ t= e; e=m; m=t;}
if(m<f){ t= f; f=m; m=t;}
Output.writeString("largest value is "+ m +"\n");
```

largest value is 23

This requires as many statements as there are variables. To order the set of numbers this sequence could be placed in a repeat loop as long as ordered output is all that is wanted: writing out the value of *m* at the end of each iteration.

```
int a= 5,b=8,c=-9,d=23,e=14,f=3,t=0;
for(int i=0;i<6;i++){
    int m= Integer.MIN_VALUE;
    if(m<a){ t= a; a=m; m=t;}
    if(m<b){ t= b; b=m; m=t;}
    if(m<c){ t= c; c=m; m=t;}
    if(m<d){ t= d; d=m; m=t;}
    if(m<e){ t= e; e=m; m=t;}
    if(m<f){ t= f; f=m; m=t;}
    Output.writeString("largest value is"+ m +"\n");
}
```

largest value is23
largest value is14
largest value is8
largest value is5
largest value is3
largest value is-9

If a new list of ordered variables, is wanted then this selection sequence will have to be duplicated for each value, or included in a list of procedure calls to this selection code structured as the sub program *selectTheLargest.*

```
int X1= selectTheLargest();  Output.writeString(" X1 is "+ X1 +"\n");
int X2= selectTheLargest();  Output.writeString(" X2 is "+ X2 +"\n");
int X3= selectTheLargest();  Output.writeString(" X3 is "+ X3 +"\n");
int X4= selectTheLargest();  Output.writeString(" X4 is "+ X4 +"\n");
int X5= selectTheLargest();  Output.writeString(" X5 is "+ X5 +"\n");
int X6= selectTheLargest();  Output.writeString(" X6 is "+ X6 +"\n");
```

X1 is 23
X2 is 14
X3 is 8
X4 is 5
X5 is 3
X6 is -9

To make this work it is necessary to make the variables *a* to *f* static global variables.

```
static int selectTheLargest(){
    int m= Integer.MIN_VALUE;
    for(int i=0;i < 6;i++){
        if(m<a){ t= a; a=m; m=t;} if(m<b){ t= b; b=m; m=t;} if(m<c){ t= c; c=m; m=t;}
        if(m<d){ t= d; d=m; m=t;} if(m<e){ t= e; e=m; m=t;} if(m<f ){ t= f; f=m; m=t; }
    } return m;
}
```

The solution to this potentially massive duplication of the same code pattern is the use of array naming. The collection of numbers is treated as a single object and is given a group name. Within the group the individual elements are identified by a second, variable name holding an index value. This is implemented by storing the collection of numbers in neighbouring locations in memory. The array name is then associated with the base-address of the first location in memory used for the block of data, and the index is used to give the offset from this position for each element in the collection by adding the index value to the base-address of the array. This allows a short single program to handle different sized arrays of numbers with the same code.

```
int[] a = new int[]{5,8,-9,23,14,3};
int t;
for(int i=0;i<a.length;i++)
    for(int j=1;j<a.length;j++)
        if(a[j]>a[j-1]){ t=a[j]; a[j]=a[j-1]; a[j-1]=t;}
for(int i=0;i<a.length;i++)
    jOutput.writeString(" a["+i+"] is "+ a[i] + "\n");
```

```
a[0] is 23
a[1] is 14
a[2] is 8
a[3] is 5
a[4] is 3
a[5] is -9
```

If many arrays are defined in a program, and they are placed next door to each other in memory: they cannot be increased in size. This becomes a limitation for example if more data are entered into the program than space has been allocated for them. Another limitation of this particular form of group naming is that an array has to be a collection of elements of the same type. It is often useful to have a name for a bundle of elements that are of different types. In C, Pascal and various, other programming languages this possibility has been catered for by providing *"structure"* names for collections of differently typed variables. In order to access these individually a different naming convention has evolved. The name of the group is followed by a period, followed by the name of the individual variable. The structure is not of great use by itself but duplicated it provides a building block for more flexible linked data structures to handle collections of data that vary in size: dynamic data structures.

In an array, neighbouring elements can be obtained by adding offsets from a current index value. Stepping one by one through an array's index values allows all its elements to be processed. This depends on the data being stored in adjacent memory locations. An alternative approach to this task of storing data collections, allows individual elements to be stored anywhere in memory, so that collections can be incrementally built up or reduced in size as required. To do this reference information must be included with each data-element to allow its neighbours to be located. *"structure"* data types proved particularly suited to this construction. Such a structure can be made up from variables for holding the primary data, along with variables of the type of the structure itself. These structure variables hold the location addresses of neighbouring structures in a collection. In Java these structures will be objects containing link variables of their own type, as references to neighbours.

Arrays allow lists of elements to be stored. An equivalent dynamic data structure can be built up from *structures* of the type *ListElement* defined in the following way:

```
class ListElement{
    public ListElement left=null, right=null;
    public Object object = null;
    public boolean comparable = true;
    ListElement(){}
    ListElement(boolean comparable){ this.comparable=comparable; }
}
```

This class definition is a template, which defines the data framework that each *ListElement* must have. It also provides two procedures, which create new structures of this type called constructors. A list is a sequence of these objects linked by the references *left* and *right*.

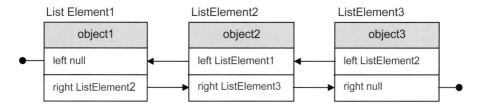

Figure 2.7 Double linked list

Although this list does not have a name for the data collection as a whole, the linked list structure allows all objects in its length to be processed using code of the form:

```
reference = firstElement;
while(reference != null){
    output(reference.object);
    reference = reference.right;
}
```

The list is accessed through a variable of type *ListElement*. Multiple lists can be set up and accessed by holding their leftmost element in a *ListElement* variable. This is the approach found in C, Pascal, Modula2 and similar languages. The only problem is that it is possible to build a variety of linked list structures using the *ListElement* as the building block. Since different accessing functions need to be used with these data structures, more information needs to be stored with each construction than a simple *ListElement* reference. In other words these list structures are of different types, and to handle them in a consistent way they should be given individual variable names with a type that reflects the kind of list-object that they reference.

Object Oriented languages such as Java not only allow these more complex data collections to be constructed but also to be given names as single objects. As objects of the same type, it will be possible to operated on them in common ways, and these operations implicitly define the type of object they are. The class definition of a *List* provides a reference to the first *ListElement* in the list and the last *ListElement*. More than this it includes the methods for operating on these structures in the following way:

```java
class List{
    public int length = 0;
    public boolean comparable = true;
    public ListElement start = null,  finish = null;

    List( ){ };
    List(boolean c){this.comparable = c;};

    public List makeNewList(List lst){
        List lst0 = new List(true);  ListElement ref = lst.start;
        while(ref!=null){ lst0.append(ref.object);  ref=ref.right; }
        return lst0;
    }

    public void setComparable(boolean c){comparable = c;}

    public ListElement push(Object n){
        if (n==null) return null;
        ListElement m = new ListElement(comparable);
        m.object = n;   m.right = this.start;  m.left = null;
        if (this.start == null){this.finish = m;} else{ this.start.left = m;}
        this.start = m;  this.length = this.length + 1;
        return m;
    }
    public Object pop(){
        if (this.start == null)return null;
        Object m = this.start.object;  this.start = this.start.right;
        if(this.start != null) this.start.left = null;  else this.finish = null;
        this.length = this.length - 1;
        return m;
    }
    public ListElement append(Object n){
        if (n==null) return null;
        ListElement m = new ListElement(comparable);   m.object = n;
        if(this.finish == null){ this.finish = this.start = m;}
        else{ this.finish.right = m;  m.left = this.finish;  finish = m; }
        this.length = this.length + 1;
        return m;
    }
```

```
public Object remove(){
    if(this.finish == null)return null;
    Object m = this.finish.object;
    if(finish.left == null){ this.start= this.finish = null;}
    else{ this.finish.left.right = null;  this.finish = this.finish.left; }
    this.length = this.length - 1;
    return m;
}
public Object delete(ListElement m){
    if(m == null) return null;
    else if (m.left == null) return this.pop();
    else if (m.right == null) return this.remove();
    else{ m.left.right = m.right;  m.right.left = m.left;
        this.length = this.length - 1;
    } return m.object;
}
public ListElement insertBefore(ListElement after,Object n){
    ListElement m=null;
    if (n==null) return null;
    if(after == null) m=this.append(n);
    else if(after.left == null) return this.push(n);
    else{
        m = new ListElement(comparable);  m.object = n;
        ListElement before = after.left;  before.right = m;  m.left = before;
        m.right = after;   after.left = m;  this.length = this.length + 1;
    } return m;
}
public ListElement insertAfter(ListElement before,Object n){
    ListElement m=null;
    if (n==null) return null;
    if(before == null) m=this.push(n);
    else if ( before.right == null)return this.append(n);
    else{
        m = new ListElement(comparable);  m.object = n;
        ListElement after = before.right;  after.left = m;  before.right = m;
        m.left = before;  m.right = after;  this.length = this.length + 1;
    } return m;
}
public List joinTo(List b){
    if(this.start==null)return b;
    if(b.start==null)return this;
    List a = new List();
    a.start = this.start;  a.finish = b.finish; this.finish.right= b.start;
    b.start.left = this.finish;
    return a;
}
}
```

Lists and Trees

Compared with an array a list has a major draw back. Finding an element in a list involves following the links from one end of the list to the other searching for the required element. The same is true for an array where elements are stored in any order. However, if values are stored in order in an array, elements can be found by dividing the array into two halves selecting the half containing the target and then recursively subdividing the new reduced sub-array in the same way until the target is "*found*" or determined to be "*not present*". Instead of taking '*n*' steps for a list of '*n*' elements long this process requires '*log(n)*' steps. However adding values to an ordered array of values will involve moving entries along to make room for a new member, which on average still adds a serious overhead to the work.

In contrast the tree data structure allows a fast "*find*" operation to be applied while at the same time providing a fast insertion method. Tree data structures can be constructed from *ListElement* objects in the way shown in Figure 2.8 merely by employing a different linking strategy. Like bit patterns, the data element, building blocks, are the same but the overall type is determined by the permitted operations on the data. These will be provided by the Tree class methods. The only draw back is that tree building operations based on inserting new elements can distort the balanced shape of the tree, which is the property that makes fast '*find*' operations work.

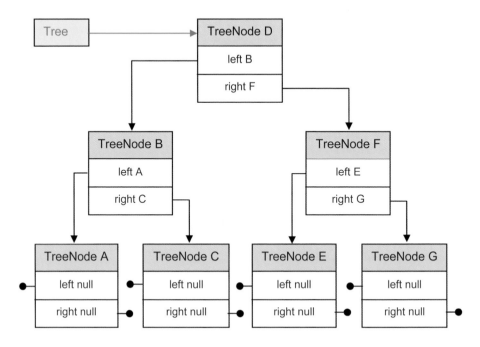

Figure 2.8 ListElements linked as a tree data structure

The tree like the list can be accessed and used through recursive procedures. Accessing elements in trees can be done in various ways depending on the application, however there are three simple tree traversal algorithms, which are used again and again. These can be expressed using the same three code statements but arranged in three different orders. Where the tree is used to store an ordered list of values shown in Figure 2.9, the output order for the prefix traversal will be D, B, A, C, F, E, G. The output order will be A, B, C, D, E, F, G for the infix traversal, and A, C, B, E, G, F, D for the postfix traversal.

```
abstract class Tree{
    static TextWindow IO = null;
    public boolean comparable = true;   public ListElement root = null;
    public Tree(){ };
    public Tree(TextWindow IO){Tree.IO = IO;};
    public abstract ListElement insert(ListElement ls, ListElement nw);
    public abstract void output(Object o);
    public void prefixTraversal(ListElement tree){
        if(tree==null)return;
        output(tree.object);
        prefixTraversal(tree.left);
        prefixTraversal(tree.right);
    }
    public void infixTraversal(ListElement tree){
        if(tree==null)return;
        infixTraversal(tree.left);
        output(tree.object);
        infixTraversal(tree.right);
    }
    public void postfixTraversal(ListElement tree){
        if(tree==null)return;
        postfixTraversal(tree.left);
        postfixTraversal(tree.right);
        output(tree.object);
    }
}
class StringTree extends Tree{
    public StringTree(TextWindow IO){Tree.IO = IO;};
    public ListElement insert(ListElement ls,ListElement ln){
        if(ls==null)ls= ln;
        else{String o1 = (String) ls.object, o2 = (String)ln.object;
            int test = o1.compareTo(o2);
            if (test > 0)   {ls.left  = insert(ls.left, ln);}
            else if(test < 0){ls.right = insert(ls.right,ln);} // new string matches existing string
        }return ls;
    }
    public void output(Object o){IO.writeString((String)o+" ");}
}
```

```
public static void main( String[ ] args ){
    ListElement root = null;String str=" ";
    StringTree stree = new StringTree(IO);
    IO.writeString("please enter 7 strings: ");
    for(int i=0; i<7;i++){
        str= IO.readTextString( );
        ListElement treeNode = new ListElement();
        treeNode.object=str;
        root= stree.insert(root, treeNode);
    }
    stree.infixTraversal(root);   IO.writeLine();
    stree.prefixTraversal(root); IO.writeLine();
    stree.postfixTraversal(root);IO.writeLine();
    }
}
```

Text Window   _ □ ×

please enter 7 strings: david brian fred alfred carl edward george
alfred brian carl david edward fred george
david brian alfred carl fred edward george
alfred carl brian edward george fred david

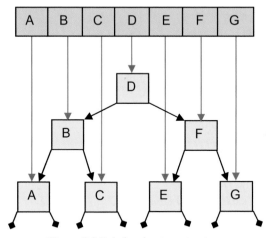

Figure 2.9 Rebalancing tree structures

If an already ordered sequence of input values is used to build a tree, using these procedures will create a single list, not a tree. There are several more sophisticated tree-building algorithms designed to keep the resulting tree reasonably well balanced as it is constructed. These can be found presented in detail in books on data structures and algorithms. However, an alternative, simpler approach to this problem is to record the number of levels in each tree-object, and when it becomes too unbalanced restructure the tree.

In practice it is often useful to switch the data structures used to implement data collections when processing different steps in a task. An example of the way this can be done is provided by tree data structures. An array can be used to rebalance a linked list tree structure built from an ordered list of values. If the linked list is traversed in infix order and the output is placed in an array of the appropriate size. The entries in the array will, by design, be in value order. If these are then accessed using a recursive binary subdivision of the array, selecting the middle value, and entering it back into a new linked list tree, the resulting tree will be balanced in the way shown in Figure 2.9.

A tree structure can be implemented either using an array or a double linked list. The tree given in Figure 2.8 could be stored in an array in the way shown in Figure 2.10, and output in infix-order A, B, C, D, E, F, G produced using the following code:

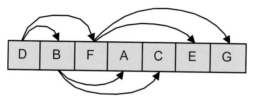

Figure 2.10 A tree in an array

```
public void infixTraversal(char[ ] tree, int index){
    if( index >= tree.length)return;
    infixTraversal(tree, index*2+1);
    output(tree[index]);
    infixTraversal(tree, index*2+2);
}
```

This arrangement is used to implement an alternative set of tree data types called *heap* structures: access is still based on the order of values but not in the same way.

Stacks, Queues and Deques

The same data structures are used again and again in different contexts, so it is natural that they should be considered as objects from the same class. However there are a series of common, dynamic, list based data structures that are essentially different types, that can be built up from the same *ListElement* units. The most general structure is the *List*, with the accessing methods given in the class listed above. However, a list that can only be accessed at one end is called a *"stack"* or a LIFO (last in first out list), similarly a list which has inputs at one end and outputs at the other is called a *"queue"* or a FIFO (first in first out list), and a queue that can go forwards or backwards is called a *"deque"* or double ended queue.

To identify these specialised lists as distinct types in Java can be achieved by grouping the methods that define these types in a similar construct to a *"class"* called an *"interface"*. The methods can all be implemented in the same class, however access to them can be limited by declaring variables of the types defined by the interface names. The implementation class is then set up to "*implement*" these interfaces.

```
interface Stack{
    public ListElement push(Object n);
    public Object pop();
}
interface Queue{
    public ListElement append(Object n);
    public Object pop();
}
class List implements Stack, Queue{
    //  as above
}
```

This allows Stack and Queue objects to be generated by statements of the form:

```
Stack stack1  =  new List( );
Queue queue =  new List();
```

In these two cases *stack1* and *queue* can only use the methods in their interface definitions, not all the methods available in the List class.

Although Java provides a library of such structures in the *Collections* package, it is often necessary to build purpose-built linked list structures for graphics applications. An important example is producing an ordered threaded list of a polygon boundary. A list of coordinates representing a polygon boundary cannot be changed without losing the structure of the boundary however it is often necessary to arrange the coordinates in sequential order.

Although it is possible to create two coordinate lists each with its own order, they are of little use unless they can be cross-linked. This requires the *ListElement* structure to be extended to allow list elements to be linked together where they refer to the same coordinate object. The extended *ListElement* structure used in later examples is defined in the following way. The reference names link1 and link2 are provided to support this cross-reference between different list structures.

```
class ListElement{
    public ListElement left=null, right=null, link1=null, link2=null;
    public Object object = null;
    public int tag = 0;
    public String name = "";
    public boolean comparable = true;

    ListElement(){ }
    ListElement(boolean comparable){ this.comparable=comparable; }
}
```

Sets, Abstract Data Types and Encapsulation

What emerges is a collection of types, which can be implemented as linked lists, indexed arrays, or as tree structures without changing their external overall behaviour. If a "black box" approach is taken to defining a data type, then only the operations needed to define the correct behaviour of objects of the type need to be made visible to the user. All that needs to be known is the correct output that can be expected from an operation, generate from a given set of inputs, without the need to specify any mechanism for turning one into the other. This introduces *abstraction* and *hierarchy*.

Set objects provide a good example of this kind of data type where there are a variety of ways that they can be implemented. One way in which the set operations of union, intersection difference and symmetric difference can be implemented is based on ordered lists and a merge operation outlined in the merge-sort algorithm illustrated in Figure 2.5. The set must be represented by an ordered list of objects where no element is duplicated. The union of two sets can then be implemented by processing the two lists sequentially the smallest values first. The two lists are merged by comparing the two "next" elements from each list, and outputting the smaller if they are different, but outputting only one, and discarding the other, if they are the same. This gives a new list that conforms to the structure of the set, having no duplicate elements and holding an ordered series of values. The intersection of the sets can be implemented by only outputting one copy of any elements that match, discarding the rest. Clearly the implementation can use linked lists or arrays, and the user does not need to know which.

If the programmer wishes to work with the abstract properties of a polygon without having to consider its implementation at a "lower" level, the true representation of a polygon can be hidden in a polygon class. When the class is implemented in a program, a particular data structure to represent the polygon can be chosen, for example a list of vertex co-ordinates. A method called area can then be written to calculate the area of a polygon modelled in this way. However, the user may provide information in various ways through different class constructors to define the polygon. The system will then have to generate the internal representation of the polygon as a list of co-ordinates from the information given to the constructors. If a particular polygon has been given the variable name *polygon1* then the area of the polygon can be returned by the "area" method using a statement of the form *polygon1.area( )*, and the real data structure need never be referred to.

Java allows objects like polygons to be treated in the same way that numbers are treated in simpler languages. Their type is defined by a class definition, which includes the operations that are permitted on the objects from the class of that type. The class definition also contains the data structures used by the sub-programs (called methods), to implement the operations on objects of the class.

If the polygon class is implemented well, then a polygon object can be worked with in much the same way that an integer object can be worked with, in expressions, relationship-tests and the like. The rules governing these operations will be different and often more complex for "higher level" objects, but a more unified programming environment results.

The object-oriented approach also allows a more natural use of names to be adopted so that program code reflects the objects, which are being worked on, like polygons, in a more direct way, than was possible in previous programming languages. In Java the main program building blocks become the class and interface definitions. These define the data structures necessary to implement objects of the class, and to manage the set of objects generated by the class. Each class has the potential mechanism to generate data structures to represent or "instantiate" multiple objects of the type that the class defines. These objects can be worked with using variables of the type the class defines, which can be used in programs, very much like variables holding numbers, by assigning object to them.

Hierarchy Inheritance and Abstraction.

This process of hiding implementation details is called encapsulation and makes many programming tasks much clearer. An important aspect of the hierarchy supported by the class structure is that one class can be defined as a refinement or modification of another class. The new class "inherits" much of its structure from its parent class, but has properties and methods of its own. This makes it possible to implement programs in a way that minimises duplication, which is very important in maintaining programs, so that changes can be carried out in as few locations as possible. It also allows template classes to be defined where the full implementation is left unfinished to be implemented by inherited classes. These are called Abstract classes illustrated by the *Tree* class given above. For example a *StringTree* class can inherit from the *Tree* class its general tree traversal methods but must provide the *output* and the *insert* procedures in the specialised form required by String objects for their *write* and *compareTo* methods.

In Java direct inheritance is permitted from only one "super" class. In real life, objects can be thought of belonging to many classificational sets. A car is a "vehicle" and also a "manufactured object". The flexibility this demands in a programming language, however, can lead to ambiguities and complex errors. Consequently, Java provides the different construction called an Interface illustrated in the case of List objects. This extends the methods and names, which can be applied to objects in one class as though they were objects from a different class of a different type

The next step is to provide equivalent input output facilities for graphic objects to that provided for text. One of the operations necessary for spatial objects such as polygons is presenting them in graphic displays. In the next chapter a simple display window class is introduced which will allow basic display operation to be executed.

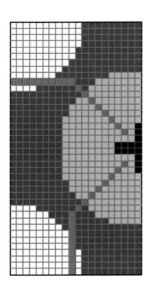

3
Programming Development Environment for Graphics

Introduction

In this chapter an outline is given of the way in which the JAVA language system will be used to illustrate modelling and display algorithms. Again a basic knowledge of the language has to be assumed. Although many aspects of the language will be discussed where they affect the way algorithms have been implemented, full details will need to be found elsewhere in the many Java Language programming and reference texts already published.

One of the many benefits that a language system such as Java provides is access to an extensive and growing library of supporting software. These libraries of classes, Java being an object oriented language system, range in application from input-output facilities, through graphic display support, to providing standard data-structure objects such as arrays, lists, trees and tables. These libraries also provide support for distributed and concurrent processing; support for databases; and eventually promise to provide much more besides, as further developments continue to evolve and be included in the original system.

The core of the language: by which is meant the grammar of its "structured programming" constructs: assignments, conditionals, repeats and function-calling statements, is virtually identical to those of the 'C' programming language. To this base has been added exception or error handling statements, a different treatment of pointers and indirect references, and the whole package is contained within an "Object Oriented" framework or harness that extends and type-checks the data-types, and data-operators that the programmer can employ. This improvement to type checking, in particular, is a major advance over earlier 'C' programming

A. Thomas, *Integrated Graphic and Computer Modelling*,
DOI: 10.1007/978-1-84800-179-4_3, © Springer-Verlag London Limited 2008

environments. It seems to be relatively easy to get underway with the Java programming language following the route laid out by books such as "On To Java" by P. Winston and S. Narasimhan; and established programmers can make the transition to standard text level programming relatively painlessly. However, getting underway with graphics and the graphics user interface facilities, can be much more difficult. This is because a more detailed knowledge of the class libraries and in particular, their inter-relationships, is necessary to obtain full control over displayed results. This is not to say that there are not many arrangements of the graphics interface, set up in a cookbook manner that are not altogether satisfactory for many purposes. However, understanding why they work and how to change them is a more serious undertaking.

Getting Started

Several generations of programmers have cut their programming teeth on the "Hello World" program. It is simple, and it gives that tremendous feedback-satisfaction of having a working program almost immediately. However, although there seems to be a backlash noted in some recent publications where this programming icon from the 70's is mocked! There is more to this starting code than merely giving a novice a psychological boost. It has been the common starting point for many seriously complicated new programs: the seed from which many larger working systems have been systematically developed.

Once a system has been analysed and designed there is still the task of coding, implementing, and testing it. A powerful, if not essential, strategy for this stage of work is to maintain a working program at all costs through all stages of system building. This must be achieved by planning the implementation route so that a working code is established, and is maintained in working order, through out development. If only small incremental changes are permitted between tests, so that the program is operational at all times, then it is possible both to maintain support and localise error checking, in a systematic way.

In other words "life" is maintained at all stages in the process! The analogy of the successful manner, in which planting a seed and nurturing it as it grows, succeeds: compared with the way that a "Frankenstein" approach of putting together a total body from dead parts then expecting to shock or electrocute it into life in one "foul", science fiction act does not, is worth noting. This is particularly relevant in discussions of reusable software! A distinction or difference between transplants and resurrection from the dead must be made! In the former case there has to be a manageable amount of damage or miss match in tissue for the operation to take, in the latter case where there are too many unknown defects and incompatibilities, resuscitation becomes improbable.

With these ideas in mind, in an attempt to provide the equivalent to the "Hello World" start up for the graphics programmer, a simple interactive interface has been pre-programmed to provide an accessible starting point for the illustrations and examples presented in the following pages. Figure 3.1 illustrates this environment, which provides interactive access to a text window and a display window, and also provides access to mouse-to-screen-pointer control inputs. The minimum needed to make progress. Setting up a *"TextWindow"* and a *"GraphicWindow"* at the beginning of a program, and then using the following methods for text input and output, for basic display operations and for interactions using the mouse, provides this program development environment.

```
class TextWindow {
    TextWindow(int col, int row,int width, int height){}    // Constructor

    public String readTextString(){}
    public String readString(){}
    public char readCharacter(){}
    public void readSpaces(){}
    public String readLine(){}
    public byte readByteInteger(){}
    public short readShortInteger(){}
    public int readInteger(){}
    public long readLongInteger(){}
    public float readReal(){}
    public double readLongReal(){}

    public void writeString(String str){}
    public void writeLine(){}
    public void writeCharacter( char ch){}
    public  void writeByteInteger(byte number, int align){}
    public  void writeShortInteger(short number, int align){}
    public  void writeInteger(int number, int align){}
    public  void writeLongInteger(long number, int align){}
    public void writeReal(float number, int align,int frac){}
    public void writeLongReal(double number, int align,int frac){}
    public void newLine(){}

    public void quit()
}
```

```
class GraphicWindow{
    public Point getCoord()
    public Rpoint  getRealCoord(CoordinateFramework b)

    public void plotPoint(Point p)
    public void plotPoint(Point p,Color cc)
    public void plotPoint(RPoint pp,CoordinateFramework b)
    public void plotPoint(RPoint pp,CoordinateFramework b,Color cc)
    public void plotLine(Point p1,Point p2)
    public void plotLine(Point p1,Point p2,Color cc)
    public void plotTriangle(Point p1,Point p2,Point p3,Color cc, Color c)
    public void plotRectangle(Point p1,Point p2,Color cc)
    public void quit()
}
```

These methods can be accessed using an object of type *GraphicWindow* in the way shown below. Notice that a reference for the *"TextWindow"* object is passed to the *"GraphicWindow"* object to allow error messages to be output.

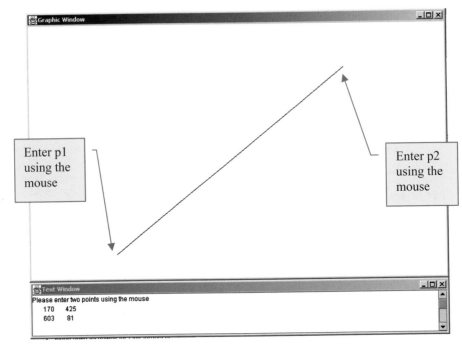

Figure 3.1 Text and graphic windows

```
public class Program3_1{
    static TextWindow IO = new TextWindow(0,500,800,100 );
    static GraphicWindow dW =
                    new GraphicWindow(IO,0,0,800,500,Color.white);

    public static void main(String[] args){
        IO.writeString("Please enter two points using the mouse");
        IO.writeLine();
        Point p1 = dW.getCoord();
        IO.writeInteger(p1.xi, 10);
        IO.writeInteger(p1.y, 10);
        IO.writeLine();
        Point p2 = dW.getCoord();
        IO.writeInteger(p2.x, 10);
        IO.writeInteger(p2.y, 10);
        IO.writeLine();
        dW.plotLine(p1,p2,Color.black);
    }
}
```

Three other classes are named and highlighted in these initial lists of methods. These are *CoordinateFramework{}*, *Point{}* and *RPoint{}*. These classes are provided to allow different floating-point co-ordinate systems to be mapped onto the integer screen co-ordinate system, which is provided as default, by the computer system. To do this it is necessary to set up a *"co-ordinate framework"* object, and to set up the scaling between it and the screen. Providing a corresponding pair of diagonal points, the first in screen co-ordinates, followed by a matching pair in the CoordinateFramework co-ordinates does this. If these co-ordinates are passed to the method *setScales(p1, p2, pa, pb)*, so that *p1* matches *pa* and *p2* matches *pb*, then screen co-ordinates can be converted to work-space co-ordinates using *scaleStoW(p)*, and work-space co-ordinates can be converted back to screen co-ordinates using *scaleWtoS(p)*.

```
class RPoint{
    public double w = 1.0, x = 0.0, y = 0.0;
    public Point toPoint();
    public RPoint copy();
    public RPoint copy(RPoint p0);
}

class Point{
    public int w =1, x = 0, y = 0;
    public RPoint toRPoint();
    public Point copy();
    public Point copy(Point p0);
}

class CoordinateFramework{
    public void   setScales(RPoint pa,RPoint pb,RPoint p1,RPoint p2)
    public Point  scaleWtoS(RPoint p1);
    public RPoint scaleStoW(Point p1);
}
```

In this arrangement *Point* co-ordinates have x and y values as integers, while *RPoint* x and y values are stored as floating point values. Care has to be taken to use the appropriate value types. Methods such as *toPoint* from *RPoint* and *toRPoint* from *Point* are provided to simplify this problem of conversion. Notice in Figure 3.2 that the screen co-ordinate y-axis is positive going down the screen while the work-space y-axis is assumed to be positive, going up the screen. This is taken into account by entering the diagonal points shown in Figure 3.2. For the working space to have a y-axis matching the display screen the top left and the bottom right corners would have to be captured instead of the pair of points shown.

The copy methods are provided in these two classes because it is often necessary to generate independent copies of co-ordinates during the course of a calculation. Merely assigning the *Point* or *RPoint* references to new point variable names can lead to errors when data for the original point is modified but the second reference is still expected to refer to the original point data. If the same approach is adopted for three-dimensional co-ordinates, then two more classes have to be defined Point3D and Rpoint3D, with similar methods for copying and changing type.

This arrangement, however, though it is clear and easy to understand in simple applications, produces fairly cumbersome code. This is because new *Point* objects have to be created to obtain graphic output from working co-ordinates, which are usually handled as real numbers. Also many operations on co-ordinates presented later are programmed using a vector or n-tuple representation treating the co-ordinate as an indexed list of real values, in other words an array of real numbers. Java makes this use of different accessing structures for the same data less easy than some other languages. The compromise solution, which has evolved from the initial approach outlined above, starts by defining a point as an n-tuple of real values, and then providing different ways of accessing these values.

```java
class NTuple {
    IOWindow tW = null;
    public double[] n = null;
    public int dimension = 0;

    NTuple(){}
    NTuple(int dim){ dimension=dim; n    = new double[dimension];}

    public void setTextWindow(IOWindow tW) {this.tW=tW;}
    public NTuple c( String st, NTuple v) {    //  copy operations
        for(int i=0;i<dimension;i++){
            if(st.equals("<-"))     {this.n[i]= v.n[i];}
            else if(st.equals("->")) {v.n[i]= this.n[i];}
        }
        if(st.equals("<-"))         return this;
        else if(st.equals("->"))   return v;
        else                 return null;
    }
    public boolean b( String st, NTuple v) {    //  boolean operations
        boolean returnValue = true;
        for( int  i=0;  i<dimension;  i++){
            returnValue = true;
            if    (st.equals("=="))  {if (this.n[i]!= v.n[i]) return false;}
            else if(st.equals("<="))  {if (this.n[i]>  v.n[i]) return false;}
            else if(st.equals(">="))  {if (this.n[i]<  v.n[i]) return false;}
            else returnValue = false;
            if(st.equals("!="))       {if (this.n[i]!= v.n[i])  return true;}
            else if(st.equals("<" ))  {if (this.n[i]<  v.n[i])  return true;}
            else if(st.equals(">" ))  {if (this.n[i]>  v.n[i])  return true;}
        }
        return returnValue;
    }
}
```

```
class Point extends NTuple{
    public int dimension

    Point()
    Point(int dim)
    Point(String st,Point p)
    Point(int w,int x,int y)

    public int wi()            public double wd()
    public int xi()            public double xd()
    public int yi()            public double yd()
    public int zi()            public double zd()

    public int w(String st,int v)    public double w(String st,double v)
    public int x(String st,int v)    public double x(String st,double v)
    public int y(String st,int v)    public double y(String st,double v)
    public int z(String st,int v)    public double z(String st,double v)

    public Point c(String st,Point v)
    public boolean r(String st,Point v)
    public Point homogenise()
}
```

In this new definition of the class *Point*, the co-ordinate is set up as an array of four double values as an n-tuple. This is then available for matrix operations and other vector operations. Conventional access to elements of the co-ordinate can no longer be made by using x and y variables, but has to be made using accessing functions. This in turn no longer permits the use of x and y variables on the left of assignment statements to change their content, and this operation will also have to be implemented in a functional form. Conventionally, this would be done in Java by using so called "getter" and "setter" functions: *p.getX()* or *p.setX(newx)*.

One of the advantages of the simple x, y notation is its use in formulae and expressions. The longer form of *p.getX()* over *p.x* makes this more cumbersome and in many cases, can obscure the form of an expression, and therefore makes the code more difficult to read. Even direct access to the *NTuple* object element *p.n[1]* instead of *p.x* only works for the real values, and makes distinguishing the coordinate axes more difficult. The integer value would need the *(int)p.n[1]* construction which again is longer and less clear. A working compromise giving reasonable flexibility and clarity is that shown above where the function *p.xi()* gives the integer value for the x co-ordinate, and *p.xd()* gives the real value. In order to convert assignment statements into this functional form and allow them to be nested in other expressions in the same way that Java permits for normal assignments -- functions of the form *p.x("=",value)*, have been defined. This, in a slightly artificial way, matches the structure of the conventional assignment statement, but unifies the treatment of the point co-ordinates under a single class definition. This redefinition of the *Point* class either requires *GraphicWindow,* its methods and related classes to be redefined or for clarity a new class *DisplayWindow* used to replace it.

```
class DisplayWindow{
    public Point  getCoord()
    public Point  getCoord(CoordinateFrame b)

    public void plotPoint(Point p)
    public void plotPoint(Point p,Color cc)
    public void plotPoint(Point pp,CoordinateFrame b)
    public void plotPoint(Point pp,CoordinateFrame b,Color cc)

    public void plotLine(Point p1,Point p2)
    public void plotLine(Point p1,Point p2,Color cc)
    public void plotLine(Point p1,Point p2, CoordinateFrame b)
    public void plotLine(Point p1,Point p2,Color cc, CoordinateFrame b)

    public void plotTriangle(Point p1,Point p2,Point p3,Color cc, Color c)
    public void plotTriangle
            (Point p1,Point p2,Point p3,Color cc, Color c, CoordinateFrame b)
    public void plotRectangle(Point p1,Point p2,Color cc)
    public void plotRectangle
            (Point p1,Point p2,Color cc, CoordinateFrame b)
    public void quit()
}
```

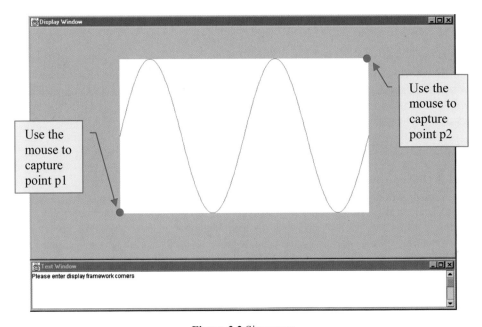

Figure 3.2 Sine wave

The example, in Figure 3.2 shows program code in which a working area is defined on the screen, which is then given the co-ordinate framework necessary to plot a sine wave from -2π to 2π. The screen co-ordinates of the required rectangle are obtained using the mouse, and the *getCoord()* method. These are integer x and y values, which must be treated as floating point numbers. These two points are then matched with the working values at the corners of the rectangular area and the scaling factors for this sub-area of the screen are then set up using a '*CoordinateFrame*' object.

```
class CoordinateFrame{
    public void   setScales(Point pa,Point pb,Point p1,Point p2)
    public Point  scaleWtoS(Point p)
    public Point scaleStoW(Point p)
}
```

The co-ordinates for the *Sine* wave graph are then calculated and plotted within this framework in the way shown by the following code segment, the output of which is shown in Figure 3.2.

```
public class Program3_2{
    static TextWindow IO = new TextWindow(0,600,1024,134 );
    static DisplayWindow dW = new DisplayWindow(IO,0,0,1024,600);

    public static void main(String[] args){
        double PI = 3.1415962;
        IO.writeString("Please enter display framework corners");
        Point pa = dW.getCoord();  Point pb = dW.getCoord();
        dW.plotRectangle(pa,pb,Color.white);
        Point pc = new Point(2);  Point pd = new Point(2);
        pc.x("=", -2.0* PI);   pc.y("=", -1.0);
        pd.x("=",  2.0* PI);   pd.y("=",  1.0);
        CoordinateFrame frameWork  = new CoordinateFrame();
        frameWork.setScales(pa,pb,pc,pd);
        int count = 200;
        double dx = PI/50.0;
        Point pLagging = new Point(2); Point pLeading = new Point(2);
        Point pl  = new Point(2); Point pn  = new Point(2);
        pLagging.x("=",pc.xd());
        pLagging.y("=",Math.sin(pLagging.xd()));
        for (int i = 0; i< count; i++){
            pLeading.x("=", pLagging.xd() + dx);
            pLeading.y("=",Math.sin(pLeading.xd()));
            pl = frameWork.scaleWtoS(pLeading);
            pn = frameWork.scaleWtoS(pLagging);
            dW.plotLine(pl,pn,Color.red);
            pLagging.x("=", pLeading.xd());
            pLagging.y("=", pLeading.yd());
        }
    }
}
```

```
public class Tiles{
    public Color[][] tileColour = null;  // array of colours
    public int rows=0, cols=0;           // number of rows and columns
    public Tiles(){ }
    public Tiles(Color cc){
        cols = 1; rows = 1;
        tileColour = new Color[cols][rows]; tileColour[0][0] = cc;
    }
    public Tiles(int numberOfColumns,int numberOfRows){
        cols = numberOfColumns; rows = numberOfRows;
        if ((cols < 1)||(rows < 1)) {
        cols = 1; rows = 1; tileColour = new Color[cols][rows];
        tileColour[0][0] = Color.white;
        }else tileColour = new Color[cols][rows];
    }
    public Tiles(int numberOfColumns,int numberOfRows,Color cc){
        cols = numberOfColumns; rows = numberOfRows;
        if ((cols < 1)||(rows < 1)) {
            cols = 1; rows = 1;
            tileColour = new Color[cols][rows]; tileColour[0][0] = cc;
        }else{
            tileColour = new Color[cols][rows];
            for (int i=0; i<cols;i++)
                for(int j=0;j<rows; j++)tileColour[i][j] = cc;
        }
    }
}
```

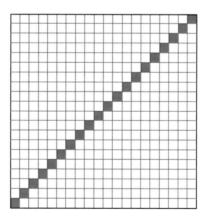

Figure 3.5 Working directly with the Tiles array

If the computer model of the display is an array of cell values then simple patterns can be defined using the standard *structured programming* constructs: expressions, conditionals, repetitions, and function calls, in the way shown in Figure 3.5, working directly on the contents of the array.

```
Tiles T = new Tiles(20,20,Color.white);
for(int j = 0; j < T.rows; j++){
    for(int i = 0; i < T.cols; i++){
        if(i == j) T.tileColour[i][j]= Color.red;
    }
}
T.display(pl);
```

An object oriented language makes it convenient for "tiles" to be set up as objects and standard tile patterns to be created by object methods such as *setBorders()* illustrated in Figure 3.6.

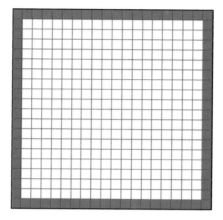

Figure 3.6 Displays defined by arrays of values set up by "pattern" methods

```
Tiles T = new Tiles(20,20,Color.white);
T.setBorders(Color.red);
T.display(pl);
```

Once a tile pattern has been created as an object it can be operated on by methods of its class of objects. One example of this results from using a tile **join** command. This is an analogous operation to the **concatenation** command, common in computer language and text processing systems. Two strings labelled A and B can be joined together to give a new string labelled C by a command of the form C = A+B. The same approach can be used to define a tile pattern C by joining together two smaller tile patterns A+B, or using a class function *A.join(B)* .

Object oriented computer languages are consequently particularly suited to the task of working with pre-made patterns. Though it is reasonably simple to concatenate characters and strings using standard language data structures, the extension into two dimensions makes the equivalent task more difficult for tiles. The entities being joined require more complex data structures, and the facilities to police operations on these structures, becomes more important if the system is to remain robust and well behaved. A complex repetitive tile pattern can be defined using this approach, by code of the form:

```
Tiles A = new Tiles(Color.green);
Tiles B = new Tiles(Color.white);
Tiles C = new Tiles(Color.red);
Tiles D = (D=(B.join(B)).clockwise()).join(D);
Tiles H = (H=(A.join(A)).clockwise()).join(H);
Tiles J = (J=(C.join(C)).clockwise()).join(J);
Tiles E = A.join(C).anticlock();
Tiles F = C.join(C).anticlock();
Tiles G = E.join(F).clockwise();
Tiles K = H.join(G).anticlock();
Tiles L = G.join(J).anticlock();
Tiles N = K.join(L).clockwise();
Tiles Q = (Q=(D.join(D)).clockwise()).join(Q);
Tiles O = Q.join(N).anticlock();
Tiles P = Q.join(Q).anticlock();
Tiles R = (R = P.join(O).clockwise()).join(R.reflected());
Tiles S = R.anticlock().join(R.clockwise());
Tiles T = (T = S.join(S).anticlock()).join(T);
T.setBorders(Color.gray);
```

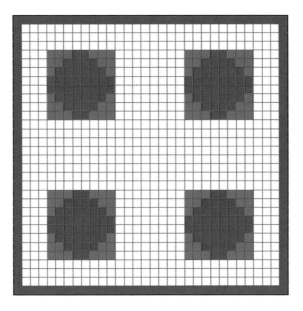

Figure 3.7 Repetitive tile pattern **T** generated by hierarchical join operations

Although the code in this example looks relatively complicated, and the object-based syntax for the join statement loses some of its potential 'expression-like' simplicity, When applied in the appropriate way, this approach saves a lot of time, space and effort, 19 lines of code in this case defining 1000 tile colours and locations. In this program segment an alternative *Tiles* constructor is illustrated which defines single cells to give unitary red, green and white tiles. The method *setBorders(Color.gray)* is used to set the edge cells of the final tile pattern grey. Also the *"Tiles" join()* method can be seen in use: linking simpler tile patterns together.

Notice that since this *'join'* operation is only defined in the horizontal direction, its 'operand' tile patterns have to be oriented correctly before they can be connected together. Also the join command can only succeed if the tile patterns being joined, have the same number of elements along the two sides being brought together. These spatial rearrangements are carried out by the three *"Tiles"* methods, *clockwise()*, which rotates a pattern a quarter circle in a clockwise direction, *anticlock()* which rotates a pattern a quarter circle in an anti- clockwise direction and *reflected()* which as its name suggests reflects a pattern horizontally about a vertical line.

These re-orientation operations can be implemented by rewriting the arrays used to represent the tile patterns in a relatively simple way, but care has to be taken working with the array indexes to maintain their correct relationship to the position of the tile pattern in a display space.

A Clockwise Quarter Turn Operation

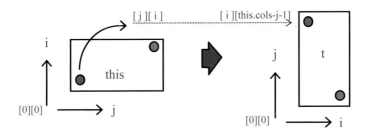

```
public Tiles clockwise(){
    Tiles t = new Tiles(this.rows,this.cols);
    for (int i=0; i<this.rows; i++){
        for(int j= 0; j<this.cols; j++){
            t.tileColour[i][this.cols-j-1]=this.tileColour[j][i];
        }
    }return t;
}
```

An Anti Clockwise Quarter Turn Operation

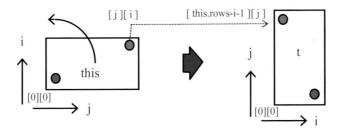

```
public Tiles anticlock(){
  Tiles t = new Tiles(this.rows,this.cols);
  for (int i=0; i<this.rows; i++){
    for(int j= 0; j<this.cols;j++)
    { t.tileColour[this.rows-i-1][j] = this.tileColour[j][i];}
  }return t;
}
```

A Reflection Operation

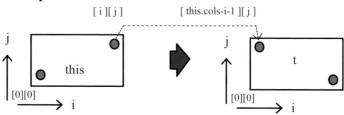

```
public Tiles reflected(){
  Tiles t = new Tiles(this.cols,this.rows);
  for (int j = 0; j < this.rows; j++){
    for(int i = 0; i < this.cols; i++)
    { t.tileColour[this.cols-i-1][j] = this.tileColour[i][j];}
  }return t;
}
```

A Join Operation

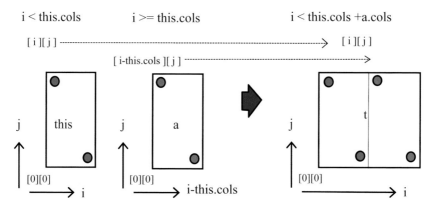

```
public Tiles join(Tiles a){
Tiles t = new Tiles(this.cols+a.cols,this.rows);
    for (int j=0; j<t.rows; j++){
      for(int i= 0; i < this.cols + a.cols; i++){
        if(i<this.cols) t.tileColour[i][j]= this.tileColour[i][j];
        else t.tileColour[i][j]= a.tileColour[i-this.cols][j];
      }
    }return t;
}
```

Combining these two approaches using an asymmetric tile pattern illustrates the need for: both rotation and reflection, reorientation-commands to obtain all possible configurations in the resulting tile layouts:

```
Tiles T = new Tiles(20,20,Color.white);
for(int j=0;j< 20; j++){
    for(int i=0; i<20; i++){
        if (((T.cols-i)*(T.cols-i)+(T.rows-j)*(T.rows-j))>144)
            T.tileColour[i][j]= Color.blue;
        if((i*i)+(j*j)< 144)T.tileColour[i][j]= Color.cyan;
        if((i<10)&&(j<10)&&(i>1)&&(j>1)&&(i==j))
            T.tileColour[i][j]= Color.red;
        if((j>9)&&(i==10))T.tileColour[i][j]= Color.red;
        if(((i<4)&&(j<4))&&((i==0)||(j==0)))
            T.tileColour[i][j]= Color.black;
    }
}
```

A bent arrow pattern Tile T is set up by the code given above for a 20 by 20 array. This is then joined with different re-orientations of itself to give the sequence of tile patterns X, Y, Z and finally a much larger pattern A in Figures 3.8 to 3.11.

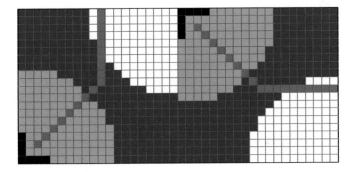

Figure 3.8 Clockwise rotation and join of a tile pattern: **X = T.join(T.clockwise( ));**

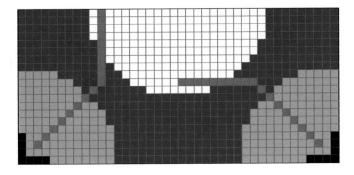

Figure 3.9 Anti-clockwise rotation and join of a tile pattern **Y = T.join(T.anticlock( ));**

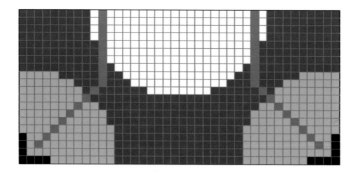

Figure 3.10 Reflection and join of a tile pattern: **Z = T.join(T.reflected( ));**

When working with complex tile patterns it is not always that easy to produce the desired result, just by rotating reflecting and joining tiles, at least without careful prior planning. If the target is to produce a predefined pattern such as that shown in Figure 3.11, then the simplest approach, realising that the target pattern has to be constructed by joining simpler patterns together, is to systematically subdivide it. Generating identical sub-tiles, where possible, will reduce the reconstruction work.

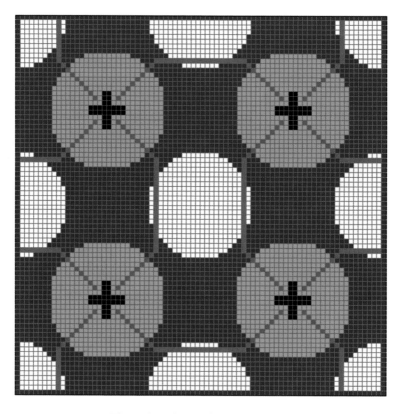

Figure 3.11 Composite Tile pattern A

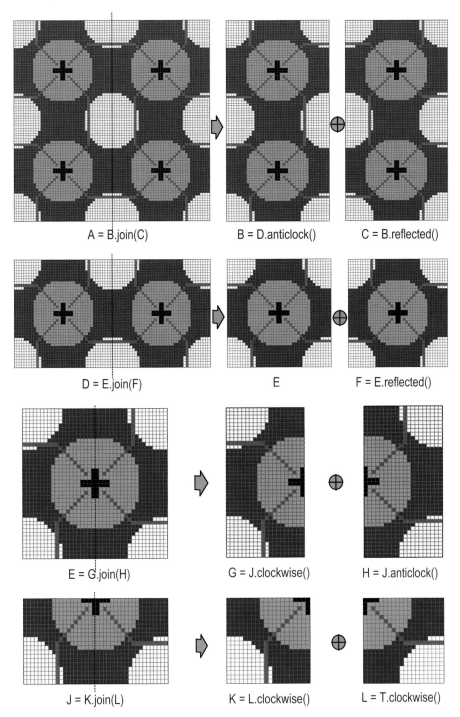

Figure 3.12 Tile pattern A constructed from Tile pattern T

The sequence building tile A from tile T is obtained by subdividing A in the systematic way illustrated in Figure 3.12. The final tile pattern shown in Figure 3.11 shows the addition of a border and introduces the next step in this examination of the ways in which arrays of colour values can be combined under computer language control to create graphic models. In this case two patterns are overlaid on each other. The blue of the border having priority over what was already on tile A. Using different combination commands can vary the way in which two overlaid patterns are combined.

Rectangle Set on Rectangle Set

An overlay task that needs to be carried out very often in an interactive window system is establishing the interrelationship of window rectangles as they are moved around the screen relative to each other. In early systems with limited memory the way this was done was critical for keeping memory usage within bounds and to allow the operation to be carried out fast enough to support real time interactive work.

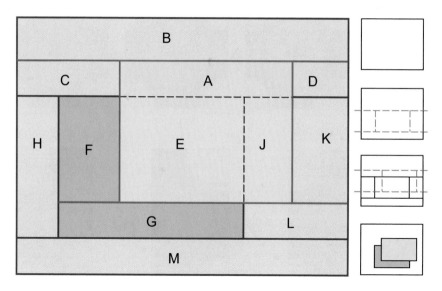

Figure 3.13 Rectangular partitioning of overlapping windows

The key operation is that shown in Figure 3.13. The process is one of partitioning the display space into rectangles. This allows fast memory transfer *"bitblt"* operations to be used to update the display, merely by repainting modified rectangles. Block memory transfer operations are primitive operations very similar to the tile operations, which can be implemented at the hardware or firmware level in the display system. If display memory is part of the computers main memory then the array of pixel values has to be accessed fast enough to keep up with the refresh cycle of the TV monitor. In early systems this could not be achieved without the special hardware outlined in chapter 5. The key task is to access the pixel values from the image array a row at a time and pass the pixels in sequence to the display system at the required speed. If the image array consists of w columns and h rows, and the data is stored in a contiguous block of memory cells starting at address *base*, then a procedure of the following form needs to be executed.

```
int address = base,  width = w, height = h, row = 0, col = 0, step = 1;
while(row<height){
    while(col<width){
        displayPixel(displayMemory[address]);
        address=address+1;  col=col+1;
    }
    address=address+step; row=row+1; col=0;
}
```

If a rectangle within this pixel array needs to be modified then this block accessing code can be extended in the following way. Given a rectangular block of width w1, and height h1 with its first pixel located at image-array index-position [x, y]: the pixel locations for this block can be accessed in the following way to paint the rectangle red.

```
Int width = w, height = h, row = 0, col = 0, step =w-w1, blockHeight = h1, blockWidth =w1;
int address = base + width*y+ x;
while(row< blockHeight){
    while(col< blockWidth){
        displayMemory[address] = Color.red;;
        address=address+1;  col=col+1;
    }
    address=address+step; row=row+1; col=0;
}
```

Variations of these block memory operations allow rectangular images to be overlaid on existing images, or combined with them in various ways. One useful operation is combining existing pixel data with new pixel data using an exclusive or operation. If this is repeated then the initial image is returned to its original state. Drawing cursors and construction lines using this technique can be very helpful when executing many interactive tasks where an element of trial and error is needed.

Another operation at the pixel level in a block memory transfer operation is combining pixel values as though an image is translucent or transparent. Including the transparency of a pixel as a component of an image pixel value allows these fast block memory transfer commands to support a range of real time image manipulation tasks that greatly extends the scope of much interactive work.

Transparent and Translucent Overlays

A traditional way for graphic designers to work employs overlapping transparent or partially transparent media so that images from different sources or generated in different ways, can be combined together. Even in oil painting, glazes, which are transparent or translucent overlays, are used to modify a lower layer pattern or image. The most direct application of this approach can be found in the cell animation employed in cartoon films.

Each layer of a scene can be manipulated independently from frame to frame allowing a whole range of movement effects to be created with very few "basic operation" types. Technical drawings are also usefully handled in this layered way, for example, allowing different engineering service plans to be seen related to each other showing how plumbing and air conditioning ducts relate to electrical circuits. This allows information in pictures to be selectively added in or left out, as their use requires it. Geographic information systems, from early on in their development, have used the *map model* concept in which different geographical distributions are constructed and stored as separate overlays for the same base map, so they can be displayed separately, together or in any functional combination required.

The traditional technical roles for the graphics model: data analysis and design, communication, and storage are now being split up and reallocated. The roles of analysis and storage are passing to the computer model, leaving the graphic model primarily for communication and interactive design. How far is the array of colours, used to create the display, also able to act as a spatial model for analysis? The next step is to investigate its use for other purposes than just display. The alternative is to find spatial models suitable for design, analysis and storage purposes, which can when necessary be converted into arrays for display purposes.

Interactive Tile Pattern Definitions

An interesting sequence of modelling issues is posed by the classical task of route finding through a maze. The array model will allow a maze to be represented by using one colour for walls and another colour for open space. The question which will be examined in the next section, is what more is needed to allow this spatial layout model to be used to support route finding from a start cells to a finish cell beyond the support the colour-array, display-model can already provide.

Figure 3.14 Hand edited maze layout

The maze pattern in Figure 3.14 was obtained by interactively modifying a basic tile pattern, using the mouse. In order to display this pattern it was necessary to convert the tile colour-array into a display pixel array. This was done using a ***Grid*** object to act as a bridge between the colour array model of the tile pattern and the pixel grid of the display screen. A simple way to do this is to multiply up each colour array element by an integer number, so that, for example, each cell becomes a block of 16 by 16 pixels. This has the drawback however, that it sets a fixed size to the resulting display.

Since display systems can have a variety of different resolutions, a more flexible solution is to take the index ranges of the tile grid and treat these as floating point values. For example an array of 10 by 20 cells can be treated as a continuous space from 0.0 to 10.0 for *x* and 0.0 to 20.0 for the *y* values. These can then be used to set up a *CoordinateFrame* object *cF* by relating them to a convenient working rectangular area on the screen, using the mouse, in the way already described for the drawing of the sine function shown in Figure 3.2. The *dW.getCoord(cF)* function can then be used to get the real value co-ordinates of a point: the *x* and *y* values of which, will, when truncated to integers, give the indexes to access the colour-array cell corresponding to the tile or grid-square being pointed to by the mouse.

In order to display tile patterns in a standard way a display method has been included in the *Tiles* class. This sets up a display *grid* object which requests a rectangular window to be defined on the screen into which it paints the tile pattern, scaled to fit. The *display* command has to be passed a reference to the "*TextWindow*" in order to have access to a text window to print out interactive messages. The code for the *Tiles* class is:

```
class Tiles{
    public Color[][] tileColour = null;
    public int cols=0, rows=0;

    public Tiles(){ }

    public Tiles(int numberOfColumns,int numberOfRows,Color cc){
        cols = numberOfColumns; rows = numberOfRows;
        if ((cols <1)||(rows<1)) {
            cols =1; rows =1;
            tileColour =new Color[cols][rows]; tileColour[0][0] = cc;
        }else{
            tileColour = new Color[cols][rows];
            for (int i=0; i<cols;i++)
                { for(int j=0;j<rows; j++) tileColour[i][j] = cc;  }
        }
    }
    public Tiles(int numberOfColumns,int numberOfRows){
        cols = numberOfColumns; rows = numberOfRows;
        if ((cols <1)||(rows<1)) {
            cols = 1; rows = 1;
            tileColour = new Color[cols][rows];  tileColour[0][0] = Color.white;
        }else { tileColour = new Color[cols][rows]; }
    }
}
```

```java
public Tiles(Color cc){
    cols = 1; rows = 1;
    tileColour = new Color[cols][rows];
    tileColour[0][0] = cc;
}

public void setBorders(Color cc){
    for(int i=0;i<this.cols;i++)
        { this.tileColour[i][0]= cc; this.tileColour[i][rows-1]= cc; }
    for(int i=0;i<this.rows;i++)
        { this.tileColour[0][i]= cc; this.tileColour[cols-1][i] = cc; }
}
public Tiles clockwise(){
    if (this.tileColour == null)return null;
    Tiles t = new Tiles(this.rows,this.cols);
    for (int i=0; i<this.rows; i++){
        for(int j= 0;j<this.cols; j++)
            { t.tileColour[i][this.cols-j-1] = this.tileColour[j][i]; }
    }return t;
}
public Tiles anticlock(){
    if (this.tileColour == null)return null;
    Tiles t = new Tiles(this.rows,this.cols);
    for (int i=0; i<this.rows; i++){
        for(int j= 0; j<this.cols;j++)
            { t.tileColour[this.rows-i-1][j] = this.tileColour[j][i];}
    }return t;
}
public Tiles reflected(){
    if (this.tileColour == null)return null;
    Tiles t = new Tiles(this.cols,this.rows);
    for (int j = 0; j < this.rows; j++){
        for(int i = 0;i<this.cols; i++)
            { t.tileColour[this.cols-i-1][j] = this.tileColour[i][j];}
    } return t;
}
public Tiles join(Tiles a){
    if ((this.tileColour == null)||(a == null)
                    ||(a.tileColour == null)||(this.rows != a.rows))return null;
    Tiles t = new Tiles(this.cols+a.cols,this.rows);
    for (int j=0; j<t.rows; j++){
        for(int i= 0; i<this.cols+a.cols;i++){
            if(i<this.cols)t.tileColour[i][j]=this.tileColour[i][j];
            else t.tileColour[i][j]=a.tileColour[i-this.cols][j];
        }
    }return t;
}
```

```
public void display(TextWindow f){
    Grid d = new Grid(f,this.tileColour,this.cols,this.rows);
    d.paintGridArray(); d.drawGridLines(Color.black,Color.gray);
}
}
```

The *Grid* class provides a series of tile-based operations based on scaling the rows and columns of the tile pattern to fit a user-defined rectangle on the screen. The display grid class sets up a co-ordinate framework in this rectangle that generates co-ordinates to match the row and column indexes used to reference tiles in the grid. A "*CoordinateFrame*" object provides this scaling, in the following way.

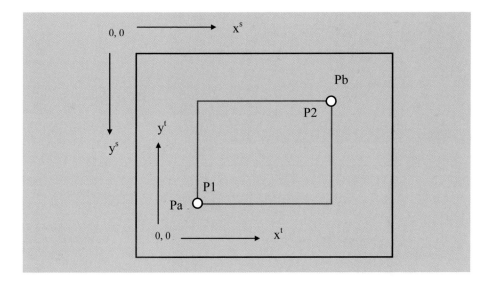

Figure 3.15 Co-ordinate framework: screen and tiles co-ordinates

The basic co-ordinate system of the display system is the row and column, integer, index pairs used to reference pixels on the display screen. These indexes may relate directly to the hardware defining the memory addresses of the pixels in the frame buffer memory. Or in the Java environment it may relate to a pixel based co-ordinate with its origin defined by the upper left-hand corner of a *Container* such as a *Canvas,* which is displayed inside a window.

The task for the *CoordinateFrame* object is to map a different co-ordinate system onto this underlying pixel array. The only assumption or constraint is that the corresponding axes from the two systems are parallel, so the mapping operation has to handle only scale and origin differences. The mapping functions are worked out from the two diagonal corner points generally provided to define the working area on the screen. These points are represented in both systems so that the relationship between them can be calculated, allowing subsequent points to be transformed to match. In Figure 3.15 the two points are P1 and P2 in screen space corresponding to

Pa and Pb in the tile space. The values of the screen co-ordinates and the tile pattern co-ordinates are (x^s_1, y^s_1) matching (x^t_a, y^t_a), and (x^s_2, y^s_2) matching (x^t_b, y^t_b). Since the x and y axes are orthogonal relationships along the x axis and the y axis can be treated independently. If the x axis values are analysed first, then it is possible to show the linear relationship between the two co-ordinate systems by labelling the screen x- values "u", and the tile x-values "v".

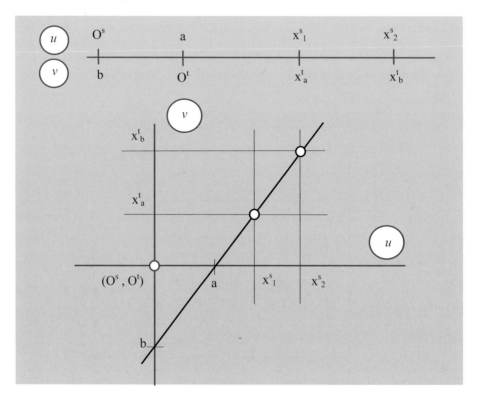

Figure 3.16 CoordinateFrame scaling X axes

This allows the relationship between the two to be drawn up as the line graph shown in Figure 3.16. From this the equation of the line can be written in the intercept form:

$$\frac{u}{a_x} + \frac{v}{b_x} = 1$$

cross multiplying gives: $0 = a_x.v + b_x.u + c_x$ $(c_x = -a_x.b_x)$

the slope of this line becomes: $\dfrac{dv}{du} = \dfrac{-b_x}{a_x}$

the coefficients of this equation can be calculated by the code

```
ax = p2.x-p1.x;
bx = pb.x-pa.x;
cx = pa.x*p2.x - p1.x*pb.x;
```

The same treatment can be applied to the y axes:

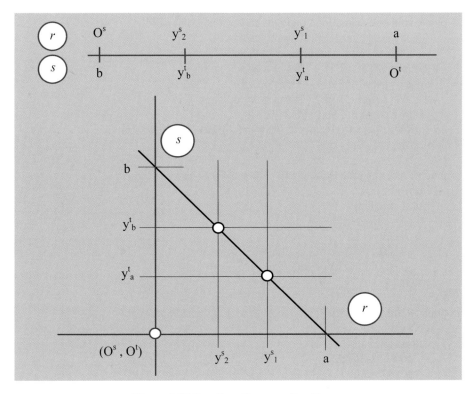

Figure 3.17 CoordinateFrame scaling Y axes

Again the equation of the line can be written in the form: $0 = a_y.s + b_y.r + c_y$
The coefficients of this equation can be calculated by the code:

```
ay = p2.y - p1.y;
by = pb.y - pa.y;
cy = pa.y*p2.y - p1.y*pb.y;
```

Once these coefficients are set up in a *CoordinateFrame* object *q* by the method *q.setScales(..)* then any point obtained from the screen with the mouse using *getCoord()* can be converted using the method *q.scaleStoW(..)* or can be obtained directly using the *getCoord(q)* which converts the screen co-ordinates to *CoordinateFrame q* co-ordinates.

Similarly any points calculated in the context of a *CoordinateFrame q*, can be converted to screen co-ordinates using the method *q.scaleWtoS(..)*.

```
class CoordinateFrame{
    public boolean scaleSet = false;
    public double ax, bx, cx, ay, by, cy;


    public CoordinateFrame(){ }

    public void setScales(Point pa,Point pb,Point p1,Point p2){
        if((pa==null)||(pb==null)||(p1==null)||(p2==null))
            { this.scaleSet = false; return;}
        this.ax = p2.xd()-p1.xd();    this.bx = pb.xd()-pa.xd();
        this.cx = pa.xd()*p2.xd() - p1.xd()*pb.xd();
        this.ay = p2.yd()-p1.yd();    this.by = pb.yd()-pa.yd();
        this.cy = pa.yd()*p2.yd() - p1.yd()*pb.yd();
        if((ax==0.0)||(bx==0.0)||(ay==0.0)||(by==0.0)) this.scaleSet = false;
        else this.scaleSet = true;
    }
    public Point scaleWtoS(Point p1){       //  scale work-space to screen
        Point p0 = null;
        if (this.scaleSet){
            p0 = new Point(2);
            p0.n[0] = 1.0;
            p0.n[1] = ((this.bx*p1.xd() + this.cx)/this.ax);
            p0.n[2] = ((this.by*p1.yd() + this.cy)/this.ay);
        }return p0;
    }
    public Point scaleStoW(Point p1){       //  scale screen to work-space
        Point p0 = null;
        if (this.scaleSet){
            p0 = new Point(2);
            p0.n[0] = 1.0;
            p0.n[1] = (this.ax*p1.xd() - this.cx)/this.bx;
            p0.n[2] = (this.ay*p1.yd() - this.cy)/this.by;
        } return p0;
    }
}
```

Although it is possible to work directly with a *CoordinateFrame* object it is convenient when working with grids and tile patterns to set up a new class, *Grid* to bridge between the programming level and the display level of work. The *Grid* variables and constructors are:

```
class Grid extends CoordinateFrame {
    public int width, height;  private TextWindow f=null;
    public DisplayWindow dW=null; public int rows = 0,cols = 0;
    private Color[][] array = null;  private Color[][] dualArray = null;
    private double[][] values = null;
    private Point pa=null, pb=null, pc=null, pd=null;
```

```
Grid(){}
Grid(TextWindow t, DisplayWindow dW ,Color[][] a,int cols,int rows){
    int width = cols; int height = rows; this.f = t; this.dW = dW;  this.array = a;
    if((a==null)||(cols<1)||(rows<1)) f.writeString("Null array for display \n");
    }else setGridDisplay(a,cols,rows,width,height);
}
Grid(TextWindow t,DisplayWindow dW,
                        Color[][] a,int cols,int rows,int width,int height) {
    this.f = t; this.dW = dW;
    if((a==null)||(cols<1)||(rows<1))  f.writeString("Null array for display \n");
    else   setGridDisplay(a,cols,rows,width,height);
}
public void setGridDisplay(Color[][] a,int cols,int rows,int width,int height){
    this.array=a;  this.rows=rows;  this.cols=cols;
    dualArray=new Color[cols+1][rows+1];
    for(int i=0;i<cols+1;i++)
        for(int j=0;j<rows+1;j++){dualArray[i][j]=Color.white;}
    f.writeString("Please Enter Diagonal Corners of Display Grid \n");
    pa = dW.getCoord();  pb = dW.getCoord();
    if (pa.xd()>pb.xd())
        {double save = pa.xd(); pa.n[1] = pb.xd(); pb.n[1] = save;}
    if (pa.yd()>pb.yd())
        {double save = pa.yd(); pa.n[2] = pb.yd(); pb.n[2] = save;}
    pc  = new Point(2);  pc.n[1] = 0.0;              pc.n[2] = (double)height;
    pd  = new Point(2);  pd.n[1] = (double)width;   pd.n[2] = 0.0;
    setScales(pa,pb,pc,pd);
}
```

This arrangement allows direct access, using the mouse, to modify the colour array model and sets up a third way of generating tile patterns: *interactive composition by the system user*. Methods from the *Grid* class provide different ways for manipulating tile images and their colour array models. An example of the way these facilities can be used to manually modify a display pattern: to set up a 20 by 30 tile pattern for the display of a maze layout made up from black tiles on a white setting, is as follows:

```
int Cls = 30, Rws = 20;
Tiles T = new Tiles(Cls,Rws,Color.white);
T.setBorders(Color.black);
Grid d = new Grid(IO, dW, T.tileColour, Cls,Rws,Cls,Rws);
d.paintGridArray();
d.drawGridLines(Color.black,Color.gray);
int j=0, k=0; boolean test = false;
IO.writeString("Use the mouse pointer to enter wall cells"); IO.writeLine();
IO.writeString("To finish click mouse outside the maze"); IO.writeLine();
do{
    Point p = d.getCell();
```

```
            k = p.xi(); j = p.yi();
            double r = p.xd(); double s = p.yd();
            double cl = (double)Cls;  double rw = (double)Rws;
            if(test =(((cl-r)*(-r)<0)&&((rw-s)*(-s)<0))){
                if(d.getCellColor(k,j) == Color.black){
                    d.paintInnerCell(k,j,0,Color.white);
                    }else{ d.paintInnerCell(k,j,0,Color.black);}
            }
        }while(test);
```

In order to support maze-solving applications in the next chapter, using a grid based display to draw mazes -- various methods were added to this class. Different ways for painting in the grid cells showing search paths and backtracking paths in distinguishable ways needed to be provided.

```
public Point getCell ( )
{    Point p1 = dW.getCoord();  Point p = scaleStoW(p1);  return p;  }
```

The procedure *getCell()* returns a Point co-ordinate of type double in the number range defined by the grid array indexes. For example if the mouse pointer is pointing inside cell [5][6], the return values might be [5.4003][6.3145], in which case to use these values they will have to be truncated to their corresponding integer values, before they can be used to access the colour array.

```
public boolean contains(Point p){
    if((p.xi()>0)&&(p.xi()<=cols) && (p.yi()>0) && (p.yi()<=rows)) return true;
    else return false;
}
```

The *contains* method allows a point to be tested to see if it lies inside the rectangle of the grid. A method for determining the existing colour of a particular grid cell and one for setting a grid cell to a new colour was essential for interactive work and was written as follows:

```
public Color getCellColor(int i, int j) { return array[i][j]; }
public void setCellColor(int i, int j, Color cc) { array[i][j] = cc; }
```

where *getCellColor(..)* and *setCellColor(..)* are two methods that can be used if the colour array is not directly accessible to the user.

```
public void paintArrayCell(int i, int j, Color cc){
    Point p1 = new Point(2);  Point p2 = new Point(2);
    p1.n[1] = (double) i;  p1.n[2] = (double) j;
    p2.n[1] = p1.xd() + 1.0;  p2.n[2] = p1.yd() + 1.0;
    Point pa = scaleWtoS(p1);  Point pb = scaleWtoS(p2);
    array[i][j] = cc;
    dW.plotRectangle(pa,pb,cc);
}
```

The *paintArrayCell(.)* method fills in the cell indexed *[i][j]* with the colour defined by *cc.* The *paintGridArray()* procedure paints in the whole grid, directly from its object's colour array.

```
public void paintGridArray() ){
   for(int i =0;i<cols; i++){
      for(int j = 0; j<rows; j++)
         { paintArrayCell(i,j,array[i][j]); }
   }
}
```

The *paintInnerCell(.)* method gives a more versatile cell painting procedure which leaves a frame *k* pixels wide round the edge of the painted cell area.

```
public void paintInnerCell(int i, int j, int k, Color cc){
   Point pq1 = new Point(2);     Point pq2 = new Point(2);
   pq1.n[1] = (double) i;        pq1.n[2] = (double) j;
   pq2.n[1] = pq1.xd() + 1.0;    pq2.n[2] = pq1.yd() + 1.0;
   Point p1 = scaleWtoS(pq1);  Point p2 = scaleWtoS(pq2);
   if(k>0){
      if((Math.abs(p1.xd()-p2.xd())<Math.abs(2*k))
                          ||(Math.abs(p1.yd()-p2.yd())<Math.abs(2*k))){
         if(Math.abs(p1.xd()-p2.xd())<Math.abs(p1.yd()-p2.yd()))
            k = Math.abs(p1.xd()-p2.xd())/2;
         else k = Math.abs(p1.yd()-p2.yd())/2;
      }
   }
   p1.n[1] = p1.xd()+1+k;       p1.n[2] = p1.yd()-k;
   p2.n[1] = p2.xd()-k;         p2.n[2] = p2.yd()+1+k;
   array[i][j] = cc;
   dW.plotRectangle(p1,p2,cc);
}
```

The *drawGridLines(..)* procedure draws in the grid lines separating the separate coloured cells. It permits the outer boundary line to be of colour *"boundary"* distinguishing it, if required, from the colour *"cc"* of the remaining grid lines.

```
public void drawGridLines(Color boundary,Color cc){
   Point pq1=null, pq2=null;
   pq1 = new Point(2);
   pq2 = new Point(2);
   for(int i =1;i<cols; i++){
      pq1.n[1] = (double) i; pq1.n[2] = (double) 0;
      pq2.n[1] = (double) i; pq2.n[2] =(double) rows;
      Point p1 = scaleWtoS(pq1); Point p2 = scaleWtoS(pq2);
      dW.plotLine(p1,p2,cc);
   }
```

```
for(int j = 1; j<rows; j++){
    pq1.n[1] = (double) 0;      pq1.n[2] = (double) j;
    pq2.n[1] = (double) cols;  pq2.n[2] = (double) j;
    Point p1 = scaleWtoS(pq1); Point p2 = scaleWtoS(pq2);
    dW.plotLine(p1,p2,cc);
}
for(int i=0;i<=cols;i=i+cols){
    pq1.n[1] = (double) i;      pq1.n[2] = (double) 0;
    pq2.n[1] = (double) i;      pq2.n[2] =(double) rows;
    Point p1 = scaleWtoS(pq1); Point p2 = scaleWtoS(pq2);
    dW.plotLine(p1,p2,boundary);
}
for(int i=0;i<=rows;i=i+rows){
    pq1.n[1] = (double) 0;      pq1.n[2] = (double) i;
    pq2.n[1] = (double) cols;  pq2.n[2] =(double) i;
    Point p1 = scaleWtoS(pq1); Point p2 = scaleWtoS(pq2);
    dW.plotLine(p1,p2,boundary);
}
}
```

The procedure *drawPath(.)* has been included specifically for maze drawing, allowing the current cell , to be linked to the next cell in the direction defined by the variable *direction*, by a line segment joining the centres of the cells. Where the current cell is defined by a Point object *p2* in which the *p2.x* and *p2.y* values correspond to the first and second indexes of the cell's position in the colour array. The use of this method will be illustrated in the next chapter

```
public void drawPath(int direction, Point p, Color cc)
    Point p1 = new Point(2);      p.c("t>",p1);
    Point p2 = new Point(2);      p.c("t>",p2);
    switch (direction){
        case 0:  p2.x("=",p2.xd()+1.0); break;
        case 1:  p2.y("=",p2.yd()+1.0); break;
        case 2:  p2.x("=",p2.xd()-1.0); break;
        case 3:  p2.y("=",p2.yd()-1.0); break;
    }
    p1.x("=",p1.xd()+0.5);        p1.y("=",p1.yd()+0.5);;
    p2.x("=",p2.xd()+0.5);        p2.y("=",p2.yd()+0.5);;
    Point pa = scaleWtoS(p1);  Point pb = scaleWtoS(p2);
    dW.plotLine(pa,pb,cc);
}
```

Figure 3.14 illustrates a maze pattern constructed by defining a *Tiles* object, 30 cells wide by 20 cells high, setting up its boundary in black, and then hand editing in the remaining pattern using the code given above. Automatically searching a maze for its finish point is discussed in the next chapter as an introduction to computer based problem solving.

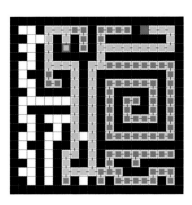

4
Conditional Action
- Spatial Searching
& Problem Solving

Representing a Maze

If the pattern generating process described in the previous chapter is taken as a starting point, then it is clear that a graphic model of a maze can be prepared using an "***array of colours***" as its corresponding computer model. Working out a solution to the maze by hand would involve marking up the route from the start point to the finish point on a picture of the maze, and this would generally include redrawing a few false exploratory attempts.

Unconditional Commands

If this marking up process is to be supported by the computer system then it will be necessary to represent the walls by one colour the open space by another and the route by either a third colour or by a line showing the path followed. Marking up the route in Figure 4.1 could then be implemented by a sequence of simple move and turn commands, which select the cells in the array model which need to be modified to show the path taken, in the following way.

```
M.setStart(11,3);
M.setFinish(21,10);
M.turnAntiClockwise();
M.move(3);
M.turnAntiClockwise();
M.move(5);
M.turnClockwise();
M.move(10);
M.turnClockwise();
M.move(6);
```

A. Thomas, *Integrated Graphic and Computer Modelling*,
DOI: 10.1007/978-1-84800-179-4_4, © Springer-Verlag London Limited 2008

```
M.turnClockwise();
M.move(3);
M.turnClockwise();
M.move(3);
M.turnAntiClockwise();
M.move(3);
M.turnAntiClockwise();
M.move(12);
```

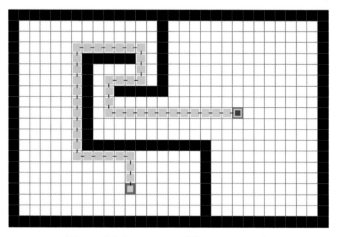

Figure 4.1 A sequence of direct commands

This operation merely automates the display task. It is clear that in order to do it, a set of adjacent cells, which are not wall cells, that link the start position to the finish position, have to be located. This has to be done by looking at the colours of the cells in the display, and then specifying the sequence of computer commands needed to mark up the selected route.

The next step is to see whether the representation used to create the display can be used to automatically select the cells, which were chosen manually in this example, by using an appropriate sequence of computer language statements. It is clearly possible to test a cell to see if it is a "wall cell" by comparing its colour with the colour selected to represent a wall. The question arises whether by testing these colour values in a systematic way, the route finding operation can be automated in a general-purpose search algorithm looking for the "finish" cell.

Conditional Commands

Examining the maze solving problem step by step: if simple direct instructions are employed, such as *move( )* or *turnClockwise( )*, then only by viewing the maze and selecting an appropriate sequences of commands, can the cursor be moved so that it does not hit a wall. In maze M0, in Figure 4.2 a move of 3 places, stays in open space. If in contrast a move of 30 steps is attempted in the way illustrated in the Figures 4.3 then a wall is hit!

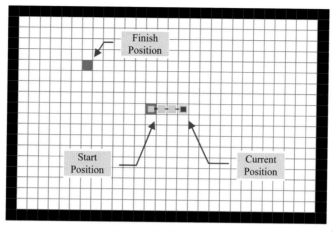

Figure 4.2 M0.move(3)

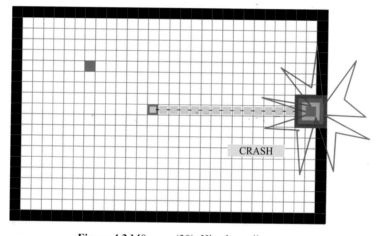

Figure 4.3 M0.move(30): Hits the wall

However, if a move is made one place at a time, testing the contents of the next cell before moving in to it, which can be done because the colour array representing the maze will have different colour codes for walls and for open space, then a test *wallAhead()* can be programmed and used in the following way:

```
if ( ! M.wallAhead()) M.move(); else M.turnClockwise();
```

"If the next cell is open space then move", is a conditional statement that can be repeated safely using a *for(;;){}* loop without fear of hitting a wall, and also can be extended to act differently, for example by turning clockwise, if a wall is found in the next cell. However, if this command is repeated 30 times the result is that shown in Figure 4.4. What is needed is a repeat operation, which will continue until a test indicates that the finish or target cell has been found. Such a test can be provided as a Boolean function *finished()*, by comparing the current position, array index-pair, with

that of the predefined finish position index pair. Such a test can be used to terminate a conditional repeat loop when or if a move command locates the finish cell.

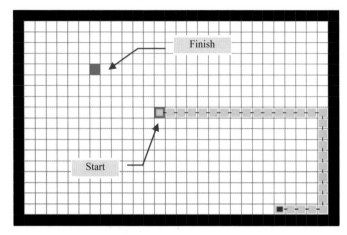

Figure 4.4 (move if the next cell does not contain a wall, otherwise turn) 30 times

```
for(int i=0;i<30;i++) {
    if(M.wallAhead()) M.turnClockwise(); else M.move();
}
```

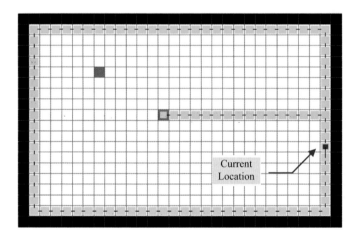

Figure 4.5 An infinite loop where the target will never be reached

```
while(!M.finished()){
    if (M.wallAhead())M.turnClockwise();else M.move();
}
```

In the case of maze M0, in Figure 4.5, simply replacing the *for* loop with a *while* loop still does not help. The program shown simply follows the outer wall forever. At this point the maze-solving problem must be examined separately from the graphics display task. If the finish position is to be found a systematic search of the total space of the maze is required.

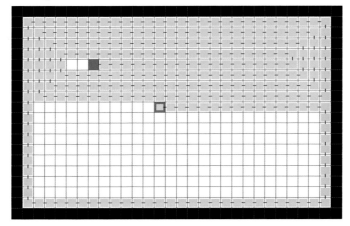

Figure 4.6 M0, a simple space filling search algorithm finding the target

```
while( ! M.finished() ){
    if( M.wallAhead() ) M.turnClockwise();
    else if( M.alreadyVisited() ) M.turnClockwise();
    else M.move();
}
```

If a cell has already been visited then clearly it is desirable not to visit it again if it can be avoided. This can again be achieved using the colour codes stored in the maze colour array to indicate that they have been visited, then setting up and using an *alreadyVisited()* test in the conditional move command, in the way shown above.

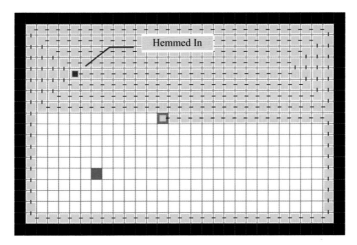

Figure 4.7 M1, a space filling search algorithm, hemmed in, in an infinite loop

Figure 4.9 Maze M2 : the general maze solving algorithm

Figure 4.9 illustrates the general maze solving procedure in action. An interesting variation on its coding which makes the relationship between *hemmedIn()* and *wallAhead()* and *alreadyVisited()* more explicit, can be obtained by observing that there are three tests, and yet there are only three actions. If the full decision table for the three tests is laid out there are, potentially, eight actions that can be discriminated from each other. However, only five test combinations are possible, and two pairs require identical actions. This suggests the use of a switch statement. If the truth values of the separate tests are combined together as a single decimal number *j* shown in the fourth row of the decision table it can be used as a switch variable. This switch variable can be generated either by Boolean bit-wise operations or a matching use of arithmetic operations in the way illustrated in the following code sequence:

alreadyVisited()	1	true			0	false		
wallAhead()	1	true	0	false	1	true	0	false
hemmedIn()	1 true	0 false	1 true	0 false	1 true	0 false	1 true	0 false
j	7	6	5	4	3	2	1	0
ACTIONS	*	*	Back Track	Turn	Back Track	Turn	*	Move

```
while(!M.finished()){
    int j=0;
    if (M.hemmedIn())      j=j+1;
    if (M.wallAhead())     j=j+2;
    if (M.alreadyVisited()) j=j+4;
    switch(j){
        case 3:case 5: M.backTrack();    break;
        case 2:case 4: M.turnClockwise(); break;
        default:M.move();
    }
}
```

Having generated a solution to the maze-searching problem using a "bottom up" exploratory approach, the next stage is to see if there are any ways in which the resulting algorithm can be refined. A direct approach to this task is to layout as many different ways of representing the algorithm's structure as possible. One application of a decision table analysis is to rearrange the order of tests in nested conditional statements to group common actions together to simplify code. In this case there are not many possibilities however one pair of actions can be merged in the following way. If the *wallAhead()* and *alreadyVisited()* tests are combined as the single test *obstructed()*

```
public void obstructed() { return (this.wallAhead()|| this.alreadyVisited());  }
```

this will support the code:

```
while(!M.finished()){
        int j=0;
        if (M.hemmedIn())      j=j+1;
        if (M.obstructed())    j=j+2;
        switch(j){
            case 3: M.backTrack();      break;
            case 2: M.turnClockwise();  break;
            default:M.move();
        }
}
```

Switch or case statements are very effective constructs for handling a range of simple geometrical programming problems. There are many cases where a discrete number of patterns characterise a particular task while the variations in the problem space are continuous. If tests can be set up to classify the patterns involved then the switch statement allows a function table to be set up; each function able to handle the particular patterned interpretation of the otherwise continuously varying data. These patterns can be considered to be states of the continuous problem, so allowing it to be turned into a finite state problem by the appropriate classification and naming of parts. The maze-solving algorithm turns into the finite state machine given in Figure 4.10.

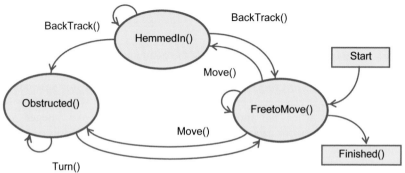

Figure 4.10 Maze-solving automata

The Maze Class

Maze objects were set up from Tiles objects by passing the colour array of a tile object and converting it into a new maze "integer" array model *mazeArray*. This was done by converting wall colours to wall codes '-1', open space from white space coloured cells to codes of '0', and then setting up start and finish cells in the colour array part of the maze model as red and magenta cells respectively.

```
class Maze{
    private TextWindow f;
    private Grid d = null;
    private Color[][] tileArray = null;
    private int delay = 100000, maxValue = 0;
    private boolean singleStep = false;
    public int[][] mazeArray = null;
    public int cols = 0, rows = 0, direction = 0, count = 1;
    public Point start = null, finish = null, cursor = null;

    Maze(){}

    Maze(Tiles t, Grid g, TextWindow h){
        this.f = h; this.d = g; this.cols = t.cols; this.rows = t.rows;
        this.tileArray = t.tileColour;                // link to a tile colour array
        if ((this.cols <1)||(this.rows<1)){
            this.cols = 0; this.rows = 0;this.mazeArray = null;
        } else{
            this.mazeArray = new int[this.cols][this.rows];
            for (int i=0; i<this.cols;i++){
                for(int j=0;j<this.rows; j++){
                    if(t.tileColour[i][j] == Color.black) this.mazeArray[i][j] = -1;
                    else if(t.tileColour[i][j] == Color.gray) this.mazeArray[i][j] = -1 ;
                    else this.mazeArray[i][j] = 0;
                }
            }
        }
    }    /* end of maze constructors */
```

The maze methods were built to provide the commands and tests discussed above based on this array model of the maze. The direct commands were *move()*, *turnClockwise()*, *turnAntiClockwise()*, and *backTrack()*. In order for these commands to be implemented it was necessary to define the direction in which any movement would take place when the move command was issued. An integer variable 'direction' was defined for each maze object along with a current location point for the cursor. The direction variable was allowed to hold one of four values 0, 1, 2, 3. to represent East, North, West, and South in corresponding order. This made it possible to define the turn clockwise operation as that of subtracting one from the 'direction', modulo 4, and turn anticlockwise operation as adding one to the 'direction', modulo 4.

```
public void turnClockwise()
{   direction = direction -1; if (direction < 0) direction = 3;}
public void turnAntiClockwise()
{   direction = direction +1; if (direction > 3) direction = 0;}
```

The direction variable could then be used in a switch statement to select the code that would implement the appropriate modification to the current cursor position for the move commands in the following way:

```
public void move(){
    int i = this.cursor.xi(); int j = this.cursor.yi();
    switch (direction){
      case 0: this.cursor.x("=",i+1); break;
      case 1: this.cursor.y("=",j+1); break;
      case 2: this.cursor.x("=",i -1 ); break;
      case 3: this.cursor.y("="j -1 ); break;
    }
    this.mazeArray[i][j] = this.count++;
}
```

However, testing the next adjacent cell is done repeatedly, and is most neatly implemented using a separate index-increment method controlled by the direction, saving a switch statement in every procedure that wishes to look ahead:

```
public int nextl(int i,int k,int direction){
    int dir; if((dir = (direction+k)%4) == 1 )return ++i;
    if(dir == 3)return --i; else return i;
}
```

This simplifies the move method to give:

```
public void move(){
    int i = this.cursor.xi(); int j = this.cursor.yi();
    this.cursor.x("=", this.nextl(i,1,direction));
    this.cursor.y("=", this.nextl(j,0,direction));
    this.mazeArray[this.cursor.xi()][this.cursor.yi()] = this.count++;
    /* Mark this cell yellow*/
}
```

```
public void backTrack(){
    this.maxValue = 0; int i = this.cursor.xi(); int j = this.cursor.yi();
    checkDirection(i+1,j,0); checkDirection(i,j+1,1);
    checkDirection(i-1,j,2); checkDirection(i,j-1,3);
    /* Mark this cell green */
    this.cursor.x("=", this.nextl(i,1,direction));
    this.cursor.y("=", this.nextl(j,0,direction));
}
```

The backtrack command is more complicated in that the appropriate direction to move has to be determined by comparing the surrounding tag values in the maze array cells. The direction for the backtracking move is set up by a separate procedure *checkDirection(..)*

```
private boolean checkDirection(int i,int j,int dir){
    if(!outOfBounds(i,j) && (mazeArray[i][j]>this.maxValue)){
        if(mazeArray[this.cursor.xi()][this.cursor.yi()]>mazeArray[i][j]) {
            this.maxValue = mazeArray[i][j];
            this.direction = dir;  return true;
        }
    }
    return false;
}
```

The various tests are relatively simple comparisons of either index values or next door neighbour, array cell values in the following way:

```
public boolean finished(){
    if((finish.xi()==cursor.xi())&&(finish.yi()==cursor.yi())) return true;
    return false;
}
```

```
public boolean wallAhead(){
    int i = this.cursor.xi(); int j = this.cursor.yi();
    i = this.nextI(i,1,direction); j = this.nextI(j,0,direction);
    if(outOfBounds(i,j)||(mazeArray[i][j]<0)) return true;
    else return false;
}
```

```
public boolean alreadyVisited(){
    int i = this.cursor.xi();      int j = this.cursor.yi();
    i = this.nextI(i,1,direction);  j = this.nextI(j,0,direction);
    if(!outOfBounds(i,j) && (mazeArray[i][j]>0))  return true;
    else return false;
}
```

```
public boolean hemmedIn(){
    int i = this.cursor.xi();  int j = this.cursor.yi();
    if    (openSpace(i-1,j))return false;
    else if (openSpace(i+1,j))return false;
    else if (openSpace(i,j-1))return false;
    else if (openSpace(i,j+1))return false;
    /* Mark the cell green */
    return true;
}
```

These procedures developed step by step as the maze problem was analysed. The final step in this sequence was to take the finite state machine shown in Figure 4.10 and attempt to simplify its structure. The inputs to trigger the transitions between the states are relatively complex tests on the data in the array representing the maze, round the current cell. Figure 4.11 illustrate a reconstruction of this state diagram to give a different state transition system where these tests are simplified to identifying if the next cell directly ahead in the direction being travelled is free to be moved into. This still leaves the backtrack operation as a relatively complex task, needing to interrogate more than one cell's contents to determine its implementation. This information can be gathered in the *Obstructed* states ready for use if the system enters the *HemmedIn* state.

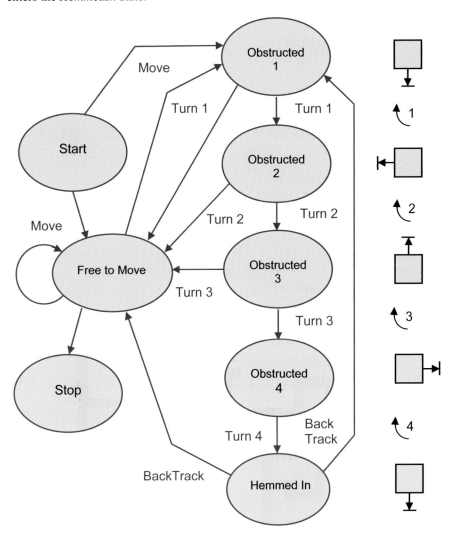

Figure 4.11 State machine for maze solving

```
public int x,y,nx,ny,X,Y, state=0, savevalue=0, savedir=0, dir=0, cellcount=1;
public int[ ] xincr = new int[]{1,0,-1,0};    public int[ ] yincr = new int[]{0,1,0,-1};
public void finiteStateMachine(){
    x= start.xi(); y= start.yi(); X= finish.xi(); Y= finish.yi();
    dG.paintInnerCell(x,y,0,Color.magenta);
    while((x!=X)||(y!=Y)){
        switch(setstate()){
            case 0:move(Color.yellow,2); mazeArray[x][y]=cellcount++; break;
            case 1:case 2:case 3:case 4:  checknext();dir=(dir+1)%4;break;
            case 5:dir=savedir;  move(Color.green,4);break;
            }nx = x+xincr[dir]; ny = y+yincr[dir];
        }dG.paintInnerCell(x,y,0,Color.cyan);
}
private void move(Color cc,int b){
    int xx=x,  yy=y;
    x=x+xincr[dir];  y= y+yincr[dir];
    dG.paintInnerCell(xx,yy,b,cc);
    Point pa = new Point(2); Point pb= new Point(2);
    pa.x("=",x); pa.y("=",y); pb.x("=",xx); pb.y("=",yy);
    dG.drawPath(pa,pb,Color.black);
}
private int setstate(){
    if(mazeArray[nx][ny]==0)state=0;
    else if(state<5){state++; return state;}
    else state =1;
    savevalue=0; savedir=dir; return state;
}
private void checknext(){
    if((mazeArray[nx][ny]<mazeArray[x][y])&&(mazeArray[nx][ny]>savevalue))
            {savevalue=mazeArray[nx][ny];  savedir = dir;}
}
```

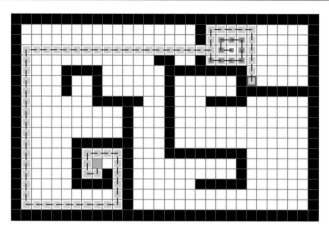

Figure 4.12 Finite state algorithm

This way of evolving a programming solution to a problem by trying out different ways of combining primitive operations and tests, and building up towards a solution is called *Bottom Up* design. It can be contrasted with the subdivision method employed in the case of the *Tiles* examples, which by starting with the whole problem was able to work from the top downwards to a solution: an approach called *Top Down* design, a form of structured reverse engineering!

Once an initial algorithm has been put in place it is possible to evaluate it and refine it. The observation made earlier that a tree-structured search results from the "footsteps in the snow" strategy, opens up several important possibilities. However, firstly it is necessary to check that this assertion is valid. If it is then a more efficient and terse program can be written to achieve the same result that the "game playing", exploratory-approach, has achieved so far.

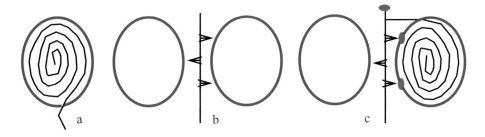

Figure 4.13 Search patterns

A maze is assumed to be made up from a set of connected cells. By this is meant that a route exists from any cell to any cell through adjacent cells, if the appropriate sequence is chosen. There are no disjoint sets of cells in the original collection and every cell is adjacent to at least one other.

If a path is taken through a maze from cell to adjacent cell, but only entering cells that have not previously been visited, then two outcomes are possible. In the first case all cells in the set are visited, case (a) in Figure 4.13. In the second case only some of the cells are visited before a cell is reached from which no progress can be made, but where there are other cells in the initial set that still have to be visited. Figure 4.13(b).

At this point consider the collection of unvisited cells. They may be one connected set, or they may be divided into several internally connected subsets disjoint from each other. Even though separated these subsets must be connected through the earlier search path because of the initial condition that they start out as one connected set. The separation is by cells that have already been visited. Since the search path is linear, it is clear that backtracking along it will potentially visit all these subsets in order. If each sub-area that is located in this back track sequence is searched fully then at the end of the back track the whole area of the maze will have been searched.

The search problem created by entering one of these sub-areas is identical in nature, to the initial search problem. It can therefore be treated recursively. This was the property, which led to the assumption that the search path would be a tree structure. To ensure that this is the case it is necessary to show that each sub area can be totally searched and that the search path returns to the entry point used for accessing each sub-area, when backtracking. The latter is necessary to ensure that the backtrack-route can fully search its initial search path, for all links to sub-areas, and cannot bye pass any links to sub-areas. For example, by entering an area at point 1 on its route but exiting back onto its search path at point 2, in between which links to other areas might lie.

Although a sub area may have several potential links to an initial search path, they will be inaccessible to normal move commands because they will already have been visited as shown in Figure 4.13 (c). This means that the only route out of an area is by using a backtrack operation, which by definition must exit by the route used for entry. The areas will be fully searched because unless the target cell is found in a sub area, this algorithm will hierarchically continue to subdivide it until all the sub-areas it finds can be fully explored. This argument allows the following program to be written:

```
public boolean mazeSearch(int i,int j){
    delays();
    Point p = new Point();  p.x("=",i);   p.y("=",j);
    for(int k =0; k<4; k++){
        int ii = nextl(i,1,k);
        int jj = nextl(j,0,k);
        if((finish.xi()==ii)&&(finish.yi()==jj)) return true;
        if (mazeArray[ii][jj]==0){
            mazeArray[ii][jj] = 1;
            /* mark cell [ii][jj] yellow */
            if ( mazeSearch(ii,jj))return true;
        }
    }
    /* mark cell [i][j] green */
    return false;
}
```

In this algorithm when a cell has been visited it is marked with a single value 1. This means this approach can be employed using the original colour-array model of the maze, since a single colour can be used in the same way as this integer value. The reason why this is possible is because this method is recursive. Each call of the procedure will place the local variables i and j onto the program stack, effectively modelling the route using the *"ball of string"* approach, which makes the marking up of a maze array model to record the track, unnecessary.

The algorithms explored so far in this chapter will search for a target cell in a closed maze. Once such a cell has been located, which in the worst case may have meant visiting all the cells in the maze at least twice, the next task to consider, is finding the best route from the start of the maze to its finish point. This optimisation process will be discussed in the next section.

Backtracking and Optimum Route Finding

Looking at the display in Figure 4.14 an interesting pattern is evident. The cells that have not been backtracked over provide an almost direct route from the beginning to the end of the maze. The places where they do not are rectangular areas of yellow, like that on the right hand side of the display. This suggests the next stage in the evolution of this program.

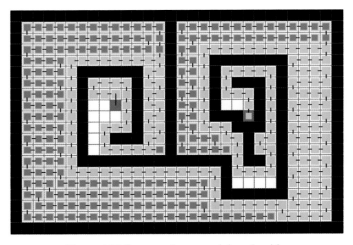

Figure 4.14 Tree search maze solving algorithm

If as before the cells are marked with ascending integer values as the search is carried out, but cells that are backtracked over are set back to open space, then the yellow cells will be in ascending order from the start to the finish. Once the finish point is found the method *backTrace()*, by moving to the neighbouring cell with the smallest marked value, can trace out the direct route, and because of the ordering of cells in rectangular areas of yellow cells, will select a boundary route through them, as shown in Figure 4.15.

```
public boolean mazeSearch (int i,  int j){
    for(int k =0; k<4; k++){
        cursor.x("=",( int ii = nextl(i,1,k));  cursor.y("=",(int jj = nextl(j,0,k));
        if((finish.xi()==ii) && (finish.yi()==jj)){
            mazeArray[ii][jj] = this.count++;
            return true; }
        if (mazeArray[ii][jj]==0){
            mazeArray[ii][jj] = this.count++;
            /* mark cell [ii][jj] yellow */
            if ( mazeSearch(ii,jj))return true;
        }
    }
    /* mark cell [ii][jj] green */
    mazeArray[i][j]= 0; return false;
}
```

```
public void backTrace(){
    int savei=0, savej=0, i=cursor.xi(), j=cursor.yi(),
    minvalue = mazeArray[i][j];
    for (int k = 0; k<4; k++){
        int ii = nextl(i,1,k); int jj = nextl(j,0,k);
        if ((mazeArray[ii][jj]>0) && (minvalue > mazeArray[ii][jj]))
        {   savei = ii; savej = jj;   minvalue = mazeArray[ii][jj]; }
    }
    cursor.x("=",savei); cursor.y("=",savej);
    /* mark cell [savei][savej] red */
}
```

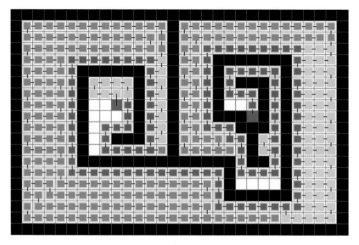

Figure 4.15 Tree search and back trace algorithm

At first sight a shortest route algorithm seems to be appearing, however when the mazes shown in Figure 4.16 are compared it can be seen that though it is an optimised route from tree pruning, it is not the shortest route. In each row, in the two examples, the target position is placed above and below the mid-point of the intervening wall. The search path in yellow, in each case fills the search space from the top, because of the arbitrary initialising conditions in the program. Consequently the optimised route goes above the wall even when the shortest route should take a path below the wall. It is clearly unable to do this because the search has found the target working from the top before any of the lower cells are visited.

The tree algorithm given, implements what is in effect a depth first search strategy. A route is followed as far as possible until, being blocked, the search process is forced to backtrack. As has already been observed, where a set of cells has several entry points from an initial pathway, only one will be explored. This exploration will then block the other routes by labelling the alternative access cells as having "*already been visited*". To find the shortest route it is necessary to explore the paths that might be generated by following these alternative access points.

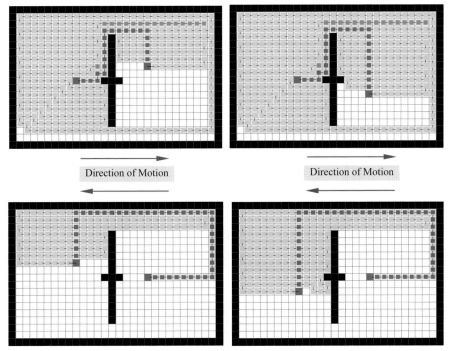

Figure 4.16 Depth first tree search and back trace algorithm

This can be done, by implementing the tree search algorithm in a different way. In the previous approach once a cell had been visited then it was blocked from subsequent visits. If this constraint is relaxed then the tree search will visit every cell many times from all possible routes. To find the shortest route it is necessary to do two things: one is to reduce unnecessary visits in this recursive process, secondly, as part of this task, to keep an account of the minimum distances travelled.

```
public void branchAndBound(int i, int j, int cnt){
    if(this.maxCount<cnt ) return;
    mazeArray[i][j] = cnt;
    if((finish.xi() == i)&&(finish.yi() == j)){
        this.maxCount = cnt;  cursor.x("=",i); cursor.y("=",j);  return;
    }
    for(int k =0; k<4; k++){
        int ii = nextl(i,1,k); int jj = nextl(j,0,k);
        if (mazeArray[ii][jj]==0){
            /*  paint cell  yellow */
            branchAndBound(ii,jj,cnt+1);
        }
        else if (mazeArray[ii][jj]>(cnt+1)) branchAndBound(ii,jj,cnt+1);
    }
    return;
}
```

All possible routes through a grid of cells where each route is defined by visiting most of the cells for each route definition will involve an unreasonably large amount of computation, if it is not limited in some way. Each call of the tree searching method can keep a value for how far it has travelled by keeping a count of how many times it has called itself. In the recursive revisits to cells if it finds that it has travelled further than some other path has taken to get to the same point, then it can stop and return.

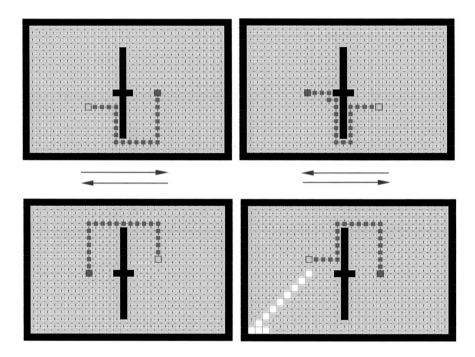

Figure 4.17 Branch and bound determination of the shortest route

To allow this to operate the minimum travel distance to each cell so far found needs to be recorded in the maze cell as an integer number. Another limitation that can be set on researching new routes, which is associated with the name given to this approach, *"branch and bound"*, is that once a route has been found to the target cell there is no point in pursuing other routes that are longer than this solution. As shorter routes are found they can be used to update this minimum search path distance, so reducing subsequent searches.

Experiments with this algorithm are shown in Figure 4.17 for the test maze layouts used before. The result of using the back trace algorithm to select the optimal route generates a shortest *"Manhattan"* route in each case. The problem with this approach is that, even with the constraints on unnecessary searching it is still fairly time consuming. Not only are the whole areas of the mazes shown in Figure 4.17 visited in the search pattern they are visited many times, as all the alternative routes are explored.

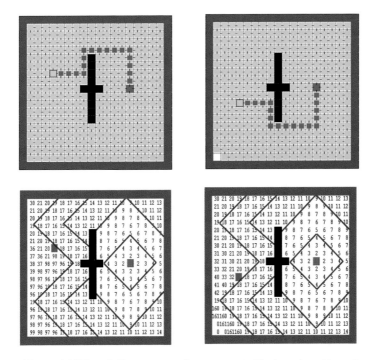

Figure 4.18 Travel-distance wave fronts generated by branch and bound

In Figure 4.18 the result of applying the branch and bound algorithm on the maze-array tag-distance values is shown. The distribution of values is the same for the same start position, wherever the target is placed. The branch and bound tree search generates a wave fronts of equal travel distance values in rings round the start cell which produces the V shape shown, where the wave front folds round the wall and meets up on its other side. It is this, which enables the back trace procedure to find the shortest way back to the origin.

If the distance values in the yellow shaded area are examined, this approach becomes similar to steepest ascent or steepest descent, optimising algorithms. The initial search path creates rings round the source point each "higher" than its immediate predecessor, which means that the back trace path simply has to follow the steepest route down to get to the origin by the shortest route.

If the general algorithm for finding the shortest or fastest route through a network shown in Figure 4.19 is taken as the starting point, then it can be seen that a wave front approach is also employed, followed by backtracking to locate the successful route.

Working out from the start box S, in Figure 4.19 the first step is to fan out to all the directly adjacent cells. Each of these cells will collect the time or distance cost of getting to them. The next step is to fan out from the cell with the lowest values in this case from A to D. The cost of getting to D will be 4, this is still less than the cost of getting to B or C, so continue fanning out from D. This will take the wave front to cells G and B. This will lower the cost of getting to B from seven to six.

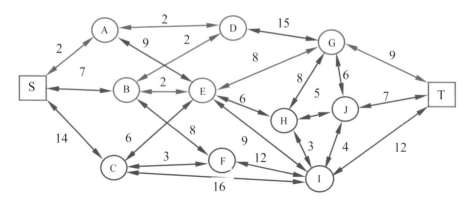

Figure 4.19 Shortest route or critical path

From B the next step is to E, where the cost goes up to 8, which is still less than any of the other routes from A. The next step from E is to C, F, G, H and I. The cost of getting to H is 14, the same as that getting from A to C, and from B to F, therefore the next tree structured links fan out from C, F and H to G, J, and I.

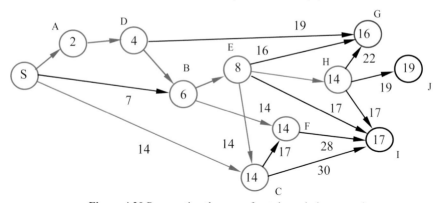

Figure 4.20 Propagating the wave front through the network

The cell with the minimum distance of 16 is now G. Fanning out from G gives the first link to the terminal box T, however there may be other shorter routes. Fanning out from C and F merely links to I. Fanning out from H links to J and I. Completing the tree-structured expansion following the minimum distance cells gives the final links to the end cell T, in the way shown in Figure 4.21. Backtracking through the links, which have contributed to the minimum score of 25 at the terminal cell defines the shortest route shown coloured red.

Working with a grid is simpler than this general network because there are only unit steps from cell to cell. Notice that each grid cell is four-way connected to its neighbours. However, if a sequential wave front expansion is generated then these links are divided to give one entry link followed by three, two, one or no exit links, depending on whether neighbours are walls or have already been visited.

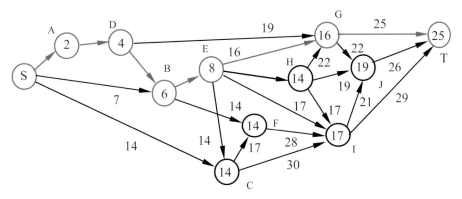

Figure 4.21 A tree search through the network followed by a backtrack

In the case of the grid it is not necessary to find the minimum distance taken to reach the cell in the way required in the general algorithm, since this will be one more than the cell providing its entry link.

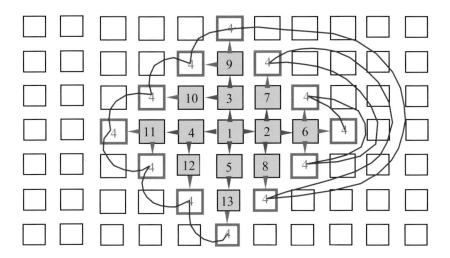

Figure 4.22 Tree structured wave front generation

These links created by the wave front expansion give a tree structure in the way shown in Figure 4.22. The numbers in the yellow cells indicate one possible ordering in which they might be visited by the search algorithm. The red cells indicate the newly generated wave front, and the purple line indicates the threaded list set up to link the wave front cells together. The order of the cells in this list is controlled by the order of the visits to the neighbouring cells made by the search algorithm and the nature of the linked list used.

One implementation of the algorithm is two nested iterative loops. The inner loop searches all the neighbouring cells, not already visited, adjacent to the current wave front path, labels them with the wave front number shown in red and creates from them a new wave front list. The outer loop processes successive wave fronts until the target cell is located or the maze space has been totally explored. Given the tree structure shown, all the cells in each wave front lie at the same depth in the tree. This is the reason for classifying this approach as a breadth first tree search.

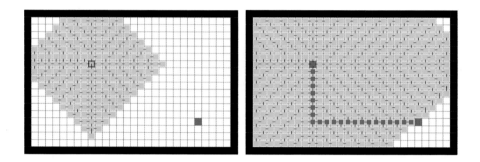

Figure 4.23 Expanding wave-front search followed by back trace

In Figure 4.23 the outward flow of the wave front is shown followed by the back tracing operation to give the shortest route. In Figures 4.24 the corresponding maze layouts to those shown in Figure 4.16 are given to illustrate the selection of the true shortest route.

The original search area covers alternative routes from the start to the finish so the *back tracing* will select the shortest one, depending on the tag values it finds in the neighbouring maze cells. However it must be noted that there are still alternative layouts for this path because the shortest route is defined as a "*Manhattan*" distance.

The selections made in Figure 4.24 are based on the *backTrace()* algorithm which selects the smallest neighbouring tag value starting in the Easterly direction and then rotating round through North, West and finishing in a Southerly direction.

The same maze relationships are shown in Figure 4.25 but in this case using the original *backTrack()* procedure to find the shortest route. This employs a selection operation where the order for testing neighbouring cells is West, East, South then North, which gives the alternative shortest route layouts shown, for comparison with the output from the *backTrace()* algorithm. Having made this distinction, Figure 4.26 shows that the route selected will also differ in detail if the start and finish locations are reversed.

One advantage that these tree-structured searches provide is to give methods for handling the same task in three or more dimensions. Working out a systematic search path in three dimensions is not an easy problem however extending this tree structured search to enter five neighbouring cells in a three dimensional grid provides a relatively easy starting point.

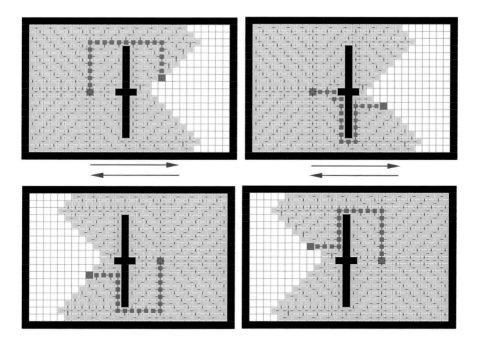

Figure 4.24 Breadth first tree search and **backTrace()** method

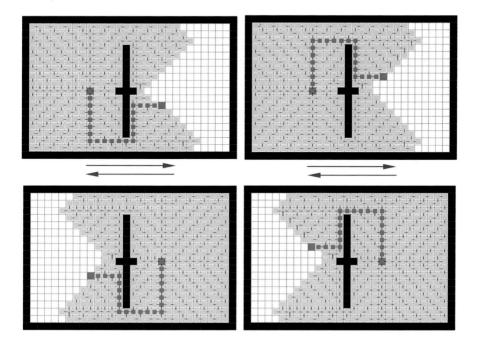

Figure 4.25 Breadth first tree search and **backTrack()** method

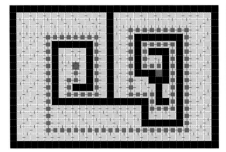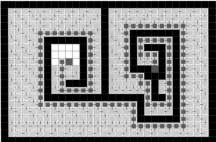

Figure 4.26 Forward and backward shortest routes

```
class MCell{
   public int i=0,j=0; MCell nxt = null;
   MCell() { }
}

class Maze{
...
   public void ShortestRoute(){
      int i = this.cursor.xi(); int j = this.cursor.yi(); int ii=0,jj=0;
      MCell lst = null, cll=null;  MCell cell = null;
      boolean finished = false;  mazeArray[i][j] = 1; int level = 2;
      cell = new MCell(); cell.i = i; cell.j=j;
      while(!finished){
         while((cell != null)&&(!finished)){
            i = cell.i;   j = cell.j;   p.x("=",i);   p.y("=",j);
            for(int k = 0; k<4; k++){
               ii = nextl(i,1,k); jj = nextl(j,0,k);
               if (mazeArray[ii][jj]==0){
                  /* mark cell [ii][jj] yellow */
                  tileArray[ii][jj] = Color.yellow;
                  mazeArray[ii][jj] = level;
                  cll = new MCell();   cll.i=ii; cll.j=jj; cll.nxt = lst; lst = cll;
               }
               if((finish.xi() == ii)&&(finish.yi() == jj)) {finished = true; break;}
            }cell = cell.nxt;
         }level = level+1; cell = lst;count = level;lst=null;
      }
      this.cursor.x("=",ii); this.cursor.y("=",jj);
      while(mazeArray[cursor.xi()][cursor.yi()]>1){
         backTrack();                                        // or backTrace();
         /* mark cell [cursor.x][cursor.y] red */
      }
   }
...
}
```

It is instructive to visualise the way in which both the hill climbing, and steepest descent, shortest-route finding algorithms work, using a three-dimensional display of the search distances to each cell, which can also be constructed from these maze arrays in a reasonably simple way.

If the tag values generated by the search path in the maze array are projected upwards, perpendicular to the maze plan, then a three-dimensional structure is generated. If each of these cell values is scaled and converted into a vertical bar then a simple drafting technique can be used to create a projected drawing of the resulting block model.

The first step is to rotate the maze grid through 45 degrees. If the cells are considered to be square then the geometry of the orthogonal grid allows a secondary grid to be set up, shown coloured blue in Figure 4.27. The bars can then be defined as six triangles with co-ordinates determined by the layout shown on the left of Figure 4.27. This is a parameterised shape, which adjusts its vertical size depending on the *value* used to define its height.

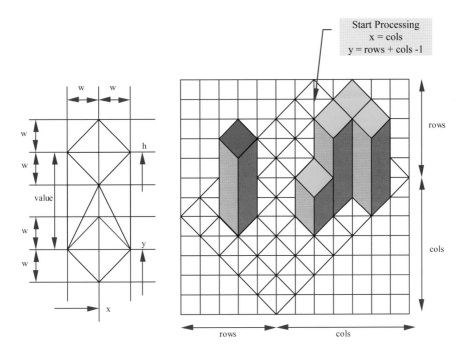

Figure 4.27 Axonometric drawing

If bars are processed from the back of the grid working forward then each new bar will overwrite the previous bars to give an image of the form shown in Figure 4.27. In the examples given the vertical dimensions are determined by the rectangle entered into the display window using the mouse. The parameterised shape is then scaled to fit. This gives the apparently less distorted images shown in Figure 4.28.

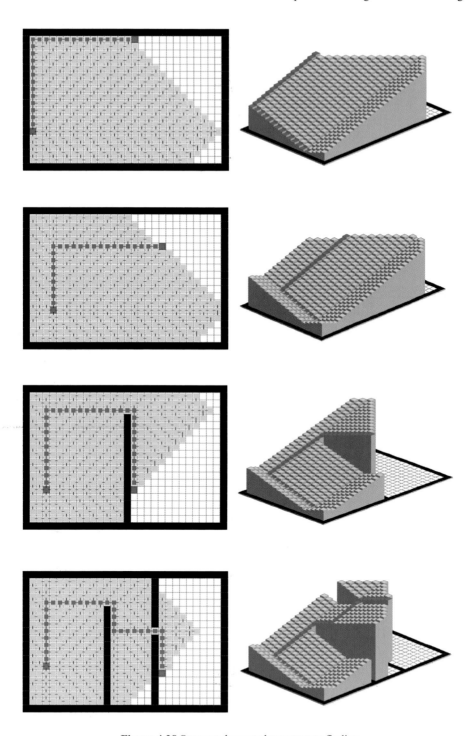

Figure 4.28 Steepest descent shortest route finding

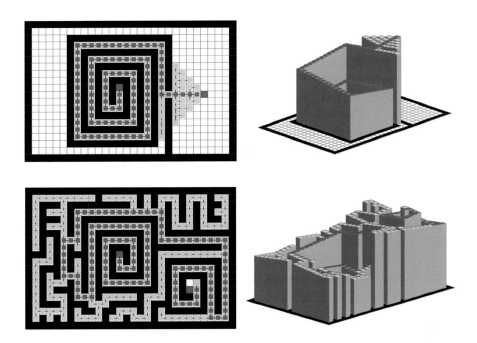

Figure 4.28 Steepest descent shortest route finding

```
public void bar(double x,double y,double w,double h,Color cc){
    Point p0 = new Point(2);
    Point p1 = new Point(2);     Point p2 = new Point(2);
    Point p3 = new Point(2);     Point p4 = new Point(2);
    Point p5 = new Point(2);     Point p6 = new Point(2);

    p0.x("=",x);       p0.y("=",h+w);
    p1.x("=",x-w);     p1.y("=",h);   p2.x("=",x);      p2.y("=",h-w);
    p3.x("=",x+w);     p3.y("=",h);   p4.x("=",x-w);    p4.y("=",y);
    p5.x("=",x+w);     p5.y("=",y);   p6.x("=",x);      p6.y("=",y-w);

    Point pa0 = scaleWtoS(p0);
    Point pa1 = scaleWtoS(p1); Point pa2 = scaleWtoS(p2);
    Point pa3 = scaleWtoS(p3); Point pa4 = scaleWtoS(p4);
    Point pa5 = scaleWtoS(p5); Point pa6 = scaleWtoS(p6);

    dW.plotTriangle(pa0,pa1,pa3,cc,Color.gray);
    dW.plotTriangle(pa1,pa2,pa3,cc,cc);
    dW.plotTriangle(pa2,pa1,pa4,Color.gray,Color.gray);
    dW.plotTriangle(pa2,pa4,pa6,Color.gray,Color.gray);
    dW.plotTriangle(pa2,pa5,pa3,Color.lightGray,Color.lightGray);
    dW.plotTriangle(pa2,pa6,pa5,Color.lightGray,Color.lightGray);
}
```

```
public void displayBlock(){
  double value;
  double x = (double)cols; double y = (double)(cols+rows);
  double range = x+y;
  for(int i =0; i<rows; i++){
    double xx =x, yy=y;
    for (int j=0;j<cols; j++){
      if(values[cols-1-j][rows-1-i]>0)
        value = (double)values[cols-1-j][rows-1-i];
      else value = 0.5;
      dG.bar(xx,yy,1,value+yy,array[cols-1-j][rows-1-i]);
      xx=xx-1; yy=yy-1;
    }
    x=x+1; y=y-1;
  }
}
```

These two procedures can be called from the main program in the following way to create the displays shown in Figure 4.28.

```
M.ShortestRoute();
dW.clearScreen(Color.white);
int width = T.cols+T.rows; int height = width*2;
dG = new Grid(f,T.tileColour,T.cols,T.rows,width,height);
M.setDisplayGrid(dG);
M.displayBlock();
```

The method used to generate the block model as a three dimensional bar graph can be extended to model the surface of a function calculated as an array of values located on the nodes of a regular grid in the way illustrated for the Sync function, using the *displayFunction()* method given below.

The construction is carried out in a similar way to the previous block model. The tile grid is rotated by 45° to obtain the secondary grid drawn in blue. In this case, however, instead of the tile boundaries, shown dashed, the dual grid shown in heavy dark grey lines, linking the centres of the tiles is taken as the main data grid. This is because instead of the bars being projected up from each tile, only the centre point is projected upwards to act as the corner value for triangular patches in the way shown in Figure 4.29. Each square cell gives two triangular patches. The values on the dual grid are processed cell-by-cell working from the back, towards the front, as before, to allow nearer triangles to overwrite those further from the viewing point. This gives the simplest hidden area removal process called the *painter's algorithm*, which will be discussed in a latter chapter. Although all the geometric dimensions are calculated using the orthogonal blue grid as a framework, the final drawings are made more realistic and less distorted by the foreshortening which occurs when these images are scaled into a rectangular rather than a square display window by the *DisplayGrid* method *displaySurface()*.

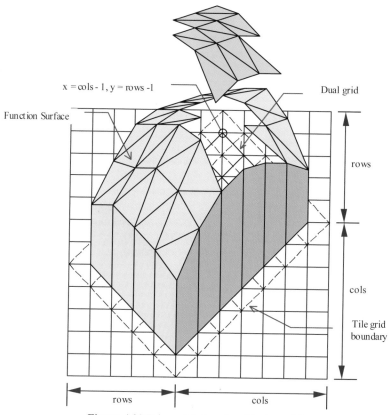

x = cols - 1, y = rows -1

Dual grid

Function Surface

rows

cols

Tile grid boundary

rows

cols

Figure 4.29 Triangulated surface block model

The Sync function $y = sin(x)/x$ rotated about the y axis gives the Mexican hat surface shown in Figure 4.30.

```
public void displayFunction(){
    double PI = 3.1415962;
    values = new double[this.cols][this.rows];
    double c = ((double)this.cols)/2.0; double d = ((double)this.rows)/2;
    for(int j=0; j<rows;j++){
        for(int i=0;i<cols;i++){
            double x = (double)i; double y = (double)j;
            double r = Math.sqrt((c-x)*(c-x)+(d-y)*(d-y));
            double X = 3.0*PI*r/c;
            double sinx = Math.sin(X);
            if(X==0)values[i][j] = 1.0;
            else values[i][j] = sinx/X;
        }
    }
}
```

```
public void displaySurface(){
   double value;
   double x = (double)cols;              double y = (double)(cols+rows);
   double h = (double)rows*2.0;          double range = x+y;
   for(int i = 0; i<rows-1; i++){
      double xx =x, yy=y;
      for (int j=0; j<cols-1; j++){
         Point p1 = new Point(2);      Point p2 = new Point(2);
         Point p3 = new Point(2);      Point p4 = new Point(2);
         p1.x("=",xx); p1.y("=",yy + h*values[cols-1-j][rows-1-i]);
         p2.x("=",xx+1.0); p2.y("=",yy - 1.0 + h*values[cols-1-j][rows-2-i]);
         p3.x("=",xx); p3.y("=",yy - 2.0 + h*values[cols-2-j][rows-2-i]);
         p4.x("=",xx-1.0); p4.y("=",yy -1.0 + h*values[cols-2-j][rows-1-i]);
         Point pa = this.scaleWtoS(p1);   Point pb = this.scaleWtoS(p2);
         Point pc = this.scaleWtoS(p3);   Point pd = this.scaleWtoS(p4);
         dW.plotTriangle(pa,pb,pd,Color.lightGray,Color.black);
         dW.plotTriangle(pb,pc,pd,Color.lightGray,Color.black);
         xx=xx-1; yy=yy-1;
      }
      x=x+1; y=y-1;
   }
}
```

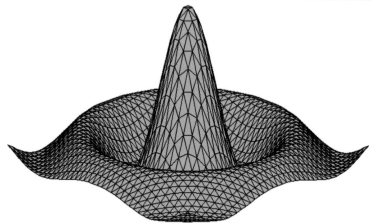

Figure 4.30 Rotated Sync function

These developments introduce the concept of a "graphics primitive", which will be discussed further in the next chapter. The primitive operation in this case is displaying a triangle. The *DisplayWindow* class generates and displays the triangles needed for these examples. Two triangles were needed to shade the top of vertical bars in the block-models, (in their simplest forms as bisected squares), and triangles were also needed to shade the sides of the bars. The Sync function surface was entirely constructed from an array of triangles where the triangles were of both variable size and variable shape.

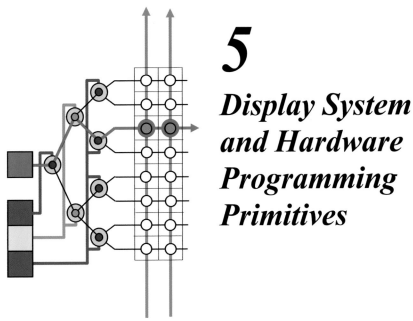

5

Display System and Hardware Programming Primitives

Introduction

In this section the "primitive" operations are reviewed, which the computer system provides the programmer for generating graphic displays. Because the idea of a primitive operation is a relative concept in the hierarchy of a modern computing system, the idea is to start at the lowest level in the hierarchy, with the hardware, and then briefly demonstrate how each level has emerged to handle more and more complex tasks.

In the previous chapter two "primitive" operations were used to compose a range of more complex images. The first primitive operation was displaying a rectangle, the second displaying a triangle. The *Grid* class generated rectangles and the *DisplayWindow* class generated triangles. However, the display hardware consists of a surface divided up into a fine mesh of grid points where the colour and brightness of each point is controlled by a number stored in a matched array of memory cells in the computer. Consequently these "primitive" elements the rectangle and the triangle have to be approximated by the appropriate selection of picture elements (pixels), which in turn are controlled by the values stored in their corresponding memory cells.

The relationship between these two levels of working illustrates a key aspect of the problem facing the computer graphics programmer. The rectangle and the triangle are used as abstract geometrical objects that have different properties to the set of point values used to display them on a grid. Lines are used as primitives for constructing drawings in a similar way. To construct complex images using programs it is not only convenient to use these abstract geometrical elements, it becomes an essential step if the content of a display is to be defined in a flexible way.

A. Thomas, *Integrated Graphic and Computer Modelling*,
DOI: 10.1007/978-1-84800-179-4_5, © Springer-Verlag London Limited 2008

In traditional graphics work, geometrical, analogue processes are employed to handle the interactions between elements. For example locating the intersection point between two lines is a matter of measuring where they cross in a drawing. Changing scale, zooming in on such an intersection point can be done photographically. The physical recording medium models the required result.

In contrast if the same operation is carried out using a grid of display points to represent the two lines, then it is not possible, using this data representation, to "zoom in on the intersection point" to see the point with higher accuracy. It is necessary to apply the abstract mathematical properties of a line to recalculate the display points approximating the lines on a new grid at a higher resolution, before the common point where they intersect can be defined more precisely. This example illustrates that computer models exist in two modes shown diagrammatically in Figure 5.1: The functional or implicit model and the data or explicit model. The explicit model poses particular difficulties for graphical or spatial entities using point data. This is because representing a "spatial-continuum" would require an "infinite set of points" which is a practical impossibility using a fixed-size computer memory.

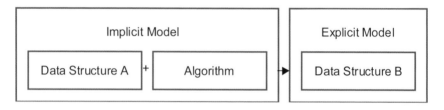

Figure 5.1 Computer models

At best, in Figure 5.1, data set B, the explicit model can only be a sampled set, such as a grid of point values. In most cases it is necessary to work with an implicit model using a finite data set A, and a corresponding algorithm to give the required data set B also as a finite set of values. Lines will consequently have to be treated as implicit models except for the explicit forms used for grid based display data, and interactions between lines will be between implicit model data sets. A simple example is the calculation of the crossing point between a pair of lines. This will start with an *implicit* model of the point calculated from two sets of line equation coefficients, from which the appropriate algorithm will calculate the *explicit* co-ordinate defining the crossing point.

Although the behaviour of the human perception system allows a high-resolution array of pixels to convince viewers they see a continuous object. It is desirable that the objects they portray should be represented in the computer system in a form that allows their properties to be modelled in a more accurate way. In a software system models of the basic form shown in Figure 5.1 can be chained together, and the flow of data from a user into the system and out again, back to the user, will have the structure shown in Figure 5.2.

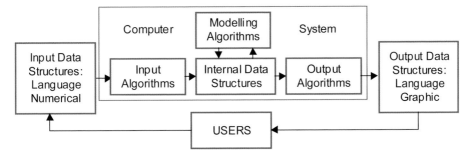

Figure 5.2 System data flow

The input to this chain must be in a form that can be understood by the user. At the minimum, this will be in a high-level, computer language for programs, and formatted numbers for numerical values. Output will be in similar highly structured presentations usually through a graphics user interface (GUI), and the software system has to be built hierarchically to support these facilities.

It is necessary to provide a bridge between the desirable forms of input and output, and the primitive operations supported by display and computing hardware. To handle the complexity of this task in a manageable way, the computer system is usually divided into the following layers:

- Application level tasks
- Programming levels
- Hardware level primitives

Although it is possible to design special purpose hardware for each application task, this loses the advantage of having a general purpose, programmable device. A program dedicates the computer to a particular task, which means building one type of machine can service many different requirements simply by changing programs.

In much the same way it would be possible to program application tasks directly in dedicated machine code programs. However the complexity of this task quickly puts a practical upper threshold on the approach. It also involves a considerable amount of duplication of the many intermediate tasks that are common in most application programs. Consequently, it is efficient to build libraries of "sub-programs" that can be reused in different applications to provide these common operations or functions. Using libraries of prewritten functions provides the programmer with a more powerful working context. In a sense the programmer then faces a new, "virtual" machine with an extended instruction set or programming language. This view has led to an alternative classification of the layers in a computing system by the "language" used to write programs at each level.

However, the nature of what is practical to build as superstructure in this way often depends on limitations imposed by the basic elements in the system. Very simple devices will support general-purpose computing and therefore any functionality required from the superstructure, however, the architecture and design of the computing and display system's hardware ultimately determines its cost, size, speed and efficiency and therefore its practical usefulness in different application systems.

Display Technologies

There are a variety of mechanisms and physical properties that can be used to generate displays. One division is between on-line interactive systems where the displays are ephemeral and disappear when the machine is switched off, and hardcopy systems, which as their name implies create permanent graphic products such as drawings or photographs. Another division is between line display systems and raster based pixel array systems. Yet another is given by the delivery technology, projection, off screen, or direct-view systems.

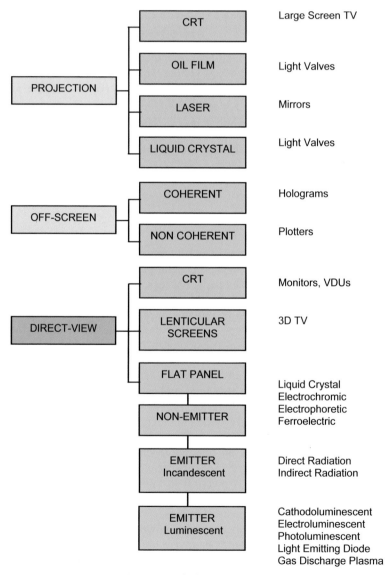

Figure 5.3 Display technologies

The line plotter and Cathode Ray Tube (CRT) were the earliest dedicated output devices used to create graphic displays. The plotter was developed to automate the production of technical drawings for the automobile industry and the aerospace industries. CRT displays initially were used to create interactive point and line displays, but later by using a different electron beam deflection technology, moved from the expensive but more accurate, electrostatic control, to the cheaper but less accurate magnetic control used in TV systems. Eventually this permitted the development and wide spread use of raster display systems. Refresh line display systems were expensive and before TV technology matured, storage tubes represented an affordable compromise for many applications through the 1970s.

The nature of display devices depends on physical systems either holding changes in colour from the application of ink or paint, or changing colour from the application of light such as photographic paper and in a sense TV cameras, or changing colour from the application of an electrical field. Alternatively display devices can be built from elements that give out light when electrically stimulated, such as phosphors in CRT and TV tubes, LED s - light emitting diodes and plasma panel cells.

Although there are many delivery mechanisms, there is a limited number of ways in which images can be generated and transferred to these varying display surfaces. The main division is between continuous and discrete methods. These in turn mostly come down to line or point based systems. However, it is convenient to use the following headings:

1. Free form line drawing systems
2. Raster systems, scan line based systems
3. Raster systems, block fill systems
4. Addressed cellular structures.

Drawing systems generate lines as continuous elements from which the display is constructed. Line plotters give hard copy while vector CRT systems give interactive refresh displays.

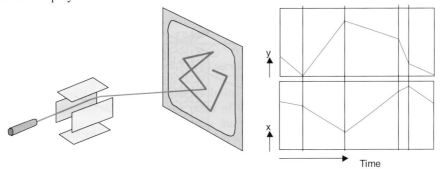

Figure 5.4 Refresh line drawing systems

The deflection of the electron beam or pen plotter turret is controlled by time based x and y signals of the general form shown in Figure 5.4.

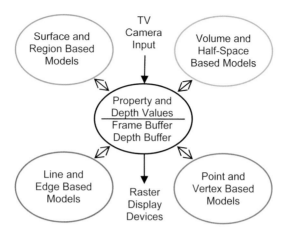

Figure 5.14 Common base for display generation

The speed of scene-content generation being de-coupled from the refresh scanning of memory needed to keep the display set up and flicker free supported a new level of computer graphics. The separation between the input and output tasks meant that the output had to be serviced first if a coherent display was to be maintained. The rubber banding effect was lost at the basic level, since lines had to be erased and redrawn in the frame buffer, in order to move them in the display.

In order to explore this topic further it is necessary to examine the implementation of both the computer system and display buffers at a more detailed level.

The Basic Digital Hardware System

Historically the automation of computing can be traced back to mechanical calculators. However, its roots go back further to the calculating table and the abacus where the position of stones or beads in a frame allowed calculating procedures to develop with highly "automata" like characteristics. In other words the rules for applying the calculating process were simple and easy to operate even if the theory behind why they worked were more difficult to understand. Consequently, adopting the Arabic notation for numbers, because it derived from the states of the abacus represents the first substantial step towards mechanising arithmetic.

Leibniz and others developed mechanical devices for adding and subtracting numbers based on this notation. The developments made by Charles Babbage and Lady Lovelace extended these into a mechanism, which could carry out a sequence of such operations under the control of a program, coded on a series of punched cards. This work established the main components of the modern computer system. These were a "store" to hold numbers; a "mill" to carry out calculations, "input" and "output" devices, and a "control unit" to execute the sequence of commands in the program. All that was lacking was a satisfactory technology to implement the scheme. Only with the development of electronics has the information processing revolution been able to take off! The problem with mechanical systems was the transfer of values from one place in the computing machine to another. The difference in the way that values were stored using the new technology, by active electrical components, made these difficulties vanish. All that was needed was to link devices together by setting switches in a communication network, and the active nature of electronic components allowed values to be transferred at electronic speeds as electric current flowed from one device to the other. The early computers were electromechanical, and still represented numbers in a decimal notation. A digit was represented by the state of a wheel, which rotated in ten steps, matching the construction of mechanical calculators. However, the basic nature of electronic switching devices made two-state systems simpler to implement, consequently, numbers became easier to represent and handle as binary codes.

For the purposes of this book, it is not necessary to explore the electronic level of computing systems in any great depth. However it is useful to know how certain key parts of the system, at this level, operate. The key advantage offered by electronic components was that they were active, in the sense that they generated different voltages to represent the state they were in. A voltage difference could be used to transfer electrical energy to a receiving unit linked to it by a wire, which could in turn be used to control subsequent dependent changes in the state of the receiving unit.

This can be contrasted with the state of a mechanical register being the position of number wheels, which is a passive representation. Transferring its state needs the positions of the wheels to be sensed and some way of replicating its state at a different location to be implemented. The output voltage of an electronic element, by signalling its state, can be used as a direct input to a second element. It can also be amplified and transferred at electronic speed over long distances, again over linking wires to many receiving elements. This has given increases in speed of many orders of magnitudes over previous computing devices, and at the same time has resulted in a technology that it has proved possible to reduce in size by matching orders of magnitude.

Switch Circuits and Logic Functions

The basic electronic component is the field effect transistor. This can be considered to be a switch in the way illustrated in Figure 5.15. When the control C, input voltage is low the switch is off and no current can pass through the transistor, when the control input voltage is high the transistor conducts current and the switch is on.

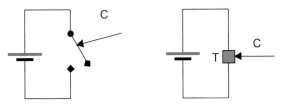

Figure 5.15 Transistor switch

A large variety of circuits can be built up using switches or relays. However, it is the logic gate, which is the building block that is the primitive component for building most digital electronic components. The easiest way to visualise how these units operate is given in the diagrams in Figures 5.16 to 5.18. These circuits are based on the ability of an electric current to activate an electro-magnet solenoid. When the current is on then the magnet is active and can be used to control the setting of a switch. If a switch is controlled by a solenoid, which is activated by a second switch, then if this control switch turns the second switch off, as in Figure 5.16 b, then the relationship between the settings of the two switches is defined by a *not* truth table.

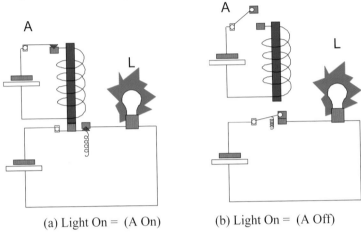

(a) Light On = (A On)   (b) Light On = (A Off)

Figure 5.16 Coupled switching circuits

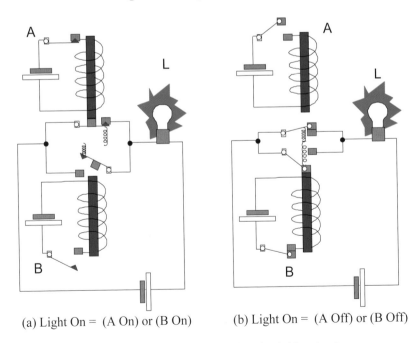

(a) Light On = (A On) or (B On)   (b) Light On = (A Off) or (B Off)

Figure 5.17 Coupling parallel pairs of switching circuits

Where the second circuit is activated, this can be shown by a light bulb being switched on. Figure 5.16 shows the two forms of coupling (a) and (b), which the two kinds of switches, active on and active off, permit. If more than one controlled switch is used to couple circuits in the way shown in Figures 5.17 and 5.18, then the way they affect the light can be defined in input output tables that correspond to various logic truth tables.

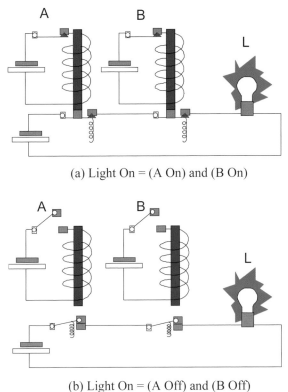

(a) Light On = (A On) and (B On)

(b) Light On = (A Off) and (B Off)

Figure 5.18 Serial pairs of coupled switching circuits

Transistors and Logic Gates

The basic digital electronic component is the logic gate. This is a unit that can be considered to process signals in two states: {low, high}, {0, 1} or {false, true}. Although gates operate on continuously varying electrical values of voltage and current, operationally these are masked from the user (in most cases). The gate is an element that combines these binary values as input signals in different ways depending on its function, to give a binary output value. So different gates exist to provide the logical operations {*and, or, not, nand, nor, exclusive-or*} on binary input signals to give binary outputs.

These elements can be presented as black boxes defined by their input output tables. *and, or* and *not* are defined in Chapter 2, *nand* is *not-and*: !(A&B), the *exclusive-or* is (A&(!B)) | ((!A)&B) and *nor* is *not-or*: !(A|B). However, using the transistor switch it is possible to illustrate how these gate circuits can be constructed. These circuits serve two purposes. They generate the various logical functions from their inputs, but they can also amplify the output signal, or at a minimum prevent it from dying away as it passes from one unit to the next.

Gates have a directional behaviour in contrast to the transistor, which acts as a simple switch in other words an open or closed link in a wire.

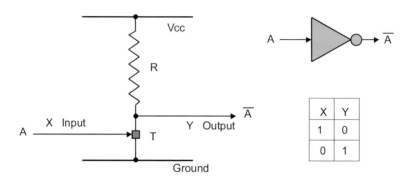

Figure 5.19 Inverting, not gate

The basic circuit for an inverting gate is given in Figure 5.19. The resistor R creates a voltage gradient between the Ground, the Output and the Vcc line, while the transistor switch T is on, which occurs when the Input is high. This makes the Output low. Conversely when the voltage applied to the Input goes low, the transistor switch is turned off, and the voltage at the Output goes high as current flows through the resistor until the Output voltage reaches Vcc. The relationship between Input and Output is the same as the logic function *not*.

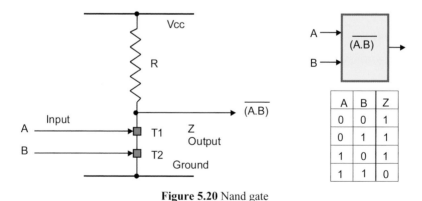

Figure 5.20 Nand gate

Two transistor switches in series produce a *nand* gate circuit in the way shown in Figure 5.20. In a similar way two transistor switches in parallel as shown in Figure 5.21 produce a *nor* gate.

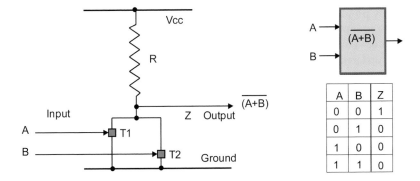

Figure 5.21 Nor gate

The basic digital electronic primitive is the logic gate. For the purposes of this discussion a second primitive element will be defined. This is the memory cell. In practice a memory cell can be constructed from gates using feedback paths from one gate to another.

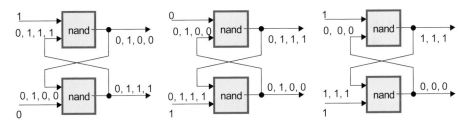

Figure 5.22 Latch using nand gates

In figure 5.22 the sequence of changes in outputs for two *nand* gates given different inputs is shown. Where the two external inputs are opposite, then the circuit settles down to a stable output, shown by the sequence of outputs and feed back values. Where one of these stable outputs is matched with an input of two 1s then there is no change the outputs are held as they were. If the two inputs are changed to two 0s then the output becomes unstable. A similar set of results is obtained for a pair of *nor* gates shown in Figure 5.23, in this case the two 0s hold the value in memory, whereas the two 1s cause instability.

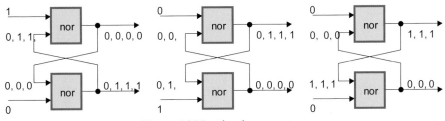

Figure 5.23 Latch using nor gates

The only problem with these circuits is they take a certain amount of time to settle with changed inputs. This creates difficulties when passing a value down a chain of these memory cells. The visualisation of the problem is having a row of people each holding a different coloured baton in their hand, being asked to transfer the baton to their immediate neighbour on their left. Chaos is the likely result! The solution is to make the operation a two-step process. Step one is to start with the baton in the right hand, then pass the baton to the other hand. Step two is then for each person to pass the baton from their left hand to their neighbour's right hand. Repeating this cycle will safely transfer all batons down the line. This introduces the master-slave memory cell, shown for *nand* gates in Figure 5.24.

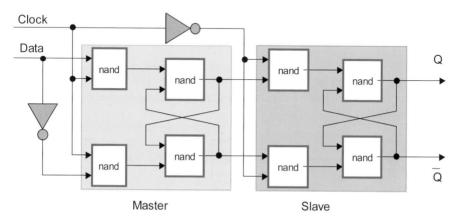

Figure 5.24 Master slave memory cell

This also introduces the idea of a synchronising clock signal to control the alignment of actions in a large circuit. The transfer of batons down the line would fail even with the second strategy, if the transfers were not synchronised in time so the action of each person matches with his or her neighbour. In Figure 5.24 when the clock is high or 1, then data is allowed to enter the master stage latch. When the clock signal goes low or 0, then the first latch is held in memory mode and its outputs are allowed to enter the second stage into the second latch. When the clock goes high again then the second latch is switched to memory mode holding its values on its output lines. This arrangement corresponds to using two hands to pass a baton down a line.

The clock signal allows time for circuits that need time to settle to reach a stable state before entering values into memory cells. Where there are feed back paths in a circuit this can be critical. The feedback links in the latches themselves are local and their design can be arranged to avoid incorrect behaviour. Where long wires are encountered or complex logic functions, which take indeterminately different amounts of time, depending on data, then the use of a synchronising clock signal makes the task much simpler to manage correctly.

A similar use of a two-stage memory unit can be constructed using transistor circuits in the following way. Two clock signals are set up where each clock goes high in turn, to switch on pass transistors, i.e. switches to let signals pass through. Each clock signal only goes high when the other one is low. This produces the same two-step storage sequence as the master slave unit. Each stage consists of an inverter, and they are arranged in a feedback loop separated by pass transistors each controlled by one of the two clock signals {Φ1, Φ2}. This arrangement means that during the first stage, when the inputs to the first inverter are isolated, the charge, currently controlling the setting of the inverter-input, holds the output to the second stage steady. The second pass transistor in this stage is on and allows this signal into the follow on inverter, charging up its input transistor. During the second phase the second clock pass-transistor switches off, so the charge on the second inverter's transistor is isolated holding its output steady, which is switched through the first clock pass transistor which is on during the second stage, allowing it to recharge the first inverter's transistor.

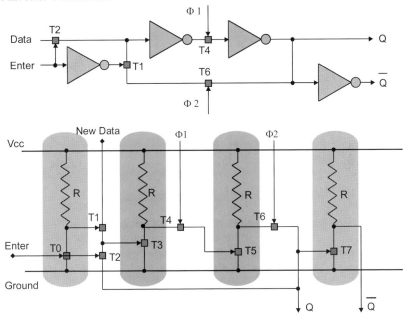

Figure 5.25 NMOS memory cell

Although it is useful to have some idea how memory units work, there are several alternatives, often using the physical properties of their storage medium, in special ways. It is consequently, easier to treat memory units as a second form of black box when constructing more complex components.

The memory cell holds a value or stores it, until it receives an *"enter"* signal. The *"enter"* signal causes whatever is waiting on the memory cell's input wire to enter the unit and be stored. This value is then held constant on its output wire until a new enter signal is received to change it.

Placing memory units between functional blocks provides a robust way of breaking down complex tasks into sequences of more manageable operations. It is the basis for pipelined and systolic processing, which is a topic, which will be returned to in a later chapter discussing more advanced display processor concepts.

Memory, Function Blocks and Micro Programming

A single memory cell is of limited usefulness, however a large collection of memory cells poses the problem of accessing the desired cell. What is needed is a circuit, which allows a specific cell to be addressed, either to input a new value, or to obtain its contents as output. These input and output values will need to be transferred onto common communication wires, so the task is to route the input to the target memory cell from the wire carrying it, or to link the output from the target cell to the desired output wire. However, any one of the memory cells may need to be linked to the same input or output wire. One solution is an accessing switch on the input and the output of each cell. This can be a physical switch in the sense that it breaks the conducting path of the wire, or it can be a logic gate. In the latter case either *and* or *or* gates can be used depending on which value {0, 1} is taken as the off state. In the examples given below {0} is taken as the off value, where it is needed.

Logic Switching Circuits

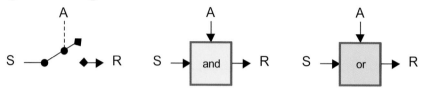

Figure 5.26 Switching circuits using logic gates

The problem remains how to select which memory access-switches should be on and which off. This can be done using an addressing circuit. If the switches are all controlled by a switching signal S: all the addressing circuit has to do is to route it to the appropriate cell. The black box definition of this unit is given in Figure 5.27.

Addressing Circuits

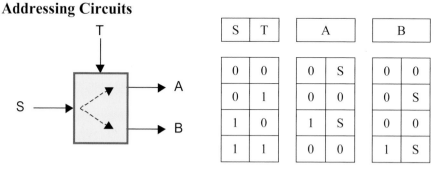

S	T	A		B	
0	0	0	S	0	0
0	1	0	0	0	S
1	0	1	S	0	0
1	1	0	0	1	S

Figure 5.27 A basic routing cell

This unit can also be implemented by the appropriate combination of gates in the way shown in Figure 5.28. The logic functions that give the same behaviour as the black box are given at the top of each output column under the output labels A and B. Notice in this case the use of the {0} value as the off value. Equivalent circuits can be set up where the off value is taken to be {1}.

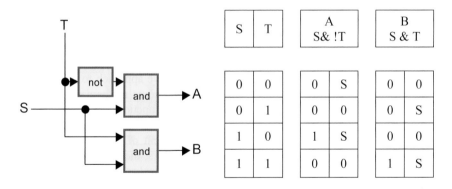

Figure 5.28 A basic routing cell implemented using gates

A signal T is used to select one of two output-wires onto which the value of the switch signal S will be transferred. The implication in this arrangement, having off as 0, is that only when a 1 is received, will the receiving unit be switched on. The application of this unit becomes apparent when it is used in layers to select from a larger number of output wires. This gives the tree structure shown in Figure 5.29 addressing a larger block of memory cells.

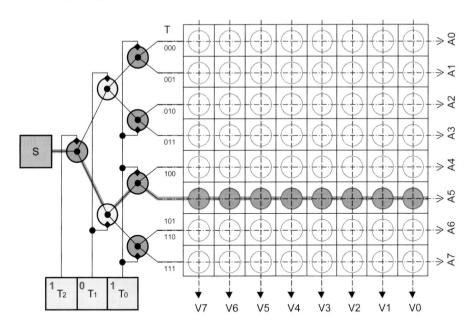

Figure 5.29 Routing cells used in an addressing tree

If the routing units (circles) are arranged in the tree structure shown in Figure 5.29 and each column of routing units in the tree is switched on by the same signal. In this case there are three routing values T0, T1, T2 stored in a register T. If the values stored in T are entered as a binary number they will sequentially access the addresses

Selection Circuits

The tree structured addressing circuit routes a selection signal through to a set of memory cells. It also can be used to select which memory circuit is switched onto output wires. Selecting a signal from a set of signals on several wires to pass to a single wire is an alternative approach to obtaining a single output from a block of memory. A two-way selector is shown in Figure 5.38, and a logic gate implementation is given in Figure 5.39. Selection circuits can be built up in much the same way as the addressing tree. The selection signal T can be extended to multiple bits, which again can be treated as an address for the selected line.

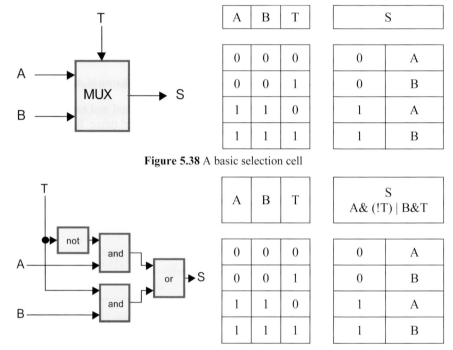

A	B	T		S
0	0	0	0	A
0	0	1	0	B
1	1	0	1	A
1	1	1	1	B

Figure 5.38 A basic selection cell

A	B	T	S $A\&(!T)\mid B\&T$	
0	0	0	0	A
0	0	1	0	B
1	1	0	1	A
1	1	1	1	B

Figure 5.39 A basic selection cell implemented using gates

Sequential Processing

There are limits to the look up table approach for evaluating many functions. The sorting task discussed in Chapter 2 highlights the limitations. Many tasks can be handled more effectively by a sequence of simpler tasks carried out in steps. The sorting process consisted of sequentially selecting and removing the largest value from a list until the list was empty. In a similar way the adder, shown in Figure 5.30, divides the adding task down to a sequence of bit position operations where the carry input, for each bit position, is the carry output from the operation to the right. In this case the whole task is limited by the time it takes for the sequence of carry calculations to ripple through the whole circuit. An alternative approach to this task is, like the sorting process, to build up the answer in time interval steps. This requires a single bit-position adding-circuit with a memory cell to hold the carry bit from one step to the next. This allows the two numbers being added together to be processed serially by the circuit shown in Figure 5.40.

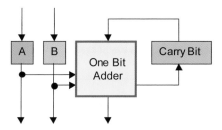

Figure 5.40 Sequential adding circuit

This circuit is an example of "sequential" circuits, which use feedback to implement a step-by-step evaluation of functions.

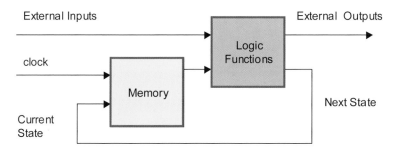

Figure 5.41 Generic sequential circuit

Display Processor Development

The development of the display systems using raster data evolved in the same way that the line displays had before them, to include more complex display functions delegated to the display system to take the processing load off the host computer.

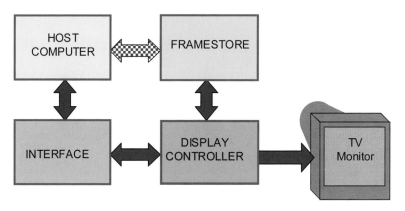

Figure 5.42 Schematic raster display system

The key elements are shown in Figure 5.42. The frame store unit is a large block of memory capable of holding the image. In some systems the frame store is an independent block of memory, in others it is part of the main memory of the computer system. There is a variety of ways in which the memory can be structured. As memory has become cheaper and available in larger volumes so more versatile and powerful raster display systems have emerged.

Frame Store Architectures

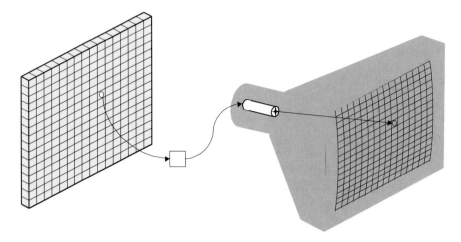

Figure 5.43 Single bit plane black and white

The simplest raster display system is that shown in Figure 5.43 a single bit plane capable of holding one bit per pixel: representing an array of black and white points on a regular grid.

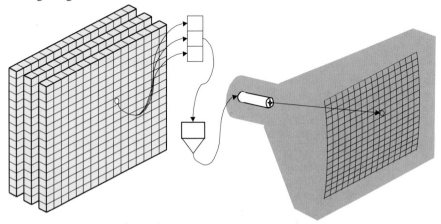

Figure 5.44 Multiple bit plane grey scale

Where the resolution of the display is high enough, this is a replacement for the line or vector display. However, certainly in its early days the resolution was far from sufficient to do this well. The visible steps in lines represented by points on a grid, where the lines are nearly vertical or horizontal introduced the term jaggies to denote this kind of undesirable, display-system artefact.

Introducing more bit-planes i.e. more bits for each pixel location and a digital to analogue converter, and allowing the beam intensity to be varied: gave the ability to display greyscale images. This allowed the anti-aliasing algorithms which model the way lines are displayed in standard camera captured TV images, to be implemented.

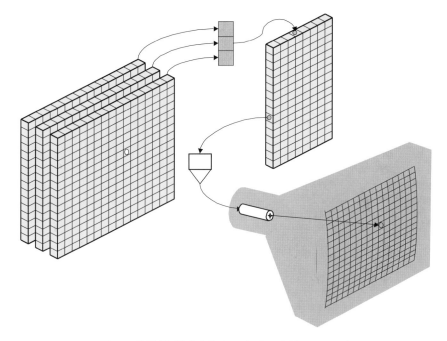

Figure 5.45 Multiple bit plane lookup table grey scale

However, even this was not a total solution. Once numbers were used to represent different levels of grey, it was discovered that the perceived grey scale of the display produced from a linear sequence of numbers was not itself linear. A (*gamma*) correction function was needed to map the linear digital values to a linear visually perceived greyscale on the screen. This was possible using an intermediate lookup table mapping the input value as the address to the appropriate output value to give the required greyscale in the display in the way shown in Figure 5.45.

Once memory was cheap enough to have multiple bit planes they could also be grouped to give red, green and blue values for each pixel position in the way shown in figure 5.46. The simplest mode of colour display was to have a single bit per colour channel. This gave eight colours to work with on screen. Again the boundaries were pushed forward as memory was made more accessible, and each colour channel could be represented by a range of the primary colours in a system in the way shown in Figure 5.47.

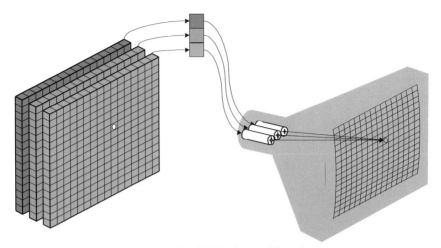

Figure 5.46 RGB bit planes eight colours

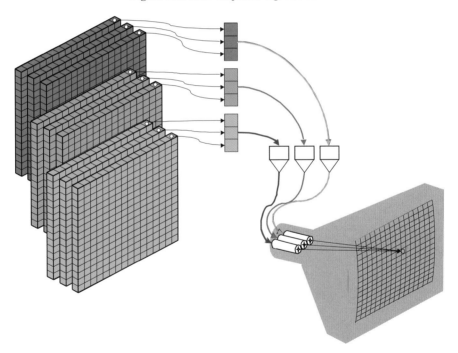

Figure 5.47 A basic RGB display system

The lookup table was found to provide a new service in this context. The simple combination of the primary colours was found to give a very uneven distribution in the perceived colour space. So again a lookup table translation to provide a better spread of colours as the input numbers were varied could be provided. This scheme also allowed the lookup table to have many more bits for each colour, refining the accuracy of the colour specification.

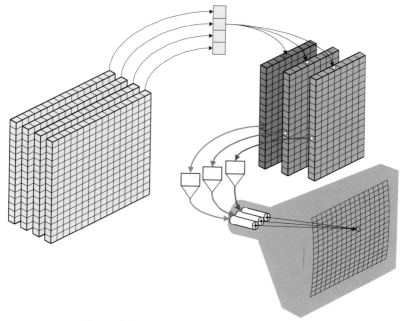

Figure 5.48 Four bit addressing of RGB lookup tables

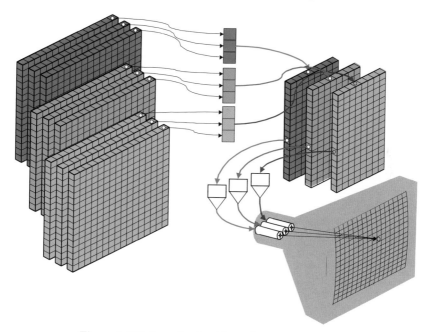

Figure 5.49 Independently addressed RGB lookup tables

This introduced the idea of a colour pallet where the number of bit planes gives the number of colours that can be defined, but the colours themselves can be separately specified in the lookup table. This provided the analogue of the artist's working pre-mixed set of colours, pre-prepared on his pallet ready for painting. In Figure 5.48 16 colours can be defined, where each colour is made up from a mixture of red green and blue primaries, each specified with eight bits giving a linear range of 256 values for each of the three colours.

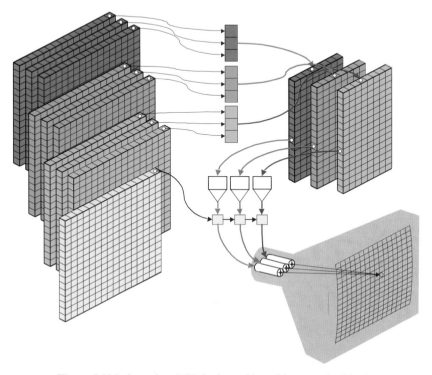

Figure 5.50 Independent RGB lookup tables with an overlay bit plane

With these components many variations become possible, for example the eye is far less sensitive to blue than it is to red and green so fewer bits in the look up tables need be provided to define the blue component of a pixel's colour. Multiple images could be placed in the same memory block and mixed at the pixel level. Overlays containing text and annotations could be provided by extra circuitry of the form shown in Figure 5.50.

As memory became cheap enough to support a geometric model of a scene, a depth buffer was added to the basic frame store containing image data. By including the depth of surfaces making up a scene at the pixel level, rather like a terrain model used in cartography a series of scene composition and display operations became much simpler to implement. In particular the depth buffer algorithm evolved from the painters algorithm to remove the difficulties generated by objects inadvertently being made to overlap in interactive work. Similarly, extended pixel data defining the transparency of a surface allowed more sophisticated display operations based on existing manual collage and overlay techniques to be developed.

TV Cameras and Image Processing

At the same time that it became practical to store TV images in a digital form, the technology of the TV camera also evolved in a similar direction. Initially images were captured as analogue electronic signals, and these had to be passed through analogue to digital conversion units before they could be stored in frame-store based display systems. However the conversion operation to a digital format has now been built into new TV cameras so their output is provided directly in a digital form. These advances make it easier for synthesised and captured images to be merged within the same system. In particular it makes a series of "image-processing" operations that need to be applied to captured-images simpler. Operations that can also be carried out on data held in a display processor's image memory. The specialisation to handle display operations of this kind along with the flexible allocation of memory to depth buffers, alpha buffers and multiple image buffers, expands the role of the display controller from a slave refresh process to a fully autonomous computing system in its own right.

Display Processor Control Unit

In practice there are many ways of constructing a frame store, depending on how the pixel values are mapped into memory packages. One of the problems with early memory based frame stores apart from the volume of memory needed was the speed of access. Even as low a resolution as 250 by 250 pixels requires 625,000 values to be read each refresh cycle. If this is $1/25^{th}$ of a second this requires a read time of:

$$\frac{1,000,000,000}{25 \times 250 \times 250} = 640 \text{ nano seconds per pixel}$$

Where memory access was still mille-seconds this would not support a realistically useful system.

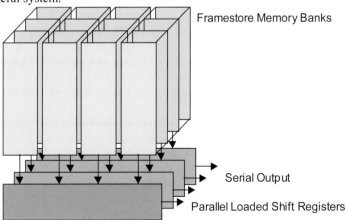

Figure 5.51 Frame store data output

However, the serial-order in which data needs to be supplied to the TV monitor to match the scanning sequence of the electron beam makes it possible to speed up the delivery to the monitor. If adjacent pixel values are stored in separate banks of memory they can be addressed and accessed in parallel. This allows several adjacent pixel values to be read in one memory cycle into a fast shift register, which can then supply the individual values at the required refresh rate. In Figure 5.51 the output is four times the limit set by the memory units.

This structure however, limited the freedom to enter data into the frame store memory whenever and wherever a display program might wish to. If the speed of the output from the memory perfectly matched the TV monitors input needs, then there would appear to be no time for data entry into the frame store. Fortunately the standard TV has gaps in its refresh cycle when the electron beam is switched off. This occurs when each line is complete and the beam flies back to start the next line and at the end of each field of the frame when the beam flies back from the bottom of the screen to the top to start again.

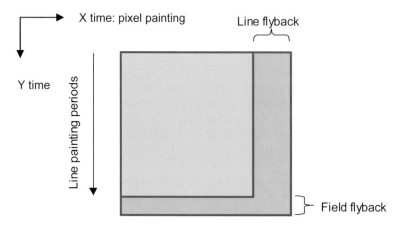

Figure 5.52 TV display: field-timing diagram

A diagram to illustrate the use of time can be set up in the way shown in Figure 5.52. The scan line refresh cycles in each image frame refresh cycle are shown scaled down the y-axis, and the number of pixel refresh periods in each line cycle is scaled along the x-axis. The pale green area then shows the time actively used to display an image, and the orange area the fly back time left to write new values into the display memory.

Updating a display within this restricted time-window can be a messy process, and is generally inadequate to support animated moving images. Luckily the serial TV scan pattern can be further exploited, to improve the situation. Where there is an interleaved display, each field can be stored in separate blocks of memory, so allowing one to be written while the other is being displayed. With non-interleaved memory the same result can be obtained using two memory blocks alternately reading into one and displaying the other. However, dividing the memory into blocks to service horizontal bands across the display can reduce the memory requirements for an equivalent system. Again one band can be outputting data for display while its neighbour is receiving new inputs. The only restriction this imposes is on which section of the image can be refreshed at any moment in time. Lines vertically scanning the image need to be partitioned appropriately. However, when panel displays are employed the over all scanning sequence can be broken up in different ways. The multi-TV wall display can be implemented in principle at any scale as parallel processing tile arrays.

As integrated circuit technology advanced and memory chips became larger and larger and faster and faster then some of these initial frame store, design difficulties disappeared. However, some of them became more difficult to resolve because the subdivided structure of the frame store became more difficult to implement efficiently using the larger units. The partitioning into input blocks and output blocks to make input-output simpler to handle became more difficult. However, the expense and size of a fully double-buffered system became less prohibitive. The rise of the personal computer led to a much larger market for display systems, this in turn justified specialised memory for refresh display devices. A major development was to take the structure shown in Figure 5.51 and place it on a dual port memory integrated circuit, which allowed virtually free address based input to memory while maintaining the high serial flow of data out needed to refresh the TV system.

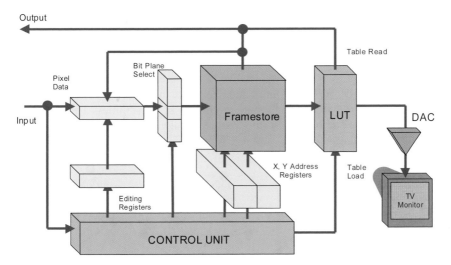

Figure 5.53 Schematic display processor structure

Developments in IC technology led to a similar evolution in display processors. Managing the growing complex range of alternative configurations fell to the display control unit. Initially this was built as a custom logic circuit. However as more and more flexibility became necessary, so the controller became a programmable system in its own right.

It is possible to configure the components shown in Figure 5.53 to carry out a variety of display operations, depending on the way data is routed from element to element. Commands sent from the host computer to select a particular operation would need to set the appropriate switches in the network to implement the desired function. A simple way to implement this is to make the command code an address to a lookup table holding the different switch setting for the various configurations.

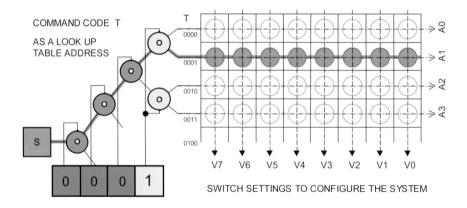

Figure 5.54 Commands as lookup table addresses for switch settings

With this arrangement it is clear that the design of the switch network is the factor that determines how flexible the overall system can be made. Linking every element to every other element in the circuit will provide maximum flexibility. Complete networks corresponding to the "connection-machine" architecture can be represented by graphs of "simplexes" in the way shown in Figure 5.54. A system of n nodes can be set up with $y = n.(n-1)/2$ links and the same number of switches y. If all configurations were required then there would be 2^y possibilities. This value gets large very quickly, and like the sorting problem, only makes lookup table selection viable for small values of n.

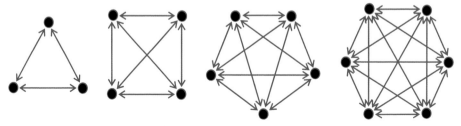

Figure 5.55 Simplex graphs: complete linkage networks

However, there are several ways in which the number of possibilities is reduced. Firstly the number of switch settings, which provide useful configurations, will usually be a fraction of all the possible settings. Secondly the number of switching arrangements can be further reduced by careful use of common lines of communication such as busses: and then selectively switching elements on and off them. Thirdly replacing parallel one-shot operations by serial sequences of operations can greatly reduce the number of necessary components, and consequently the switching complexity. However this does incur a time cost.

In practical systems the number of configurations will still be smaller than the number of possible settings for the switches needed to implement them. This makes a lookup table an effective approach. Also by changing the contents of the lookup table, the same command codes can be made to activate different operations. This introduces the lowest level of programming and a small extension to this approach allows sequential operations to be set into motion from a single instruction.

Display controllers made from dedicated circuits could be set up as "finite state machines" to execute serial operations. There are two forms for these mechanisms: Mealy machines and Moore machines shown in Figure 5.56. They both depend on feedback: making their action depend on current state variables stored in memory. The current state and new inputs determine the next state. The Moore machine makes the output depend on just the current state, whereas the Mealy machine makes the output depend on the current state and the current inputs to the system.

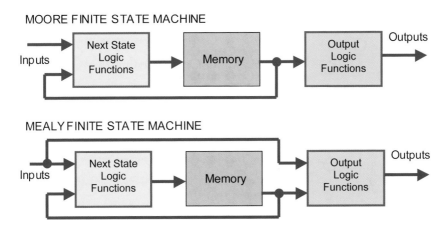

Figure 5.56 Finite state machines

Both these systems can be implemented using a lookup table, in the way outlined in Figure 5.57. The state is determined by switch settings but the state is represented by the address of the settings in the lookup table. The feedback consists of the address of the next state, switch settings for the machine.

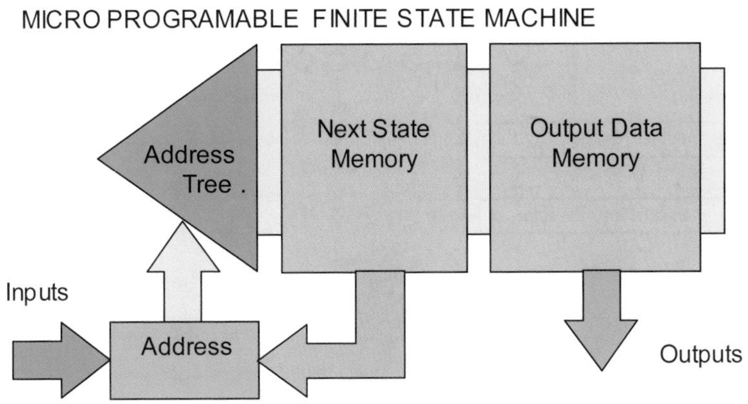

MICRO PROGRAMABLE FINITE STATE MACHINE

Figure 5.57 Finite state machine using look up tables

However to make this system implement a set of different sequential operations requires the feedback to also provide a selection signal to allow a new external address to be entered to replace the feedback address. Take as an example eight main external commands requiring 3bits to select them and an overall set of 32 switch settings requiring 5 bits to select them. If the three low order bits are used to select the external command addresses, and the top two bits in these cases are set to zero then the first switch setting for each of these commands will lie in the top eight rows of the lookup table in Figure 5.58 labelled A to H.

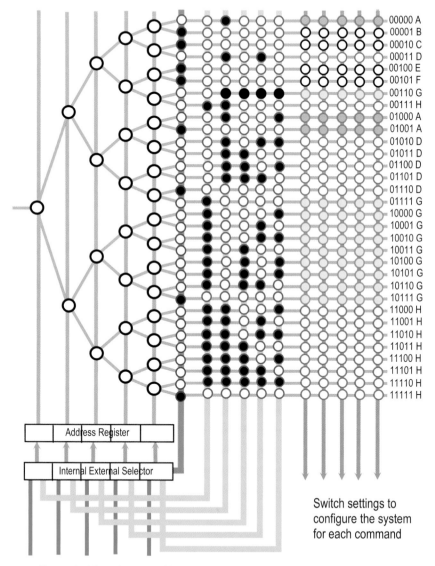

Figure 5.58 Micro-program to execute eight instructions

In Figure 5.58 the commands labelled A, D, G and H all require a series of steps to execute them but the commands labelled B, C, E and F only require one step. What is interesting about the arrangement shown in Figure 5.58 is that though each of the first steps of the eight commands are located in adjacent locations at the top of the lookup table, in the case of A, D, G and H the subsequent steps are separated and are in contiguous blocks in the remainder of the lookup table. This is clearly the result of using a five-bit address to access the internal states but only the low order three-bits of the address to identify the starting state of the eight externally defined commands. If the external commands were defined by a five bit address then they could start anywhere, what is more they could then be arranged so all the steps for each command were in blocks of adjacent addresses in the lookup table. However the codes for the external commands would no longer be neighbouring numbers.

What is implemented in Figure 5.58 is a linked list structure linking the series of switch settings that a command needs to implement it. This gives total flexibility to locate each switching-set in any address location. However it is an unnecessarily powerful technique for this application. If the start addresses were placed in a counter register then all that would be needed to access a series of contiguous addresses would be to increment the counter between each step. The only feedback would be the single bit identifying when the end of a series of commands had been reached and a new external address needed to be entered into the counter-register to locate the next command.

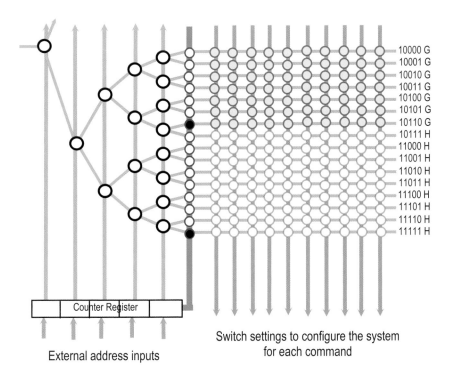

Figure 5.59 Micro-program to execute instructions G and H using a counter

The linked list approach is useful when the steps needed to find space in the lookup table do not determine the address for a command but where the address is set externally. This occurred when a family of computers of different speeds and power were given the same machine code, so programs could run on all of the computer family though with different levels of performance. An example of this is given by the multiply operation. In the fast top of the range system special hardware could implement this operation in one step. However on simpler machines it could be implemented using a series of shift and add commands.

If an addition is implemented as a hardware function, and a shift left operation moving all the bits in a register one place to the left, is also implemented at that level. Because the shift operation is equivalent to multiply by two, a series of shift and add operations can implement binary multiplication in a series of steps, in the way that conventional long multiplication for decimal arithmetic is carried out by hand.

Machine Language Programming

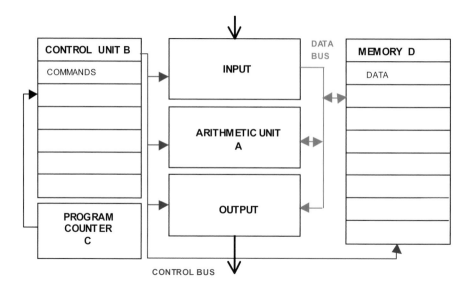

Figure 5.60 A schematic computing system

Externally defined programs could be executed in the scheme shown in Figure 5.59, however there would be a great deal of duplication if serial operations such as multiplication, implemented using shift and add, needed to be repeated within a program. The microprogramming unit in Figure 5.59 can be used in a control unit for a more general computing system in the way shown in Figure 5.60, by placing the external program in the control unit's memory also accessed by a counter, and having commands in the program that can change the address in this counter when the flow of control needs to be rearranged.

Allowing the commands stored in the control unit, shown in Figure 5.60, to be placed into the main memory D, gives the general stored program computer discussed in the next chapter.

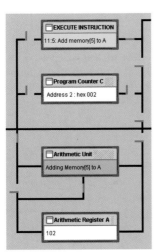

6

Computer Hardware and Low Level Machine Language Programming

The Stored Program Computer

In the previous chapter the "primitive" hardware facilities needed to display images were introduced. The main form of current display hardware consists of a surface divided up into a fine mesh of grid points where the colour and brightness of each point is controlled by a number stored in a matched array of memory cells in the computer system. Computer graphic displays are generated by writing programs to enter the appropriate values into these data arrays in memory.

In this section a "minimal" computer is presented to illustrate the nature of machine language programming, and the way it activates the electronic building blocks, which make up the computer system. The aim is to present machine language programming, in as simple a manner as possible and to provide a visualisation to help understand the dynamic behaviour of a computer system and the programs running it. Real systems follow the same principles, though they are usually more complex in order to give greater speed, flexibility or overall performance. However a simplified presentation should go a long way to clarifying the constraints that are imposed on computer graphics by the programming process, at the same time, hopefully, demystifying some of the "black magic" aspects of the subject.

A primitive, schematic, computing system was shown in Figure 5.60. A program in the "control-unit" contains a list of machine language commands, which are sequentially applied to the system. Each command sets the switches in the system routing data to or from memory, to or from output and input units, or directs data to or from the arithmetic unit. The sequence of commands is a computer program, which can be changed depending on the task. The commands are stored as blocks of binary digits which when they are entered into an "instruction register" set switches in the overall computing system to execute the commands they represent. However,

A. Thomas, *Integrated Graphic and Computer Modelling*,
DOI: 10.1007/978-1-84800-179-4_6, © Springer-Verlag London Limited 2008

the data that such a program is designed to operate on, also has to be stored in fixed sized blocks of binary digits in the main memory of the system. It is therefore a small step to make this system more flexible by storing the program in the same location as the data in the computer's main memory. Potentially this allows the instructions to be processed as data and therefore modified under program control. This gives the stored-program computer-architecture, attributed to Von Neumann, shown diagrammatically in Figure 6.1 that is the subject of the rest of this chapter. This more flexible system operates on two levels. The first level is a cycle repeatedly fetching and executing program-instruction from memory, while the second level handles the actions defined by the program itself.

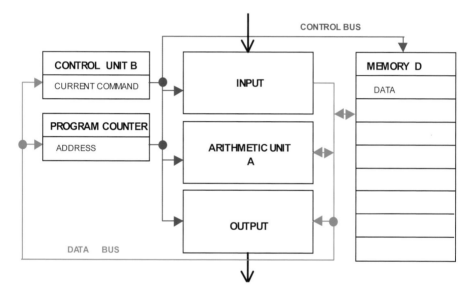

Figure 6.1 A store program computer system

In the final system discussed in this section the blocks, or "words" of memory, will be 16 bits long. If this bit pattern is interpreted as a decimal number it can hold integers approximately in the range of –32,000 to +32,000. However if it is interpreted as a character it will hold one Unicode, Java character code, or two ASCII eight-bit character codes packed together. If it is interpreted as a command what is done will depends on the structure of the electronics in the computer's control unit registers.

Machine Language Programming

Machine language programming depends on the instruction set provided by a particular computer. At its most basic level it consists of entering a series of binary number codes into the machines memory and then setting the system to access these values as a sequence of instruction codes. In this chapter Java will be used to simulate a series of simplified machines to illustrate important aspects of this task.

The idea of using numbers to represent instructions is first illustrated by building a system that uses decimal command codes. This is then extended to handle hexadecimal codes as these directly match the binary values used in the hardware for machine programming: each hexadecimal character corresponding to four binary digits.

Conceptually these simulator programs can be constructed in two or three parts depending on how the task is viewed. Firstly there is the program that simulates the behaviour of the machine, secondly there is the display generating program that illustrates the way it is operating and acts as an interface to the system user, and the third section is a control program that sets up and links these first two units together. In order to develop the simulators incrementally the implementation that is discussed below is initially evolved as one program. However, it can be refactored to use Java's "graphic-model" approach so its component parts become separate "threads" loosely coupled so actions in one are correctly reflected in the behaviour of the others. Building these simulators allows the application of some of the graphics facilities provided in the Java support libraries to be explored: in particular the window system used to display the computer system in Figure 6.3.

A fairly natural starting point for building a computer simulator is to write a program to model the behaviour of the target system. A graphic model to provide the appropriate visualisation of the system in action can then be added to improve the program's user interface. The system shown in Figure 6.1 is implemented in Figure 6.3 as a decimal coded machine where the instruction set for the system consists of ten command codes: each command being made up from a single decimal digit representing its action followed by a four decimal digit memory location address, on which the action is to be applied. This allows 10000 memory locations to be addressed, and numerical values in the range −99999 to 99999 to be used as data in calculations.

Decimal Instruction set

0	Read a new value into the addressed memory location
1	Output a value from the addressed memory location
2	Load the Arithmetic Register from the addressed memory location
3	Add the addressed value to the Arithmetic Register
4	Subtract the addressed value from the Arithmetic Register
5	Multiply the Arithmetic Register by the addressed value
6	Divide the Arithmetic Register by the addressed value
7	Store the Arithmetic Register in the addressed memory location
8	Jump to the instruction in the addressed memory location
9	Jump to the addressed location if the Arithmetic Register is negative

Most of the ten commands are direct and can be assembled into simple sequences. However to make intelligent programs it is necessary to provide conditional operations that depend on relationships found in the data. This flexibility is provided by the last command in this set of operations, which changes the program counter only when the value in the arithmetic register is negative. This allows tests to be set up as arithmetic calculations the outcome of which can then be used to select the next program code-sequence to execute. The two "jump" commands permit the program counter to be loaded with a new address to change the sequence otherwise obtained by the program counter automatically incrementing after each instruction.

A program to implement this computer's behaviour can be set up as follows:

```java
import java.awt.*;  import javax.swing.*;  import javax.swing.table.*;
import javax.swing.event.*;import java.awt.image.*;  import java.awt.event.*;
import java.util.*;  import java.text.*; import java.lang.*;
public class MicroComputer{
    static  int programCounter = 0,address = 0;
    static  int command = 0,instruction = 0,accumulator = 0;
    static  boolean finished = false;
    static int[] memory = new int[10000];
    static TextWindow IO = new TextWindow(20,70,300,300 );

    MicroComputer (){ }
    public static void main(String[] args){
        memory[0]=1;
        while(!finished){
            instruction = fetchInstruction(programCounter);
            programCounter++;
            command = instruction / 10000;
            address = instruction % 10000;
            if(command>9)finished=true;
            else executeInstruction(instruction,command,address);
        }
    }
    static int fetchInstruction(int address){
        return memory[address];
    }
    static void executeInstruction(int instruction,int command,int addr){
        switch(command){
            case 0: memory[addr]=getInput(); break;                 // read
            case 1: putOutput(memory[addr]); break;                 // write
            case 2: accumulator=memory[addr]; break;                // load
            case 3: accumulator= accumulator+memory[addr]; break;   // add
            case 4: accumulator= accumulator-memory[addr]; break;   // subtract
            case 5: accumulator= accumulator*memory[addr]; break;   // multiply
            case 6: accumulator= accumulator/memory[addr]; break;   // divide
            case 7: memory[addr]=accumulator; break;        // store the accumulator
            case 8: programCounter = addr; break;           // unconditional jump
            case 9: if (accumulator<0) programCounter = addr; break;        // jump if...
            default: finished = true;
        }
    }
}
```

The *"main"* procedure executes an infinite loop. Within this loop it fetches the contents of the memory cell addressed by the program counter, which it then splits into a command code and an address. As long as the command code is in the range 0 to 9, the program counter is incremented and the instruction executed, otherwise the loop is exited. Where the command is a jump the contents of the program counter is overwritten directly by the instruction's address field. The *fetchInstruction()* method is a simple memory access operation. Whereas the *executeInstruction()* method uses the operation code in a *switch* statement to select the *case* which implements the action it represents on the data in the addressed memory location, accompanying it.

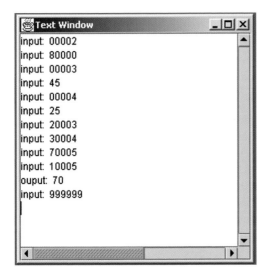

Figure 6.2 Test input and output

Most actions consist of moving data between memory locations and the various registers in the control and arithmetic units. However data input and output need facilities for the program to interact with the user. Figure 6.2 shows a simple input and output window to allow a decimal coded program to be entered into the simulator and the result output. Since all the operations in programs written for this system are defined as operations on data in addressed memory locations -- including the input and output commands. This means that programs have to be entered into the computer's memory before they can be run.

The minimum facility needed to support this task is provided by pre-setting the first memory location to 00001: the command to read a new value to the second memory location 0001. If the program counter is initialised to address this value in location 0000 then the first operation will be to read the new value into location 0001. If this first value is a command to read a value into location 0002, directly followed by the jump command 80000, a self-loading process results. Once 80000 has been read into location 0002, the program counter increments from 0001 to 0002, accessing it as a command and executes it as a jump instruction, returning to location 0000, and the read command 00001. This gives a loop, which will read in and execute the following-on sequence of instructions shown in Figure 6.2: reading in two numbers 45 and 25 adding them together, and finally outputting the result: 70.

The methods used in this simulation to input and output values to a *TextWindow* are *putOutput()* and *getInput()*. Because these methods only work with character-string inputs and outputs, a method to convert such a digit string into the internal representation of a number also has to be provided:

```
static void putOutput(int value){ IO.writeString("ouput:  "+value+"\n");}
static int getInput(){
    output.writeString("input: ");
    String str = IO.readString();IO.readLine();
    return number(str);
}
static int number(String str){
    int num=0;
    for(int i=0;i<str.length();i++){
        int digit = (int)str.charAt(i)-(int)'0';
        num = num*10+digit;
    }return num;
}
}
```

Building a Micro Processor Simulator in Java

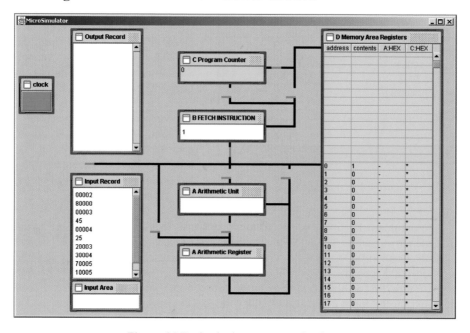

Figure 6.3 Decimal microcomputer simulator

Although this program simulates the way the target computer works and produces the right answers, it does not present the user with any indication how it does so. The simplest way to do this is to link a graphic model of the computer systems to the simulator. Setting up a program that runs a computer model of the hardware system and a coupled computer graphic model will allow its behaviour to be seen in a step by step manner. This allows the user to follow what is going on as a sequence of instructions is executed in a program. The *"main"* procedure needs to set up the graphic model, the operational computer model and initialise the system variables such as the contents of the program counter and the instruction and arithmetic registers as well as the memory. As a computer's program-instructions are executed they also have to generate the appropriate changes in the graphic display to represent the actions taking place in the physical hardware of the system.

Once the basic computer model is working it can be linked to a display of the form shown in Figure 6.3. In Figure 6.3 the control unit consists of two registers, the first called the instruction register in which the current command word is stored. This register, labelled the B-register, would be coupled to the circuitry that sets the switches to implement the command. A second register is associated with it, holding the address in memory of the next instruction from the program that is being executed. This register is called the program counter or the C-register. In this scheme, this conveniently allows the arithmetic unit to be labelled the A-register, and the Memory or data storage registers to be labelled the D registers.

Constructing a graphic model of the computer system in the way shown in Figure 6.3 depends on using objects from the Java libraries, in particular the *AWT* and the *Swing* libraries. These provide display objects such as *windows* and *frames* that act as graphic *containers* for other image building elements. In the *TextWindow* used in Figure 6.2 this context is provided by a *JFrame* from the *Swing* classes to which has been added a *TextArea* object to handle the input and output of text. In Figure 6.3 the display is also built up within a *JFrame*. Each of the working elements is generated in an *JInternalFrame container* object that is added to the main *JFrame*.

Each of the registers is modelled by a *JinternalFrame* containing a *Textfield* to display its contents. The clock button is a *JinternalFrame* to which has been added a *Jbutton* object.

This graphic model can then be used in the way shown in Figures 6.4-6 to display the step-by-step execution of the following program to add two numbers together.

Order	Command	coded	decimal
0:	Read A	rds 3	0-0003
1:	A		25
2:	Read B	rds 4	0-0004
3:	B		45
4:	Load A	lds 3	2-0003
5:	Add B	add 4	3-0004
6:	Store C	str 5	7-0005
7:	Write C	wrt 5	1-0005

The first step is to enter the self-loading-loop instructions into the computer's memory. Clicking the mouse pointer on the red clock-button triggers the execution and display of each fetch and execute step in running the program in the way shown in the following sequence of screen shots.

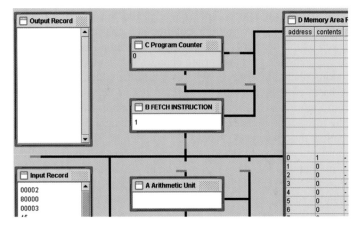

Figure 6.4a Obtain the next instruction from memory location 0

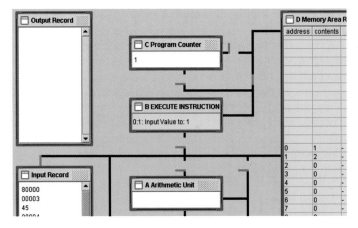

Figure 6.4b Execute instruction:read next input to location 1

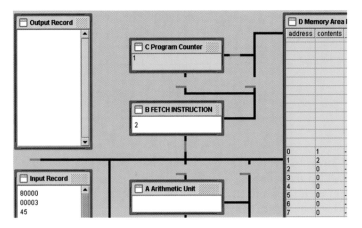

Figure 6.4c Obtain the next istruction from location 1

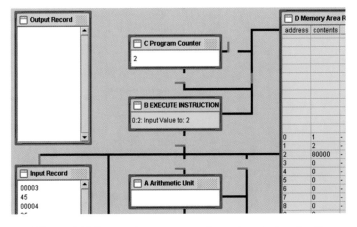

Figure 6.4d Execute instruction: read next input to location 2

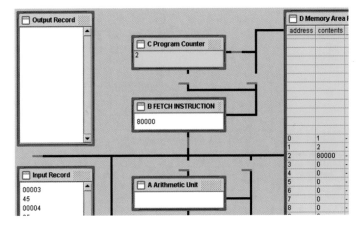

Figure 6.4e Obtain the next istruction from location 2

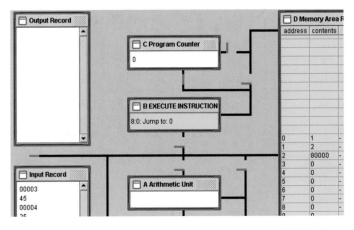

Figure 6.4f Execute instruction: reset program counter back to 0

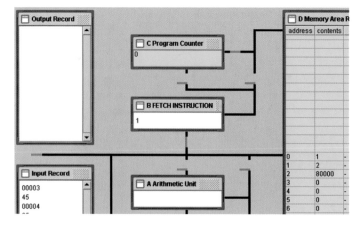

Figure 6.5a Obtain the next istruction from location 0

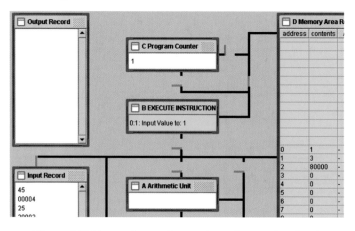

Figure 6.5b Execute instruction: read next input to location 1

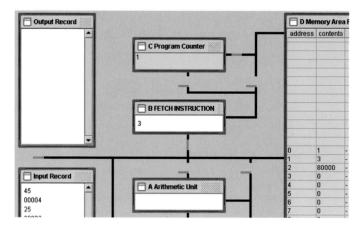

Figure 6.5c Obtain the next istruction from location 1

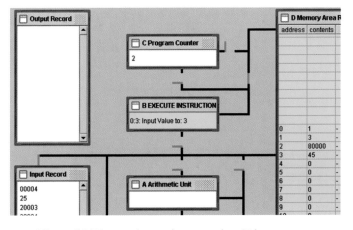

Figure 6.5d Execute instruction: enter data 45 into memory

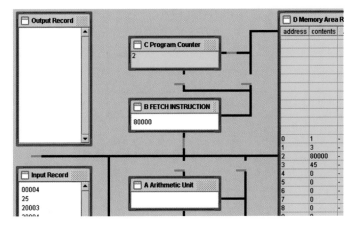

Figure 6.5e Obtain the next istruction from location 2

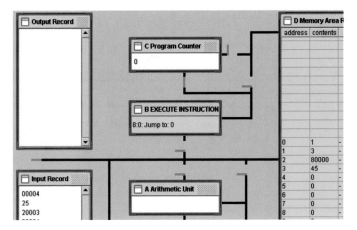

Figure 6.5f Execute instruction: reset program counter back to 0

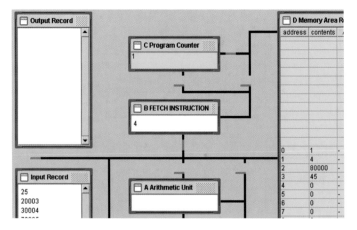

Figure 6.6a Obtain the next istruction from location 1

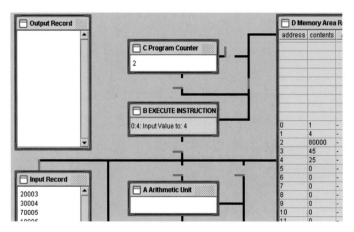

Figure 6.6b Execute instruction: enter data 25 into memory

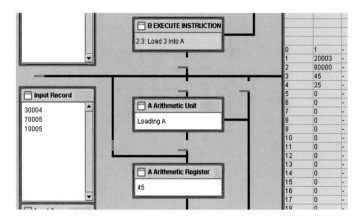

Figure 6.7a Executing the program from Figure 6.3: load the accumulator

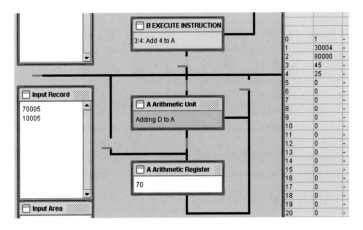

Figure 6.7b Executing the program from Figure 6.3: add to the accumulator

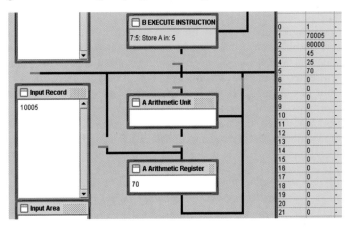

Figure 6.7c Executing the program from Figure 6.3: store the accumulator

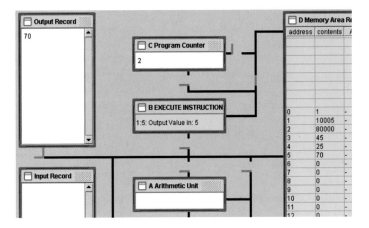

Figure 6.7d Executing the program from Figure 6.3: output the answer

This visualisation sequence is given at length to illustrate the way the simulator program can be used to show the switch settings that implement each command, and the way that the data is routed through the various units of the computer system in order to execute the whole program.

The system in Figure 5.60 was simplified by having the program code separated from the data it operates on. This made the way a program is executed easier to describe. However the system gains in generality and flexibility if the program code is stored in memory as data. This is possible because commands are represented by numerical codes. This makes the task of repeating command sequences merely a matter of revisiting the same command codes in memory and potentially allows program instructions to be treated as data and modified as a program is being run.

Hexadecimal Operation Codes

The basic way a minimal computer system might be designed to work is illustrated above using decimal number codes. However, electronic components operate on binary coded data and instructions. The simulator can be brought closer to a real system by exchanging the decimal number system by the hexadecimal number system. Where numbers have to be worked with as bit patterns, it is easier to work with a hexadecimal representation of the number, because each four bits of the 16 bits can be presented as one hexadecimal digit {0-9, A-F}, consequently the 16 bits will match four hexadecimal characters. The equivalent decimal representation of a binary number obscures the structure of the binary data and hence the way the number can be partitioned into fields that represent an instruction code. The decimal version of the number however remains much easier to work with when discussing arithmetic operations.

0:	nop	0: A null command, used as a continue statement for labels
1:	rds	1: Read a number into memory D[i]
2:	wrt	2: Write a value out in decimal format from memory D[i]
3:	wrh	3: Write value out in hexadecimal format from memory D[i]
4:	wrc	4: Write a character out from memory D[i]
5:	jmp	5: Load the program counter: register C, with address i.
6:	jng	6: If register A is negative load register C with address i.
7:	jez	7: If register A is zero load register C with address i.
8:	stc	8: Store register C back into memory unit D[i]
9:	lds	9: Load Arithmetic Unit register A from memory D[i]
A:	str	A: Store Arithmetic Unit register A back to memory D[i]
B:	add	B: Add value from memory D[i] to register A
C:	sub	C: Subtract memory value D[i] from register A
D:	mul	D: Multiply register A by value from memory D[i]
E:	div	E: Divide register A by value from memory D[i]
F:	mod	F: Reminder of register A divided by memory D[i]

In the simplified systems presented in this section, instructions consist of a command code followed by a single memory address. Each command therefore represents an action related to a specified memory cell. If the first hexadecimal digit is used to code the command then the remaining three digits can be used to represent the address. This gives sixteen possible commands and allows 4096 cells or words to be addressed in the memory. An instruction set for a simulator system can be set up as follows, where D[i] is the memory word addressed by the instruction *f(i)*. The control unit again consists of a single register called the instruction register in which the current command word is stored. This register, again labelled the B-register, is coupled to the circuitry that sets the switches to implement the command. A second register is associated with it, which holds the address in memory of the next instruction to be executed. This register is called the program counter or the C-register. This allows the arithmetic unit to be labelled the A-register, and the Memory or data storage registers again to be labelled the D registers.

Most of these commands are direct and can be arranged in simple sequences. However, again to make intelligent programs it is necessary to provide conditional operations, which depend on relationships found in the data. In this extended set of operations this flexibility is provided by two commands which change the program counter either if the value in the arithmetic register is negative, or if it is zero. This allows tests to be set up as arithmetic operations, the outcome of which can be used to select the next program code sequence to be executed.

```
public class MicroComputer{
    static int pc = 0, address = 0, command = 0, instruction = 0;
    static int accumulator = 0;
    static boolean finished = false;
    static int[] memory = new int[4096];
    static InputWindow input = new InputWindow(20,370,300,70 );
    static OutputWindow output = new OutputWindow(20,70,300,300 );
    MicroComputer (){}
    public static void main(String[] args){
        memory[0] =4097; memory[2] = 20480;
        int instruction = fetchInstruction(pc);
        while((instruction<16*16*16*16)&&(!finished)) {
            instruction = memory[pc];
            address= instruction%4096;
            pc = executeInstruction(instruction);
            address = pc;
        }
        output.writeString("program completed \n\n");
    }
    static int fetchInstruction(int address){ return memory[address];}
    static int executeInstruction( int instruction){
        command = instruction/4096; address = instruction%4096;
        switch(command){
            case  0: if(address>0){
                output.writeString("to finish enter y \n");
                if(input.readString( ).equals("y"))finished = true;}
                break;                                          // nop
```

```
        case  1: memory[address]=getInput();break;              // rds
        case  2:  putOutput(memory[address]);break;             // wrt
        case  3: case  4: break;              // wrh wrc not implemented
        case  5: pc  = address;return pc;                       // ldc
        case  6: if(accumulator<0){ pc=address;}
                 else pc= pc+1;return pc;                       // lnc
        case  7: if(accumulator==0){pc=address;}
                 else{pc=pc+1;}return pc;                       // lec
        case  8: memory[address]= pc;break;                     // stc
        case  9: accumulator = memory[address];break;           // lds
        case 10: memory[address]= accumulator;break;            // str
        case 11: accumulator= accumulator+memory[address];break;   // add
        case 12: accumulator= accumulator-memory[address];break;   // sub
        case 13: accumulator= accumulator*memory[address];break;   // mul
        case 14: accumulator= accumulator/memory[address];break;   // div
        case 15: accumulator= accumulator%memory[address];break;   // mod
        default: output.writeString("error end \n");finished=true;
    }pc=pc+1; return pc;
}
static void putOutput(int value){
    output.writeString(
        "output decimal: "+value+ " hexadecimal: "+hexString(value,4)+"\n");
}
static int getInput(){
    output.writeString("input:  ");
    String str = input.readString(); output.writeString(str+" \n");
    return hexNumber(str,4);
}
static int hexNumber(String program,int len){
    int val=0;
    for(int j=0;j<len;j++){
        char ch = program.charAt(j);
        if((ch>='0')&&(ch<='9')){val = val*16+ (int)ch - (int)'0';}
        else if((ch>='A')&&(ch<='F')){val = val*16+ 10+(int)ch - (int)'A';}
    }return val;
}
static String hexString(int val,int len){
    String str ="";
    if(val<0)return "****";
    for(int j=0;j<len;j++){
        int digit = val%16;
        val = val/16;
        if((digit>=0)&&(digit<=9)){str = ((char)(digit+((int)'0')))+str;}
        else if((digit>=10)&&(digit<=15))
            {str = ((char)(digit-10+((int)'A')))+str;}
    }return str;
}
```

```
static int number(String str){
    int num=0;
    for(int i=0;i<str.length();i++){
        int digit = (int)str.charAt(i)-(int)'0';
        num = num*10+digit;
    }return num;
}
}
```

The "jump" commands permit the program counter to be loaded with a new address. This allow repeated sequences of code to be set up as sub-programs, or subroutines. A program can jump to and return from these blocks of code where ever necessary.

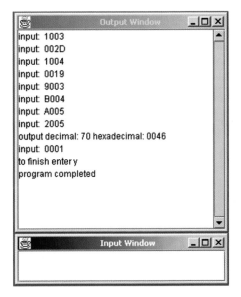

The state of the overall machine can be defined as the contents of the memory and the setting of the arithmetic register, and the program counter. However, if the memory contains a program, its state at any point in its operation can be defined merely by the values of the arithmetic register and the program counter. If two programs are stored in memory at the same time, it is possible to change from one program to the other merely by switching the arithmetic register value and the program counter value for each program. This context switching allows independent program threads to be interleaved, to make more efficient use of the system, if, for example, one program thread is waiting for input data.

Figure 6.8 Test hexadecimal program

The classes *InputWindow* and *OutputWindow* set up the windows to enter and output data. These two classes extend the facilities provided by the *JFrame* class provided by the Java *Swing* library. This class provides the window objects, which allow output and the input obtained from the keyboard to be displayed on the screen. The system facilities that support these actions are separate program threads that run independently of the simulator program, sharing the computer facilities with the simulator program in an interleaved way under the control of the Java language environment. The only links that need to be set up between these programs occur when the input and output data is transferred between them. This is achieved by an "event" driven communication scheme.

The objects that handle input and output text data are obtained from the *TextField* and *TextArea* classes. However these objects are *Container* objects that have to be added to the *JFrames*. The *TextField* provides methods for working with a simple line of text, whereas the *TextArea* is more complex and supports multiple lines of text in a document format. In particular they provide the bridge to the system routines that manage text inputs from the keyboard through event-generator and event listener procedures. An *ActionListener* for the *TextField* is set up to receive events generated when a carriage return character is input at the keyboard. When an *ActionListener* object is added to a *TextField* object it is registered with the keyboard system methods, so that they can notify the listener when the appropriate event – the carriage return -- occurs.

The list of listeners which the system holds registered for an event are all notified when the event occurs by calling the procedure, in each listener set up to handle the event. This is achieved by writing a method with the standard name *ActionPerformed( )* to to carry out the action required in its local context . In the code given below: the action taken on receiving such an event notification, is to get the text string from the system after each carriage return, and copy it to a String variable available to the rest of the simulator program. A *TextArea* provides similar event-linked communication with the Java system, but is more complex to use.

For this reason the input has been channelled through a *TextField*, but this means that every time new text is read into the simulator through the *readString()* method of the *InputWindow* it needs to be output to the *TextArea* in the *OutputWindow* by the *getInput()* method of the simulator program, to keep track of an overall input sequence in the way shown in Figure 6.8.

```
class OutputWindow extends JFrame{
    public JTextArea ta = null;

    OutputWindow(int col, int row,int width, int height){
    super("                         Output Window");
    this.ta = new JTextArea();
    JScrollPane jsp = new JScrollPane(ta);
    jsp.setVerticalScrollBarPolicy(JScrollPane.VERTICAL_SCROLLBAR_ALWAYS);
    this.getContentPane().add(jsp);
    this.setSize(width,height);
    this.ta.setBackground(Color.white);
    this.ta.setEditable(false);
    addWindowListener(
        new WindowAdapter (){
            public void windowClosing(WindowEvent e){ System.exit(0);}
        }
    );
    this.setLocation(col,row);
    this.setVisible(true);
    }
    public void writeString(String str) { this.ta.append(str);}
}
```

```
class InputWindow extends JFrame{
    public JTextField tf = null;
    private String text = null;
    private boolean noInput = true;
    InputWindow(int col, int row,int width, int height){
        super("                    Input Window");
        this.tf = new JTextField();
        this.tf.setBackground(Color.white);
        this.getContentPane().add(tf);
        this.tf.addActionListener(
            new ActionListener(){
                public void actionPerformed(ActionEvent e){
                    text = tf.getText();
                    tf.setText("");
                    noInput= false;
                }
            }
        );
        this.setSize(width,height);
        addWindowListener(
            new WindowAdapter (){
                public void windowClosing(WindowEvent e){ System.exit(0);}
            }
        );
        this.setLocation(col,row);
        this.setVisible(true);
    }
    public String readString(){
        noInput = true; this.tf.setBackground(Color.yellow);
        while (noInput){Dummy.dummy();}
        this.tf.setBackground(Color.white);
        return text;
    }
}
```

```
class Dummy{
    public Dummy(){ }
    static void dummy(){for(int i=0; i<8; i++){}}
}
```

Because the simulator program and the system keyboard programs run independently of each other, the *readString()* method, when it is called, could transfer what ever the last input string was, that previously had been entered into the system, back to the simulator program,. When the *readString()* method is called what is wanted is usually a response to an input request. To ensure what is received is produced in the right order, as soon as *readString()* is called it sets a Boolean variable *noInput* to be true and then while it is true executes a "busy-wait" loop until the next text input is received. As soon as the *ActionPerformed()* procedure receives a new text *String* it sets *noInput* false so allowing the *readString()* method to escape the busy-loop and return the new text-string.

The call to the *dummy()* method in the *Dummy* class is a programming fix found necessary to allow the Java system to recover control during the busy-wait loop. This enables the system to interleave the necessary keyboard and other methods that are being run as separate program threads, concurrently with the simulator program thread. Otherwise the system would stay locked in the simulator's infinite busy-loop, if the appropriate external action of these other methods were not permitted to happen. There are more sophisticated facilities in Java for managing the task of waiting for an event, but this is a simple, minimal way, that works in this context to achieve the required behaviour.

Running A Simple Hexadecimal Program

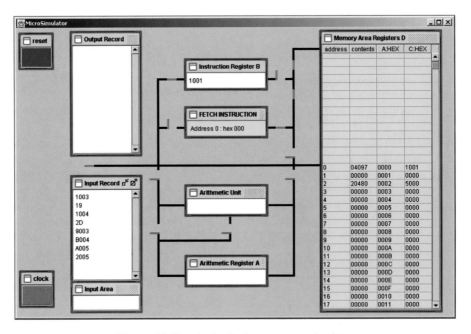

Figure 6.9 Hexadecimal microcomputer simulator

The hexadecimal system can be set up to run in the same way that the decimal machine was set up to run. If the 0 address in memory is initialised to *rds 1*, the next place left empty and the address position 2 initialised with *jmp 0*, then an infinite fetch execute loop is defined. This can be set going by initialising the program counter to address the first location in memory, address 0. This will allow the program code shown in the table given below to be entered serially into the system with the required data values in the appropriate location to be executed correctly; and then allow the system to return safely to read in the next program. A program to add two numbers together, in this system will consist of the machine code to execute the sequence of operations given in the following table.

A program to add two numbers together

Command	coded	decimal	hex	Binary Machine Code			
0: Read A	rds 3	1 – 3	1- 003	0001-	0000	0000	0011
1: A		25	19	0000	0000	0001	1001
2: Read B	rds 4	1 – 4	1- 004	0001-	0000	0000	0100
3: B		45	2D	0000	0000	0010	1101
4: Load A	lds 3	9 – 3	9- 003	1001-	0000	0000	0011
5: Add B	add 4	11– 4	B- 004	1011-	0000	0000	0100
6: Store C	str 5	10– 5	A- 005	1010-	0000	0000	0101
7: Write C	wrt 5	2 – 5	2 - 005	0010-	0000	0000	0101

In the initial column the program is defined using variable names for the values. In the second column these have been converted into address locations in memory, and the commands have been converted into the codes obtained from the table of commands given earlier. In the third column these commands have been translated into decimal numbers for the command field of a machine code instruction and the address field of the instruction. Converting these numbers to their hexadecimal format gives the code in the fourth column, and this can then be converted directly to its binary format shown in the final column.

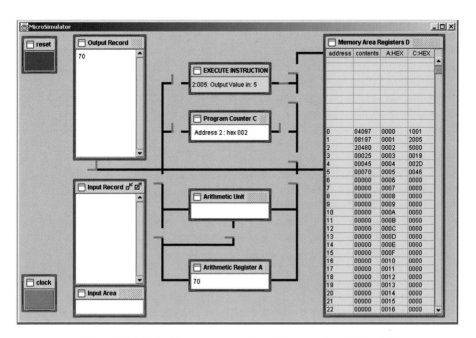

Figure 6.10 Output from a program to add two numbers: 25 and 45

Loading a Program

The problem with the approach given above is that only one line of the program is in the system at one time and once it has been executed it is lost. This prevents the reuse of code for example in repeat loops. What is needed is a way of entering the previous program into a position in memory and then changing the program counter to address this code and execute it.

A program to add two numbers together

command	coded	decimal	hexa-decimal	binary			
0: Read A	rds 6	1 – 6	1- 006	0001-	0000	0000	0110
1: Read B	rds 7	1 – 7	1- 007	0001-	0000	0000	0111
2: Load A	lds 6	9 – 6	9- 006	1001-	0000	0000	0110
3: Add B	add 7	11– 7	B- 007	1011-	0000	0000	0111
4: Store C	str 8	10– 8	A- 008	1010-	0000	0000	1000
5: Write C	wrt 8	2 – 8	2 - 008	0010-	0000	0000	1000

If the code given in the table above is located in address locations 0 to 5 and the program counter is initialised to 0, then five *fetch-execute* cycles would read in two numbers, add them together and output the answer. When the program counter gets to 5 the program will be complete, however there is nothing to stop the system incrementing the address for the next instruction to memory location 6, the contents of which would then be read as though it were an instruction. This would result in data being executed as program code probably with disastrous results. There needs to be a way to stop once a program has finished running, or at least a way to set up and run another program.

Also manually placing program code directly into memory in this way would require special hardware to do it, and would be a slow and tedious task. Some way of automating the process is desirable. In fact almost nothing new is needed to achieve this. All that is required is the short, program loading loop already demonstrated for the decimal machine, to be used in a slightly different way.

This can be achieved within the basic infinite fetch-execute-loop of the running machine by interleaving each main program command with a loading command to place it at some required location in memory. The only extension necessary is to end the overall input sequence with a jump command. This must pass control to the program wherever it has been placed in memory once it has been completely loaded. Also an extra command must be placed before this, at the end of the actual program code sequence, to jump back to the reading loop at location 0, in the way shown in the table given below.

A self loading program to add two numbers together

Read 7 Read A	rds 7 rds 4	1 – 7 1 – 4	1007 1004
Read 8 Read B	rds 8 rds 5	1 – 8 1 – 5	1008 1005
Read 9 Load A	rds 9 lds 4	1 – 9 9 – 4	1009 9004
Read 10 Add B	rds 10 add 5	1 – 10 11 – 5	100A B005
Read 11 Store C	rds 11 str 6	1 –11 10 – 6	100B A006
Read 12 Write C	rds 12 wrt 6	1 – 12 2 – 6	100C 2006
Read 13 Jump 0	rds 13 jmp 0	1 – 13 1 – 0	100D 5000
Jump 7	jmp 7	5 – 7	5007

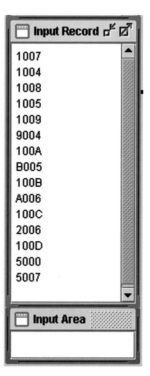

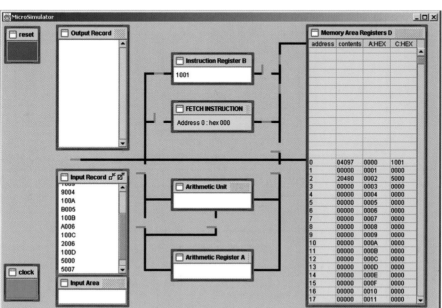

Figure 6.11 A self loading program to add two numbers ready for input

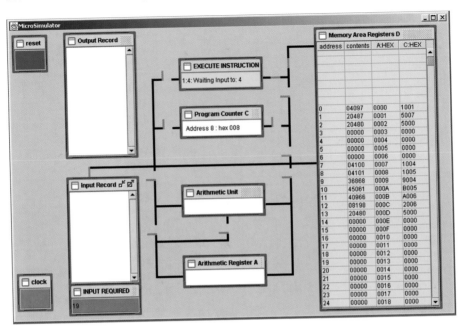

Figure 6.12 Input first value

Figures 6.11 to 6.15 illustrate the self loading of a simple program to add two numbers together and write out the answer.

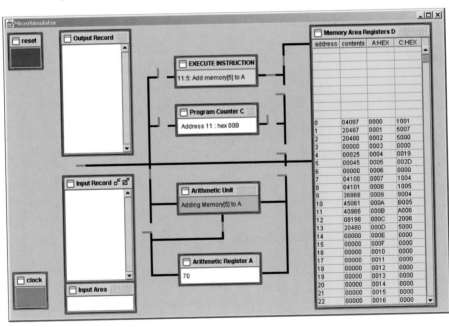

Figure 6.13 Adding two numbers

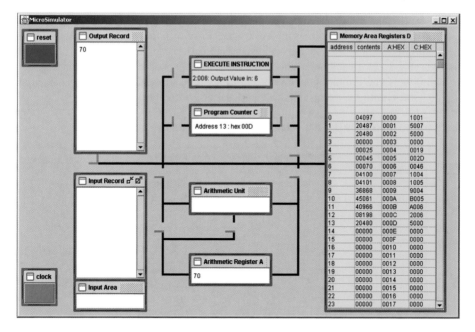

Figure 6.14 Output from a program to add two numbers: 25 and 45

Decimal 25 or hexadecimal 19 is added to decimal 45 or hexadecimal 2D to give an output decimal value of 70.

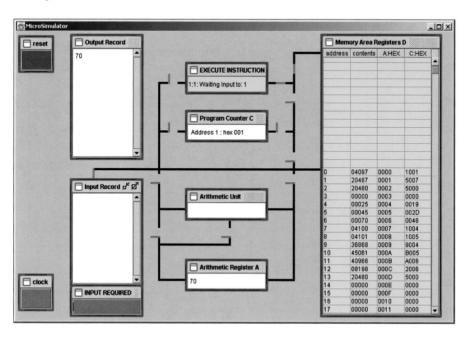

Figure 6.15 Ready for the next program

The system which has been developed to this point still has serious limitations. It is not clear at first sight, for example, how an array of numbers can be entered into neighbouring memory addresses, or how the collection can be processed if the array is of varying size. If each value has to be identified by a single explicit address, in a program, the same limitations would seem to apply to those faced in chapter 2 using simple names, until the *array* construct and *for* loop were introduced.

A Program to Load and Run Other Programs

Location	Labels	Program	Comment	
1003	"Repeat"	101B	Read to "Start"	
1004		101C	Read "Length"	
1005		901D	Load "ReadOp"	
1006		B01B	Add "Start"	
1007		A00A	Store "ReadOp"	
1008		B01C	Add "Length"	
1009		A01E	Store "ReadEndOp"	
100A	"ReadOp"	0000	Read next command	
100B		900A	Load "ReadOp"	
100C		B01F	Add "One"	
100D		A00A	Store "ReadOp"	
100E		C01E	Subtract "ReadEndOp"	
100F		7011	If Zero Jump to "End"	
1010		500A	Jump to "ReadOp"	
1011	"End"	9020	Load "StoreOp"	
1012		B01B	Add "Start"	
1013		B01C	Add "Length"	
1014		A016	Store "StoreEndOp"	
1015		9021	Load "JumptoRepeat"	
1016	"StoreEndOp"	0000	Store at "Code End"	
1017		9022	Load "JumpOp"	
1018		B01B	Add "Start"	
1019		A01A	Store "JumptoStartOp"	
101A	"JumptoStartOp"	0000		
101B	"Start"	0000		
101C	"Length"	0000		
101D	"ReadOp"	1000		
101E	"ReadEndOp"	1000		
101F	"One"	0001		
1020	"StoreOp"	A000		
1021	"JumptoRepat"	5003		
1022	"JumpOp"	5000		
5033				

Figure 6.16 Loading program

Fortunately there is a way round this difficulty. The solution results from the fact that each instruction is a numeric code that can be treated as an arithmetic value. This means it can be entered into the arithmetic unit and modified, for example by adding one to it so incrementing the address section of the command. This makes it

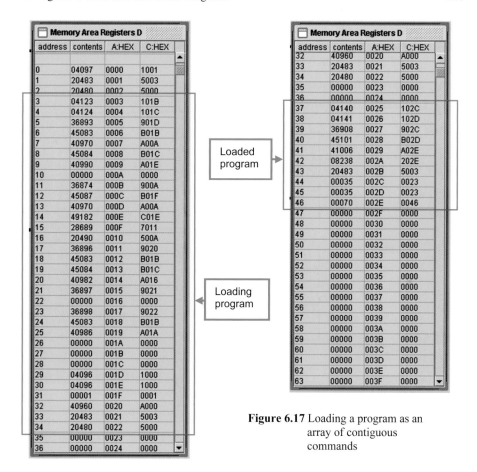

Figure 6.17 Loading a program as an array of contiguous commands

possible to take a command from a program sequence, modify it, and then store it back in its operational place. If the code is constructed as a loop then the same command can be applied to consecutive memory locations by incrementing its address in this way as part of the loop-code. It is also possible to set a limit value to such a modified command so that if subtracting the limit value from it at the end of each cycle results in zero, an appropriate jump command can exit the loop and pass control to following commands. The way this can be used to write a program to load a new program into a predetermined memory space is shown above in Figure 6.16 and is illustrated in use by the simulation system shown in Figures 6.17 and 6.18.

As before this program (in the yellow column) has to be loaded using the primitive loader and a sequence of locating read-commands (in the blue column). However once the main program is in memory it will take over the task of loading subsequent programs into memory. All that needs to be added to each new program is its start address and its length. The new program will then be read into the required space and be executed on control being passed to it. A return jump-command must be added at the end of each new program to return control to the loader when the newly entered program has run, ready for the next program. Figure 6.17 shows the loading program correctly located in memory and a short program that it has entered into the space starting at hex address 25. This program runs to give the output shown in Figure 6.18 before returning ready to enter the next program.

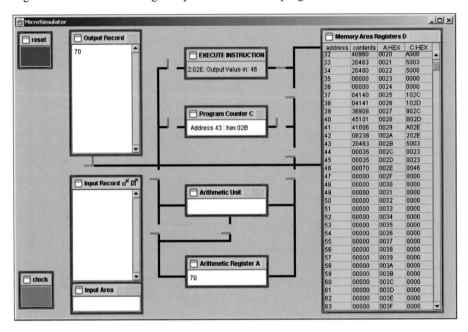

Figure 6.18 Running the loaded program

Processing Arrays

Sequentially processing numbers in an array, within a loop, requires a long sequence of commands to reset the addresses of the code within the loop to allow it to be repeatedly executed in the desired way. Take as an example the problem of locating and outputting the largest value in an array of numbers. In Java the code would be a simple loop followed by a write statement:

```
int temp=0;
int [ ] array = new int[ ]{23,45,2, 12,99};      // init
for(int i=0; i<array.length; i++ ){              // setop
    if(array[i-1]>array[i]){                     // test
        temp= array[i-1];                        // swap
        array[i-1]= array[i];                    // swap
        array[i]=temp;                           // swap
    }
}
write(array[array.length-1]);                    //out
```

The inner statements in this loop can be implemented in machine code by a short instruction sequence. However each command on an array value in this sequence has to have its address modified before the next cycle of the loop can be executed.

Location	Labels	Program	Comment	
2				
3		"9031",	Load array "start address"	init
4		"A035",	Store in "current"	init
5		"B034",	Add "one"	init
6		"A036",	Store in "next"	init
7		"B032",	Add "length-1"	init
8		"B020",	Add "loadcurrentOP"	init
9		"A037",	Store "endtest"	init
A	"cycle"	"901C",	Load "loadnextOP"	setop
B		"B036",	Add "next"	setop
C		"A01C",	Store "loadnextOP"	setop
D		"A022",	Store "loadnextOP2"	setop
E		"C037",	Subtract "endtest"	test
F		"702A",	Jump if equal to zero to "end"	test
10		"9025",	Load "storenextOP"	setop
11		"B036",	Add "next"	setop
12		"A025",	Store "storenextOP"	setop
13		"9020",	Load "loadcurrentOP"	setop
14		"B035",	Add "current"	setop
15		"A020",	Store "loadcurrentOP"	setop
16		"901D",	Load "subtractcurrentOP"	setop
17		"B035",	Add "current"	setop
18		"A01D ",	Store "subtractcurrentOP"	setop
19		"9023",	Load "storecurrentOP"	setop
1A		"B035",	Add "current"	setop
1B		"A023",	Store in "storecurrentOP"	setop
1C	"loadnextOP"	"9000",	Load "next"	test
1D	"subtractcurrentOP"	"C000",	Subtract "current"	test
1E		"6020",	Jump if negative to "loadcurrentOP"	test
1F		"5026",	Jump to "resetincrements"	test
20	"loadcurrentOP"	"9000",	Load "current"	swap
21		"A033",	Store in "temp"	swap
22	"loadnextOP2"	"9000",	Load "next"	swap
23	"storecurrentOP"	"A000",	Store in "current"	swap
24		"9033",	Load "temp"	swap
25	"storenextOP"	"A000",	Store in "next"	swap
26	"resetincrements"	"9034",	Load "one"	rep
27		"A035",	Store in "current"	rep
28		"A036",	Store in "next"	rep
29		"500A",	Jump to "cycle"	rep
2A	"end"	"902E",	Load "writeOP"	out
2B		"B032",	Add "length-1"	out
2C		"B031",	Add "start address"	out
2D		"A02E",	Store "writeOP"	out
2E	"writeOP"	"2000",	Write Largest Value in the Array	out
2F		"5000",	Jump to Loader	stop
30		"0000",	Null command	
31	"start address"	"0038",		var
32	"length-1"	"0004",		var
33	"temp"	"0000",		var
34	"one"	"0001",		var
35	"current"	"0000",		var
36	"next"	"0000",		var
37	"endtest"	"0000",		var

Figure 6.19 Program to select the largest value in an array

38	"0005",	data
39	"0001",	data
40	"0009",	data
41	"0004",	data
42	"0008"	data

Figure 6.20 Test data array

Selecting the largest value in an array can be programmed using the same strategy used to implement the loading program. Each array operation is set up with a base address of 0. These operation codes are then modified by adding on an address-offset to give the address to which they are to be applied. This group of commands loading the initial command, adding on the offset and then storing the result back into the program sequence where it applies, allows the code to be initialised to target the first element in an array, and then by changing the offset being added to the previous operation code, can be used to increment the addresses to apply to the neighbouring sequence of locations in the array. Figure 6.19 shows a bubble sorting algorithm used to sweep the largest value in an array to the highest index position. The inner loop is identified by the swap labels in the right hand column. The operations needed to modify these instruction to handle successive locations in the array, are shown labelled setop. Though the inner loop appears relatively simple the overhead to apply it to an array is clearly very large. Even the output statement for the last element in the array, is made up from five commands.

The importance of array based operations is one reason why better ways of executing these kinds of algorithms evolved. One approach is to provide hardware support for a base-displacement addressing operation. The program shown in the micro-processor simulator memory in Figure 6.21 can be considerably reduced using this kind of extension to the addressing circuitry in the computer system's hardware.

Memory Area Registers D			
address	contents	A:HEX	C:HEX
2	20480	0002	5000
3	36913	0003	9031
4	41013	0004	A035
5	45108	0005	B034
6	41014	0006	A036
7	45106	0007	B032
8	45088	0008	B020
9	41015	0009	A037
10	36892	000A	901C
11	45110	000B	B036
12	40988	000C	A01C
13	40994	000D	A022
14	49207	000E	C037
15	28714	000F	702A
16	36901	0010	9025
17	45110	0011	B036
18	40997	0012	A025
19	36896	0013	9020
20	45109	0014	B035
21	40992	0015	A020
22	36893	0016	901D
23	45109	0017	B035
24	40989	0018	A01D
25	36899	0019	9023
26	45109	001A	B035
27	40995	001B	A023
28	36864	001C	9000
29	49152	001D	C000
30	24608	001E	6020
31	20518	001F	5026
32	36864	0020	9000
33	41011	0021	A033
34	36864	0022	9000
35	40960	0023	A000
36	36915	0024	9033
37	40960	0025	A000
38	36916	0026	9034
39	41013	0027	A035
40	41014	0028	A036
41	20490	0029	500A
42	36911	002A	902F
43	45106	002B	B032
44	45105	002C	B031
45	00000	002D	0000
46	41007	002E	A02F
47	08192	002F	2000
48	20480	0030	5000
49	00056	0031	0038
50	00004	0032	0004
51	00000	0033	0000
52	00001	0034	0001
53	00000	0035	0000
54	00000	0036	0000
55	00000	0037	0000

Figure 6.21 Loaded program

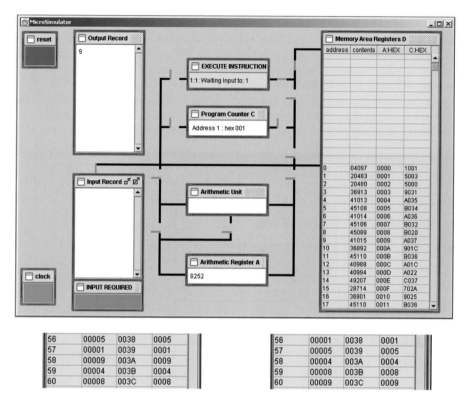

56	00005	0038	0005
57	00001	0039	0001
58	00009	003A	0009
59	00004	003B	0004
60	00008	003C	0008

56	00001	0038	0001
57	00005	0039	0005
58	00004	003A	0004
59	00008	003B	0008
60	00009	003C	0009

Figure 6.22 Input array and final array

The output from the program in Figure 6.21 is shown in Figure 6.22 along with the memory showing the initial data array and the memory showing the reordered array.

Extending the Java Micro Processor Simulator

If the address used to access memory is calculated by adding together a base address held in one register and the address held in an instruction address field then the program given in Figure 6.19 can be rewritten in a different way. The extension to the simulator to support this is shown in Figure 6.23. However, this change in the hardware requires an extension to the instruction set needed to run the system.

Once further registers are added to the system the number of instructions to control the alternative settings of the system which might be needed can became larger than the sixteen which are supported by one hexadecimal character. This could involve changing the 16 bit word size that is the basis of this simulator system's architecture. However, the objective in this chapter is not to explore the many design options which might be used for computer processors but to illustrate the key ideas underlying the way that computer systems have evolved, employing a minimum number of changes to the initial simulator system. To this end two steps have been taken, the first is to include a mode or state variable for the simulator system. This can be used to extend the addressable microcode range of commands. For example one bit potentially gives 32 alternative instructions if needed. The second step is to make the registers accessible using memory addresses. This is called memory mapping. In this first example the addresses of the extended set of registers are located at the high end of the memory. The program counter is accessed as memory location FFF, and the associated base address register is located at position FFE. This allows the existing instruction-set to operate on these registers without creating new instructions for each register. The state variable (0,1) selects the register set used to provide the next address.

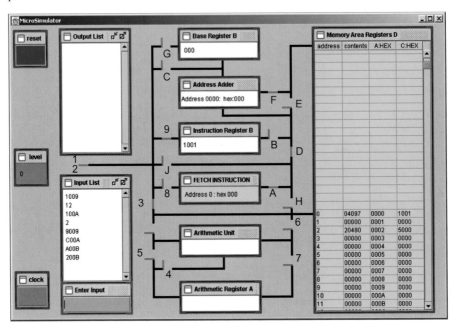

Figure 6.23 Read in two numbers and subtract the second from the first

0:	nop	0: A null command, used as a continue statement for labels
1:	rds	1: Read a hexadecimal format number into memory D[i+j]
2:	wrt	2: Write a value out in decimal format from memory D[i+j]
3:	lvl	3: Set Base Registers: k=0 absolute, k=1 relative addressing
4:	sta	4: Store Arithmetic Unit register A in absolute location D[i]
5:	jmp	5: Load the program counter: register C, with address i.
6:	jng	6: If register A is negative load register C with address i.
7:	jez	7: If register A is zero load register C with address i.
8:	lda	8: Load Arithmetic Unit register A from absolute location D[i]
9:	lds	9: Load Arithmetic Unit register A from memory D[i+j]
A:	str	A: Store Arithmetic Unit register A back to memory D[i+j]
B:	add	B: Add memory value D[i+j] to register A
C:	sub	C: Subtract memory value D[i+j] from register A
D:	mul	D: Multiply register A by value from memory D[i+j]
E:	div	E: Divide register A by value from memory D[i+j]
F:	mod	F: Remainder of register A divided by memory value D[i+j]

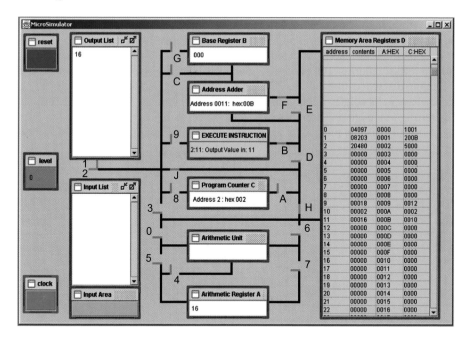

Figure 6.24 Output the result of subtracting one number from another

Table 6.1 Micro-code switch settings for the extended simulator

```
private static int [ ][ ] swS = new int[ ][ ]{              // command switch settings
                //0 1 2 3 4 5 6 7 8 9 A B C D  E F G H  J    // switch labels
/*nop 0*/       {0,0,0,0,0,0,0,0,0,0,0,0,0,0,0,0,0,0,0 },    //0: no operation
/*rds 1*/       {0,0,1,0,0,0,0,0,0,0,0,1,0,0,0,1,0,1,1 },    //1:read a number
/*wrt 2*/       {0,1,0,0,0,0,0,0,0,0,0,1,0,0,0,1,0,1,1 },    //2:write a decimal number
/*lvl 3*/       {0,0,0,0,0,0,0,0,0,0,0,0,0,0,0,0,0,0,0 },    //3:set state register
/*sta 4*/       {0,0,0,0,0,0,1,1,0,0,0,1,0,0,1,0,0,0,0 },    //4:store acc absolute
/*jmp 5*/       {0,0,0,0,0,0,0,0,1,0,0,1,0,1,0,0,0,0,1 },    //5:direct jump
/*jng 6*/       {0,0,0,0,0,0,0,0,0,0,1,0,1,0,0,0,0,0,1 },    //6:if acc negative jump
/*jez 7*/       {0,0,0,0,0,0,0,0,0,0,1,0,1,0,0,0,0,0,1 },    //7:if acc zero jump
/*lda 8*/       {1,0,0,0,0,1,0,0,0,0,0,1,0,0,1,0,0,0,0 },    //8:load acc absolute
/*lds 9*/       {1,0,0,0,0,1,0,0,0,0,0,1,0,0,0,1,0,0,0 },    //9:load acc relative
/*str A*/       {0,0,0,0,0,0,1,1,0,0,0,1,0,0,0,1,0,0,0 },    //A:store acc relative
/*add B*/       {1,0,0,0,1,0,0,1,0,0,0,1,0,0,0,1,0,0,0 },    //B:add to acc relative
/*sub C*/       {1,0,0,0,1,0,0,1,0,0,0,1,0,0,0,1,0,0,0 },    //C:subtract acc relative
/*mul D*/       {1,0,0,0,1,0,0,1,0,0,0,1,0,0,0,1,0,0,0 },    //D:multiply the acc relative
/*div E*/       {1,0,0,0,1,0,0,1,0,0,0,1,0,0,0,1,0,0,0 },    //E:divide the acc relative
/*mod F*/       {1,0,0,0,1,0,0,1,0,0,0,1,0,0,0,1,0,0,0 },    //F:modulo the acc relative
                //0 1 2 3 4 5 6 7 8 9 A B C D E F G H  J    // switch labels
/*fch 10*/      {0,0,0,1,0,0,0,0,0,1,1,0,0,1,0,1,0,0,0 },    //*:fetch-mode relative
/*fch 11*/      {0,0,0,1,0,0,0,0,0,1,1,0,0,1,1,0,0,0,0 }     //*:fetch-mode absolute
                };
```

The array shown in table 6.1 defines the switch settings used to connect the registers in the simulator to implement the different instructions. All the operations on data are implemented using relative addresses calculated by adding a base address and an offset address (the address contained within each command), with the exception of two commands to load and store the accumulator using the offset address as an absolute value. This allows the accumulator to act as a way of transferring values to and from the registers. Two settings are possible for the fetch cycle. Initially just the relative fetch sequence is used, allowing the setting of the base register to zero to provide an absolute addressing mode.

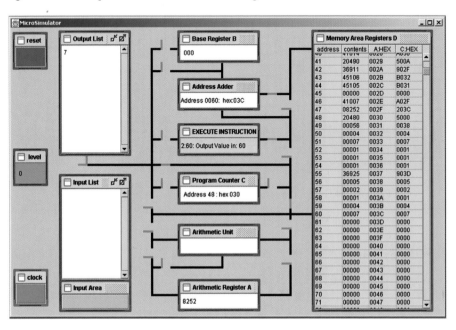

Figure 6.25 Selecting the largest value from an array

The program shown in Figure 6.19, can still be executed in this extended system in the way shown in Figure 6.25. However the same algorithm can be rewritten to take advantage of the new registers in the following way: If the address of the working location in the array is held and modified in the base address register but offsets from these values, used to compare neighbouring values in the array, are held in the program instructions, then the program can be considerably reduced in length in the way shown in Figure 6.26, with the output illustrated in Figure 6.27.

Program Libraries

Providing these extension to the addressing hardware makes the related task of relocating a program a simpler exercise. Changing the location of the program requires a base-address register to be used in the fetch cycle. Setting up the new value in this base-address register immediately changes the next address being operated on. This will need the program counter to be modified at the same time

Input Program

3		"9019",	Load Start
4		"A01C",	Store Index
5	Loop:	"4FF8",	Set BaseR
6		"9000",	Load Index
7		"C001",	Subtr Index+1
8		"600F",	If –ve ->Incr
9		"9001",	Load Index+1
A		"401B",	Store Temp
B		"9000",	Load Index
C		"A001",	Store Index+1
D		"801B",	Load Temp
E		"A000",	Store Index
F	Incr:	"801D",	Load Zero
10		"4FF8",	Set BaseR
11		"901C",	Load Index
12		"B01E",	Add One
13		"A01C",	Store Index
14		"C01A",	Subtr Length
15		"C019",	Subtr Start
16		"701F",	If 0 -> Output
17		"901C",	Load Index
18		"5005",	Jump Loop
19	Start:	"0025",	
1A	Length:	"0005",	
1B	Temp:	"0000",	
1C	Index:	"0000",	
1D	Zero:	"0000",	
1E	One:	"0001",	
1F	Output:	"901C"	Load index
20		"4FF8"	Set BaseR
21		"2000"	Write index
22		"801D"	Load zero
23		"4FF8"	Set BaseR
24		"7000"	If 0 ->restart
25		"0007",	Array
26		"0009",	Array
27		"0000",	Array
28		"0002",	Array
29		"0004",	Array
2A		"0003"	Array

Program after Running .

address	contents	A:HEX	C:HEX
0	04097	0000	1001
1	20483	0001	5003
2	20480	0002	5000
3	36889	0003	9019
4	40988	0004	A01C
5	20472	0005	4FF8
6	36864	0006	9000
7	49153	0007	C001
8	24591	0008	600F
9	36865	0009	9001
10	16411	000A	401B
11	36864	000B	9000
12	40961	000C	A001
13	32795	000D	801B
14	40960	000E	A000
15	32797	000F	801D
16	20472	0010	4FF8
17	36892	0011	901C
18	45086	0012	B01E
19	40988	0013	A01C
20	49178	0014	C01A
21	49177	0015	C019
22	28703	0016	701F
23	36892	0017	901C
24	20485	0018	5005
25	00037	0019	0025
26	00005	001A	0005
27	00003	001B	0003
28	00042	001C	002A
29	00000	001D	0000
30	00001	001E	0001
31	36892	001F	901C
32	20472	0020	4FF8
33	08192	0021	2000
34	32797	0022	801D
35	20472	0023	4FF8
36	28672	0024	7000
37	00007	0025	0007
38	00000	0026	0000
39	00002	0027	0002
40	00004	0028	0004
41	00003	0029	0003
42	00009	002A	0009

Figure 6.26 Select the array's largest value using base displacement addressing

if chaos is to be avoided. One way of achieving this switch to new addressing values is to have a duplicate pair of registers and while one is being used, allow the other set to be reset with new values. If the switch to the new values is done in one command then the modified instruction set given above can be made to work using code loaded anywhere into memory -- merely by changing its base address value.

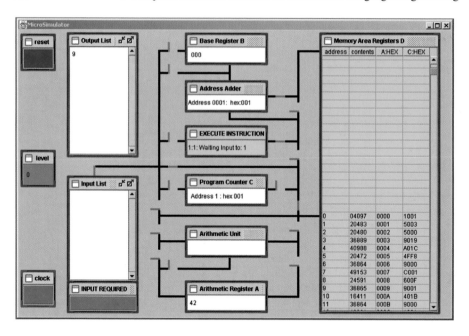

Figure 6.27 Selecting an array's largest value using base displacement addressing

Relocating Programs

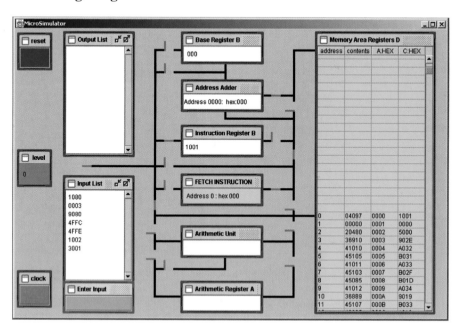

Figure 6.28 Relocating a program: written to operate from address 000
but being run from address 003 without code changes

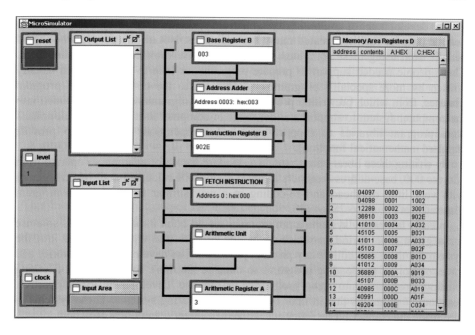

Figure 6.29 Relocating a program: written to operate from address 000 but being run from address 003 without code changes

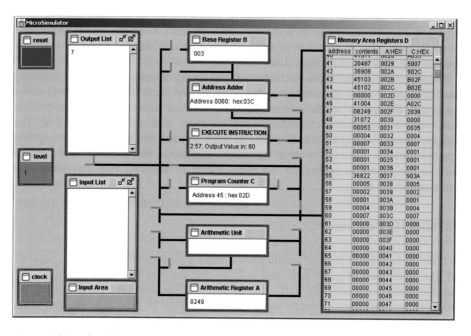

Figure 6.30 Relocating a program to select the largest value in an array: written to operate starting from address 000, but being run from address 003

back to to the original entry. A simple version of a *StringTable* class can be set up to support this name table function in the following way.

```java
class StringTable{
    private String[] table = new String[100];
    public int[] val= new int[100]; private int size = 0;

    StringTable(){ }
    StringTable(String[] str){size=str.length; table=str;}
    public int locate(String str){
        for(int i=0; i<size; i++){if(str.equals(table[i]))return i;}  return -1;}
    public int add(String str){
        int index = locate(str);
        if(index== -1){index = size; table[index]= str; size=size+1;} return index;
    }
    public String getElement(int i){
        if((i>=0)&&(i<size))return table[i]; else return "";}
    public boolean contain(String str){
        int k= locate(str);
        if (k<0)return false;  else return true;
    }
    public int getSize(){ return size;}
    public void setTable(String[] tbl){ table = tbl;size= tbl.length;}
}
```

In this program translation example, a separate table for the operation-code names, ordered so the index of each code name corresponds to its machine code, is set up at the beginning of the translation program. Once the potential operation codes placed in *programElements* in the second column, are confirmed to be correct, by locating them in this operation-code table, the indices of their positions in this table provide the numerical command codes for their equivalent machine code instructions.

By processing the third column of *programElements* last, the variable names, not used as labels, can be added to the name table after the labels. These names refer to data storage spaces. If spaces are allocated for these variables at the end of the program, the addresses that refer to them can be generated and entered into their undefined value-array spaces (not already set: as the label locations will be), merely by counting on from the last line of the *machineProgram* code generated from the second column of the *programElements* array. Once the value array in the name table has been completed the third column of *programElements* containing variable and label names can be used to look-up the address that each name must be replaced by, and which can then be added to the machine code address field, to complete the machine program in the way shown in Figure 7.1.

```java
import java.awt.*; import javax.swing.*; import javax.swing.text.*;
import javax.swing.table.*; import javax.swing.event.*; import java.awt.image.*;
import java.awt.event.*; import java.util.*; import java.text.*; import java.lang.*;
public class Assembler{
```

```
static TextWindow jInput= null;  static TextWindow jOutput = null;
static boolean notFinished = true;
static int i = 0, stringIndex =0, recordIndex =0,  index =0, labelCount = 0;
static String str = "", inputString ="";
static int[] record = new int[500];
static TextWindow IO = new TextWindow(20,170,800,600,"Assembler" );

public static void main(String[] args){
    StringTable names = new StringTable(), commands = new StringTable();
    commands.setTable(
                new String[]{"rds","wrt","lds","add","sub","mul","div","str","jmp","jng"});
    String [] program = new String [100];
    int [] machineProgram = new int[100];
    String [][] programElements = new String[100][3];
    for(int i=0; i<100;i++) machineProgram[i]= -1;
    int i=0; int m=0;
    IO.writeString("Please enter the source program: \n");
    String str = IO.readLine();
    while (!str.equals("stop")){
        program[i++]= str+" "; str = IO.readLine();}          // READ IN THE NEXT PROGRAM
    int programLength = i;

    for(int j=0; j < programLength;j++){
        int k=0,  n=0;  str="";
        char ch = program[j].charAt(k++);
        while(ch ==' '){ch = program[j].charAt(k++);};                  // SPACES
        if((ch>='a')&&(ch<='z'))
            while(((ch>='a')&&(ch<='z'))|| ((ch>='0')&&(ch<='9'))){
                str= str+ch;
                ch = program[j].charAt(k++);
        }
        while(ch==' '){ch = program[j].charAt(k++);};                   // SPACES
        if(ch==':'){
            ch = program[j].charAt(k++);
            programElements[j][0]= str;                                 // LABELS
            while(ch==' '){ch = program[j].charAt(k++);};               // SPACES
            str = "";
            while((ch >= 'a')&&(ch <= 'z')){
                str= str+ch; ch = program[j].charAt(k++);}
        }
        programElements[j][1] = str;                                    // OP CODES
        str="";
        while(ch==' '){ch = program[j].charAt(k++);};                   // SPACES
        while((ch>='a')&&(ch<='z')){
            str= str+ch; ch = program[j].charAt(k++);}
        programElements[j][2]=str;                                      // VARIABLES
    }
```

```
        int n=0;                                                       //  GENERATING MACHINE CODE
        for(int j=0;j < programLength;j++)                                    //  LABELS
            if(programElements[j][0] != null){
                n = names.add(programElements[j][0]);   names.val[n]= -( j+1);
        }
        int numberOfLabels=names.getSize();
        m=0;
        for(int j=0;j<programLength;j++){                              //  OPERATION CODES
            if((m=commands.locate(programElements[j][1]))>=0){
                machineProgram[j]= m*10000;}
        }
        for(int j=0;j<programLength;j++){                                    //  VARIABLES
            n = names.add(programElements[j][2]);
            if (names.val[n]>= 0)names.val[n]= programLength+n-numberOfLabels;
            else names.val[n]= -names.val[n]-1;
            machineProgram[j]= machineProgram[j]+ names.val[n];
        }
        IO.writeLine();  IO.writeString("output program : \n");        // OUTPUT MACHINE CODE
        for(int j=0;j<programLength;j++){
            if(j<10)IO.writeString("00"+j+": ");
            else if(j<100)IO.writeString("0"+j+": ");
            else IO.writeString(j+": ");
            if (machineProgram[j]<0) {IO.writeString("00000"); machineProgram[j]=0;}
            else if(machineProgram[j]<10)IO.writeString("0000");
            else if(machineProgram[j]<100)IO.writeString("000");
            else if(machineProgram[j]<1000)IO.writeString("00");
            else if(machineProgram[j]<10000)IO.writeString("0");
            IO.writeString(machineProgram[j] + " : "+ program[j]+"\n");
        }
        int jj = programLength;
        for(int j=numberOfLabels;j<names.getSize();j++){  // OUTPUT VARIABLE SPACES
            if(jj<10)IO.writeString("00"+jj+": ");
            else if(jj<100)IO.writeString("0"+jj+": ");
            else IO.writeString(jj+": ");
            jj++;
            IO.writeString("variable "+ names.getElement(j)+"\n");
        }
    }
}
```

The program that translated the assembly program shown in Figure 7.1 is given
above. However, the resulting machine code is set up to operate in the memory space
starting from address location 0000. This has to be relocated to a final working
position. In the case of this system of commands the relocation can be done, simply
by adding the new start address value to all the address fields of the program's
instructions. Clearly the data spaces set aside for variables must be left unmodified,
which is important if literals are to be added to the assembly code.

This translation program can be extended to generate hexadecimal code, but also to compose sequences of machine code from single new assembly commands. A simple example of this, that is very convenient in the case of this basic computer simulator, is to allow numerical values to be built into program codes as literals so they do not need to be read into the system as the value of variables. This can be done by extending the code definition in the following way.

<Operation Code>   <Variable Name> ';'

<Label> ':' <Operation Code>   <Variable Name> ';'

<Operation Code> ' " ' <Integer Number> ' " ' ';'

<Label> ':' <Operation Code> ' " ' <Integer Number> ' " ' ';'

This allows statements like *add "25";* to be included in a program. In order to support this extension, the resulting analysis of the input strings must create an entry in the name table to create a variable space for the associated integer value and allow repetitions of the same value to reference the same memory space.

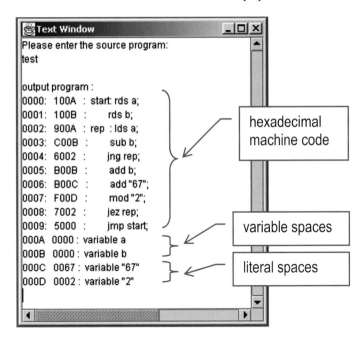

Figure 7.2 Hexadecimal program containing literals numbers

A more complex example of an extension that can be made to the assembly language is a command to load a program to a predetermined memory location.

"load" <address> ';'

If an explicit load command is provided in the assembly language then it has to be expanded as a sequence of commands. This process is called macro expansion, where the single line of code such as *load "24"* is replaced by a sequence of machine codes to implement the operation. However it is possible to use the first version of the assembler to translate a loader written in assembly code to obtain the machine code to add to subsequent programs to produce self-loading units from a new bootstrapped assembly language translation system.

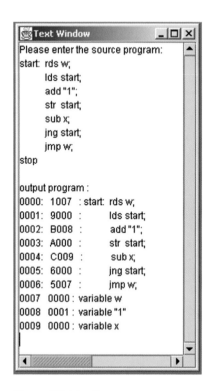

A loading program written in assembly code is given in Figure 7.3. This consists of a read command *"start: rds w;"* to repeatedly place a given number of program instructions into neighbouring positions in memory and then transfer control to them. In order to do this the value *w* has to be set to the start address for the new program. This command then has to be loaded as data into the arithmetic unit and incremented by one before being replaced in its original location to provide the repetition needed to read the main program statements into neighbouring memory locations. The value of *x* has to be set to the *"start"* address, added to the *length* of the program then added to the *read code* to give the test to terminate this reading loop. This occurs when subtracting x from the read command, in the accumulator, no longer gives a negative value. The assembler translation program has to be extended to add this loading code with the appropriately modified command to precede the translated machine program. The machine code for the loader given in Figure 7.3 still has to be modified both to load itself, and then to load the follow on target-program code.

Figure 7.3 A loading program

```
static void writeLoader(int codeLength,int[] exeCode){
    int[] loaderCode = new int[]{
        4099,        4107,        // 1003,    100B,    -- 4107,      rds 11;
        4100,        1,           // 1004,    0001,    -- 1,         1;
        4101,        4107,        // 1005,    100B,    -- 4107,      rds 11;
        4102,        36869,       // 1006,    9005,    -- 36869,     lds 5;
        4103,        45060,       // 1007,    B004,    -- 45060,     add 4;
        4104,        40965,       // 1008,    A005,    -- 40965,     str 5;
        4105,        49155,       // 1009,    C003,    -- 49155,     sub 3;
        4106,        24581,       // 100A,    6005,    -- 24581,     jng 5;
        20485 };                  // 5005
    loaderCode[1]= 4107+ codeLength;  // SET UP THE PROGRAM LENGTH TEST
    for(int i=0;i<17; i++){
        exeCode[i]=loaderCode[i]; }   // ADD THE LOADER TO THE NEW OUTPUT CODE
}
```

A version of this loader setup to place the main program directly after itself in memory is given above. By placing its data space for 'x' and "1" before the loading code the final jump command can be removed to give an eight-line program that will drop through to 'w' the start of the main program. This eight-line program has to have input location commands added to enter it, and a jump to its start command, also the main program has to finish with a jump back to the primitive loader.

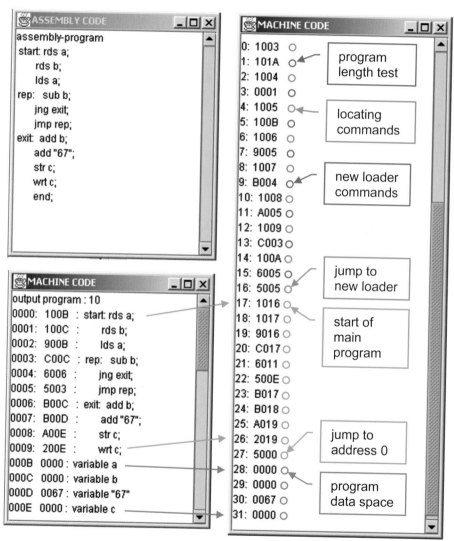

Figure 7.4 Self-loading code

The final machine code given in Figure 7.4 will self load and run to give the output shown in Figure 7.5. The program given in Figure 7.4 reads in two values and subtracts the second from the first. If the result is negative the second value is added back. Hexadecimal 67 is then added to the result.

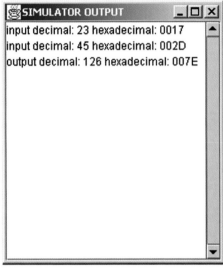

Figure 7.5 Simulator output

This gives 7E in the example shown in Figure 7.5, by adding 23 or hexadecimal 17 to hexadecimal 67. In decimal notation 23 is added to 103 giving decimal 126. The program to translate the assembly program into machine code that will self load and run on the Micro-simulator program developed in chapter 6, is given below. Two separate window classes *Input* and *Output*, replacing the *TextWindow* class used in earlier programs, provide the input and output to this program. The *Simulation* class calls up a machine program execution method and an assembly code translation method from the *MicroComputer* and *Assembler* classes respectively given below.

```java
import java.awt.*;  import javax.swing.*;  import javax.swing.text.*;
import javax.swing.table.*;  import javax.swing.event.*;  import java.awt.image.*;
import java.awt.event.*;  import java.util.*;  import java.text.*;  import java.lang.*;
public class Simulation{
    static String str = "";
    static  Input jInput = new Input(12,412,1000, 100,"INPUT");
    static  Output jOutput0 = new Output(12,512,1000, 100,"OUTPUT");
    static  Output jOutput2 = new Output(262,12,250, 400,"ASSEMBLY CODE");
    static  Output jOutput3 = new Output(512,12,250, 400,"MACHINE CODE");
    static  Output jOutput4 = new Output( 762,12,250, 400,"SIMULATOR OUTPUT");

    public static void main(String[] args){
        int width = 300, height = 100; int [] machineProgram = null;
        String[] program = null; String str = jInput.readString();
        while(!str.equals("stop")){
            if(str.equals("assembly-program")){
                jOutput2.writeString(str +"\n");
                jOutput0.writeString("Please enter the assembly program: \n");
                Assembler ass = new Assembler();
                machineProgram = ass.program(
                                 null,0, jInput, jOutput0, jOutput2, jOutput3);
                MicroComputer.execute(machineProgram,jOutput4,jInput);
            }
            str = jInput.readString();
        }
    }
}
```

```
class Assembler{
    static  Input jInput = null;
    static  Output jOutput0 = null;
    static  Output jOutput2 = null;
    static  Output jOutput3 = null;

    public Assembler(){}

    public int[] program(String[] program,int length,
                    Input jln, Output jOut,Output jOut2, Output jOut3){
        jInput = jln; jOutput0 = jOut; jOutput2 = jOut2; jOutput3 = jOut3;
        int programLength = 0; int[] exeCode=null;
        boolean finished = false, errors = false;
        StringTable names = new StringTable();
        int [] machineProgram = new int[200];
        for(int i=0; i< machineProgram.length;i++)  machineProgram[i]= -1;
            int i=0;
            if(program == null){
                program = new String[200];
                String str="";
                boolean test= true;
                str = jInput.readString();
                if(!str.equals("test")){
                    while (! str.equals("end")){          // enter a program from the keyboard
                        program[i] = str+" \n";  i++;
                        jOutput2.writeString(str+" \n");
                        str = jInput.readString();
                    }
                }else{ i = testData(program)-1;}          // enter a prewritten test program
                programLength = i;
            }else{ programLength = length-1;}
            int codeLength=0;
            if((codeLength = translate(programLength,machineProgram,program,names))>0){
                exeCode = new int[codeLength+17];
                writeLoader(codeLength,exeCode);          // add on self loader
                for(i= 0;i<codeLength;i++){
                exeCode[i+17]=machineProgram[i];
            }
            for(i= 0;i<codeLength+17;i++){
                jOutput3.writeString("\n");
                jOutput3.writeString(i + ": ");
                writeHex(exeCode[i], 4);
            }
            jOutput3.writeString("\n");
        }
        return exeCode;                                   // return assembly code program
    }
```

```
static int translate (int programLength,int[] machineProgram,
                                    String[] program,StringTable names){
    StringTable commands = new StringTable();
    commands.setTable(new String[]
        {"nop","rds","wrt","wrh","wrc","jmp","jng","jez","stc","lds",
                            "str","add","sub","mul","div","mod","end"});
    String [][] programElements = new String[100][3];
    jOutput0.writeString("assembler : "+ programLength +" \n");
    for(int i=0; i< machineProgram.length;i++) machineProgram[i]= -1;
    int i = 0; int m = 0;
    char[] c = new char[1];   c[0]=' ';  String[] strr = new String[1];
    for(int j=0; j < programLength;j++){  // separate labels opcodes variables
        int k=0,  n=0;
        k= name(program[j],k, strr, c);
        if(c[0] ==':'){
            programElements[j][0]= strr[0];                   //  LABELS
            k= name(program[j],++k,strr,c);
            programElements[j][1] = strr[0];              // OP CODE
        }
        else programElements[j][1] = strr[0];             // OP CODE
        if((c[0]>='a')&&(c[0]<='z')){
            k= name(program[j],k,strr,c);
            programElements[j][2]= strr[0];
        }else if(c[0]==""){                           // HEX LITERALS
            String str=""+c[0]; k++;  c[0] = program[j].charAt(k++);
            while(((c[0]>='0')&&(c[0]<='9'))||((c[0]>='A')&&(c[0]<='F'))){
                str= str+c[0]; c[0] = program[j].charAt(k++);
            }
            if (c[0]!=""){str="";} else{str=str+"";k++;}
            programElements[j][2]= str;
        }else if(c[0]=='^'){                          // DECIMAL LITERALS
            String str=""+c[0]; k++; c[0] = program[j].charAt(k++);
            while((c[0]>='0')&&(c[0]<='9')){
                str= str+c[0];  c[0] = program[j].charAt(k++);
            }
            if (c[0]!='^'){str="";} else{str=str+'^';k++;}
            programElements[j][2]= str;
        }
    }
    int n=0;                  // process labels opcodes and variables to give machine code
    for(int j=0; j < programLength; j++){                    //  LABELS
        if(programElements[j][0] != null){
            n = names.add(programElements[j][0]);
            names.val[n]= -(j+1);                   // LABELS Flagged
        }
    }
    int numberOfLabels = names.getSize();
```

```
m=0;
for(int j=0;j<programLength;j++){              //  OPERATION CODES
        if((m = commands.locate(programElements[j][1])) >= 0){
              machineProgram[j]= m*16*16*16;
        }
}
for(int j=0; j<programLength; j++){
      if(machineProgram[j]<16*16*16*16){                  //  VARIABLES
            int gg=0, vals=0; char ch= ' ';
            n = names.add(programElements[j][2]);
            if (names.val[n]>= 0) {
                  gg = programLength + n - numberOfLabels + 1;
                  machineProgram[j] = machineProgram[j] + gg;
                  int k=0;
                  if(programElements[j][2].length()>1)
                      if(programElements[j][2].charAt(k)==""){
                          k++; vals=0;
                          ch = programElements[j][2].charAt(k++);
                          while(((ch >='0')&&(ch<='9'))||((ch>='A')&&(ch>+'F'))){
                              if((ch >='0')&&(ch<='9'))
                                    vals= vals*16 + ((int)ch-(int)'0');
                              if((ch>='A')&&(ch>+'F'))
                                    vals= vals*16 + 10 +((int)ch-(int)'A');
                              ch = programElements[j][2].charAt(k++);
                          }
                      }else if(programElements[j][2].charAt(k)=='^'){
                          k++;  vals=0;
                          ch = programElements[j][2].charAt(k++);
                          while((ch >='0')&&(ch<='9')){
                              if((ch >='0')&&(ch<='9'))
                                    vals= vals*10 + ((int)ch-(int)'0');
                              ch = programElements[j][2].charAt(k++);
                          }
                      }
                  machineProgram[gg] = vals;
            } else {
                  gg = - names.val[n]-1;
                  machineProgram[j] = machineProgram[j] + gg;
            }
      }
} jOutput3.writeString("output program : "+ programLength+"\n");
for(int j=0;j < programLength; j++){
    writeHex(j,4); jOutput3.writeString(":  ");
    writeHex(machineProgram[j],4);
    jOutput3.writeString("  : "+ program[j]);
    machineProgram[j] = machineProgram[j]+11;
}
```

```
        machineProgram[programLength]= 20480;   // add jump to primitive loader
        int jj = programLength+1;
        for(int j = numberOfLabels;j < names.getSize(); j++){
            writeHex(jj,4);
            jOutput3.writeString("   ");
            writeHex(machineProgram[jj],4);
            jOutput3.writeString(" :  variable "+ names.getElement(j)+"\n");
            jj++;
        }return jj;
    }
```

```
static int name(String program, int k,String[] strr,char[] ch){
    char c = program.charAt(k);
    while(c==' '){c = program.charAt(++k);}
    String str= ""; strr[0]=str;
    if((c>='a')&&(c<='z')){
        while(((c>='a')&&(c<='z'))||((c>='0')&&(c<='9'))){
            str = str + c;  k++;  c = program.charAt(k);
        }
    }
    strr[0]= str;
    while(c==' '){c = program.charAt(++k);}
    ch[0]=c;
    return k;
}
```

```
static void writeLoader(int codeLength,int[] exeCode){
    int[] loaderCode = new int[]{
    4099,4107,          // 1003,   100B,    -- 4107,  rds 11;
    4100,1,             // 1004,   0001,    -- 1,     1;
    4101,4107,          // 1005,   100B,    -- 4107,  rds 11;
    4102,36869,         // 1006,   9005,    -- 36869, lds 5;
    4103,45060,         // 1007,   B004,    -- 45060, add 4;
    4104,40965,         // 1008,   A005,    -- 40965, str 5;
    4105,49155,         // 1009,   C003,    -- 49155, sub 3;
    4106,24581,         // 100A,   6005,    -- 24581, jng 5;
    20485};             // 5005
    loaderCode[1]= 4107+codeLength;
    for(int i=0;i<17; i++){exeCode[i]=loaderCode[i];}
}
```

```
static int spaces(String program,int k,char[] ch){
    ch[0] = program.charAt(k);
    while(ch[0] ==' '){k++; ch[0] = program.charAt(k);};
    return k;                                              // SPACES
}
static void writeHex(int num, int size){
```

```
        if(num>16*16*16*16-1)return; //{jOutput3.writeString("STOP");return;}
        if(num<0){ jOutput0.writeString("write hex integer error \n");return;}
        if(num<=0)for(int i=0;i < size;i++)jOutput3.writeString("0");
        else{
            int sz = size-1;
            int digit[]= new int[size+1];
            for(int i=0;i<=size;i++) digit[i]=0;
            while(num>0){ digit[sz--] = num%16; num = num/16;}
            for(int j=0;j<size; j++){
                if(digit[j]<10)jOutput3.writeString(""+digit[j]);
                else jOutput3.writeString(""+(char)((int)'A'+digit[j]-10));
            }
        }
    }
}
```

```
    static int testData(String[] program){                    // test program
        program[0] =        " start:    rds a;\n";
        program[1] =        "           rds b;\n";
        program[2] =        "           lds a;\n";
        program[3] =        " rep:      sub b;\n";
        program[4] =        "            jng exit;\n";
        program[5] =        "           jmp rep;\n";
        program[6] =        " exit:     add b;\n";
        program[7] =        "           add \"67\";\n";
        program[8] =        "           str c;\n";
        program[9] =        "           wrt c;\n";
        program[10] =       "           end;\n";
        for(int i=0;i<11;i++){jOutput2.writeString(program[i]);}
        return 11;
    }
}
```

```
class MicroComputer{
    static  int pc = 0, address = 0, command = 0, instruction = 0;
    static  int accumulator = 0;
    static  boolean finished = false;
    static int[] memory = new int[4096];
    static Input input  = null;
    static Output output = null;

    MicroComputer (){}
        private static int codeindex = 0;private static int[] code;
    public static void execute(int[] program,Output out,Input in){
        ....
    }                                                // replaces main in chapter 6
}
```

Wherever repeating patterns of code occur the programmer will tend to use a shorthand psuedo code to represent it when analysing and designing a program. In the case of this simple system an example of this can be found in the case of statements such as a=b+c. The single-address machine-code requires this and similar arithmetic operations to be expanded into the three commands *lds x; add y; str z;* where a short hand of the sort *add(x y z);* gives a template that can be applied to each of the binary arithmetic operations replacing *add* by, *sub, mul* or *div* as required. This kind of short hand can be added to the assembly language and it can be translated automatically by "macro" substitution before translation to machine code proper in the way shown in Figure 7.6.

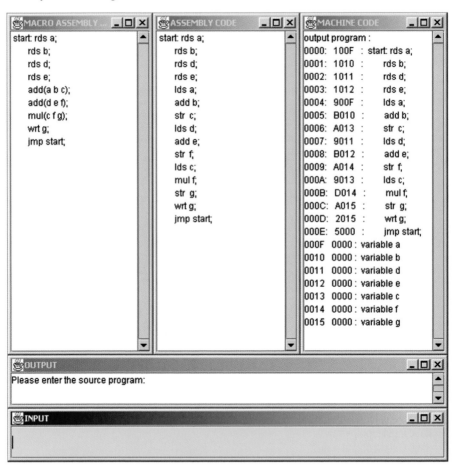

Figure 7.6 Macro substitution in the assembly translation process

The program to execute the macro expansion in the case of this example is given below. Where a pattern of code is used repeatedly and is relatively short this inline substitution for a shorthand command generally makes code easier to read and simpler to write. However, where the substitute code is long an alternative treatment can be more efficient.

```
public class MacroAssembler{
static  Input jInput      = new Input (12,562,750, 100,"INPUT");
static  Output jOutput0= new Output(12,462,750, 100,"OUTPUT");
static  Output jOutput1= new Output(12,12,250,450,"MACROASSEMBLY CODE");
static  Output jOutput2= new Output(262,12,250, 450,"ASSEMBLY CODE");
static  Output jOutput3= new Output(512,12,250, 450,"MACHINE CODE");
   public static void main(String[] args){
        boolean finished = false, errors = false;
        StringTable names = new StringTable();
        String [] program = new String [100];
        int [] machineProgram = new int[200];
        for(int i=0; i<100;i++)machineProgram[i]= -1;
        jOutput0.writeString("Please enter the source program: \n");
        boolean test= true;  int i=0;
        String str = jInput.readString();
        if(!str.equals("test")){
            while (! str.equals("stop")){
                jOutput1.writeString(str);  program[i++] = str;
                str = jInput.readString();
            }
        }else{
             i = testData(program);
             for(int j = 0; j < i; j++){ jOutput1.writeString(program[j]);}
        }
        int programLength = i;
        if((program= macroExpansion(program,programLength))!=null)
            if(translate(program.length,machineProgram,program,names)){}
}
   static String[] macroExpansion(String[] program,int programLength){
        String[] temp = new  String[3*programLength];
        int k=0;  String str="";String stt="";boolean macro=false;
        for(int i=0;i<programLength;i++){
            str = program[i];
            macro=false;
            stt=""; char ch;
            for(int j=0; j<str.length();j++){
                ch=str.charAt(j); if(ch =='(')macro=true; stt=stt+ch;
            }
            if(macro){
                String[] variables = new String[4];
                int kkk =0; String st="";
                ch = stt.charAt(kkk++);
                while(ch ==' '){ch = stt.charAt(kkk++);};              // SPACES
                while((ch >='a')&&(ch<='z')){st= st+ch; ch = stt.charAt(kkk++);}
                variables[0]=st;
                while(ch==' '){ch = stt.charAt(kkk++);};               // SPACES
                if(ch=='('){
```

```
                ch = stt.charAt(kkk++);
                for(int kk = 1; kk < 4; kk++){
                    while(ch==' '){ch = stt.charAt(kkk++);};        // SPACES
                    st = "";
                    while((ch >= 'a')&&(ch <= 'z')){
                        st= st+ch; ch = stt.charAt(kkk++);
                    }
                    variables[kk]= st;
                }
            } else { jOutput0.writeString("error \n");return null;}
            temp[k++]= "    lds "+variables[1]+";\n";
            temp[k++]= "    "+variables[0]+" "+ variables[2]+";\n";
            temp[k++]= "    str "+ variables[3]+";\n";
        }else temp[k++]= stt;
    }
    String[] prog = new String[k];
    for(int i=0;i<k;i++){
        prog[i]=temp[i];
        jOutput2.writeString(prog[i]);
    }
    return prog;
}
static int testData(String[] program){        // macro test program
    program[0] = " start: rds a;\n";
    program[1] = "    rds b;\n";
    program[2] = "    rds d;\n";
    program[3] = "    rds e;\n";
    program[4] = "    add(a b c);\n";
    program[5] = "    add(d e f);\n";
    program[6] = "    mul(c f g);\n";
    program[7] = "    wrt g;\n";
    program[8] = "    jmp start;\n";
    return 9;
}


    ... // remaining assembler procedures
}
```

The example of macro substitution given above replicates the block of code based on a template, substituting strings from the input statement to appropriate positions in the output statements: "*#1 (#2 #3 #4);*" is converted to "*lds #2;*" "*#1 #3;*" "*str #4;*". An alternative approach to handling repeating patterns of code is to set up one example of it and transfer control to it each time it is needed. This gives the sub-program or subroutine call. However this transfer and return of control involves a sequence of commands that imposes an overhead to the repeated code that is only worthwhile if the code is longer than the overhead needed to transfer and return control. Otherwise a form of macro or inline code substitution is more efficient.

The code to tranfer control to a subprogram can be set up as a macro expansion in the following way for a subroutine call:

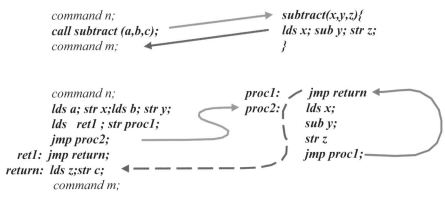

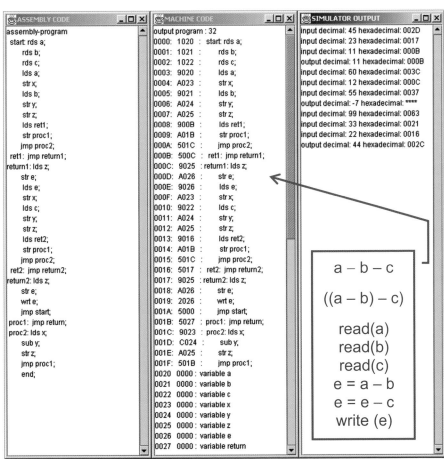

Figure 7.7 Subroutine(a,b){ return a – b;}

The appropriate jump command, to return control to the calling program can be constructed and loaded at the head of the subprogram before moving to the main code of the subprogram, in the way shown above. This return jump-command can then be returned to as soon as the subprogram completes its task. This basic transfer of control also has to be extended to include code to transfer the values of the variables that the subprogram is expected to operate on.

There are a variety of ways for making this information available to subprograms. One possibility can be set up as follows: copy the values that are to be operated on into a standard temporary location that is use by the subprogram for its parameters. This allows different values to be passed to a subprogram from different calling locations. Return values also have to be placed in similar standard but temporary locations and these can then be copied back to the appropriate local variables when control is returned to the calling program.

In practice program code can be built up from a collection of procedures in two ways. The first is a simple sequential list of procedures. The second is a nested structure or a block structured arrangement. A one level system on one hand or a hierarchical layered system on the other. The simplest program is the sequential list of procedure calls where the flow of control will simply pass from one procedure call to the next walking down the list. However even when this program runs: these procedures could pass control from one to another in a variety of more or less complicated ways. The scheme illustrated in Figure 7.7 will support a layered or hierarchical calling sequence, as long as the same routine is only called once in any such sequence. Clearly if it is not then the data giving return addresses and parameter values will be over written and the correct flow of control will be lost.

To allow program code to be revisited within a calling sequence this flow of control information has to be protected. Using a stack data structure to hold the necessary address data and parameter values can do this. Every time a program is called then its data values are placed on the stack and when control is passed back to the calling program then this data can be released or popped from the top of the stack. In essence this allows flow of control to change from a tree structure linking separate procedures, to a graph structure linking procedures that includes loops revisiting the same procedure more than once. The classical example of a "recursive" program that behaves in this way is a procedure to calculate the factorial value of a number.

```
int factorial(int k) {  if(k==0)return 1; else return k*factorial(k-1); }
```

To implement this program in assembly language or machine code, using the simple microprocessor employed above, is difficult. However, the extended system which uses base displacement addressing makes the task a little more tractable. If the assembly language is extended by four macros to define the beginning and end code for each procedure, such as: *proc factorial(k); endproc factorial();* and command sequences for transferring to and returning control from the procedure: *call factorial(x);* and *return(x);* then the program given above can be expressed in assembly code in the following way.

```
          proc main();
start:    rds a;
          call factorial(a);
          str answer;
          wrt answer;
          jmp start;
          endproc main();
          proc factorial(x);
          lds x;
          jez one;
          sub "1";
          str y;
          call factorial(y);
          mul  x;
          return();
one       lds "1";
          return();
          endproc factorial();
```

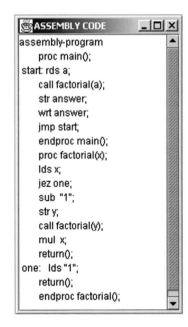

Figure 7.8 Factorial procedure

However, to expand the macros into a full assembly language program for the extended simulator still raises another difficulty when translating it into machine code. The problem is the way that constants are handled as variables in the system that has been developed so far. In the new context this means that these values will have to have new memory space allocated for them each time a program calls itself, and also the constant values will have to be copied across as if they were parameters, into the new space, which clearly involves an unnecessary overhead. It would be convenient if such data could be held in the original program-code address-space.

At first sight this would seem to require all the arithmetic commands to be duplicated to work on either program-space variables or data-space variables. Clearly further extensions of the hardware are an option. However this opens up a complex design area, which though it is important in its own right, is not critical to the main theme of this chapter. The desire is to keep to the original sixteen-instruction set, and the sixteen bit word size for this schematic machine, since the objective is to illustrate the processing stages for a high-level language program: say in Java, to a physical hardware system capable of executing it. Graphics requires specialised hardware and since computer graphics is concerned with the language based control of this hardware, illustrating a way the bridge between these two modes of communication, -- language and graphics -- can be implemented, is an important issue in this book.

The task is to present this bridge in as clear and accessible way as possible. Though recursive procedures may be possible to implement using the original simple simulator scheme the resulting code becomes so long and tortuous that its ability to illustrate the desired capability tends to be lost! As it stands the programs shown in Figures 7.7 and 6.31 are already becoming fairly opaque, and need the reader to play computer with them to be convinced that they work. Running the simulator will step through the code and show how each step operates, but anything longer than these examples ceases to serve its explanatory purpose!

A working solution to this particular problem that keeps the programs short enough to work with can be achieved by implementing the unused command "*lvl 0*" to change the mode of addressing used in the command that immediately follows it in a code sequence. In the hardware circuit diagram in Figure 6.34, this can be done by changing the selection signal *Y* for one fetch-execute cycle: to add the *PCbase* register value rather than the *IRbase* register value to the relative address from the *IR* register.

The new assembly language statements that need to be expanded as macros can be executed using the following substitutions.

jack: lds "1"; → *jack: lvl #0#; lds "1"; lds "1";* → *lvl #0#; lds "1";*

```
String str = program[i]; Boolean macro=false,constant=false;
String stt=""; char ch=str.charAt(0);
for(int j=0; j<str.length();j++,ch=str.charAt(j)){
    if(ch =='(')macro=true;
    if(ch =='\"')constant=true;
    stt=stt+ch;
}
if (constant){
    String strr="", sttr=""; int j=0;
    ch=str.charAt(j);
    while(ch == ' '){ch=str.charAt(j++);}
    if((ch >= 'a')&&(ch <= 'z'))
        while((ch >= 'a')&&(ch <= 'z')||(ch >= '0')&&(ch <= '9')){
            strr=strr+ch;  ch=str.charAt(j++); }
    while(ch == ' '){ch=str.charAt(j++);}
    if(ch == ':'){j++;
        while((ch=str.charAt(j++))!= ';'){sttr = sttr+ch;}
        temp[k++]= strr+":   lvl #0#;\n";
        temp[k++]= sttr+";\n";
    }else{
        temp[k++]= "   lvl #0#;\n";
        temp[k++]= stt;
    }
}
```

proc factorial(x); call factorial(y); return(); endproc factorial();

```
else if(macro){
    String strr=""; int j=0;
    ch=str.charAt(j);
    while(ch == ' '){ch=str.charAt(j++);}
    while((ch >= 'a')&&(ch <= 'z')){strr=strr+ch; ch=str.charAt(j++);}
```

```
if(strr.equals("proc")) {
    String st="";                                         //get name of procedure
    while(!((ch >= 'a')&&(ch <= 'z'))){ch=str.charAt(j++);}
    while((ch >= 'a')&&(ch <= 'z')){ st=st+ch;  ch=str.charAt(j++);}
    ind = procNameTable.add(st);                          // place name in procNameTable
    procNameTable.procAddress1[ind]= k+mprg;// start address of procedure
    temp[k++]= "      nop "+ " #0# "+ ";\n";     // number of local variables
    temp[k++]= "      lda "+ " #FF9# "+ ";\n";        // load register value
    temp[k++]= "      str "+ " #0# "+ ";\n";;         // store in location 0
    temp[k++]= "      lda "+ " #FFB# "+ ";\n";        // load register value
    temp[k++]= "      str "+ " #1# "+ ";\n";;         // store in location 1
    temp[k++]= "      lda "+ " #FFD# "+ ";\n";        // load register value
    temp[k++]= "      str "+ " #2# "+ ";\n";;         // store in location 2
    temp[k++]= "      lda "+ " #FFF# "+ ";\n";        // load register value
    temp[k++]= "      str "+ " #3# "+ ";\n";;         // store in location 3
}

else if(strr.equals("endproc")) {
    temp[k++]= "endproc:nop "+ " #0# "+ ";\n";//
    temp[k++]= "      lds "+ "#0#"+ ";\n";            // load old program address
    temp[k++]= "      sta "+ " #FF9# "+ ";\n";        // place in pc base reg
    temp[k++]= "      lds "+ "#1#"+ ";\n";   // load old program base address
    temp[k++]= "      sta "+ " #FFB# "+ ";\n";        // place in alt pc base reg
    temp[k++]= "      lds "+ "#2#"+ ";\n";            // load old IR address
    temp[k++]= "      sta "+ " #FFD# "+ ";\n";        // place in other IR reg
    temp[k++]= "      lds "+ "#3#"+ ";\n";            // load old IRbase address
    temp[k++]= "      sta "+ " #FFF# "+ ";\n";   // place in other IR base reg
    temp[k++]= "      lds "+ " return"+ ";\n";
    temp[k++]= "      lvl "+ " #3# "+ ";\n";          // switch all registers
    procNameTable.procAddress2[ind]= k+mprg; // end address of procedure
}

else if(strr.equals("call")) {
    String [] params = new String[10];
    while(!((ch >= 'a')&&(ch <= 'z'))){ch=str.charAt(j++);}
    String st="";                                         // get name of procedure
    while((ch >= 'a')&&(ch <= 'z')){  st=st+ch; ch=str.charAt(j++);}
    int ind2 = procNameTable.add(st);        // place name in procNameTable
    while(!((ch >= 'a')&&(ch <= 'z'))){ ch=str.charAt(j++);}
    int count=0;;
    while((ch != ')')&&(ch != ';')){                      // get parameters
        String stt1 = "";                                 //get parameter name
      if((ch >= 'a')&&(ch <= 'z'))
          while((ch >= 'a')&&(ch <= 'z')||(ch >= '0')&&(ch <= '9')){
              stt1=stt1+ch;  ch=str.charAt(j++); }
        params[count++]= stt1;
        while((ch == ' ')||(ch == ',')){ ch=str.charAt(j++);};
    }
```

```
        temp[k++]= "    lda "+ " #FFE# "+ ";\n";    // load base data address
        temp[k++]= "    lvl #0#;\n";
        temp[k++]= "    add "+ '#'+0+'#'+ ";\n";//add number of local variables
        temp[k++]= "    sta "+ " #FFF#"+ ";\n";    // store base data address
        for(int ii=0 ; ii < count; ii++){              // code  to pass parameters
            temp[k++]= "    lds "+ params[ii] + ";\n";
            temp[k++]= "    lvl "+ " #2# "+ ";\n";
            temp[k++]= "    str "+ '#'+(ii+4)+'#'+ ";\n";
            temp[k++]= "    lvl "+ " #2# "+ ";\n";
        }
        temp[k++]= "    lvl #0#;\n";
        temp[k++]= "    lds "+ "\"0\""+ ";\n";
        temp[k++]= "    sta "+ " #FF9# "+ ";\n";    // place 0 in pc other reg
        temp[k++]= "    sta "+ " #FFB# "+ ";\n";    // place 0 in IR other reg
        temp[k++]= "    lda "+ '#'+ getHex(ind2+10,3) +'#'+ ";\n";
        temp[k++]= "    sta "+ " #FFD# "+ ";\n"//store new procedure address
        temp[k++]= "    lvl "+ " #3# "+ ";\n";
    }
    // switching all registers passes control to the new procedure
    else if(strr.equals("return")) {
        temp[k++]= "    str "+ " return"+ ";\n";
        temp[k++]= "    jmp "+ " endproc "+ ";\n";
    }
```

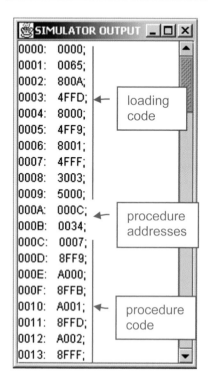

These expansions allow the factorial program in Figure 7.8 to be converted into the code shown in Figure 7.10. This runs successfully on the extended simulator shown in Figure 7.11. Figures 7.12 and 7.13 show the same scheme applied to a procedure to add two numbers together. The overhead required to handle recursive procedure calls can be seen by comparing these programs with Figure 7.7. However in both these expanded programs the code is relocatable merely by changing the base address used for each procedure. In these examples these addresses have been placed in the initialisation code added onto the beginning of the program shown in Figure 7.9 In a full system these values would have to be set up by the program loading system. If these procedures are held in a library then this task becomes an important part of the selection and placement of procedures.

Figure 7.9 Program initialisation code

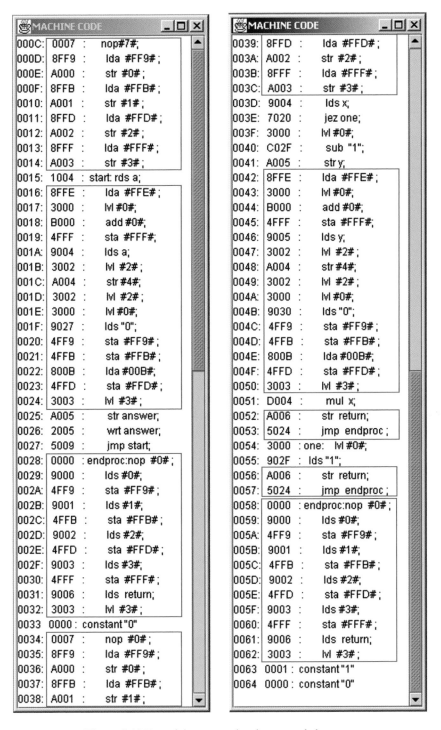

Figure 7.10 Factorial program showing expanded macros

```
public int[] prg(String[] program, int proglength){
    machineProgram = new int[2000];  procNameTable = new NameTable();
    for(int i=0; i<2000;i++) machineProgram[i]= -1;
    if (program==null){
        program = new String [1000];  prog = new String[2000];
                                        // enter source code into program[ ]
    }
    mprg=0; boolean proc = true; int begin=0, programLength=0, i=0, k=0;
    while(i<proglength){
        String str = program[i], strr=""; int j=0;
        char ch=str.charAt(j++); while(ch == ' '){ch=str.charAt(j++);}
        while((ch >= 'a')&&(ch <= 'z')){ strr=strr+ch; ch=str.charAt(j++);}
        if(strr.equals("proc")) proc=true;
        while((proc)){
            begin = i; j=0; strr="";  str = program[i++];
            ch = str.charAt(j++); while(ch == ' '){ch=str.charAt(j++);}
            while((ch >= 'a')&&(ch <= 'z')){ strr=strr+ch; ch=str.charAt(j++);}
            if(strr.equals("endproc")){proc=false;}
        }
        begin= programLength; programLength = i;
        macroExpansion(program,begin,programLength);
    }
    mprg=procNameTable.getSize()+10;
    for(int n=0; n<procNameTable.getSize(); n++){
        int end = procNameTable.procAddress2[n];
        StringTable names = new StringTable(), constants = new StringTable();
        mprg = translate(start, end, prog, names, constants, procNameTable, n);
    } for(int jj=0;jj<procNameTable.getSize();jj++)
        machineProgram[jj+10]= procNameTable.procAddress1[jj];
    machineProgram[0] = 0;
    machineProgram[1] = mprg;                   //overall program length
    machineProgram[2] = 8*4096+ 10;                     //lda 10;
    machineProgram[3] = 4*4096+ 4093;                   //sta FFD;
    machineProgram[4] = 8*4096;                         //lda 0;
    machineProgram[5] = 4*4096+ 4089;                   //sta FF9;
    machineProgram[6] = 8*4096+1;                       //lda 1;
    machineProgram[7] = 4*4096+ 4095;                   //sta FFF;
    machineProgram[8] = 3*4096+3;                       //lvl 3 ;
    machineProgram[9] = 5*4096;                         //jmp 0;
    for(int jk=0;jk<mprg;jk++){
        writeHex(jk,4,jOutput4); jOutput4.writeString(":   ");
        writeHex(machineProgram[jk],4,jOutput4); jOutput4.writeString(";\n");
    }
    int [] mp = new int[mprg];
    for(int jj=0;jj<mprg;jj++){ mp[jj]= machineProgram[jj];}
    return mp;
}
```

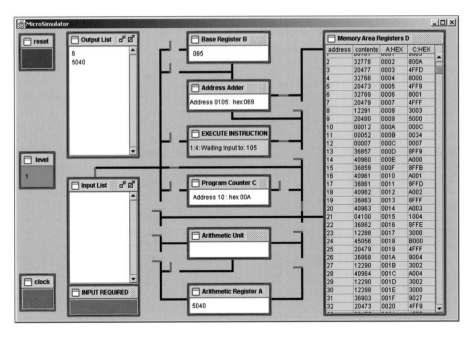

Figure 7.11 Executing the recursive factorial procedure for: 3!, 7!

Figures 7.11 and 7.12 show two assembly language programs translated to machine code and running on the extended hardware simulator.

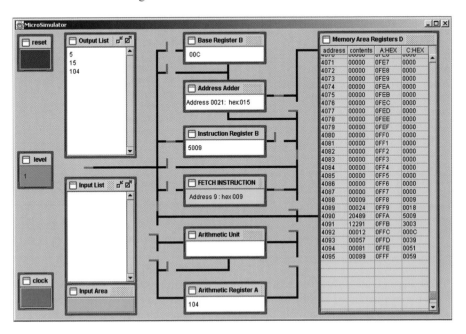

Figure 7.12 Executing the addition procedure for Hex inputs: 2+3, 7+8, 45+23

```
MACHINE CODE                              _□×
000C: 0008 :    nop#8#;
000D: 8FF9 :    lda #FF9#;
000E: A000 :    str #0#;
000F: 8FFB :    lda #FFB#;
0010: A001 :    str #1#;
0011: 8FFD :    lda #FFD#;
0012: A002 :    str #2#;
0013: 8FFF :    lda #FFF#;
0014: A003 :    str #3#;
0015: 1004 : start: rds a;
0016: 1005 :    rds b;
0017: 8FFE :    lda #FFE#;
0018: 3000 :    lvl #0#;
0019: B000 :    add #0#;
001A: 4FFF :    sta #FFF#;
001B: 9004 :    lds a;
001C: 3002 :    lvl #2#;
001D: A004 :    str #4#;
001E: 3002 :    lvl #2#;
001F: 9005 :    lds b;
0020: 3002 :    lvl #2#;
0021: A005 :    str #5#;
0022: 3002 :    lvl #2#;
0023: 3000 :    lvl #0#;
0024: 902C :    lds "0";
0025: 4FF9 :    sta #FF9#;
0026: 4FFB :    sta #FFB#;
0027: 800B :    lda #00B#;
0028: 4FFD :    sta #FFD#;
0029: 3003 :    lvl #3#;
002A: A006 :    str answer;
002B: 2006 :    wrt answer;
002C: 5009 :    jmp start;
002D: 0000 : endproc:nop #0#;
002E: 9000 :    lds #0#;
002F: 4FF9 :    sta #FF9#;
0030: 9001 :    lds #1#;
0031: 4FFB :    sta #FFB#;
0032: 9002 :    lds #2#;
0033: 4FFD :    sta #FFD#;
0034: 9003 :    lds #3#;
0035: 4FFF :    sta #FFF#;
0036: 9007 :    lds return;
0037: 3003 :    lvl #3#;
0038: 0000 : constant "0"
```

```
MACHINE CODE                              _□×
0038: 0000 : constant "0"
0039: 0007 :    nop #0#;
003A: 8FF9 :    lda #FF9#;
003B: A000 :    str #0#;
003C: 8FFB :    lda #FFB#;
003D: A001 :    str #1#;
003E: 8FFD :    lda #FFD#;
003F: A002 :    str #2#;
0040: 8FFF :    lda #FFF#;
0041: A003 :    str #3#;
0042: 9004 :    lds x;
0043: B005 :    add y;
0044: A006 :    str return;
0045: 500D :    jmp endproc;
0046: 0000 : endproc:nop #0#;
0047: 9000 :    lds #0#;
0048: 4FF9 :    sta #FF9#;
0049: 9001 :    lds #1#;
004A: 4FFB :    sta #FFB#;
004B: 9002 :    lds #2#;
004C: 4FFD :    sta #FFD#;
004D: 9003 :    lds #3#;
004E: 4FFF :    sta #FFF#;
004F: 9006 :    lds return;
0050: 3003 :    lvl #3#;
```

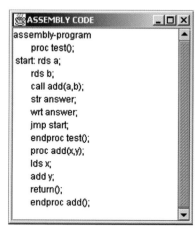

```
ASSEMBLY CODE                             _□×
assembly-program
    proc test();
start: rds a;
    rds b;
    call add(a,b);
    str answer;
    wrt answer;
    jmp start;
    endproc test();
    proc add(x,y);
    lds x;
    add y;
    return();
    endproc add();
```

Figure 7.13 Macroexpansions assembler and machine code for an addition procedure

Another operation that also requires information to be stacked in order to carry it out is the evaluation of arithmetic expressions. However it is possible in this case to rearrange the operations required to evaluate the expression into a linear sequence of instructions that can be expressed directly in the assembly language commands defined for the simple unextended microprocessor system.

Expressions and Formulae

The simple simulator developed in chapter 6 allows the user to automatically execute sequences of arithmetic operations that a person using a simple desktop calculator might employ to evaluate a more complex calculation. However many calculations are expressed as formulae or expressions that do not immediately give the correct sequence of operations needed to evaluate them. This is because expressions and formulae are generally nested structures where the first calculation required is not necessarily the first one encountered reading the expression in a standard left to right order.

The arithmetic operations in an expression have to be reordered if they are to be carried out sequentially. Also intermediate partial results need to be held while other sections of an expression are evaluated. A model of the reordering process can be provided by the much older task of reordering the carriages in a train using railway sidings, where the process has to be done step by step: the carriages have to be kept on the tracks and moved one by one.

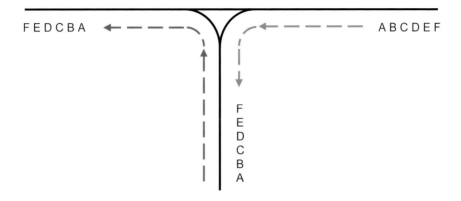

Figure 7.14 Reversing a sequence using a "siding" or a first in last out stack

Evaluating Expressions

Given an expression of the form $\dfrac{(c-a)}{(b-d)*(e-f)}$ the first step is to rewrite it replacing

the fraction notation to give $((c-a)/(b-d))/(e-f)$ or $(c-a)/((b-d)*(e-f))$, where every operator is a simple binary operator acting on the two elements to the left and right of it. The next step is to reorder the operations so they can be carried out sequentially. In the original presentation the subtraction operations would all have to be carried out before the multiply or division operations. In this example the ordering is defined by the use of brackets, however in the expression: $a-b*c$ the multiply has to be carried out first because of the convention that multiply and division takes precedence over subtract and add operations. If this were not the case the ordering would also require brackets to obtain the required result for: $a-(b*c)$.

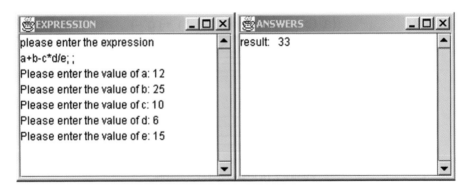

Figure 7.18 Expression processing

A calculator program to evaluate expressions in this way using two stacks can be written in the following way:

```
import java.awt.*; import javax.swing.*; import javax.swing.text.*;
import javax.swing.table.*; import javax.swing.event.*; import java.awt.image.*;
import java.awt.event.*; import java.util.*; import java.text.*;import java.lang.*;
public class Calculator{
    static String str = "";
    static Input jInput = new Input(12,412,500, 100,"INPUT");
    static Output jOutput0 = new Output(12,512,500, 100,"OUTPUT");
    static Output jOutput1 = new Output(12,12,250, 400,"EXPRESSION");
    static Output jOutput2 = new Output(262,12,250, 400,"ANSWERS");
    public static void main(String[] args){
        int width = 300, height = 100;
        int [] machineProgram = null;  String[] program = null;
        String str = jInput.readString();  jOutput0.writeString(str + "\n");
        while(!str.equals("stop")){
            if(str.equals("Calculator")){
                while(!str.equals("end")){
                    jOutput1.writeString("please enter the expression \n");
                    str = jInput.readString();
                    if(str.equals("end")) break;
                    str=str+" ;";
                    jOutput1.writeString(str+"\n");
                    AExpression expression = new AExpression(str);
                    int num = expression.evaluate(jInput,jOutput1,jOutput2);
                    jOutput2.writeString(
                        "result: "+expression.decString(num,4)+"\n");
                }
            }str = jInput.readString(); jOutput0.writeString(str+ "\n");
        }
    }
}
```

```
class AExpression{
    private Output out = null; private Output output = null;private Input in = null;
    public StringTable names = new StringTable();
    public int codeCount = 0;
    public String exprs = "";
    static CharacterTable alphaSet = new CharacterTable(new char []
    {'a','b','c','d','e','f','g','h','i','j','k','l','m','n','o','p','q','r','s','t','u','v','w','x','y','z','A','B',
    'C','D','E','F','G','H','I','J','K','L','M','N','O','P','Q','R','S','T','U','V','W','X','Y','Z'});
    static CharacterTable numericSet = new CharacterTable(new char []
        {'0','1','2','3','4','5','6','7','8','9'});
    static CharacterTable operatorSet = new CharacterTable(new char []
        {'+','-','*','/','%',';'});
    static StringStack nameStack = new StringStack();
    static StringStack operatorStack = new StringStack();
    static NumberStack valueStack = new NumberStack();
    public AExpression(){}
    public AExpression(String exprs)  {this.exprs=exprs;}

    public int evaluate(Input in, Output output1,Output output){
        nameStack = new StringStack();
        operatorStack = new StringStack();
        valueStack = new NumberStack();
        names = new StringTable();
        this.setIO(in, output1,output);
        String[] tokens = this.tokenize(exprs);
        this.setValues();
        String[] code = this.translate(tokens);
        return valueStack.pop();
    }
```

```
    public void setValues(){
        int size =names.getSize();
        for(int i=0 ; i < size; i++){
            String ss=""; int nm=0;
            if((ss = names.getElement(i)).charAt(0)==""){
                for(int j=1; j<ss.length()-1;j++){
                    int digit = (int)ss.charAt(j)-(int)'0';
                    nm = nm*10+digit;
                }names.val[i]=nm;
            }else{
                output.writeString(
                    "Please enter the value of "+names.getElement(i)+": ");
                String num= in.readString(); output.writeString(num +"\n");
                names.val[i]= number(num);
            }
        }
    }
```

```
public String[] tokenize(String exprs){
    String[] tokens = new String[200];
    int i=0, j=0;
    if(exprs.charAt(exprs.length()-1)!=';')exprs=exprs+';';
    while(i<exprs.length()){
        if (alphaSet.contain(exprs.charAt(i))){
            String str = ""+ exprs.charAt(i++);
            while((i<exprs.length())&&(alphaSet.contain(exprs.charAt(i))
                            ||numericSet.contain(exprs.charAt(i))))
                {str= str + exprs.charAt(i); i++;}
            tokens[j++]= str;  names.add(str);}
        else if(numericSet.contain(exprs.charAt(i))){
            String str= "\"";
            while((i<exprs.length())&&(numericSet.contain(exprs.charAt(i))))
                {str = str + exprs.charAt(i);i++;}
            str=str+"\"";  tokens[j++]= str;  names.add(str);}
        else if(operatorSet.contain(exprs.charAt(i))){
            String str = "";
            while((i<exprs.length())&&(operatorSet.contain(exprs.charAt(i))))
                {str = str + exprs.charAt(i);i++;}
            tokens[j++]= str;}
        else if(exprs.charAt(i)=='('){tokens[j++]= "(";i++;}
        else if(exprs.charAt(i)==')'){tokens[j++]= ")";i++;}
        else if(exprs.charAt(i)==' ')while(exprs.charAt(i)==' '){i++;}
        else if(exprs.charAt(i)=='\n'){i++;}
        else   {i++;  out.writeString("unexpected character \n");}
    }
    String[] tokenArray = new String[j];
    for(i=0;i<j;i++) tokenArray[i]=tokens[i];
    return tokenArray;
}
```

```
public void executeOperator(int operatorIndex, String[] code){
    if((operatorIndex>=0)&&(operatorIndex<=4)){
        int a = valueStack.pop(), b = valueStack.pop();
        switch(operatorIndex){ // '+','-','*','/',"%".';'
            case 0: valueStack.push(b+a); break;              // add
            case 1: valueStack.push(b-a); break;              // subtract
            case 2: valueStack.push(b*a); break;              // multiply
            case 3: valueStack.push(b/a); break;/             // divide
            case 4: valueStack.push(b%a); break;              // modulo
            default:                                          // end of expression
        }
    }else if(operatorIndex==7){
        int a=valueStack.pop();valueStack.push(-a);    // unary minus
    }
}
```

```java
public String[] translate(String[] tokenArray){
    StringTable operators = new StringTable(
    new String []{"+","-","*","/","%",",","(","~"});
    int[] precedence = new int[]{3,3,4,4,4,1,2,5};
    String[] code = new String[100];
    int topPrecedence=0, tPrecedence, operatorIndex; String operator, s;
    for(int i=0;i<tokenArray.length;i++){
        String t = tokenArray[i];
        if(names.contain(t)){                                       // variable name
            valueStack.push(names.val[names.locate(t)]);
        }else if(t.equals("(")){                                    // open bracket
            operatorStack.push("("); topPrecedence= 2;
        }else if(operators.contain(t)){                             // operator
            operatorIndex = operators.locate(t);
            tPrecedence = precedence[operatorIndex];
            while(tPrecedence<=topPrecedence){
                operator = operatorStack.pop();
                operatorIndex = operators.locate(operator);
                executeOperator(operatorIndex,code);
                int index =operators.locate(operatorStack.peek());
                if (index<0) topPrecedence=0;
                else topPrecedence= precedence[index];
            }operatorStack.push(t); topPrecedence = tPrecedence;
        }else if(t.equals(")")){                                    //close bracket
            while(!((s = operatorStack.pop()).equals("("))){
                operatorIndex = operators.locate(s);
                executeOperator(operatorIndex,code);
            }int index =operators.locate(operatorStack.peek());
            if (index<0)topPrecedence=0;
            else topPrecedence= precedence[index];
        }
    } return code;
}
```

```java
public void setIO(Input input,Output output,Output output1){
    this.output = output; this.out = output1; this.in = input;}
public void setNames(StringTable nameTable) {this.names = nameTable;}
static String hexString(int val,int len){
    String str ="";
    if(val<0)return "****";
    for(int j=0;j<len;j++){
        int digit = val%16; val = val/16;
        if((digit>=0)&&(digit<=9)){str = ((char)(digit+((int)'0')))+str;}
        else if((digit>=10)&&(digit<=15))
            {str = ((char)(digit-10+((int)'A')))+str;}
    }return str;
}
```

```
static String decString(int val,int len){ws
    String str ="";  int v=val;
    if(val<0){val=-val;}
    while(val>0){
        int digit = val%10; val = val/10;
        if((digit>=0)&&(digit<=9)){str = ((char)(digit+((int)'0')))+str;}
    }
    if(v<0){str="-"+str;}
    return "  "+str;
}
static int number(String str){
    int num=0,n=1;
    int k=0; while(str.charAt(k)==' '){k++;}
    if(str.charAt(k)=='-'){n= -1; k++;}
    for(int i=k;i<str.length();i++){
        if(numericSet.contain(str.charAt(i))){
            int digit = (int)str.charAt(i)-(int)'0';
            num = num*10+digit;
        }
    }return num*n;
}
}
```

```
class NumberStack{
    private int [] stack = new int[100];private int top = 0;
    NumberStack(){}
    public int pop(){ if (top == 0)return -9999; else {top--; return stack[top];}}
    public int peek(){ if (top == 0)return -9999; else {return stack[top-1];}}
    public void push(int ch)  {stack[top] = ch; top++;}
    public int getTopIndex() { return top;}
}
class CharacterStack{
    private char [] stack = new char[100]; private int top = 0;
    CharacterStack(){}
    public char pop(){if (top <= 0)return (char)0; else {top--;return stack[top];}}
    public char peek(){if (top == 0)return '\0'; else {return stack[top-1];}}
    public void push(char ch) {stack[top]=ch; top++;}
    public int getTop() { return top;}
}
class StringStack{
    private String [] stack = new String[100]; private int top = 0;
    StringStack(){}
    public String pop() {if (top == 0)return ""; else {top--;return stack[top];}}
    public String peek() {if (top == 0)return "";  else {return stack[top-1];}}
    public void push(String str) {     stack[top]= str; top++;}
    public int getTop() {return top;}
}
```

```
class CharacterTable{
    private char[] table = new char[100];
    int[] val= new int[100];
    private int size = 0;
    CharacterTable(){}
    CharacterTable(char[] str){ size = str.length; table = str; }
    public int locate(char ch){
        for(int i=0; i<size; i++){if(ch== table[i])return i;} return -1;}
    public int add(char ch){
        int index = locate(ch);
        if(index == -1){index = size; table[index] = ch; size++;}
        return index;
    }
    public char getElement(int i){
        if((i>=0)&&(i<size))return table[i];  else    return (char) 0;
    }
    public boolean contain(char ch){
        int k= locate(ch);
        if (k<0)return false; else return true;
    }
    public int getSize(){return size;}
    public void setTable(char[] tbl){ table = tbl;size= tbl.length;}
}

class StringTable{
    private String[] table = new String[100];
    int[] val= new int[100];
    private int size = 0;
    StringTable(){}
    StringTable(String[] str){size=str.length; table=str;}

    public int locate(String str){
        for(int i=0; i<size; i++){if(str.equals(table[i]))return i;} return -1;}
    public int add(String str){
        int index = locate(str);
        if(index== -1){index = size; table[index]= str; size=size+1;}
        return index;
    }
    public String getElement(int i){
        if((i>=0)&&(i<size))return table[i];else return "";
    }
    public boolean contain(String str){
        int k= locate(str);
        if (k<0)return false; else return true;
    }
    public int getSize(){ return size;}
    public void setTable(String[] tbl){    table = tbl;size= tbl.length;}
}
```

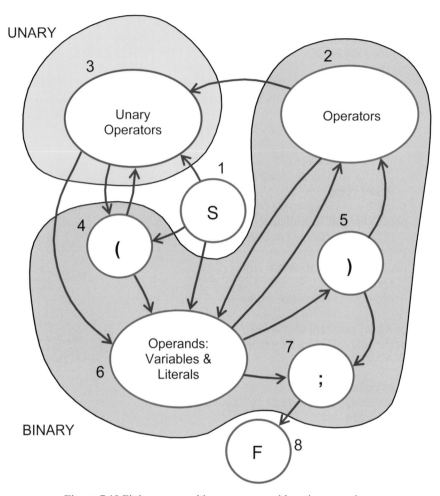

Figure 7.19 Finite state machine to process arithmetic expressions

There are two drawbacks to this program, the first is that it works correctly for a correctly structured input expression, but may or may not handle an incorrectly structured expression in a safe way. Wrong input may result in failure, however sometimes apparently correct output may result from wrong input, a situation which cannot be accepted. Where it fails then the user is aware of the problem however the latter case could be dangerous because an apparently correct but wrong result might be used in subsequent work. The second draw back is that arithmetic expressions often contain unary minuses. The first step in tackling these problems is to layout the possible sequences of operators, brackets, variable names and numbers that can occur in correctly structured arithmetic expressions and convert them into a state transition diagram of the form shown in Figure 7.19. This allows the tokenise procedure to be modified to only accept the correct sequences of elements that this diagram permits, and to identify when a minus sign is a unary not a binary operator. By changing the token used to represent the unary minus and giving it a high precedence the rest of the program will work correctly as before.

```
public String[] tokenize(String exprs){
    CharacterTable operator = new CharacterTable(new char []  {'+','-','*','/','%'});
    String[] tokens = new String[200];  int i=0, j=0;
    if(exprs.charAt(exprs.length()-1)!=';')exprs=exprs+';';
    boolean notFinished= true;
    int state =1;   char ch=';';
    String str="";
    while(notFinished){
        switch(state){
        case 0:  output.writeString("input expression error\n");
                state = 8; notFinished=false; break;
        case 1:while((ch = exprs.charAt(i))==' '){i++;};
                if (alphaSet.contain(ch)||numericSet.contain(ch))state = 6;
                else if(ch== '-')state = 3;  else state=0; break;
        case 2:  tokens[j++]= ""+ch; i++;
                while((ch=exprs.charAt(i))==' '){i++;};
                if(alphaSet.contain(ch) || numericSet.contain(ch))state=6;
                else if(ch== '-')state = 3; else state =0; break;
        case 3:  tokens[j++]= "~";i++;
                while((ch=exprs.charAt(i))==' '){i++;};
                if (alphaSet.contain(ch)||numericSet.contain(ch))state = 6;
                else if(ch=='(')state=4; else state=0; break;
        case 4:  tokens[j++]= "("; i++;
                while((ch=exprs.charAt(i))==' '){i++;};
                if (alphaSet.contain(ch)||numericSet.contain(ch))state = 6;
                else if(ch=='-')state=3; else state=0; break;
        case 5:  tokens[j++]= ")";i++;
                while((ch=exprs.charAt(i))==' '){i++;};
                if(operator.contain(ch))state=2;
                else if(ch==';')state=7; else state=0; break;
        case 6:  if (alphaSet.contain(ch)){ str=""+ch; i++;
                    while((alphaSet.contain((ch=exprs.charAt(i)))
                        ||numericSet.contain(ch))) {str= str + ch; i++;}
                    tokens[j++]= str; names.add(str);
                }else if(numericSet.contain(ch)){str= "\"";
                    while(numericSet.contain((ch=exprs.charAt(i))))
                        {str = str + ch; i++;}
                    str=str+"\""; tokens[j++]= str; names.add(str);
                }while((ch=exprs.charAt(i))==' '){i++;};
                if(ch==')') state=5;   else if(ch==';')state=7;
                else if(operator.contain(ch)) state=2;  else state=0; break;
        case 7:  tokens[j++]= ";"; notFinished = false; state=8;  break;
        }
    }String[] tokenArray = new String[j];
    for(i=0;i<j;i++) { tokenArray[i]=tokens[i]; }
    return tokenArray;
}
```

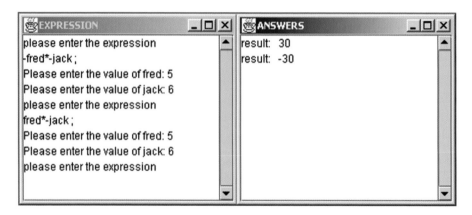

Figure 7.20 Including unary minus operators within arithmetic expressions

Translating Expressions into Assembly Language Programs.

What the calculator program does is to convert the nested operations of the input expression into a sequence of simple binary or unary arithmetic operations on values stored on the top of a stack. Binary operations can be implemented using three assembly language commands

$$a+b \;\; \to \;\; lds\ a;\ \ add\ b;\ \ str\ c;$$

And a unary minus becomes

$$-a \;\; \to \;\; lds\ \text{“0”};\ \ sub\ a;\ \ str\ c;$$

However the order necessary to support a sequence of simple binary or unary operations is based on using the stack. If the hardware provided stack based addressing operations this would be easy to implement directly. However in this case it is necessary to generate a stack-behaviour by composing names in the translation program that simulate access to a stack. As operands are obtained from the original expression they can be stored using a sequence of names s1, s2, … s.., constructed from a counting variable *index* using the *String* concatenate operator in the following way to simulate a stack:

"lds variable;" "str s" + index+";";   index++;

As soon as binary operators are obtained, in reverse Polish order, each one can be translated into a sequence of commands using the current value of the index variable, which provides the required stack based behaviour in the following way:

"lds s" + (index-2)+";" ; "add s" + (index-1)+";" ;   "str s" + (index-2)+";" ;

The unary minus operator can be implemented as follows:

"lds \" 0 \" ;" "sub s" + (index-1)+";";   "str s" + (index-1)+";";

Translating Expressions into Assembly Code

```
public String[] translate(String[] tokenArray){
    StringTable operators = new StringTable(
                        new String [){"+","-","*","/","%",",","(","~"});
    int[] precedence = new int[]{3,3,4,4,4,1,2,5};
    StringStack nameStack = new StringStack();
    StringStack operatorStack = new StringStack();
    String[] code = new String[100];
    int j=0, topPrecedence=0, tPrecedence, operatorIndex; String operator,s;
    for(int i=0;i<tokenArray.length;i++){
        String t = tokenArray[i];
        if(names.contain(t)){                              //operand
            nameStack.push("s"+j);
            code[codeCount++]="     "+"lds "+t+";\n";
            code[codeCount++]="     "+"str "+"s"+j+";\n";
            j++;
        }else if(t.equals("(")){                           //open bracket
            operatorStack.push("(");
            topPrecedence= 2;
        }else if(operators.contain(t)){                    //operator
            operatorIndex = operators.locate(t);
            tPrecedence = precedence[operatorIndex];
            while(tPrecedence<=topPrecedence){
                operator = operatorStack.pop();
                operatorIndex = operators.locate(operator);
                j = executeOperator(operatorIndex,j,code);
                int index =operators.locate(operatorStack.peek());
                if (index<0)topPrecedence=0;
                else topPrecedence= precedence[index];
            }
            operatorStack.push(t);
            topPrecedence = tPrecedence;
        }else if(t.equals(")")){                           //close bracket
            while(!((s=operatorStack.pop()).equals("("))){
                operatorIndex = operators.locate(s);
                j = executeOperator(operatorIndex,j,code);
            }
            int index =operators.locate(operatorStack.peek());
            if (index<0)topPrecedence=0;
            else topPrecedence= precedence[index];
        }
    }
    String[] assemblycode= new String[codeCount];
    for(int i=0;i<codeCount;i++)assemblycode[i]=code[i];
    return assemblycode;
}
```

```
public int executeOperator(int operatorIndex,int j, String[] code){
    if(operatorIndex==7){
    code[codeCount++]="    "+"lds "+ "\"0\";\n";
    code[codeCount++]="    "+"sub "+"s"+(j-1)+";\n";
    code[codeCount++]="    "+"str "+"s"+(j-1)+";\n";
    return j;
    }
    code[codeCount++]="    "+"lds "+"s"+(j-2)+";\n";
    switch(operatorIndex){ // '+','-','*','/','%" ':'
    case 0: code[codeCount++]=" "+"add "+"s"+(j-1)+";\n";break;        //add
    case 1: code[codeCount++]=" "+"sub "+"s"+(j-1)+";\n";break;        // subtract
    case 2: code[codeCount++]=" "+"mul "+"s"+(j-1)+";\n";break;        // multiply
    case 3: code[codeCount++]=" "+"div "+"s"+(j-1)+";\n";break;        // divide
    case 4: code[codeCount++]=" "+"mod "+"s"+(j-1)+";\n"; break        // modulo
    default:return 0;                                          // end of expression
    }
    code[codeCount++]="    "+"str "+"s"+(j-2)+";\n";
    return j-1;
}
```

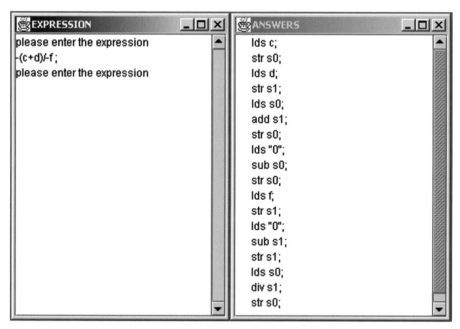

Figure 7.21 Translating an arithmetic expression into assembly code

Boolean Expression Processing

As soon as conditional operations are required it becomes necessary to use Boolean expressions. The Boolean expression will have to be evaluated using the available operations provided by the microcomputer system. In a conventional computer processing-unit, logic operations will be provided as primitive hardware operations. With this system all that is available is the set of arithmetic operations. However, these are sufficient to support the use of Boolean expressions given the appropriate translation to basic assembly language and machine code sequences. If false is represented by 0 and true is represented by 1 then the "**and**" operation can be provided by the standard arithmetic "**multiply**" operation. The "**complement**" operation can be handled by **subtracting** values **from 1**, the "**or**" operation can be implemented by **adding** the values and the **exor** by subtracting the values in the following ways.

A	1–A	!A	
0	1	1	lds "1";
1	0	0	sub A;
			X: ...

A	B	A*B	A&&B	
0	0	0	0	lds A;
0	1	0	0	mul B;
1	0	0	0	X: ...
1	1	1	1	

A	B	A+B	A||B	
0	0	0	0	lds A;
0	1	1	1	add B;
1	0	1	1	jez X;
1	1	2	1	lds "1"
				X: ...

A	B	A–B	A $\oplus$ B	
0	0	0	0	lds A;
0	1	–1	1	sub B;
1	0	1	1	jez X;
1	1	0	0	lds "1"
				X: ...

A	B	A–B	A $\otimes$ B	
0	0	0	1	lds A; sub B;
0	1	–1	0	jez X;
1	0	1	0	lds "0";
1	1	0	1	jmp Y;
				X: lds "1";
				Y: nop;

   Combination of these basic operations will support the full implementation of the often more complex Boolean expressions used in *conditional* and *while* statement tests, and they can be extended to include numerical comparison tests in the following way:

Numerical comparison tests using arithmetic operations

A	B	A − B	A<B	
5	4/5	1/0	0	lds A; sub B; jng X; lds "0";
4	5	-1	1	jmp Y; X: lds "1"; Y: nop;

A	B	B − A	A<=B	
5	4	-1	0	lds B; sub A; jng X; lds "1";
4	5/4	1/0	1	jmp Y; X: lds "0"; Y: nop;

A	B	A − B	A>=B	
5	4/5	1/0	1	lds A; sub B; jng X; lds "1";
4	5	-1	0	jmp Y; X: lds "0"; Y: nop;

A	B	B − A	A>B	
5	4	-1	1	lds B; sub A; jng X; lds "0";
4	5/4	1/0	0	jmp Y; X: lds "1"; Y: nop;

A	B	A − B	A==B	
5	5	0	1	lds A; sub B; jez X; lds "0";
4	3/5	1/-1	0	jmp Y; X: lds "1"; Y: nop;

A	B	A − B	A !=B	
5	5	0	0	lds A; sub B; jez X; lds "1";
4	3/5	1/-1	1	jmp Y; X: lds "0"; Y: nop;

```java
public String[] translate(String[] tokenArray){
    StringTable operators = new StringTable(new String []
            {"||","&&","!","<",">","==","<=",">=","!=",";","("});
    int[] precedence = new int[]{3,4,5,6,6,6,6,6,4,1,2};
    StringStack nameStack = new StringStack();
    StringStack operatorStack = new StringStack();
    String[] code = new String[100];
    int topPrecedence=0,tPrecedence, operatorIndex;
    int[] j=new int[2]; j[0]=0; j[1]=0;
    String operator,s;
    for(int i=0;i<tokenArray.length;i++){
        String t = tokenArray[i];
        if(names.contain(t)){                          //variable name
            nameStack.push("s"+j);
            code[codeCount++]="    "+"lds "+t+";\n";
            code[codeCount++]="    "+"str "+"s"+j[0]+";\n";
            j[0]++;
        }else if(t.equals("(")){                       //open bracket
            operatorStack.push("(");
            topPrecedence= 2;
        }else if(operators.contain(t)){                //operator
            operatorIndex = operators.locate(t);
            tPrecedence = precedence[operatorIndex];
            while(tPrecedence<=topPrecedence){
                operator = operatorStack.pop();
                operatorIndex = operators.locate(operator);
                j = executeOperator(operatorIndex,j,code);
                int index =operators.locate(operatorStack.peek());
                if (index<0)topPrecedence=0;
                else   topPrecedence= precedence[index];
            }
            operatorStack.push(t);
            topPrecedence = tPrecedence;
        }else if(t.equals(")")){                        //close bracket
            while(!((s=operatorStack.pop()).equals("("))){
                operatorIndex = operators.locate(s);
                j = executeOperator(operatorIndex,j,code);
            }
            int index =operators.locate(operatorStack.peek());
            if (index<0)topPrecedence=0;
            elsetopPrecedence= precedence[index];
        }
    }
    String[] assemblycode= new String[codeCount];
    for(int i=0;i<codeCount;i++)assemblycode[i]=code[i];
    return assemblycode;
}
```

In Boolean expressions each of the variables A and B in these expansions will have to be stacked variables constructed in the same way used for the arithmetic expression: s0, s1, s2 etc. depending on their place in the expression and consequently the position in which they are placed in the evaluation stack. The labels X and Y also have to be modified to make each one unique within the larger program

The simpler version of the *tokenise* procedure is adequate for processing input Boolean expressions because the unary operator '!': ***not***, does not need to be distinquished from a binary operation represented by the same character, as in the case of the arithmetic unary minus.

However, a change has to be made to the arithmetic expression *translate* procedure by providing a counter for contructing label names, similar to the counter used to create the stack-simulating names The two counters are passed to and from the *translate* and *executeOperator* methods through the array parameter *j[]* holding two values, where *j[0]* follows the top of the stack and *j[1]* gives the next free label postscript. The main extension needed to implement these sequences of commands is to replace the *executeOperator* procedure in the way illustrated below.

```
public void booleanop(
    int ca, int cb, String cc, String cd, String ce,String cf,int j[],String[] code){
    int l1 = j[1];  j[1]++;
    int l2 = j[1];  j[1]++;
    code[codeCount++] = "    "+"lds s"+ca+";\n";                    //load first variable
    code[codeCount++] = "    "+cf+" s"+cb+";\n";                    //apply second variable
    code[codeCount++] = "    "+cc + " x"+ l1  + ';' +"\n";          //jump
    code[codeCount++] = "    "+"lds "+ cd+";\n";                    //result
    code[codeCount++] = "    "+"jmp " + "x"+ l2  + ';' +"\n";       //jump
    code[codeCount++] = "x"+ l1 +":   " + "lds " + ce +';' +"\n";   //result
    code[codeCount++] ="x"+ l2 +":  "+"str s"+(j[0]-2)+";\n";       //store the result
}

public int[] executeOperator(int operatorIndex, int j[], String[] code){
    String f ="\""+"0"+"\"",t ="\""+"1"+"\"";
    switch(operatorIndex){
    case 0:booleanop(j[0]-1,j[0]-2,"jez",t,f,"add",j,code);j[0]--;break;      //"||"
    case 1:booleanop(j[0]-1,j[0]-2,"jez",t,f,"mul",j,code);j[0]--;break;      //"&&"
    case 2:j[0]++;booleanop(j[0]-2,j[0]-2,"jez",f,t,"add",j,code);j[0]--;break   //"!"
    case 3:booleanop(j[0]-2,j[0]-1,"jng",f,t,"sub",j,code);j[0]--;break;      //"<"
    case 4:booleanop(j[0]-1,j[0]-2,"jng",f,t,"sub",j,code);j[0]--;break;      //">"
    case 5:booleanop(j[0]-2,j[0]-1,"jez",f,t,"sub",j,code);j[0]--;break;      //"=="
    case 6:booleanop(j[0]-1,j[0]-2,"jng",t,f,"sub",j,code);j[0]--;break;      //"<="
    case 7:booleanop(j[0]-2,j[0]-1,"jng",t,f,"sub",j,code);j[0]--;break;      //">="
    case 8:booleanop(j[0]-1,j[0]-2,"jez",t,f,"sub",j,code);j[0]--;break;      //"!="
    }return j;
}
```

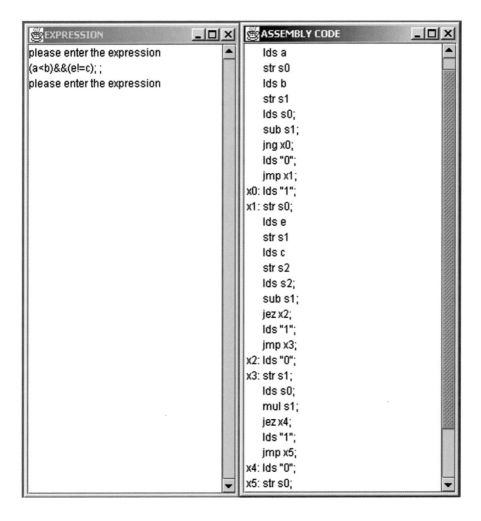

Figure 7.22 Translating a Boolean expression into assembly code

This approach works but loses the capacity to check that an expression is correctly formed. Though it is still possible to check that the structure of a Boolean expression is correct using a state transition system in the way developed for the arithmetic expression, as soon as the relationship test between two arithmetic values is extended from:

'('<variable> <relational operator> <variable> ')'

to

'('<arithmetic expression> <relational operator> < arithmetic expression > ')'

a more powerful and general technique for checking structure is required. One approach to this task is explored in the next chapter.

Assembly languages made programming early computers a much easier and less error prone activity. The move to a more natural language way of writing programs did not stop there, and the evolution of high level computer languages developed apace as better ways to automatically translate them to machine code were discovered.

The calculator program discussed in this chapter can be included in a class of language processing programs called interpreters. The arithmetic expression is the input language form that the calculator expects, and which it interprets directly, evaluating or calculating the required answer from the given variable values. In contrast the translation programs that convert the expression into equivalent programs in a simpler language structure that are eventually executed by the computer simulator, are called compilers. Notice in this case that the simulator is an interpreter, but the final code could have been machine code for a hardware system.

The benefit of the hierarchical language system is that overall it provides a more flexible versatile and compact computer system. General capability is captured at the expense of local efficiencies. However the efficiency of a complex general-purpose system depends on the statistics of its patterns of use. These can vary depending on the principal applications. One of the outcomes from testing systems in operation is that critical elements that are profitable to support in a dedicated way, for example at the hardware or firmware level of systems, often turned out to be very different from what was expected. Consequently they have to be identified by running profile tests for use in each new context.

At the same time computer languages developed, many alternative designs of computer hardware were also produced generally using different machine codes, and even with the unifying possibilities offered by microcode to give families of computers using the same machine language, alternatives multiplied. This diversification gave assembly or intermediate level languages a new role.

The need for flexibility created a demand for portable computer languages that could run on many platforms. If the high level language is translated to a common intermediate language that is much simpler in structure, it is then a much smaller task to write translators or interpreters for the simpler language to run on different hardware platforms. Both Pascal and Java were designed to translate to a standard intermediate code that could either be interpreted or translated to the native code of the machine they were being run on.

This idea extends to making multiple languages run efficiently on many platforms with minimum programming effort. If each high level language is translated into a common intermediate language, then this is translated into the machine code for the individual hardware platforms a great deal of duplication can be avoided in the programming task.

8

Higher Level Languages – Translation, Interpretation and Scripting

Introduction

The problem with the calculator and expression translator discussed in the last chapter occurs when incorrectly structured expressions are entered into the system. What is necessary is a way for checking that an expression has a correct structure before attempting to execute or translate it.

<arithmetic expression>	:=	<multiply phrase><add operator><arithmetic expression> | <multiply phrase> .
<multiply phrase>	:=	<factor> <multiply operator> <multiply phrase> | <factor>.
<factor>	:=	<variable> | <number> | <bracketed expression>.
<bracketed expression>	:=	'(' <arithmetic expression> ')'.
<add operator>	:=	"+" | "-".
<multiply operator>	:=	"*" | "/" | "%".

The correct structure of a language statement is defined by the grammar of the language. This can be expressed as a set of swapping rules of the form given above for an arithmetic expression. The way to read these rules is, starting from the top left: in order to have an *arithmetic expression* it is necessary either to find a *multiply phrase* followed by an *add operator* – either plus or minus – followed by a structure that itself conforms to that of an *arithmetic expression*; or merely to find a *multiply phrase*. This definition however transfers the problem to recognising the structure of a *multiply phrase*. The second rule gives the necessary definition to do this except that again it leaves the *factor* undefined, but the next rule takes care of this! Descending

A. Thomas, *Integrated Graphic and Computer Modelling*,
DOI: 10.1007/978-1-84800-179-4_8, © Springer-Verlag London Limited 2008

down through this hierarchy of definitions ultimately leads to entities like *keywords* *punctuation* or *variable names* and *numbers* that can be directly located in the target expression by character by character comparison tests. This approach allows a program to be written that will automatically check that a new expression has a correct structure. If an expression is entered as the *String: inputString* then processed character by character using the character array index *stringIndex* to walk through the expression, its structure can be checked using the following sequence of procedures developed directly from the grammar rules.

<arithmetic expression> := <multiply phrase><add operator><arithmetic expression>
| <multiply phrase> .

```java
public static boolean arithmeticExpression(){
    int indexSave = stringIndex;
    if(multiplyPhrase())
        if(addOperator())
            if(arithmeticExpression())return true;
    stringIndex=indexSave;
    if(multiplyPhrase())return true;
    stringIndex=indexSave;
    return false;
}
```

<multiply phrase>   := <factor> <multiply operator> <multiply phrase> | <factor>.

```java
public static boolean multiplyPhrase(){
    int indexSave = stringIndex;
    if(factor())
        if(multiplyOperator())
            if(multiplyPhrase())return true;
    stringIndex=indexSave;
    if(factor())return true;
    stringIndex=indexSave;
    return false;
}
```

<factor>   := <variable> | <number> | <bracketed expression>.

```java
public static boolean factor(){
    int indexSave = stringIndex;
    if(variable())return true;
    stringIndex=indexSave;
    if(number())return true;
    stringIndex=indexSave;
    if(bracketedExpression())return true;
    stringIndex=indexSave;
    return false;
}
```

\<bracketed expression\>   := '(' \<arithmetic expression\> ')'.

```
public static boolean bracketedExpression(){
    int indexSave = stringIndex;
    if ( nextNonSpaceCharacter()=='(')
        if ( arithmeticExpression())
            if ( nextNonSpaceCharacter()==')') return true;
    stringIndex=indexSave;
    return false;
}
```

\<add operator\>   := '+' | '-'.

```
public static boolean addOperator(){
    int indexSave = stringIndex;
    if ( nextNonSpaceCharacter()=='+') return true;
    stringIndex=indexSave;
    if ( nextNonSpaceCharacter()=='-') return true;
    stringIndex=indexSave;
    return false;
}
```

\<multiply operator\>   := '*' | '/' | % .

```
public static boolean multiplyOperator(){
    int indexSave = stringIndex;
    if ( nextNonSpaceCharacter()=='*') return true;
    stringIndex=indexSave;
    if ( nextNonSpaceCharacter()=='/') return true;
    stringIndex=indexSave;
    if ( nextNonSpaceCharacter()=='%') return true;
    stringIndex=indexSave;
    return false;
}
```

\<number\> := NUMBER.

```
public static boolean number(){
    String number = ""; char ch; int indexSave = stringIndex;
    if(((ch = nextNonSpaceCharacter())>='0')&&(ch<='9')){
        number = number + ch;
        while(((ch = nextCharacter())>='0')&&(ch<='9'))number = number +ch;
        stringIndex--;
        return true;
    }
    stringIndex=indexSave;
    return false;
}
```

Applying the Maze Grammer

Processing starts by considering the first character in the input target string. Entry into the maze is at the top into the left hand cell named *statement*. This has a single exit point to a right hand cell containing two elements. Movement through the maze is carried out by systematically following the arrows showing the links between cells until a terminal cell is reached following which the path must be backtracked. If the terminal cell contains a symbol that matches the current element in the target string then the path to this cell is flagged as successful, allowing the next element in the input string be set up for matching. Where there is no match then the current element in the input string is kept in an attempt to find an alternative route to a terminal cell that can match it. Alternative routes are found by systematically continuing to search the space of the grammar-maze.

If a successful match is found from any one of the exit routes from a pink cell then indication of this success can be taken back to the previous green cell from which it was entered. However, for a success to be taken back from a blue cell to a pink cell, all the green cells within it, taken in order, have to have successfully found a route to a terminal cell matching the input string. If any green cell is returned to with a failure then this failure immediately has to be taken back to the pink cell from which its blue cell was entered.

On failure the backtracking process has to recover the location in the input target string appropriate for the cell being returned to. This is because partial matches may be found following a successful sequence of green links from a blue cell, which have to be discarded if a subsequent green link returns a failure. The previously successful returns will have incorrectly advanced the current character in the input string, being tested for matches. When a route from the start of this maze to its exit has been found then matches for whole target string will have been found. The target string can then be accepted as having a correct grammatical structure, defined by the path that was taken in finding a match for it.

The way the grammar maze is laid out in Figure 8.1 places the cells on the left where only one alternative link needs to return successfully for the cell itself to return success. In contrast the cells on the right need to have all their links returning success for them to be able to return success. The repetitive pattern that emerges makes it possible to write a general recognition procedure that can be tailored to work with any grammar structured in this way merely by presenting the grammar in the form of a pair of tables.

The first table – the *orTable* – gives the list of alternatives for each rule: corresponding to a list of the pink cells. The second table -- the *andTable* – lists the expansions for each alternative sudivision of a rule, a list corresponding to the blue cells, gives the necessary elements in the correct order that need to be identified to satisfy each alternative. The positive integers in the *orTable* are indexes to rows in the *andTable*, and similarly the positive integers in the *andTable* are indexes into the *orTable*. The negative entries in either table reference procedures that directly match input-string characters with keyword strings -- given in an array of Strings *chs* -- or provide tests for **name** or **number** character sequences in the input string.

```
public class Expression {
    static Input jInput= null; static Output jOutput = null; static int i = 0;
    static String str = ""; static int stringIndex =0;static String inputString ="";
    static int[] record = new int[1000]; static int recordIndex=0, Index=0;
    static String[]  chs = new String[]
        {"0","1","2","3","4",",","+","-","*","/","%","(",")",",","."};
    static int[][] orTable= new int[][]{
        {0},
        {1,2},
        {3,4},
        {5,6,7},
        {-6, -7},
        {-8, -9, -10}
    };
    static int[][] andTable= new int[][]{
        {1,-5},
        {2,4,1},
        {2},
        {3,5,2},
        {3},
        {-2},
        {-3},
        {-11,1, -12 }
    };

    public static void main(String[] args){
        int width = 800, height = 100;
        jInput = new Input(12,612,400, 100," INPUT ");
        jOutput = new Output(12,12,400, 600," OUTPUT ");
        String str = jInput.readString();
        while(!str.equals("stop")){
            while((!str.equals("end"))&&(!str.equals("stop"))){
                inputString = inputString + str;
                jOutput.writeString(str +"\n");
                str = jInput.readString();
            }
            recordIndex =0; stringIndex =0; record[recordIndex]=0;
            char ch = nextNonSpaceCharacter();stringIndex--;
            if(orTest(0)){
                jOutput.writeString("\n    success \n");
                AExpression ex = new AExpression();
                String[] code= ex.expressionEvaluation(jOutput,inputString);
            }else jOutput.writeString("failed \n");
            stringIndex = 0; recordIndex =0; inputString = "";
            str = jInput.readString();
        }
    }
}
```

Annotations in diagram:

- operators keywords & separators
- + numbers indexes to *andTable* entries
- + numbers indexes to *orTable* entries
- − numbers selectors for lexical elements

```
static boolean orTest(int i){
    int[] m = orTable[i];
    int indexSave = stringIndex;  int recordSave = recordIndex;
    for(int j=0; j<m.length;j++){
        record[recordIndex] = m[j];
        recordIndex++;
        switch(m[j]){
            case -1: return true;
            case -2: if(name()) return true; break;
            case -3: if(number()) return true; break;          //integer
            case -4: if(number()) return true; break;          //real
            default: if(m[j] < 0){if(Match(chs[-m[j]]))return true;
                            }else {if(andTest(m[j]))return true;}
        }
        stringIndex=indexSave; recordIndex=recordSave;
    }return false;
}
static boolean andTest(int i){
    int indexSave = stringIndex;
    int[] n = andTable[i];
    for(int j=0; j < n.length;j++){
        switch(n[j]){
            case -1: break;
            case -2: if(name())  break; return false;
            case -3: if(number()) break; return false;          //integer
            case -4: if(number())  break; return false;          //real
            default: if(n[j] < 0){if(Match(chs[-n[j]]))break; else return false; }
                        else {if(orTest(n[j]))break; else return false;}
        }
    }return true;
}
static boolean number(){
    String number = "";  char ch;
    if(((ch = nextNonSpaceCharacter())>='0')&&(ch <='9')){
        record[recordIndex]= stringIndex-1;recordIndex++;
        number = number + ch;
        while(((ch = nextCharacter())>='0')&&(ch<='9'))
            number = number +ch;
        record[recordIndex]= stringIndex-1; recordIndex++;
        return true;
    } return false;
}
static char nextCharacter(){
    if(stringIndex==inputString.length())return (char)0;
    char ch= inputString.charAt(stringIndex); stringIndex++;
    return ch;
}
```

```
static char nextNonSpaceCharacter() {
    char ch = nextCharacter( );
    while((ch == ' ')||(ch =='\n')){ch = nextCharacter( );}return ch;
}
static boolean name(){
    String nam = ""; char ch;
    if((((ch = nextNonSpaceCharacter())>='a')&&(ch<='z'))
                                  ||((ch >='A')&&(ch<='Z'))) {
        stringIndex--; record[recordIndex]= stringIndex;recordIndex++;
        while((((ch= nextCharacter()) >='a')&&(ch<='z'))
                ||((ch >='A')&&(ch<='Z'))||((ch >='0')&&(ch<='9')))
            { nam = nam +ch;}
        stringIndex--; record[recordIndex]= stringIndex;recordIndex++;
        return true;
    }return false;
}
static boolean Match(String st){
    int i=0;char ch;
    if(nextNonSpaceCharacter()== st.charAt(i)){
        stringIndex--;
        while((i < st.length())&&(st.charAt(i) == (ch=nextCharacter()))){ i++;}
    }
    if(i==st.length()){ return true;}
    return false;
}
```

The input and output from this program is shown in Figure 8.2. The first expression entered contains an error and the result is failure. The second gives a corrected input and the result is recognised as correct and passed to the expression translator outlined in chapter 7, which generates the translated version of the program in assembly code.

In order to extend the language from simple expressions the first step is to extend the grammar rules to allow the new statements to be recognised. Using the first approach discussed above, will involve writing new procedures for each of the new rules. Using the second approach merely extending the *orTable*, the *andTable*, and the array of separators, keywords and operators will achieve the same goal. However, once the extensions have been checked and recognised they will still have to be interpreted or translated into assembly code. The equivalent to the calculator or expression translator needs to be provided for the new statements.

What is important is that this step can also be carried out based on the grammar rules used to define the new statements. In carrying out the recognition process the structure of the input statements has implicitly been identified by the route taken through the grammar maze. If this structure is recorded as the list of successful alternative rules passed through, then this list can be used to support the execution or translation stage that matches the calculator program's evaluation of the arithmetic expression. The next step is to illustrate the alternative way this can be done by executing or translating an arithmetic expression based on its grammar rules.

```java
public static void main(String[] args){
    int width = 800, height = 100;
    jInput = new Input(12,612,400, 100," INPUT ");
    jOutput = new Output(12,12,400, 600," OUTPUT ");
    String str = jInput.readString();
    while(!str.equals("stop")){
        while((!str.equals("end"))&&(!str.equals("stop"))){
            inputString = inputString + str; jOutput.writeString(str +"\n");
            str = jInput.readString();
        }
        recordIndex =0; stringIndex =0; record[recordIndex]=0;
        char ch = nextNonSpaceCharacter();stringIndex--;
        if(orTest(0)){
            for(int k=0;k<recordIndex;k++){
                if((k!=0)&&(k%15==0))jOutput.writeString("\n");
                jOutput.writeString(record[k]+" ");
            }
            jOutput.writeString("\n   success \n");
            Index=0;  MiniJCProgram();
        }else jOutput.writeString("failed \n");
        stringIndex = 0; recordIndex =0; inputString = "";
        str = jInput.readString();
    }
}
```

```java
static void MiniJCProgram(){
    Index=0;jOutput.writeString("   Program \n");
    Statement();
}
static void Statement(){
    jOutput.writeString("   Statement \n");
    switch (record[Index]){
        case 0: Index++; ArithmeticExpression(); Index++;
                jOutput.writeString("   "+chs[-record[Index]]+"\n"); break;
        default: jOutput.writeString("error 1 index = "+Index+"\n");
    }
}
static void ArithmeticExpression(){
    jOutput.writeString("   ArithmeticExpression \n");
    switch (record[Index]){
        case 1: Index++;  MultiplyPhrase();
                Index++;  AddOp();
                Index++;  ArithmeticExpression(); break;
        case 2: Index++; MultiplyPhrase(); break;
        default: jOutput.writeString("error 2 index = "+Index+"\n");
    }
}
```

```
static void MultiplyPhrase(){
    jOutput.writeString("    MultiplyPhrase \n");
    switch (record[Index]){
        case 4: Index++; Factor();
                Index++; MultiplyOp();
                Index++; MultiplyPhrase(); break;
        case 5: Index++; Factor(); break;
        default: jOutput.writeString("error 3 index = "+Index+"\n");
    }
}
static void Factor(){
    jOutput.writeString("    Factor \n");
    switch (record[Index]){
        case 5: Index++; jOutput.writeString("    (\n");
                ArithmeticExpression();
                jOutput.writeString("    )\n"); break;
        case 6: Index++; name(); break;
        case 7: Index++; number (); break;
        default: jOutput.writeString("error 4 index = "+Index+"\n");
    }
}
static void Name(){
    jOutput.writeString("    NAME \n");
    jOutput.writeString("    ");
    for(int i=record[Index++];i<record[Index];i++)
        jOutput.writeString(""+inputString.charAt(i));
    jOutput.writeString("\n"); break;
}
static void AddOp(){
    jOutput.writeString("    Add Operator \n");
    jOutput.writeString("    ");
    jOutput.writeString(chs[-record[Index]]+"\n");
}
static void MultiplyOp(){
    jOutput.writeString("    Multiply Operator \n");
    jOutput.writeString("    ");
    jOutput.writeString(chs[-record[Index]]+"\n");
}
static void Number(){
    jOutput.writeString("    NUMBER \n");
    Index++;
    jOutput.writeString("    ");
    for(int i=record[Index++];i<record[Index++];i++)
        jOutput.writeString(""+inputString.charAt(i));
    jOutput.writeString("\n"); break;
}
}
```

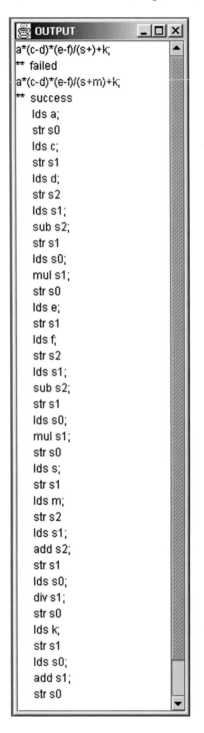

```
OUTPUT                          _□X
a*(c-d)*(e-f)/(s+)+k;
** failed
a*(c-d)*(e-f)/(s+m)+k;
** success
    lds a;
    str s0
    lds c;
    str s1
    lds d;
    str s2
    lds s1;
    sub s2;
    str s1
    lds s0;
    mul s1;
    str s0
    lds e;
    str s1
    lds f;
    str s2
    lds s1;
    sub s2;
    str s1
    lds s0;
    mul s1;
    str s0
    lds s;
    str s1
    lds m;
    str s2
    lds s1;
    add s2;
    str s1
    lds s0;
    div s1;
    str s0
    lds k;
    str s1
    lds s0;
    add s1;
    str s0
```

```
OUTPUT                          _□X
Fred*(Jack+Jim)+Alan;
0 1 3 6 0 4 -8 4 5 1 4 6 6 10 -6
2 4 6 11 14 -6 2 4 6 16 20
    success
    Program
    Statement
    ArithmeticExpression
    MultiplyPhrase
    Factor
    NAME
    Fred
    Multiply Operator
    *
    MultiplyPhrase
    Factor
    (
    ArithmeticExpression
    MultiplyPhrase
    Factor
    NAME
    Jack
    Add Operator
    +
    ArithmeticExpression
    MultiplyPhrase
    Factor
    NAME
    Jim
    )
    Add Operator
    +
    ArithmeticExpression
    MultiplyPhrase
    Factor
    NAME
    Alan
    ;
```

Figure 8.2 Translated expression   **Figure 8.3** Recognition path

In the output shown in Figure 8.3 the digit sequence following the expression gives the indexes to the rows of the *orTable* that record the successful sequence of rules found necessary to match the structure of the input expression. If as before a series of procedures matching the grammar rules are written but in the way illustrated above, then this record can be used to write out the structure found in the way also shown in Figure 8.3 following the integer sequence. It is this framework of procedures, driven by the *record[]* representing a particular expression structure, which can be set up to interpret the expression or translate it.

If a valid sentence is identified by a valid route through the grammar maze network, and each grammar rule defines the links between maze cells, then the recognition task is a route-finding spatial search looking for a path from the start cell to a finish cell that allows the target input statement to be matched. Where deadends are reached in the maze system the search has to backtrack to look for alternative options. In a general maze problem there may be a variety of routes to a finish cell. In natural language parsing this can also be the case, and this is traditionally illustrated by the two sentences:

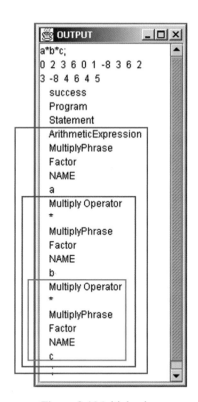

Figure 8.4 Multiply phrases

Fruit flies like a banana.

Concord flies like an arrow.

Each of these statements can be incorrectly read using the sructure that makes sense of the other. In the first example "flies" is the subject of the sentence whereas in the second "flies" is the verb! The distinction in this case cannot be made based on the structure of the strings alone but has to depend on the meaning of the words, and so stops being a purely formal analysis. The construction of the maze or search space can be arranged to prevent multiple routes to valid language strings being possible. This is done by the design of the grammar rules. In the case of standard high level computer languages the rules are specifically set up to avoid this kind of ambiguity. Once a path to a matching statement has been found in this "language space" then the route taken uniquely defines its structure, and can then be used to interpret it or translate it into an equivalent alternative language statement or sequence of statements.

As an example consider evaluating the *MultiplyPhrase* a*b*c. The sequence of procedure calls that match the analysis of this expression, are shown in Figure 8.4. If each time a variable is located its value is passed back up to the calling procedure then each multiplyPhrase will have either a single value passed back to it from the rule defining it as a factor, or two values from the alternative rule defining it as a factor followed by an operator followed by a multiplyPhrase. Applying the operator to the two values returned to it will allow it to pass back the product of the two to its calling procedure.

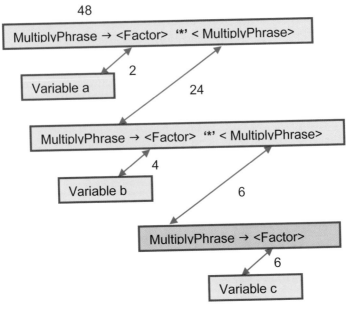

Figure 8.5 Recursive evaluation of a multiply phrase a*b*c

However, though this illustrates the principle involved, and it works in this direct way for a multiply sequence, it imposes an ordering on the application of the operators to the variables which is not appropriate for subtraction and division operations. What these procedures have to do is to construct an operator tree, which can then be traversed to give the correct order for the calculation.

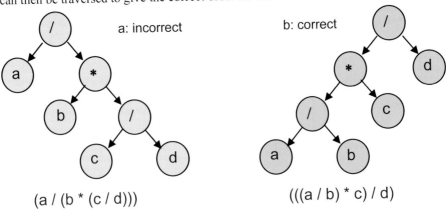

Figure 8.6 Alternative operator trees

<Arithmetic Expression> :: = <Multiply phrase>

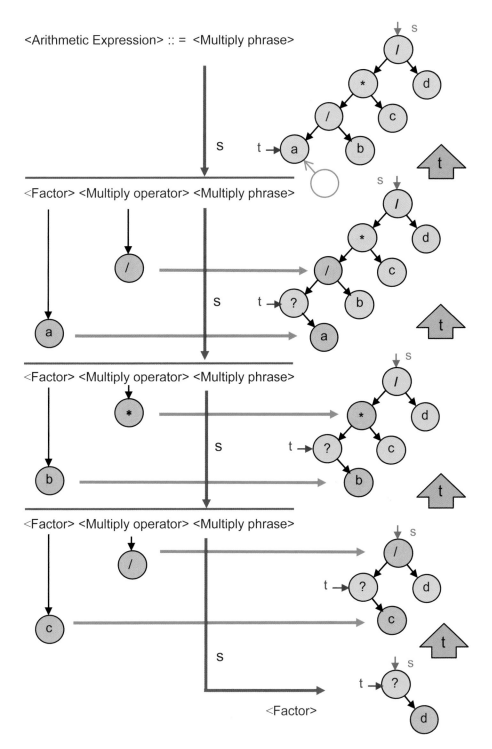

Figure 8.7 Constructing an operator tree using backward recursion

The recursive evaluation sequence shown in Figure 8.5 implements the incorrect sequence defined by the operator tree shown in in Figure 8.6a. However, the same recursive processing order can be used to produce the correct operator tree given in Figure 8.6b, in the way shown in Figure 8.7.

A program to build this tree from the record generated by the recognition program is given below.

```
static void MiniJCProgram(){
    Index=0;  jOutput.writeString("    Program \n");
    Statement(); return;
}
```

```
static void Statement(){
    switch (record[Index]){
        case 0: Index++;
                TreeNode s = new TreeNode();
                TreeNode t = ArithmeticExpression(s);
                if(t==null)s=s.right;
                else if(t.left.left==null)t.left=t.left.right;
                Index++;
                s.infixTraverse(jOutput);
                break;
        default: jOutput.writeString("error 1 index = "+Index+"\n");
    }return;
}

static TreeNode ArithmeticExpression(TreeNode s){
    TreeNode t = new TreeNode();
    int save = Index;
    Index++;
    TreeNode r = MultiplyPhrase(t);
    if(r==null)t=t.right;
    else if((r.left!=null)&&(r.left.left==null))r.left=r.left.right;
    switch (record[save]){
        case 1: Index++;
                String op = chs[-record[Index]];
                Index++;
                TreeNode p = ArithmeticExpression(s);
                TreeNode q = new TreeNode();
                if(p==null){s.name=op;s.left=q;q.right=t;return s;}
                else{ p.left.name=op; p.left.left=q; q.right=t; return p.left;}
        case 2: s.right=t;  return null;
        default: jOutput.writeString("error 2 index = "+Index+"\n");
    }
    return null;
}
```

```
static TreeNode MultiplyPhrase(TreeNode s){
    TreeNode t = new TreeNode();
    int save=Index; Index++;
    TreeNode r = Factor(t);
    if(r==null)t=t.right;
    else if((r.left!=null)&&(r.left.left==null))r.left = r.left.right;
    switch (record[save]){
        case 3: Index++;
                String op = chs[-record[Index]];
                Index++;
                TreeNode p = MultiplyPhrase(s);
                TreeNode q = new TreeNode();
                if(p==null){ s.name=op; s.left=q; q.right=t; return s;}
                else{ p.left.name=op; p.left.left=q; q.right=t; return p.left;}
        case 4: s.right=t; return null;
        default: jOutput.writeString("error 3 index = "+Index+"\n");
    }return null;
}
static TreeNode Factor(TreeNode s){
    switch (record[Index]){
        case 5: Index++;
                s.name = Name();
                return s;
        case 6: Index++;
                s.name = Number();
                return s;
        case 7: Index++;
                TreeNode r = ArithmeticExpression(s);
                return r;
        default: jOutput.writeString("error 4 index = "+Index+"\n");
    }return null;
}
static String Name(){
    String str="";
    for(int i=record[Index++];i<record[Index];i++)
        str=str+inputString.charAt(i);}
    return str;
}
static String Number(){
    String str="";
    for(int i=record[Index++];i<record[Index];i++)
        str=str+inputString.charAt(i);
    return str;
}
```

Recursing backwards creates a correct operator tree in the way shown in Figure 8.8. However a forward recursive approach gives the same result shown in Figure 8.9.

```
OUTPUT                                    _ □ ×
fred;
  success
  Program
( fred )

a+b+c+d;
  success
  Program
( ( ( ( a ) + ( b ) ) + ( c ) ) + ( d ) )

a*b*c*d;
  success
  Program
( ( ( ( a ) * ( b ) ) * ( c ) ) * ( d ) )

(a-b)*(e+f)+g*h;
  success
  Program
( ( ( ( a ) - ( b ) ) * ( ( e ) + ( f ) ) ) + ( ( g ) * ( h ) ) )

a*b*c+e/f/g+k;
  success
  Program
( ( ( ( ( a ) * ( b ) ) * ( c ) ) + ( ( ( e ) / ( f ) ) / ( g ) ) ) + ( k ) )
```

Figure 8.8 Backward recursion

```
OUTPUT                                    _ □ ×
a;
  success
  Program
( a )

a+k+h;
  success
  Program
( ( ( a ) + ( k ) ) + ( h ) )

a*n+k*y+p*q;
  success
  Program
( ( ( ( a ) * ( n ) ) + ( ( k ) * ( y ) ) ) + ( ( p ) * ( q ) ) )

(j-m-n)*(z+x+y)+p;
  success
  Program
( ( ( ( ( j ) - ( m ) ) - ( n ) ) * ( ( ( z ) + ( x ) ) + ( y ) ) ) + ( p ) )
```

Figure 8.9 Forward recursion

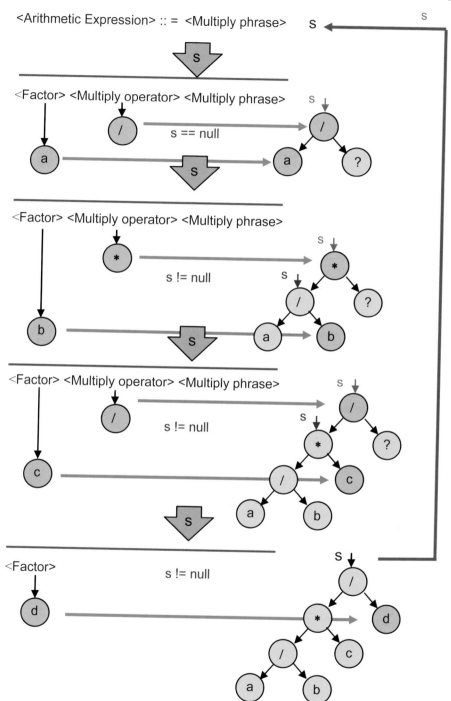

Figure 8.10 Constructing an operator tree using forward recursion

Forward recursion can be implemented by using the following modified procedures:

```
static void Statement(){
    switch (record[Index]){
        case 0: Index++;
            TreeNode t = ArithmeticExpression(null);
            t.infixTraverse(jOutput);
            break;
        default: jOutput.writeString("error 1 index = "+Index+"\n");
    }return;
}
static TreeNode ArithmeticExpression(TreeNode s){
    switch (record[Index]){
        case 1: Index++;
            TreeNode t= new TreeNode(); TreeNode r= MultiplyPhrase(null);
            if(s!=null)s.right = r ;
            Index++;  t.name = AddOp();
            if(s==null)t.left = r; else t.left = s;
            Index++;  return ArithmeticExpression(t);
        case 2: Index++;
            if(s==null)return MultiplyPhrase(null);
            else{ s.right= MultiplyPhrase(null); return s;}
        default: jOutput.writeString("error 2 index = "+Index+"\n");
    }return null;
}
static TreeNode MultiplyPhrase(TreeNode s){
    switch (record[Index]){
        case 3: Index++;
            TreeNode t= new TreeNode(); TreeNode r = Factor();
            if(s!=null)s.right = r ;
            Index++;  t.name = MultiplyOp();
            if(s==null)t.left = r; else t.left = s;
            Index++; return MultiplyPhrase(t);
        case 4: Index++;
            if(s==null)return Factor(); else{ s.right= Factor(); return s;}
        default: jOutput.writeString("error 2 index = "+Index+"\n");
    }return null;
}
static TreeNode Factor(){
    TreeNode s= new TreeNode();
    switch (record[Index]){
        case 5: Index++;  s.name = Name();return s;
        case 6: Index++;  s.name = Number();return s;
        case 7: Index++;  TreeNode r = ArithmeticExpression(null); return r;
        default: jOutput.writeString("error 4 index = "+Index+"\n");
    } return null;
}
```

Once the operator tree has been constructed a standard tree traversal algorithm will give the reverse polish order that can be used to generate assembly code as before, or to evaluate the expression. However the forward recursive procedure can be used directly without having to generate the intermediate tree structure, both to evaluate the expression and to translate it into assembly code in the following way.

```
static void MiniJCProgram(){
    Index=0;jOutput.writeString("   Program \n"); Statement(); return;
}
static void Statement(){
    switch (record[Index]){
        case 0: Index++;
            int t = ArithmeticExpression(0,0);
            Index++;
            jOutput.writeString("  result= "+t+"\n");
            break;
        default: jOutput.writeString("error 1 index = "+Index+"\n");
    }return;
}
static int ArithmeticExpression(int m, int op1){
    int save= record[Index],op2=0;
    Index++; int n= MultiplyPhrase(1,0);
    if(save==1){Index++; op2= -record[Index]; }
    if(op1==6){n= m+n;}
    if(op1==7){n= m-n;}
    if(save==1){Index++; n= ArithmeticExpression(n,op2);}
    return n;
}
static int MultiplyPhrase(int m, int op1){
    int save= record[Index],op2=0;
    Index++; int n= Factor();
    if(save==3){Index++; op2 = -record[Index];}
    if(op1==8){n=m*n;}
    if(op1==9){n=m/n;}
    if(op1==10){n=m%n;}
    if(save==3){Index++; n= MultiplyPhrase(n,op2);}
    return n;
}
static int Factor(){
    int j=0;int i = Index;Index++;
    switch (record[i]){
        case 5: if((j=names.locate(Name()))>=0)return names.val[j]; break;
        case 6: if((j=names.locate(Number()))>=0)return names.val[j]; break;
        case 7: return ArithmeticExpression(0,0);
        default: jOutput.writeString("error 4  index = "+Index+"\n");
    }return 0;
}
```

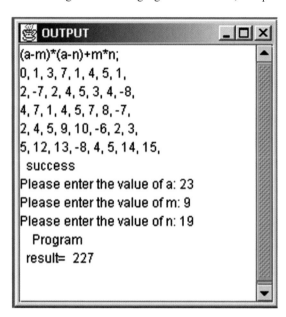

Figure 8.11 Directly evaluating an expression to give a result

```
static void MiniJCProgram(){
    Index=0; jOutput.writeString("   Program \n");
    SetValues();
    Statement();  return;
}
static int ArithmeticExpression(int m, int op1){
    int save= record[Index],op2=0;
    Index++; int n= MultiplyPhrase(1,0);
    if(save==1){Index++; op2= AddOp();}
    if(op1==6){
        n= m+n;
        code[codeIndex++] = "lds s"+(stackIndex-2)+"\n";
        code[codeIndex++] = "add s"+(stackIndex-1)+"\n";
        code[codeIndex++] = "str s"+(stackIndex-2)+"\n";
        stackIndex--;
    }
    if(op1==7){
        n= m-n;
        code[codeIndex++] = "lds s"+(stackIndex-2)+"\n";
        code[codeIndex++] = "sub s"+(stackIndex-1)+"\n";
        code[codeIndex++] = "str s"+(stackIndex-2)+"\n";
        stackIndex--;
    }if(save==1){Index++; n= ArithmeticExpression(n,op2);}
    return n;
}
```

```
static void Statement(){
    switch (record[Index]){
        case 0: Index++;
            int t = ArithmeticExpression(0,0);  Index++;
            jOutput.writeString("  result=  "+t+"\n");
            code[codeIndex++]="wrt s"+(stackIndex-1)+";\n";
            for(int i=0;i<codeIndex;i++){ jOutput2.writeString(code[i]);}
            break;
        default: jOutput.writeString("error 1 index = "+Index+"\n");
    }return;
}
static int MultiplyPhrase(int m, int op1){
    int save= record[Index],op2=0;  Index++; int n= Factor();
    if(save==3){Index++;op2 = MultiplyOp();}
    if(op1==8){
        n=m*n;
        code[codeIndex++] = "lds s"+(stackIndex-2)+"\n";
        code[codeIndex++] = "mul s"+(stackIndex-1)+"\n";
        code[codeIndex++] = "str s"+(stackIndex-2)+"\n";
        stackIndex--;
    }if(op1==9){
        n=m/n;
        code[codeIndex++] = "lds s"+(stackIndex-2)+"\n";
        code[codeIndex++] = "div s"+(stackIndex-1)+"\n";
        code[codeIndex++] = "str s"+(stackIndex-2)+"\n";
        stackIndex--;
    }if(op1==10){
        n=m%n;
        code[codeIndex++] = "lds s"+(stackIndex-2)+"\n";
        code[codeIndex++] = "mod s"+(stackIndex-1)+"\n";
        code[codeIndex++] = "str s"+(stackIndex-2)+"\n";
        stackIndex--;
    }if(save==3){Index++; n= MultiplyPhrase(n,op2);}
    return n;
}
static String Name(){
    String str="";
    for(int i=record[Index++];i<record[Index];i++)
        str=str+inputString.charAt(i);
    return str;
}
static String Number(){
    String str="";
    for(int i=record[Index++];i<record[Index];i++)
        str=str+inputString.charAt(i);
    return ('\"'+str+'\"');
}
```

```
static int Factor(){
    int j=0;String nm= "";
    switch (record[Index]){
        case 5: Index++;
            if((j=names.locate(nm=Name()))>=0){
                code[codeIndex++]= "lds "+nm+"\n";
                code[codeIndex++]= "str s"+(stackIndex++)+"\n";
                return names.val[j];
            }
        case 6: Index++;
            if((j=names.locate(nm=Name()))>=0){
                code[codeIndex++]= "lds "+nm+"\n";
                code[codeIndex++]= "str s"+(stackIndex++)+"\n";
                return names.val[j];
            }
        case 7: Index++;
            int r = ArithmeticExpression(0,0);
            return r;
        default: jOutput.writeString("error 4 index = "+Index+"\n");
    }return 0;
}
static void setValues(){
    int size =names.getSize();
    for(int i=0 ; i < size; i++){
        String ss=""; int nm=0;
        if((ss = names.getElement(i)).charAt(0)==""){
            for(int j=1; j<ss.length()-1;j++){
                int digit = (int)ss.charAt(j)-(int)'0';
                nm = nm*10+digit;
            }names.val[i]=nm;
        }else{
            code[codeIndex++] = "rds "+ss+"\n";
            jOutput.writeString("Enter the value of "+names.getElement(i)+": ");
            String num= jInput.readString();
            jOutput.writeString(num +"\n");
            names.val[i]= number(num);
        }
    }
}
static int AddOp() {return -record[Index];}
static int MultiplyOp() { return -record[Index];}
```

Extending these procedures to generate the assembly code, provides both interpreter and translator functions in the way illustrated in Figure 8.12. This demonstrates the two ways in which a specialised computer language can be set up and processed to support application requirements not directly catered for by existing general purpose, high level languages. Although most of the facilities that are needed are provided by the Java language system, a dedicated system will add considerably to the clarity with which some of the topics in later chapters can be presented. Where animation sequences need to be defined it is useful to develop a scripting language that operates at a level above a general purpose language such as Java. In order to support this it is worth extending the system to include the other basic structured programming constructs.

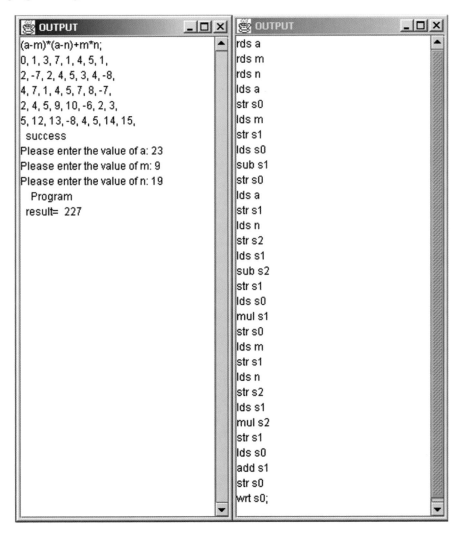

Figure 8.12 Expression interpretation and translation

The goals for the remainder of this chapter are thus two fold. The first is to implement a small high-level language: **Mini-JC** to act as a platform for later work. The second is to use this exercise to present the key ideas that underpin the hierarchical structure of larger software systems that facilitate or constrain the way that modelling and display services can be provided.

Mini-JC: High Level to Assembly Language Translator

In this section, the target is to implement the minimum set of "structured programming" language constructs needed to write general-purpose programs:

1. Statements and statement sequences.
2. Conditional or branch statements.
3. Repetition statements.
4. Subprogram calls and definitions.

Planning the Program Design

The first step is to define the rules of the language, in other words to define its grammar.

```
<program> :=       <procedure><program>|<procedure>
<programbody>:=<listofdefinitions><code>
<listofdefinitions>:=<typedefinition> ";" < listofdefinitions > | NULL
<procedure> :=     "void" "main" "(" ")" "{"<programbody> "}"
                   |<rtype><procname> <args> "{"<programbody> "}"
<typedefinition>:=<type><dimension> NAME
<type> :=          "char" | "int" |"boolean" | "double" | "geometric"
<variable> :=      NAME <matrixindex>  | NAME
<code> :=          <statement><code>  | NULL
<statement> :=     <assignment>";" |<conditional>  |<repeat> |<functioncall> ";"
                   |<return><exprs> ";"|<return> ";"
<assignment> :=    <variable> "="<exprs>
<conditional> :=   "if" "("<bexprs>")" "{"<code>"}"<elsepart>
<elsepart> :=      "else" "{"<code> "}"   | NULL
<repeat> :=        "while" "("<bexprs> ")" "{"<code> "}"
                   |"do""{"<code> "}""while" "("<bexprs>")" ";"
<functioncall> := NAME "("<parameterlist> ")"
<parameterlist>:=<exprs>","<parameterlist>  |<exprs>  | NULL
<bexprs> :=        <arithexprs><relop><arithexprs>
                   |<andphrs><orops><exprs>  | <andphrs>
<andphrs> :=       <bfactor><andops><andphrs>  |<bfactor>
<bfactor> :=       <functioncall>|<boperand>  | "!"<boperand>
<boperand> :=       "("<bexprs>")"|"true"  | "false" |<variable>
<args> :=          "(" <argslist> ")"
<arithexprs> :=    <multphrs><addops><arithexprs>  |<multphrs>
<multphrs> :=      <afactor><multops><multphrs>  |<afactor>
<afactor> :=       <functioncall>|<aoperand>  | "-"<aoperand>
<aoperand> :=      "("<arithexprs> ")"  |<variable> | INTEGER | REAL
<orops> :=         "||"  | "+"
<andops> :=        "&&"  | "."
<exprs> :=         <aexprs>  |<bexprs>
<relops> :=        "<"  | ">"  | "<="  | ">="  | "=="  | "!="
<addops> :=        "+"  | "-"
<multops> :=       "*"  | "/"  | "%"
<argslist> :=      < typedefinition >","<argslist>  |< typedefinition >  | NULL
<matrixindex> :=  "[" <arithexprs "]"<matrixindex>  | "[" <arithexprs> "]"
<dimension> :=     "[" INTEGER "]"<dimension>  | NULL
<rtype> :=         <type>  | "void"
```

This grammar definition can then be converted into the *ortable*, *andtable* and String array *chs[]* needed to work with the table driven recogniser. The second step is to build a program to test an input string to see if it conforms to these rules, and the third is to use the identified grammatical structure to carry out the translation or interpretation of the program.

```
static int [][] orTable = new int[][]{
{0,3},                          //0  <program>
{1},                            //1  <programbody>
{2,-1},                         //2  I<istofdefinitions>
{5,4},                          //3  <procedure>
{6},                            //4  <typedefinition>
{-35,-36,-37,-38,-39},          //5  <type>
{7,-2},                         //6  <variable>
{8,-1},                         //7  <code>
{33,9,10,11,12,13},             //8  <statement>
{14},                           //9  <assignment>
{16},                           //10 <conditional>
{17,-1},                        //11 <elsepart>
{18,19},                        //12 <repeat>
{20},                           //13 <functioncall>
{21,22,-1},                     //14 <parameterlist>
{34,23,24},                     //15 <bexprs>
{25,26},                        //16 <andphrs>
{20,27,28},                     //17 <bfactor>
{29,-33,-34,32},                //18 <boperand>
{49},                           //19 <args>
{35,36},                        //20 <arithexprs>
{37,38},                        //21 <multphrs>
{20,39,40},                     //22 <afactor>
{41,43,-3,-4},                  //23 <aoperand>
{-29,-11},                      //24 <orops>
{-30,-31},                      //25 <andops>
{15,30},                        //26 <exprs>
{-5,-6,-7,-8,-9,-10},           //27 <relops>
{-11,-12},                      //28 <addops>
{-13,-14,-15},                  //29 <multops>
{50,44,-1},                     //30 <argslist>
{45,47},                        //31 <matrixindex>
{46,-1},                        //32 <dimension>
{48, -41},                      //33 <rtype>
```

```
static int [][] andTable = new int[][]{
{3,0},                          //0  <procedure><program>
{2,7},                          //1  <listofdefinitions><code>
{4,-16, 2},                     //2  <typedefinition>';'< listofdefinitions >
{3},                            //3  <procedure>
```

```
{33, -2, 19,-21,1,-22},                      //4  <proc >
{-41, -44, -19,-20,-21,1,-22},               //5  <mainproc> ";"
{5, 32, -2},                                 //6  <type><dimension> NAME
{-2,31},                                     //7  <matrix variable>
{8,7},                                       //8  <statement>
{9,-16},                                     //9  <assignment>
{10},                                        //10 <conditional>
{12},                                        //11 <repeat>
{13,-16},                                    //12 <functioncall> ";"
{-40, -16},                                  //13 "return" ";"
{6, -17, 26},                                //14 <variable> "="<exprs>
{20},                                        //15 <arithexprs>
{-18,-19,15,-20, -21, 7,-22, 11},            //16 "if" "("<bexprs>")" "{"<code>"}"<elsepart>
{-23, -21, 7, -22},                          //17 "else" "{"<code> "}"
{-25, -19, 15, -20, -21,7,-22},              //18 "while" "("<bexprs> ")" "{"<code> "}"
{-24,-21,7,-22,-25,-19,15,-20,-16},          //19 "do" "{"<code> "}" "while" "("<bexprs>")" ";"
{-2,-19,14,-20},                             //20 NAME "("<parameterlist> ")"
{26, -26, 14},                               //21 <exprs> ","<parameterlist>
{26},                                        //22 <exprs>
{16, 24, 15},                                //23 <andphrs><orops><exprs>
{16},                                        //24 <andphrs>
{17, 25, 16},                                //25 <bfactor><andops><andphrs>
{17},                                        //26 <bfactor>
{18},                                        //27 <boperand>
{-32, 18},                                   //28 "!"<boperand>
{-19, 15, -20},                              //29 "("<bexprs> ")"
{15},                                        //30 <bexprs>
{13},                                        //31 <functioncall>
{6},                                         //32 <variable>
{-40, 26,-16},                               //33 "return" <exprs> ";"
{20, 27, 20},                                //34 <arithexprs><relop><arithexprs>
{21, 28, 20},                                //35 <multphrs><addops><arithexprs>
{21},                                        //36 <multphrs>
{22, 29, 21},                                //37 <afactor><multops><multphrs>
{22},                                        //38 <afactor>
{23},                                        //39 <aoperand>
{-12, 23},                                   //40 "-"<aoperand>
{-19, 20, -20},                              //41 "("<arithexprs> ")"
{13},                                        //42 <functioncall>
{6},                                         //43 <variable>
{4},                                         //44 < typedefinition >
{-27, 20, -28,31},                           //45 "[" <arithexprs "]"<index>
{-27, -3, -28, 32},                          //46 "[" INTEGER "]"<dimension>
{-27, 20, -28},                              //47 "[" <arithexprs "]
{5},                                         //48 <type>
{-19, 30, -20},                              //49 "(" <argslist> ")"
{4,-26,30},  };                              //50 < typedefinition > ","<argslist>
```

```
static String[] chs = new String []
{"0","NULL","NAME","INTEGER","REAL","<",">","<=",">=","==","!=","+","-","*","/","%",";","=","if",
"(",")","{","}","else","do","while",",", "[","]","||", "&&",",","!","true","false","char","int","boolean",
"double","geometric","return", "void", "read", "write", "main" };
```

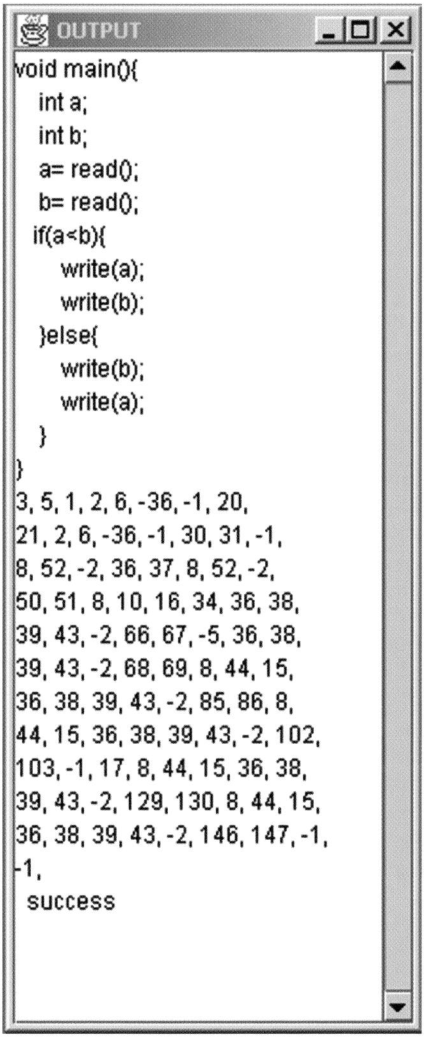

```
void main(){
  int a;
  int b;
  a= read();
  b= read();
  if(a<b){
     write(a);
     write(b);
  }else{
     write(b);
     write(a);
  }
}
3, 5, 1, 2, 6, -36, -1, 20,
21, 2, 6, -36, -1, 30, 31, -1,
8, 52, -2, 36, 37, 8, 52, -2,
50, 51, 8, 10, 16, 34, 36, 38,
39, 43, -2, 66, 67, -5, 36, 38,
39, 43, -2, 68, 69, 8, 44, 15,
36, 38, 39, 43, -2, 85, 86, 8,
44, 15, 36, 38, 39, 43, -2, 102,
103, -1, 17, 8, 44, 15, 36, 38,
39, 43, -2, 129, 130, 8, 44, 15,
36, 38, 39, 43, -2, 146, 147, -1,
-1,
 success
```

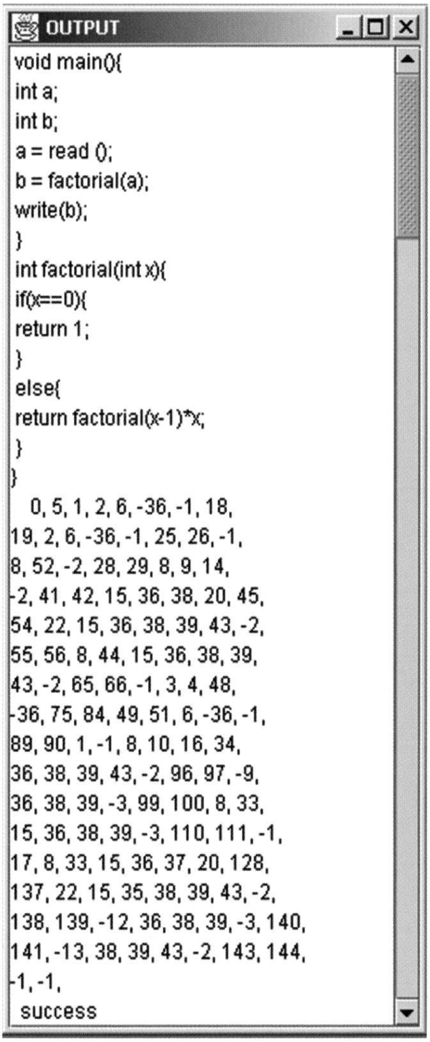

```
void main(){
  int a;
  int b;
  a = read ();
  b = factorial(a);
  write(b);
}
int factorial(int x){
if(x==0){
return 1;
}
else{
return factorial(x-1)*x;
}
}
   0, 5, 1, 2, 6, -36, -1, 18,
19, 2, 6, -36, -1, 25, 26, -1,
8, 52, -2, 28, 29, 8, 9, 14,
-2, 41, 42, 15, 36, 38, 20, 45,
54, 22, 15, 36, 38, 39, 43, -2,
55, 56, 8, 44, 15, 36, 38, 39,
43, -2, 65, 66, -1, 3, 4, 48,
-36, 75, 84, 49, 51, 6, -36, -1,
89, 90, 1, -1, 8, 10, 16, 34,
36, 38, 39, 43, -2, 96, 97, -9,
36, 38, 39, -3, 99, 100, 8, 33,
15, 36, 38, 39, -3, 110, 111, -1,
17, 8, 33, 15, 36, 37, 20, 128,
137, 22, 15, 35, 38, 39, 43, -2,
138, 139, -12, 36, 38, 39, -3, 140,
141, -13, 38, 39, 43, -2, 143, 144,
-1, -1,
 success
```

Figure 8.13 Program recognition stage

The next stage is to write the translation procedures that match the grammar structure.

```
static int[] record = new int[1000];        //record of output from the syntax tests
static int Index=0, codeIndex=0, stackIndex=0;
static NameTable array = new NameTable(); // to hold names and dimensions
static String Number(){
    String str="";
    for(int i=record[Index++]; i<record[Index]; i++)
        str=str+inputString.charAt(i);
    return ('\"'+str+'\"');
}
```

```
static void program(){                                      //0 {0,3}, program
    switch (record[Index]){
        case 0: Index++; procedure();  Index++; program();  break;
        case 3: Index++; procedure();  break;
        default: jOutput.writeString("error program index \n");
    }return;
}
static void programbody(){                                  // 1 {1},  programbody
    stackIndex=0;
    switch (record[Index]){
        case 1: Index++; listofdefinitions(); Index++; code(); break;
        default: jOutput.writeString("error programbody \n");
    }return;
}
static void listofdefinitions(){                            //2  {2,3,-1}, listofdefinitions
    switch (record[Index]){
        case 2: Index++; typedefinition(); Index++; listofdefinitions(); break;
        case -1: break;
        default: jOutput.writeString("error listofdefinitions \n");
    }return;
}
static void procedure(){                                    //3  {5,4}, procedure
    stackIndex=0;
    switch (record[Index]){
        case 4:  String argstring = "";
            Index++; rtype();  Index++; String str = Name();
            code[codeIndex] = "proc "+ str + '(' ;
            Index++; argstring= args(0,argstring);
            code[codeIndex] = code[codeIndex++]+argstring+");\n" ;
            Index++;  programbody();
            code[codeIndex++] = "endproc "+ str + "();\n";  break;
        case 5: code[codeIndex++] =  "proc main();\n";
            Index++;  programbody();
            code[codeIndex++] =  "end proc main();\n";break;
        default: jOutput.writeString("error definitions\n");
    }return;
}
static void code(){                                         //7 {8,-1}, code
    switch (record[Index]){
        case 8: Index++;statement();Index++;code();break;
        case -1:break;
        default: jOutput.writeString("error code index = "+Index+"\n");
    }return;
}
static int type(){  return -(record[Index]);}               //5 {-35,-36,-37,-38,-39}, type
static int relops(){  return -record[Index]; }              //27  {-5,-6,-7,-8,-9,-10}, relops
static int addops(){ return -record[Index];}                //28 {-11,-12}, addops
```

```
static String typedefinition(){                                    //4  {6}, typedefinition
     String st="";int size=0; int[] dim = null;
     switch (record[Index]){
         case 6: Index++; int datatype=type();
              Index++; dim=dimension(0); Index++; st= Name();
              if(dim!=null){
                   int id = array.add(st);  size= dim[0];
                   for(int k=1; k<dim.length; k++){ size= dim[k]*size;}
                   code[codeIndex++]= "  array "+st+ "("+"\""+size+"\""+")"+"\n";
                   array.dims[id]= dim;
                   return st+"[]";
              }break;
         default: jOutput.writeString("error typedefinition index = "+Index+"\n");
     }return st;
}
static void variable(){                                            //6  {7,-2}, variable
     String str = "";
     switch (record[Index]){
         case 7: Index++; str=Name();
              int id= array.add(str); Index++; matrixindex(0, array.dims[id]);
              code[codeIndex++] = "     "+ "str " +"s"+(stackIndex-1)+";\n";
              code[codeIndex++] =
                   " load "+ str +"(s"+(stackIndex-1)+", s"+(stackIndex-1)+");\n";
              break;
         case -2: Index++;str = Name();
              code[codeIndex++] = "     "+"lds "+str+"\n";
              code[codeIndex++] = "     "+"str s"+(stackIndex)+"\n";
              stackIndex++; break;
         default: jOutput.writeString("error variable\n");
     }return;
}
static void statement(){                                //8  {33,9,10,11,12,13}, statement
     switch (record[Index]){
         case 52:Index++;Index++; str = Name();
              code[codeIndex++] = "     "+"rds "+str+";\n";
         stackIndex++;break;
         case 33:Index++; exprs();
              code[codeIndex++] = "     str "+ " return"+ ";\n";
              code[codeIndex++] = "     jmp "+ " endproc "+ ";\n";  break;
         case 9: Index++;assignment();break;
         case 10:Index++;conditional();break;
         case 11:Index++;repeat();break;
         case 12:Index++;functioncall();break;
         case 13:code[codeIndex++] = "    jmp "+ " end "+ ";\n"; break;
         default: jOutput.writeString("error statement \n");
     }return;
}
```

```
static void assignment(){                              //9  {14}, assignment
    String str="";
    switch (record[Index]){
        case 14:Index++;
            switch (record[Index]){
                case -2: Index++; str = Name();
                    Index++; exprs();
                    code[codeIndex++] = "      "+"lds s"+(stackIndex-1)+";\n";
                    code[codeIndex++] = "      "+"str "+str+";\n";
                    stackIndex--; return;
                case 7:Index++; str = Name();
                    int id= array.add(str);
                    Index++; matrixindex(0, array.dims[id]);
                    code[codeIndex++] = "      "+"str s"+(stackIndex-1)+";\n";
                    int saveStack=(stackIndex-1);
                    Index++; exprs();
                    code[codeIndex++] =
                        " store "+ str +"(s"+(stackIndex-1)+", s"+saveStack+");\n";
                    stackIndex = saveStack; return;
                default: jOutput.writeString("error assignment\n");
            }
        default: jOutput.writeString("error assignment\n");
    }return;
}
static void conditional(){                             //10  {16},conditional
    switch (record[Index]){
        case 16: Index++;bexprs(0);
            int l1 = labelCount; labelCount++;
            code[codeIndex++] = "      "+"jez " + "x"+ l1 + ';' +"\n";
            Index++; code();Index++;
            if(record[Index]==17){
                int l2=labelCount;labelCount++;
                code[codeIndex++] = "      "+"jmp " + "x"+ l2  + ';' +"\n";
                code[codeIndex++] = "x"+l1+":   "+ "nop" +';' +" \n";
                Index++; code();;
                code[codeIndex++] = "x"+ l2 +":   "+ "nop" + ';' +"\n";
            }else{ code[codeIndex++] = "x"+ l1 +":   "+ "nop" + ';' +"\n";}
            return;
        default: jOutput.writeString("error conditional \n");
    }return;
}
static int orops(){                                    //24  {-29,-11},orops
    switch (record[Index]){
        case -29: case -11: return 29;
    default: jOutput.writeString("error orops\n");
    }return 0;
}
```

```
static int andops(){                                                    //25  {-30,-31}, andops
    switch (record[Index]){
        case -30:case -31:return 30;
        default: jOutput.writeString("error andops\n");
    }return 0;
}
static String parameterlist(String str){                    //14 {21,22,-1}, parameterlist
    int save = record[Index];
    switch (record[Index]){
        case 21: Index++; exprs(); Index++;
            str= str+","+ parameterlist(str); return str;
        case 22:Index++; exprs();  return " s"+(stackIndex-1)+" ";
        case -1:  return "";
        default: jOutput.writeString("error parameterlist \n");
    }return "";
}
static void repeat(){                                                   //12  {18,19}, repeat
    switch (record[Index]){
        case 18:int l1 = labelCount;  labelCount++;
            code[codeIndex++] ="x"+ l1 +":    " + "nop" + ';' +"\n";
            Index++; bexprs(0);
            int l2= labelCount;  labelCount++;
            code[codeIndex++] ="     "+"jez " + "x"+ l2  + ';' +"\n";
            Index++; code();
            code[codeIndex++] ="     "+"jmp " + "x"+ l1  + ';' +"\n";
            code[codeIndex++] ="x"+ l2 +":    " + "nop" + ';' +"\n"; break;
        case 19: l1 = labelCount;  labelCount++;
            code[codeIndex++] ="x"+ l1 +":    " + "nop" + ';' +"\n";
            Index++;code();  Index++; bexprs(0);
            l2= labelCount;  labelCount++;
            code[codeIndex++] ="     "+"jez " + "x"+ l2  + ';' +"\n";
            code[codeIndex++] ="     "+"jmp " + "x"+ l1  + ';' +"\n";
            code[codeIndex++] ="x"+ l2 +":    " + "nop" + ';' +"\n"; break;
        default: jOutput.writeString("error repeat \n");
    }return;
}
static void booleanop(int ca, int cb, String cc, String cd, String ce,String cf){
    int l1 = labelCount;  labelCount++;  int l2 = labelCount;  labelCount++;
    code[codeIndex++] = "     "+"lds s"+ca+";\n"; //(stackIndex-2)
    code[codeIndex++] = "     "+cf+" s"+cb+";\n"; //(stackIndex-1)
    code[codeIndex++] ="     "+cc + "x"+ l1  + ';' +"\n"; //"jng"
    code[codeIndex++] ="     "+"lds "+ cd+";\n"; //bfalse
    code[codeIndex++] ="     "+"jmp " + "x"+ l2  + ';' +"\n";
    code[codeIndex++] ="x"+ l1 +":    " + "lds " + ce +';' +"\n"; // btrue
    code[codeIndex++] ="x"+ l2 +":    "+"str s"+(stackIndex-2)+";\n";
    stackIndex--;
}
```

```
static void functioncall(){                              //13  {20}, functioncall
    int count =0;String  strpar="";String str ="";
    switch (record[Index]){
        case 20: Index++; str = Name();
            Index++; strpar = parameterlist(strpar);
            if(str.equals("read")){
                code[codeIndex++] = "   "+"rds s"+(stackIndex)+";\n";
                stackIndex++;
            }else if(str.equals("write")){
                code[codeIndex++] = "    "+"wrt "+strpar+";\n";
            }else{ code[codeIndex++]= "call "+ str + '('+ strpar +");\n";}
            break;
        default: jOutput.writeString("error functioncall index = "+Index+"\n");
    }return;
}
static void arithexprs(int op1){                          //20  {35,36}, arithexprs
    int save= record[Index],op2=0;
    Index++; multphrs(0);
    if(save==35){Index++; op2= AddOp(); }
    if(op1==11)numericop("add");
    if(op1==12)numericop("sub");
    if(save==35){Index++; arithexprs(op2);}
    return;
}
static void bfactor(){                                    //17  {20,27,28}, bfactor
    switch (record[Index]){
        case 20: functioncall();
            code[codeIndex++] = "     "+"str s"+(stackIndex++)+";\n"; break;
        case 27:Index++;boperand();break;
        case 28:Index++;boperand();  stackIndex++;
            booleanop(stackIndex-2, stackIndex-2, "jez ", bfalse, btrue,"mul");
            break;
        default: jOutput.writeString("error bfactor\n");
    }return;
}
static void afactor(){                                    //22  {20,39,40}, afactor
    switch (record[Index]){
        case 20: functioncall();
            code[codeIndex++] = "     "+"str s"+(stackIndex++)+";\n"; break;
        case 39:Index++; aoperand();break;
        case 40:Index++; aoperand();
            code[codeIndex++] = "     "+"lds "+'\"'+"0"+'\"'+";\n";
            code[codeIndex++] = "     "+"sub s"+(stackIndex-1)+";\n";
            code[codeIndex++] = "     "+"str s"+(stackIndex-1)+";\n";break;
        default: jOutput.writeString("error afactor \n");
    }return;
}
```

```
static int multops(){ return -record[Index]; }          //29  {-13,-14,-15}, multops
static void bexprs(int op1){                             //15  {34,23,24},  bexprs
    String bfalse ='\"'+"0"+'\"',btrue='\"'+"1"+'\"';
    int save= record[Index], op2=0;
    switch (record[Index]){
        case 34:Index++;arithexprs(0);
            Index++; int op = relops();
            Index++;arithexprs(0);
            int st1= stackIndex-1,st2 = stackIndex-2;
            switch(op){
                case 5: booleanop(st2, st1, "jng ", bfalse, btrue,"sub"); break;
                case 6: booleanop(st1, st2, "jng ", bfalse, btrue,"sub"); break;
                case 7: booleanop(st1, st2, "jng ", btrue, bfalse,"sub"); break;
                case 8: booleanop(st2, st1, "jng ", btrue, bfalse,"sub"); break;
                case 9: booleanop(st1, st2, "jez ", bfalse, btrue,"sub"); break;
                case 10:booleanop(st1, st2, "jez ", btrue, bfalse,"sub"); break;
            }break;
        case 23: case 24:
            Index++; andphrs(0);
            if(save==23)Index++; op2= orops();
            if(op1==29)
                booleanop(stackIndex-2, stackIndex-1, "jez ", btrue, bfalse,"add");
            if(save==23){Index++; bexprs(op2); }
            break;;
        default: jOutput.writeString("error bexprs \n");
    }
    return;
}
static void multphrs(int op1){                           //21  {37,38},  multphrs
    int save= record[Index],  op2=0;
    Index++; afactor();
    if(save==37){Index++; op2 = multops();}
    if(op1==13)numericop("mul");
    if(op1==14)numericop("div");
    if(op1==15)numericop("mod");
    if(save==37){Index++; multphrs(op2);}
    return;
}
static void andphrs(int op1){                            //16  {25,26}, andphrs
    int save= record[Index],  op2=0;
    Index++; bfactor();
    if(save==25){Index++; op2 = andops();}
    if(op1==30)
        booleanop(stackIndex-2, stackIndex-1, "jez ",btrue, bfalse, "mul");
    if(save==25){Index++; andphrs(op2);}
    return;
}
```

```
static void numericop(String ca){
    code[codeIndex++] = "     "+"lds s"+(stackIndex-2)+";\n";
    code[codeIndex++] = "     "+ca+" s"+(stackIndex-1)+";\n";  //eg "add"
    code[codeIndex++] = "     "+"str s"+(stackIndex-2)+";\n"; stackIndex--;
}
static void aoperand(){                                //23  {41,43,-3,-4}, aoperand
    switch (record[Index]){
        case 41:Index++;arithexprs(0);break;
        case 43:Index++;variable();   break;
        case -3:case -4:  Index++;  String str = Number();
            code[codeIndex++] = "     "+"lds "+str+";\n";
            code[codeIndex++] = "     "+"str s"+(stackIndex)+";\n";
            stackIndex++; break;
        default: jOutput.writeString("error aoperand \n");
    }return;
}
static void boperand(){                                //18  {29,-33,-34,32}, boperand
    switch (record[Index]){
        case 29:Index++;bexprs(0);break;
        case -33:code[codeIndex++] = "     "+"lds "+btrue+";\n";
            code[codeIndex++]  = "     "+"str s"+(stackIndex)+";\n";
            stackIndex++; break;
        case -34:code[codeIndex++] = "     "+"lds "+bfalse+";\n";
            code[codeIndex++]  = "     "+"str s"+(stackIndex)+";\n";
            stackIndex++; break;
        case 32:Index++; variable();break;             // matrix or name
        default: jOutput.writeString("error boperand \n");
    }return;
}
static void exprs(){                                   //26  {15,30}, exprs
    switch (record[Index]){
        case 15:Index++;arithexprs(0);break;
        case 30:Index++;bexprs(0);break;
        default: jOutput.writeString("error exprs \n");
    }return;
}
static String argslist(int count,String alist){        //30  {50,51,-1}, argslist
    int save=record[Index];
    switch (record[Index]){
        case 50: case 51:
            Index++; String st = typedefinition(); count++;
            if(save==50){ Index++; alist = argslist(count,alist); return alist;}
            return st;
        case -1: return "";
        default: jOutput.writeString("error argslist index \n");
    }return "";
}
```

```
static void matrixindex(int count,int[] dim){              //31  {45,47}, matrixindex
    switch (record[Index]){
        case 45:Index++;arithexprs(0);
            int saveStack =(stackIndex-1);
            Index++; matrixindex(count+1,dim);
            code[codeIndex++] = "    "+"mul "+"\""+dim[count]+"\""+";\n";
            code[codeIndex++] = "    "+"add s"+saveStack+";\n";
            stackIndex= saveStack+1;
            break;
        case 47:Index++;arithexprs(0);
            code[codeIndex++] = "    "+"lds s"+(stackIndex-1)+";\n";
            stackIndex--;
            break;
        default: jOutput.writeString("error matrix index\n");
    }return;
}
static int[] dimension(int count){                         //32  {46,-1}, dimension
    int[] dim = null;
    switch (record[Index]){
        case 46:Index++;String snum = Number();
            int num = number(snum);
            Index++; dim = dimension(count+1);
            if(dim==null){ dim = new int[count+1]; }
            dim[count] = num;
            return dim;
        case -1:break;
        default: jOutput.writeString("error dimension\n");
    }return null ;
}
static void rtype(){                                       //33  {48, -41}, rtype
    switch (record[Index]){
        case 48:Index++;type();break;
        case -41:break;
        default: jOutput.writeString("error rtype\n");
    }return;
}
static String args(int count,String alist){                //19  {49} args
    if(record[Index]==49){ Index++; return argslist(count,alist);}
    else  jOutput.writeString("error args \n");
    return alist;
}
static String Name(){
    String str="";
    for(int i=record[Index++];i<record[Index];i++)
        str=str+inputString.charAt(i);
    return str;
}
```

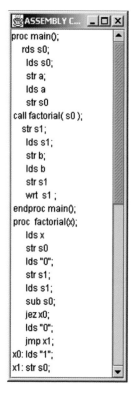

```
proc main();
  rds s0;
  lds s0;
  str a;
  lds a
  str s0
call factorial( s0 );
  str s1;
  lds s1;
  str b;
  lds b
  str s1
  wrt s1 ;
endproc main();
proc factorial(x);
  lds x
  str s0
  lds "0";
  str s1;
  lds s1;
  sub s0;
  jez x0;
  lds "0";
  jmp x1;
x0: lds "1";
x1: str s0;
```

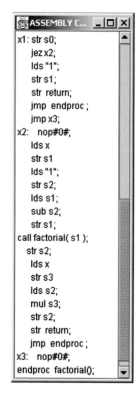

```
x1: str s0;
  jez x2;
  lds "1";
  str s1;
  str return;
  jmp endproc ;
  jmp x3;
x2: nop#0#;
  lds x
  str s1
  lds "1";
  str s2;
  lds s1;
  sub s2;
  str s1;
call factorial( s1 );
  str s2;
  lds x
  str s3
  lds s2;
  mul s3;
  str s2;
  str return;
  jmp endproc ;
x3: nop#0#;
endproc factorial();
```

This translation process is set up to take advantage of the macro expansions outlined in chapter 7. It is clear looking at the code in Figure 8.14 that the program can be tidied up and shortened. No attempt to optimise the assembly translation has been made. This introduces another area of specialist study that must be followed up elsewhere, since it is not essential to the theme of this book! This assembly code is translated by the assembler to give the machine code shown in Figure 8.16, and when this is run in the simulator the result is shown in Figure 8.15.

Figure 8.14 The program translation stage for a factorial function

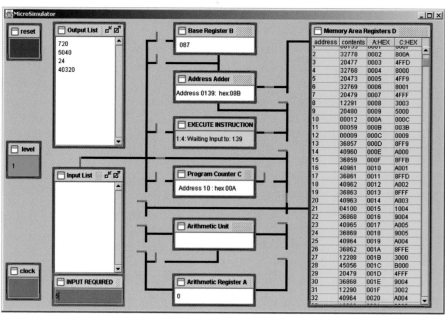

Figure 8.15 Program execution stage factorial function: 6! 7! 4! and 8!

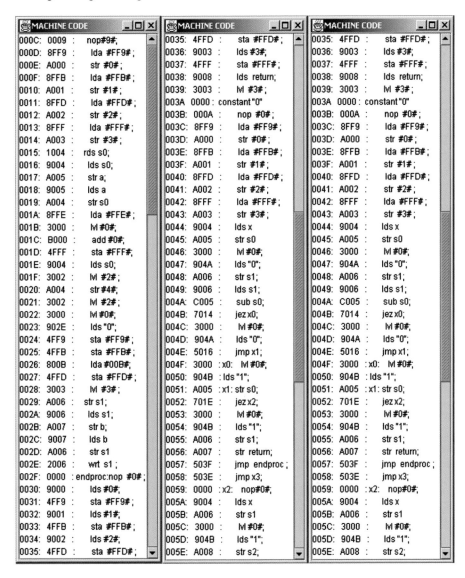

Figure 8.16 Machine code program for the factorial function

For graphics applications, processing arrays is essential. If arrays are to be included in the language one way of handling them is to also include them in the macro expansion scheme. What is needed is a way of declaring each array, and a command to load the accumulator with an element of an array and a command to store the contents of the accumulator in a defined position in the array.

array fred(100);	"array" \<name\> '(' \<size-parameter\> ')' ';'
load fred(v, i);	"load" \<name\> '(' \<variable\> ',' \<index-parameter\> ')' ';'
store fred(v, i);	"store" \<name\> '(' \<variable\> ',' \<index-parameter\> ')' ';'

```
☒ OUTPUT                    _□×
void main(){
  int i;
  int j;
  int[1 0][6] b;
  i=0;
  j=0;
  while(i<1 0){
    while(j<6){
      b[i][j] = read();
      write(b[i][j]);
      j=j+1;
    }
    i=i+1;
  }
}
```

Figure 8.17 A program to read values into an array

```
☒ OUTPUT                    _□×
proc main();
  array b("60")
// NEXT STATEMENT ;
    lds "0";
    str s0;
    lds s0;
    str i;
// NEXT STATEMENT ;
    lds "0";
    str s0;
    lds s0;
    str j;
// NEXT STATEMENT ;
// REPEAT A0 start;
x0:  nop;
    lds i
    str s0
    lds "10";
    str s1;
    lds s0;
    sub s1;
    jng x1;
    lds "0";
    jmp x2;
x1:  lds "1";
x2:  str s0;
    jez x3;
// NEXT STATEMENT ;
// REPEAT A4 start;
x4:  nop;
    lds j
    str s1
    lds "6";
    str s2;
    lds s1;
    sub s2;
    jng x5;
    lds "0";
    jmp x6;
x5:  lds "1";
x6:  str s1;
    jez x7;
// NEXT STATEMENT ;
    lds i
    str s2
    lds j
    str s3
```

```
☒ OUTPUT                    _□×
    lds j
    str s3
    lds s3;
    mul "10";
    add s2;
    str s2;
    rds s3;
    store b(s3, s2);
// NEXT STATEMENT ;
    lds i
    str s2
    lds j
    str s3
    lds s3;
    mul "10";
    add s2;
    str s2;
    load b(s2, s2);
    wrt  s2;
// NEXT STATEMENT ;
    lds j
    str s3
    lds "1";
    str s4;
    lds s3;
    add s4;
    str s3;
    lds s3;
    str j;
    jmp x4;
x7:  nop;
// REPEAT A4 finish;
// NEXT STATEMENT ;
    lds i
    str s3
    lds "1";
    str s4;
    lds s3;
    add s4;
    str s3;
    lds s3;
    str i;
    jmp x0;
x3:  nop;
// REPEAT A0 finish;
end proc main();
```

These three macros can be set up to handle arrays within the existing scheme. The first statement is needed to setup the storage space where the array values are to be placed. The second two statements are provided to transfer values between variables and indexed locations in the array. In Figure 8.17 these macros are being used to set up an array, read values into the array and finally to check the result by writing the array values out. Notice that the read function is treated in a special way, being translated directly into an assembler read command.

What is interesting about these macro statements is their similarity to the function-call macros, and the fact that they could be implemented in a similar way. However, making *load* and *store* into procedure calls for accessing an array space in the same way that the function calls have been implemented would involve a heavy overhead, it is simpler to place the relevant assembly instructions in-line in a larger program, rather than jumping to external procedure code for each array reference.

```
    // array macros
    if(strr.equals("array")) {
        String [ ] params = new String[1];
        String st="";
        while(!((ch >= 'a')&&(ch <= 'z'))){ch=str.charAt(j++);}
        if((ch >= 'a')&&(ch <= 'z')){
            while((ch >= 'a')&&(ch <= 'z')||(ch >= '0')&&(ch <= '9')){
                st=st+ch;ch=str.charAt(j++);}
        }
        int ind2 = procNameTable.add(st);                          // place array name in procNameTable
        while(!((ch >= '0')&&(ch <= '9'))){ ch=str.charAt(j++);}
        int num =0;                                                // get array dimension
        while((ch >= '0')&&(ch <= '9')){
            num=num*10+ (int)ch-(int)'0';
            ch=str.charAt(j++);
        }
        procNameTable.procAddress2[ind2] = num;
        procNameTable.procAddress1[ind2]= num;
        procNameTable.val[ind2]= num;                              // array dimension
    }
    if(strr.equals("load")) {
        String [] params = new String[2]; String st="";
        while(!((ch >= 'a')&&(ch <= 'z'))){ch=str.charAt(j++);}
        if((ch >= 'a')&&(ch <= 'z')){
            while((ch >= 'a')&&(ch <= 'z')||(ch >= '0')&&(ch <= '9')){
                st=st+ch; ch=str.charAt(j++);}
        }
        int ind2 = procNameTable.add(st);         // locate array name in procNameTable
        while(!((ch >= 'a')&&(ch <= 'z'))){ ch=str.charAt(j++);}
        int count=0;                                               // get parameters
        while((count<2)&&(ch != ')')&&(ch != ';')){
            String stt1 = "";
            if((ch >= 'a')&&(ch <= 'z')){
                while((ch >= 'a')&&(ch <= 'z')||(ch >= '0')&&(ch <= '9')){
                    stt1=stt1+ch;   ch=str.charAt(j++);}
            }
            params[count++]= stt1;                                 //parameter name
            while((ch == ' ')||(ch == ',')){ ch=str.charAt(j++);};
        }
        temp[k++]= "    lda "+ '#'+ getHex(ind2+10,3) +'#'+ ";\n";  // array address
        temp[k++]= "    add "+ params[1]+ ";\n";                    // add array index
        temp[k++]= "    sta "+ " #FFF# "+ ";\n";                    // store in alt IRbase register
        temp[k++]= "    lvl "+ "#2#"+ ";\n";                        // switch IR registers
        temp[k++]= "    lds "+ "#0#"+ ";\n";                        // load value from array
        temp[k++]= "    lvl "+ "#2#"+ ";\n";                        // switch IRbase registers back
        temp[k++]= "    str "+ params[0]+ ";\n";                    // store value from array
}
```

```
if(strr.equals("store")) {
    String [] params = new String[2]; String st="";
    while(!((ch >= 'a')&&(ch <= 'z'))){ch=str.charAt(j++);}
    if((ch >= 'a')&&(ch <= 'z')){
        while((ch >= 'a')&&(ch <= 'z')||(ch >= '0')&&(ch <= '9')){
            st=st+ch;ch=str.charAt(j++);}                           // name of array
    }
    int ind2 = procNameTable.add(st);                  // locate array name in procNameTable
    while(!((ch >= 'a')&&(ch <= 'z'))){ ch=str.charAt(j++);}
    int count=0;                                                     // get parameters
    while((count<2)&&(ch != ')')&&(ch != ';')){
        String stt1 = "";                                            //parameter name
        if((ch >= 'a')&&(ch <= 'z')){
            while((ch >= 'a')&&(ch <= 'z')||(ch >= '0')&&(ch <= '9')){
                stt1=stt1+ch; ch=str.charAt(j++);}
        }
        params[count++] = stt1;
        while((ch == ' ')||(ch == ',')){ ch=str.charAt(j++);};
    }
    temp[k++]= "    lda "+ '#'+ getHex(ind2+10,3) +'#'+ ";\n";      //array address
    temp[k++]= "    add "+ params[1]+ ";\n";                       // add index
    temp[k++]= "    sta "+ " #FFF# "+ ";\n";                        // store in alt IRbase
    temp[k++]= "    lds "+ params[0]+ ";\n";                        // load value
    temp[k++]= "    lvl "+ "#2#"+ ";\n";                            // switch IRbase
    temp[k++]= "    str "+ " #0#" + ";\n";                          // store value in array
    temp[k++]= "    lvl "+ "#2#"+ ";\n";                            // switch IRbase registers back
}
```

The address for the memory space ear-marked for the array can still be located in a slot at the head of the program in the same way that the addresses for procedures are stored. This makes it easy for all other references to the array in a program to identify its position in memory in a similar way that procedure calls locate their procedure code, through one standardised address location.

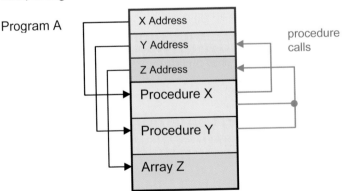

Figure 8.18 Machine language program structure

The structure of the program that has evolved in this exploration is interesting and worth discussing in the context of further high-level language extensions before returning to the use of arrays for graphics algorithms. Although in this presentation arrays have been implemented using macro expansions, the idea of implementing arrays as function calls introduces an important extension to the existing system, which is capable of providing the support needed for an object-oriented language like Java.

An array is a data-type. If a data-type is defined by the operations that it permits to be carry out on a collection of "raw" data elements, then grouping the two accessing functions and the array data-space, as a unit, gives a standardised framework for any object classified as an array. Duplicating a parameterised copy of this structure can be used to set up a new array as an "object" of type *array*. However, because procedures have been coded using base displacement addressing and program space is distinguished from data space, it means that all objects of the same type can share a single copy of the code for the procedures that define the permitted operations on the data. In Java this code is provided in a class or interface definition.

This basic definition, including both procedures and data-structures, can be treated as a template for all objects of the same data type. For the array this means new array objects only need to contain the address slots referencing the accessing procedures and the array location, along with the space for the new array, rather than the whole package. Although this would not save much memory space in the case of arrays, the framework it provides is a general one that can be used to implement a much more complex data-type and provide multiple instances of objects of that type.

A collection of objects of the same type form a set of entities which can be called by the same set or type name. Also each object in this set can be distinguished by an individual variable name having the type defined by the class or template definition of the object in question. To complete such a system, the type or class definition must include a generating function or "constructor" to create new instances of objects when they are required in other programs.

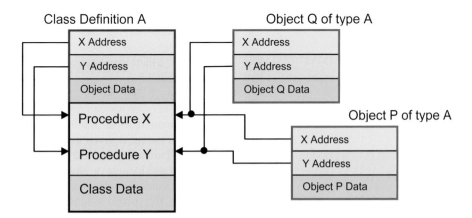

Figure 8.19 Class and object structures

This indirect reference to objects and data elements allows dynamic data structures such as linked-lists and trees to be built and demolished as required during the execution of an algorithm. The layout of program code and data space in memory can be set up in the way shown in Figure 8.20 to allow the stack space used for procedure calls and the growth and decline of dynamic data structures to occur in the most flexible way possible.

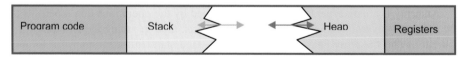

Figure 8.20 Flexible arrangment of memory space for dynamic structures

New data objects can be created by the class constructors and placed in the memory space labelled Heap in Figure 8.20. As dynamic objects are dispensed with, gaps can be created in this area of memory so there is a management task needed to ensure that memory space is not wasted. In general terms this arrangement allows the maximum expansion of stack and heap space that the available resources allow.

```
MINIJC PROGRAM                    _ □ ×
void main(){
  int [3][3] a;
  int i;
  int j;
  int b;
  int c;
  b= read();
  c= read();
  j=0;
  while(j<3){
    i=0;
    while(i<3){
      if(i==j){
        a[i][j] = b;
      }
      else{
        a[i][j] = c;
      }
      write(a[i][j]);
      i=i+1;
    }
    j=j+1;
  }
}
```

Figure 8.21 Entering the diagonal values
in a three by three array

Arrays are an important data structure for handling repetitive tasks as the previous chapters have already indicated.

Matrices and vectors are mathematical objects that can be implemented by arrays, but which need to be associated with their own set of procedures to model the behaviour of the mathematical entity that they represent.

Although the simple array is important for many graphics operations. An image as a pixel array requires many more operations than simple element-by-element access. A new object type is required which provides other array-based procedures.

Figure 8.21 shows the miniJC program to enter a set of values into the main diagonal of a small array. The simulator output is given in Figure 8.22. If the hardware of the system is extended to include a display system then this kind of program can be used to draw in the pixel values needed to create a picture.

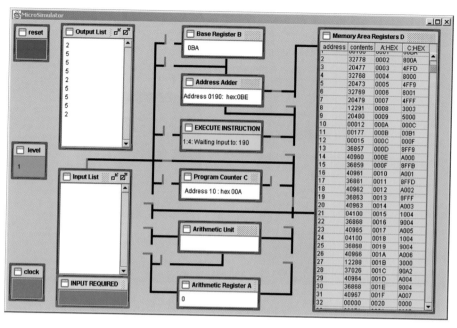

Figure 8.22 Entering values into the main diagonal of a three by three array

As the programs get longer the simulator system showing each step becomes too slow for practical purposes, even though it is still very useful for step by step debugging. The simulator with a cut down graphics user interface and with a simulated display unit added supports the final steps in this exploration linking language to graphics by running the programs but only showing the input data and the output data, along with the display screen contents.

```
public static Input jInput = null;
public static Output jOutput0 = null, jInputList = null, jOutputList = null;
public static DisplayWindow display = null;
public static CoordinateFrame cf=null;
public static Color[] colour = new Color[ ]
                    {Color.red,Color.blue,Color.green,Color.black};
public Simulator(int width,int height, Output Out){
    jOutput0=Out;
    jInput = new Input(110,501,150,101,"Enter Input");
    jInputList = new   Output(110,291,150,215,"Input List");
    jOutputList = new   Output(110,60,150,231,"Output List");
    display=new DisplayWindow(null,260, 60,518,542,Color.gray);
    Point p1= new Point(1,0,0); Point p2= new Point(1,512,512);
    Point p4= new Point(1,512,0); Point p3= new Point(1,0,512);
    display.plotRectangle(p1,p2,Color.white);
    cf= new CoordinateFrame();  cf. setScales(p1,p2,p3,p4);
}
```

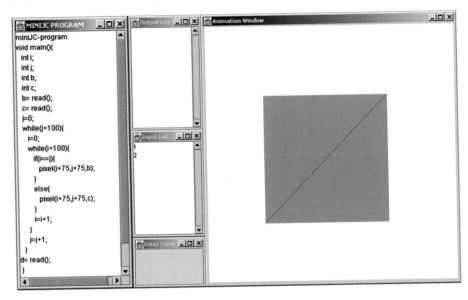

Figure 8.23 Linking a display unit into the micro-simulator

Figure 8.23 shows the result of changing the value in a display array from green to blue where the indexes are the same rather than printing out the array in the way shown in Figure 8.22. Figure 8.24 shows the program, extended to display a circular blue disk against a green background. The values entered for b and c are the indexes to a colour table contained in the display processor simulator.

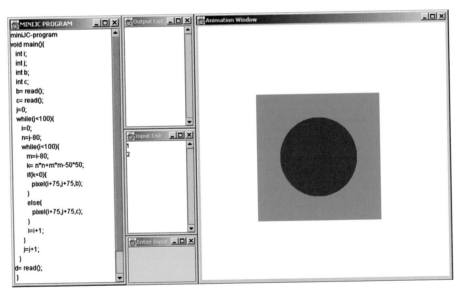

Figure 8.24 Linking a display unit into the micro-simulator

The size of the simulator's memory limited its use as a display array. When vector graphics systems were developed all that needed to be transferred to the display hardware was a list of coordinate points giving the ends of line segments. However even for a small TV raster based display a much larger block of memory is required. Two approaches to this problem evolved. The first was to extend the memory in the main computer system and allow the display system dual access to the display array in memory. The second was to output each pixel display value and its array index-positions to an independent display system with its own memory space, dedicated to a raster display system. The second approach is the simplest to implement without a major change to the existing simulator, and the language processors driving it.

A new output procedure *"pixel" '('<x>', '<y>', '<colour index>')'* is defined for the miniJC language, and a corresponding new macro: *pixel(x,y,c)* is included in the assemby language to support a display unit within the simulator. This macro uses an extension to the *lvl* command to allow three values to be passed to the display sub-system simulated using a *DisplayWindow* object described in previous chapters. The third value "c" the index of the pixel colour defined in a colour array in the display sub-system is loaded into the accumulator the second value is the y or row-value for the display array and is stored in a variable operated on by the *lvl* command. This passes this value to the display device and modifies the computer's state so the following store command not only provides the x or column value but also passes the value held in the accumulator to the display device. The extension to the miniJC compiler involves adding to the function call procedure.

```
static void functioncall(){                                          //functioncall
    int count =0,savestack=0;String strpar="";String str ="";
    switch (record[Index]){
        case 20: Index++; str = Name();
            savestack=stackIndex;
            Index++; strpar = parameterlist(strpar);
            if(str.equals("read")){
                code[codeIndex++] = "   "+"rds s"+(stackIndex)+";\n";
                stackIndex++;
            }else if(str.equals("write")){
                code[codeIndex++] = "   "+"wrt "+strpar+";\n";
                stackIndex=savestack;
            }else if(str.equals("pixel")){
                code[codeIndex++]= "pixel "+ '('+ strpar +");\n";
                stackIndex=savestack;
            }else {
                code[codeIndex++]= "call "+ str + '('+ strpar +");\n";
                code[codeIndex++] = "   "+"str s"+(stackIndex)+";\n";
                stackIndex++;
            }
            break;
        default: jOutput.writeString("error functioncall index = "+Index+"\n");
    }return;
}
```

The change to the assembler requires the macro expansion program to be extended.

```
else if(strr.equals("pixel")) {
    String [] params = new String[3]; int count=0;
    while((count<3)&&(ch != ')')&&(ch != ';')){ String stt1 = "";
        if((ch >= 'a')&&(ch <= 'z'))                        //parameter names
            while((ch >= 'a')&&(ch <= 'z')||(ch >= '0')&&(ch <= '9')){
                stt1=stt1+ch; ch=str.charAt(j++); }
            params[count] = stt1;  count++;
            while((ch == ' ')||(ch == ',')){ ch=str.charAt(j++);};
    }
    temp[k++]= "   lds "+ params[2]  + ";\n";        // pixel index colour
    temp[k++]= "   lvl "+ params[1]+ ";\n";          // pixel y value
    temp[k++]= "   str "+ params[0] + ";\n";         // pixel x value
}
```

The *lvl* and *str* commands in the hardware simulator also have to be modified.

```
case 3:                                                           // lvl
    if(addr>3){ pixel=true; rowy=readMemory(relA);
    }else{ pixel=false; /* cases addr = 0..3*/;;;; } break;
case 10:                                                          // str
    if(pixel){
        Point p1= new Point(2); p1.n[1]= readMemory(relA); p1.n[2]=rowy;
        display.plotPoint(p1,cf,colour[registerA]);
        pixel=false;
    }else  writeMemory(relA,registerA,true);
    break;
```

A similar extension can be set up to access values stored in a display memory and combining them with new values from other images. This arrangement however depends on processing locations in the display array, one by one. Where a large number of pixels need to have the same operation carried out the overhead in accessing them in this way can become large. Operations on rectangular areas of the display space have been implemented at the assembler and machine level of display systems (called *bitblt* operations) to speed up interactive image manipulation. These operations allow new pixel values to be combined with existing pixel values in a variety of ways. A very useful operation is provided by the exclusive-or combination of values. Applying this operation once changes the displayed pixels, applying it again returns the pixels to their original values. This is particularly useful in interactive displays. The mouse allows a point position to be defined as a coordinate. This location can then have a pointer symbol exclusive-ored with this location in the display array. On moving the mouse, the original pointer is erased by a second application of the operation, and the new location marked by applying the pointer to the new mouse coordinate position. The last three chapters have illustrated the way a high-level computer language can be used to to create and interact with displays. The operations that can be built on this foundation form the subject of the rest of the book.

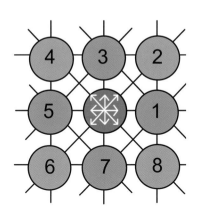

9

Primitive, Raster-Infill Operations: Line Interpolation

Introduction

The most exciting advance that resulted from including line or vector generators in a display processor was that the consequence of moving or changing a vertex point in a display file automatically moved all the lines linked to it in the display. Where this operation could be done interactively with a mouse or pointer device, it created a new form of editing allowing points to be *dragged* from one place to another, and allowing the lines linked to the points to *rubber band* into their new locations. Interactively working with a set of lines in this way was a major improvement over existing techniques, and the natural desire to drag objects around the screen in a similar way for editing or correcting a display, led to the next evolutionary step in the development of hardware display primitives.

Representing an object by a set of vertex points, reduced the data required to represent a display scene, and allowed more computation to be undertaken in each refresh, display-regenerating cycle. However, this was still a relatively low-level form of modelling. Duplicate objects in different positions in the screen still needed different line end point co-ordinate sets.

The introduction of absolute and relative co-ordinates meant that only a single instance of an object model needed to be held in store if held as a set of relative co-ordinates. These models could be copied or dragged to a new location, whenever new objects or symbols of the same type were required. By adding the *relative* co-ordinates representing the object, to *absolute* co-ordinates representing true locations in the display space, meant that multiple copies of the object could be placed all over an image at a relatively small extra computational cost in the overall display process. This calculation was akin to *base-displacement* addressing used to access blocks of storage in a computer's memory

A. Thomas, *Integrated Graphic and Computer Modelling*,
DOI: 10.1007/978-1-84800-179-4_9, © Springer-Verlag London Limited 2008

The advent of the frame store and TV monitor based display systems provided a different framework in which to generate lines. A line had to be defined using a set of points located on a regular rectangular grid or raster. Although start and end points of line segments could easily be placed on the points of a grid it is clear that most of the intermediate grid points would not lie on the line exactly and their positions would have to be approximations. Using the standard equation for a line $y = m.x + c$ gives the methods *Line(..)* and *line(..)*. For slopes from $0°$ to $45°$, the values of y are calculated for grid unit steps in x, and then rounded to the nearest grid value for y. For $45°$ to $90°$ the same is done for unit steps in y, rounding the matching x value to the nearest grid line. The output is shown in Figure 9.1.

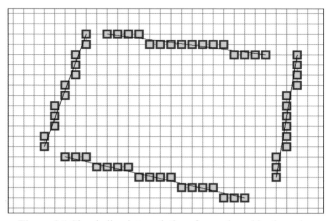

Figure 9.1 Simple line interpolation: forwards and backwards

```
public void Line(Point p1, Point p2, int wd, Color cc){
    int dx = p2.xi()-p1.xi();   int dy = p2.yi()-p1.yi();
    if (Math.abs(dx) > Math.abs(dy))
        line(p1.xi(), p1.yi(), p2.xi(), p2.yi(), wd, dx, dy, false, cc);
    else line(p1.yi(), p1.xi(), p2.yi(), p2.xi(), wd, dy, dx, true, cc);
}
private void line(int x, int y, int xend, int yend, int width,
                                    int dx, int dy, boolean steep, Color cc){
    double m=1,c=0;int kx =1;
    if(dx<0)kx= -1; else if (dx ==0) kx = 0;
    if (dx != 0){m = (double)dy/(double)dx;   c = (double)y - m*(double)x;}
    for(;;){
        if (steep) paintInnerCell(y,x,width,cc);else paintInnerCell(x,y,width,cc);
        if (x == xend) return;
        x = x+ kx;   y = (int)(m*((double)x)+c+0.5);
    }
}
```

Line interpolation needs to be as simple and as fast as possible. An improvement in speed can be obtained by removing the multiplication needed, each time, to calculate the new y value from the incremented x value. This is done by calculating the standard step size in y needed for each unit step in x. These increments are then added on to the previous co-ordinate values to give the new points. Representing a line by the steps in x and y needed to define its position on a grid produced the "Digital Differential Analyser" DDA algorithm. The output of which is shown in Figure 9.2.

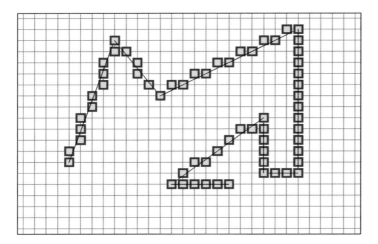

Figure 9.2 DDA line interpolation: forwards and backwards

```
public void ddaLine(Point p1,Point p2,int wd,Color cc){
    int dx = p2.xi()-p1.xi();int dy = p2.yi()-p1.yi();
    if(Math.abs(dx)>Math.abs(dy))
        dda(p1.xi(),p1.yi(),p2.xi(),p2.yi(),wd,dx,dy,false,cc);
    else dda(p1.yi(),p1.xi(),p2.yi(),p2.xi(),wd,dy,dx,true,cc);
}
private void dda(int x,int y,int xend,int yend,int width,
                                        int dx,int dy,boolean steep, Color cc){
    double m = 1, c = 0, yy = (double)y; int kx =1;
    if(dx!=0)m = (double)dy/(double)dx; //c = (double)y - m*(double)x;}
    if(dx<0){kx= -1; m = -m;} else if (dx ==0)kx = 0;
    for(;;){
        if(steep) paintInnerCell(y,x,width,cc);else paintInnerCell(x,y,width,cc);
        if (x == xend) return;
        x = x+kx;  yy = yy + m; y = (int)(yy + 0.5);
    }
}
```

This simple incrementing scheme can also be implemented by specialised hardware. Using a fixed-point representation of fractional numbers gives the circuit shown in Figure 9.3. The display part of the y value is only incremented by carries from its fractional unit. This is built from an adding circuit, which adds on the fractional increment to the y value for each unit step taken in x. The rounding operation is implicitly carried out by initialising the fractional part of the y value for the first point to 0.5, in other words, 0.10 binary.

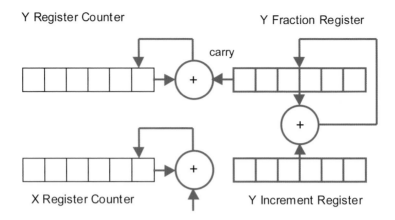

One unit step in X and one Y increment: each clock cycle

Figure 9.3 A schematic DDA line interpolation circuit

Line Following Algorithms on a Raster Grid

An alternative approach to the DDA algorithm, which avoided the use of fractional numbers, was first analysed by J. Bresenham for line plotters in the early sixties. This was based on the idea that a line could be represented by the point set obtained by walking from one point to a neighbouring point on a grid by keeping as close to the real line as possible. The task was to select, which was the best neighbouring point to move to, from the current point, in order to represent a given line most accurately. There are two possible schemes, which can be developed here depending on what the definition of neighbouring points on a grid is taken to be. The simplest arrangements can have either four or eight neighbours in a rectangular grid in the manner shown in Figure 9.4.

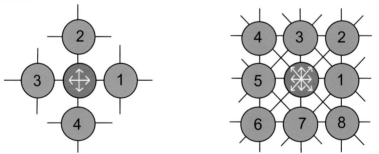

Figure 9.4 Four and eight way neighbours to a point in a grid

Four Point Neighbours

Starting with the simpler of these two possibilities, where each point is considered to have four neighbours. Each neighbouring position to the East, North, West and South of the given point, can be located by adding *or* subtracting one unit to its x *or* y *(but not both)* grid co-ordinate value for each step taken. Horizontal and vertical lines can be set up as rows and columns of grid points. Inclined lines have to be approximated by taking points in a staircase pattern in the form shown in Figure 9.5. Each step in this pattern will move along the line, either in the x or the y direction in the way that keeps the approximate line as near to the real line as possible.

One implementation of this idea can be developed in the following way. Given a line defined by its two end points (x1, y1), and (x2, y2) the slope of the line will be DY/DX where DY = y2-y1, and DX = x2-x1, and the standard equation of a line is:

$$y = m.x + c$$

Substituting m = DY/DX into the equation and multiplying out gives:

$$0 = DY.x - DX.y + C$$

From this a distance d from any point (x_a, y_a) to this line can be calculated by:

$$d = DY.x_a - DX.y_a + C$$

When a point is on the line then d = 0, when the point is above the line then d < 0 and when the point is below the line then d > 0. If the initial point is on the line and is also on a regularly spaced grid with unit intervals between rows and columns, then a step in the y direction to (x, y+1) will be at a new distance *d* from the line where:

$$d = DY.x - DX.(y+1) + C$$

However, the initial distance is:

$$d = DY.x - DX.y + C$$

The change can be obtained by subtracting: $d - d = -DX$

If this element (*d* - d) is renamed: Δd_y then:

$$\Delta d_y = -DX,$$

Similarly a move in the x direction to (x+1, y) will produce a change in d of

$$\Delta d_x = DY.$$

These distance measures and changes are particularly convenient because DX and DY as they are defined above are integer numbers. This arrangement avoids the use of fractional numbers needed to represent the slope of lines in the DDA approach. Consider the line in Figure 9.5, drawn through the origin from point (-7, -4) to (7, 4). At the start of the line at the point (-7, -4): d = 0, DX = 14 and DY = 8.

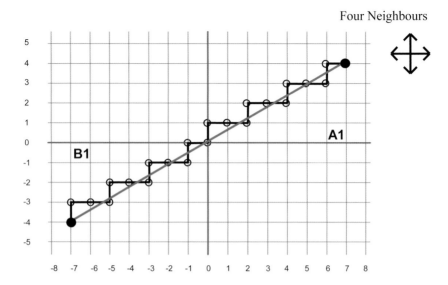

Figure 9.5 Interpolating a line from (-7, -4) to (7, 4)

From this starting point there are two options given the orientation of the line, either a step in the x direction or a step in the y direction by one unit. If a step is taken in the y direction to point (-7,3), this will give a d value of -DX, in other words the value of d becomes -14. This sign indicates a position above the line, so further steps taken in this direction will move further away from the true line.

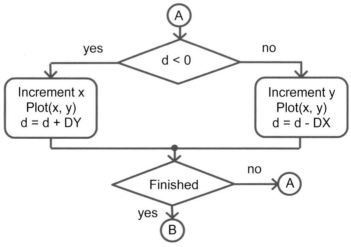

Figure 9.6 The basic line following algorithm.

To move forward, yet back towards the true line requires the next move to be taken in the x direction. A step in the x direction will involve a change in d of +8, so d becomes -6 at the point (-6,3). This is still above the line indicated by the negative value, so a further step in the same direction needs to be taken. This gives a d value of 2 at the point (-5,3). This is now below the line and increments must switch back to the y direction. A y step will immediately return to the other side of the line with the value of d changing from 2 to -12. Further steps in the x direction are then needed, until the sign of d changes again. The full sequence of points that this process generates between the point (-7, -4) and the point (7,4) is shown in Figure 9.5.

This algorithm can be expressed by the flow chart shown in Figure 9.6. A similar effect is obtained if, at the starting point, where d = 0, the initial step is taken in the x direction. In this case d becomes +8 and a move in the y direction will take place on the second step. To implement this change the increment selection test would become d<= 0 rather than d < 0 in Figure 9.6.

If this algorithm is applied to the line defined in the opposite direction so that (x1, y1) is (7, 4), and (x2, y2) is (-7, -4) then the following sequence is produced. Initially d = 0. Following the scheme in Figure 9.5 this requires an increment in the y direction, so d becomes d - DX, but DX in this case is -14 since it is the difference x2 - x1, so d becomes 14. However, this value leads to another x increment, which increases the size of d moving further from the true line, rather than nearer to it. Clearly a variation in the basic algorithm is required to correctly handle the different sign relationships that exist between DX and DY for all the line drawing directions which will be encountered.

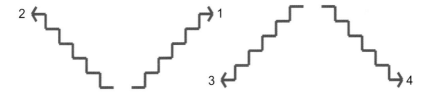

Figure 9.7 The four main line-incrementing patterns

At first sight there appear to be four possible cases, which must be considered, depending on the orientation of the line. These are shown in Figure 9.7. However if the values of the test variables that control the selection of the different incrementing patterns are laid out in a table: more possibilities emerge. A decision table based on the sign and the values that DX, DY and d can take, gives the list of the different conditions and associated actions that need to be considered.

Decision Table

								Conditions																			
DX	-ve								0									+ve									
DY	-ve			0			+ve			-ve			0			+ve			-ve			0			+ve		
d	-	0	+	-	0	+	-	0	+	-	0	+	-	0	+	-	0	+	-	0	+	-	0	+	-	0	+
									Actions																		
	C			H			B			V			P			V			D			H			A		

The four initial cases appear in this table labelled A, B, C and D. Along with these there are the vertical line and the horizontal line cases labelled V and H respectively. There is also the degenerate case where x1 = x2 and y1 = y2, labelled P. In this case the algorithm can be set up to plot a single point. From this analysis a procedure can be written with nine parts. However, if the structure of this algorithm is examined a little further some interesting properties emerge which can be used to considerably reduce the code needed to implement it.

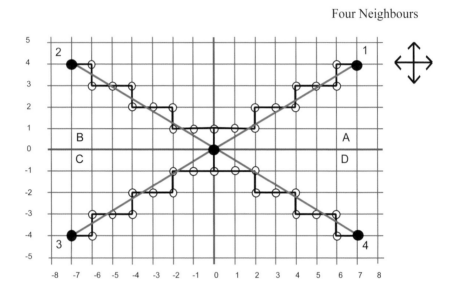

Figure 9.8 Interpolating four lines (0,0) to (7,4) to (-7,4) to (-7,-4) and to (7,-4)

Returning to the four general schemes in Figure 9.7, if the initial point of each line (x1, y1) is taken as the origin point (0, 0), these four cases correspond to lines drawn from the origin into the four different axes quadrants of the x-y plane. The signs of the values DX and DY correspond to the signs of the co-ordinates of the second point (x2, y2) in each case, as shown in Figure 9.8. This approach allows the line shown in Figure 9.5 to be two lines drawn in opposite directions from the origin, indicated by the labels B1 and A1 for the two quadrants.

The target is to find if there is a way in which these different cases can be handled by a single procedure. One approach to this task is shown in Figure 9.8. If the absolute unsigned value of the DX and DY are used to define the value of d and consequently to determine the stepping sequence, then using the correctly signed, incrementing directions for the x and y steps, will give the line patterns shown for the four quadrants. In effect the pattern of points generated in the first quadrant is reflected about the axis-lines into the three neighbouring quadrants. However, this way of amalgamating the different cases produces an obvious problem. It is necessary for consistency, that identical lines drawn in opposite directions be approximated by the same set of points. Comparing Figures 9.5 and 9.8 shows this is not the case. The line in Figure 9.5 from (-7, -4) to (7,4) does not match the two lines (0, 0) to (-7, -4) and (0, 0) to (7, 4) in Figure 9.8.

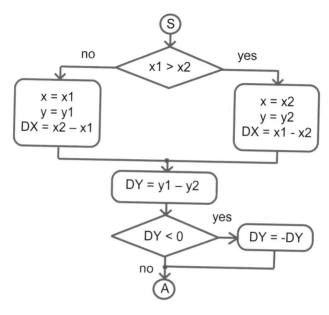

Figure 9.9 Initialising a line following algorithm for four quadrants

Matching these line patterns can be achieved by always drawing lines that should be identical, in the same direction, for example from the lesser x value to the greater x value, where necessary by reversing the order of the end point co-ordinates used to define the lines. This approach makes cases A and C the same, and cases B and D the same in the decision table. This reduces the sloping lines to two cases, which can be reduced to one case by using the unsigned value of DY when calculating changes in the value of d moving in the x direction along the normalised line, but keeping the correctly signed incrementing steps.

This "reflects" the lines in the second and fourth quadrants, (cases B and D), to the first and third quadrants, (cases A and C). To modify the algorithm shown in Figure 9.6 to work in this way requires the initialisation shown in Figure 9.9, *increment x* still means add one to the current value of x, as before. In contrast *increment y* now means add one or subtract one depending on the sign of DY.

This approach gives consistent results, but sets of points representing line segments, for example, from a polygon boundary will not be generated naturally as a continuous sequence round the polygon, which was necessary for the original application on line plotters. Such behaviour, if it is required, must be obtained by following a different strategy, which is to design a backward stepping algorithm that can create the same overall point sequence as the corresponding forward stepping algorithm.

Four Neighbours

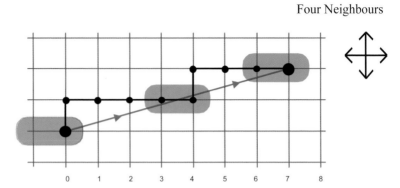

Figure 9.10 Forward stepping algorithm

The problem, which this poses, appears when lines of shallow slope are being processed. The diagram in Figure 9.10 shows the sequence, which the initial basic algorithm will generate, in the first quadrant. Most of the points are above the true line. If the same approach is adopted to generate the reverse sequence, *"by moving towards the true line until it is crossed then changing direction and repeating the process",* the first step may well move to a point corresponding to the original sequence above the true line. However, it is clear that the second step, will move across the line and this will not match the previous stepping pattern. The sequence will be of the form shown in Figure 9.11.

To get the same sequence from the backward stepping algorithm, it is necessary to test the distance value d, at the same set of positions used in the forward stepping algorithm. This is possible to achieve in the backward stepping algorithm, by *looking ahead* to the point and the test which in the forward stepping algorithm would have selected the current position in the backward stepping sequence. The critical points, which illustrate this relationship, are where a horizontal step in the interpolated line crosses the true line. The change in direction is at one end of this line segment in the forward stepping algorithm and is at the other end in the backward stepping case.

Four Neighbours

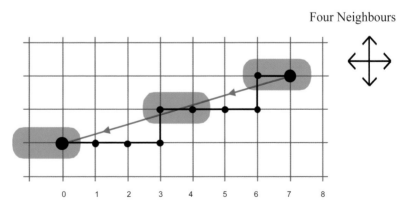

Figure 9.11 Backward stepping algorithm

These segments are high lighted in Figures 9.10 and 9.11. If a test is carried out at the other end of this critical segment in the backward algorithm it will match the forward stepping algorithm exactly.

In Figure 9.13 the algorithm for the third quadrant is given. The testing sequence, which this carries out, is illustrated in Figure 9.12. The offset value of d is calculated at the *test position* indicated by the arrow for each *current line position*. With this value, the same strategy can be applied as before: as soon as the true line is crossed, (indicated by a change of sign in the offset distance value), the incrementing direction is changed.

Four Neighbours

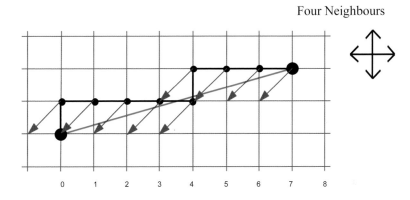

Figure 9.12 Backward stepping interpolation algorithm with look ahead

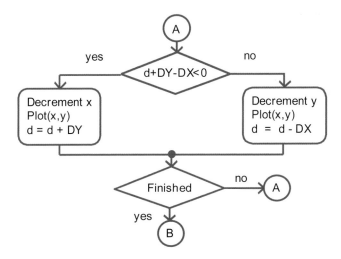

Figure 9.13 Backward stepping algorithm with look ahead test

However, a more elegant alternative is to initialise the distance value d for the backward stepping algorithm by this offset value before the line generation process is started. This can be done because the same offset step applies to all points wherever the test is being carried out. The result is the simpler selection test (d < 0) can then be applied in both forward stepping and backward stepping cases to trigger the change in stepping direction. The only exception to this has to be the condition where (d = 0), in this case: in one direction the stepping direction has to be changed in the other it must not.

If the forward stepping algorithm is labelled A1 and the backward stepping algorithm B1 then all four quadrants can be covered by reflecting the pattern of points obtained in quadrants one and three into four and two respectively, or alternatively across the other axis. The first of these alternatives is illustrated in Figure 9.14.

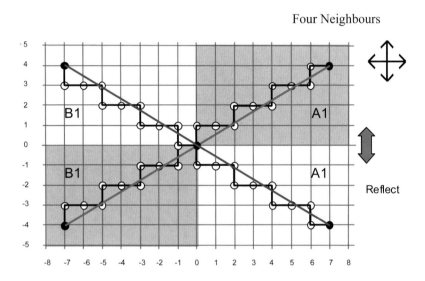

Figure 9.14 Reflecting the line increment pattern about the x- axis

If the basic selection algorithm and control variables are laid out in a table for the four quadrants, the way that such a reflection operation can be implemented is summarised reasonably clearly in the following table. Setting out these relationships in a table makes it possible to see what is fixed and what is variable. In this case it can be seen that the incrementing patterns for quadrants 1 and 2 are opposites as are 3 and 4. The diagram in Figure 9.14 indicates that the pattern in quadrant 2 could be a reflection of the line incrementing scheme produced in quadrant 3, and similarly, quadrant 1 could be reflected from quadrant 4, or vice versa. This can be implemented by treating the (dx < 0) cases as the same, but using negated **y** increments, and the two (dx > 0) cases as the same but again with negated **y** increments. For identical forward and backward drawing, the diagonal (d = 0) tests must still select opposite stepping actions, although which way round this is done appears to be arbitrary.

The simplest way to show how these relationships can be turned into working code so that all four quadrants can be handled by one procedure is to divide the task into two stages.

Quadrant 2 $+ve$	Quadrant 1 $+ve$
dx < 0 dy > 0 d = -dx - dy (Initialised value)	dx > 0 dy > 0 d = 0 (Initialised value)
d < 0: - ve : step in Y direction: d more - d = 0: zero: step in X direction: d more + d > 0: + ve: step in X direction: d more + stepX: d = d - dy, stepY: d = d - dx	d < 0: - ve : step in X direction: d more - d = 0: zero: step in Y direction: d more + d > 0: + ve: step in Y direction: d more + stepX: d = d + dy, stepY d = d - dx
Quadrant 3 $-ve$	Quadrant 4 $+ve$
dx < 0 dy < 0 d = dx - dy (Initialised value)	dx > 0 dy < 0 d = 0 (Initialised value)
d < 0: - ve : step in X direction: d more - d = 0: zero: step in X direction: d more - d > 0: + ve: step in Y direction: d more + stepX: d = d - dy , stepY: d = d + dx	d < 0: - ve : step in Y direction: d more - d = 0: zero: step in Y direction: d more - d > 0: + ve: step in X direction: d more + stepX: d = d + dy , stepY: d = d + dx

The first is to use the initial conditions and tests to generate the required sequence of X or Y steps, for example by defining a variable j so that a statement of the form:

```
if (j < 0) stepY; else stepX;
```

can be written The second is to implement *stepX* and *stepY* to increment the x and y values in the correct way for the quadrant in question. This can be done by setting up two incrementing variables kx and ky so these steps can be executed simply by $x = x + kx;$ and $y = y + ky$. These variables are initialised to +1 or -1 to correspond with the original signs of dx and dy.

To generate the X and Y stepping sequence for quadrant 1 requires d to be initialised to 0, and increments to d to be defined for X steps as $d = d + dy$, and for Y steps as $d = d - dx$. When $(d < 0)$ an X step is made otherwise a Y step is made. An identical stepping sequence to the one this will generate, however, is also wanted for quadrant 4. The simplest way to obtain such a sequence for quadrant 4 is to modify its initial conditions to correspond to working in quadrant 1. This can be done by executing the statement:

```
if (dx > 0) && (dy < 0) dy = -dy;
```

The same approach could be adopted to map quadrant 3 into quadrant 2, by reversing the sign of dy. However, the sequence for quadrant 2 can also be obtained directly from quadrant 1. If the way that the step selection *depends* on the value of d is laid out in a table, given below, it can be seen that the step selection of quadrant 2 is the opposite of that in quadrant 1.

Quadrant	dy	dx	d: -ve	d: zero	d: +ve
1	+ve	+ve	step X	step Y	step Y
2	+ve	-ve	step Y	step X	step X
3	-ve	+ve	step X	step X	step Y
4	-ve	-ve	step Y	step Y	step X

This means, if the corresponding offset-value to the value of d, required to give the correct stepping sequence in quadrant 2, is set up for quadrant 1. A sequence of X and Y steps will be generated, but with X steps for Y, and Y for X, generated by the quadrant 1 tests on d. All that is needed is to reverse these step directions appropriately, and the same pattern will then be suitable for quadrant 3 and quadrant 2. The initial offset value for d in the first quadrant is given by $d = dy - dx$. To get this and the correct increments to d requires $dx = -dx$ to map from quadrant 2 to 1, and $dx = -dx$ and $dy = -dy$ to map from quadrant 3. Combining these requirements with those required for mapping quadrant 4 to 1, can be satisfied by the following code sequence:

```
if (dy < 0){ky = -1; dy = -dy;}
if (dx < 0){kx = -1;  dy = -dy;  d = dy-dx;}
```

The switching between X and Y steps depends on the value of d, and the original sign of dx. This is retained in the sign of kx:

	$(dx >= 0)$: $(kx = 1)$	$(dx < 0)$: $(kx = -1)$
$(d < 0)$	stepX: $j = 1$	stepY: $j = -1$
$(d >= 0)$	stepY: $j = -1$	stepX: $j = 1$

This selection operation can be implemented by the code:

```
if(d < 0) j = kx; else j = -kx;
if(j < 0){d = d - dx;  y = y + ky;}
else {d = d+ dy;  x = x + kx;}
```

This gives a program, which will interpolate in all quadrants, of the form:

```
int Cls = 60, Rws = 40;
Tiles T = new Tiles(Cls,Rws,Color.white);
DisplayGrid d = new DisplayGrid(f,T.tileColour,T.cols,T.rows);
d.paintGridArray(); d.drawGridLines(Color.black,Color.gray);
f.writeString("Please enter number of line segments: ");
int number = f.readInteger();

for(int i = 0; i <number; i++){
    f.writeString("Use the mouse to enter the line vertices"); f.writeLine();
    Point pa = d.getCell();   Point pb = d.getCell();
    d.line(pa, pb, 1, Color.blue);
}
```

```
public void line ( Point p1,Point p2,int border, Color cc){
    int  kx = 1,  ky = 1,  dx,  dy,   d = 0,  j = 0;
    dx =  p2.xi() - p1.xi();   dy =  p2.yi() - p1.yi();
    if (dy < 0){ky = -1; dy = -dy;}
    if (dx < 0){kx = -1;   dy = -dy; d = dy-dx;}
    for(;;){
        paintInnerCell(p1.xi(), p1.yi(), border, cc);             // plot point
        if (p1 ("==", p2))return;
        if(dy!=0){
            if(d < 0)j = kx; else j = -kx;
            if(j < 0){d = d - dx; p1.y("=", p1.yi() + ky);}
            else {d = d+dy; p1.x ("=", p1.xi() + kx);}
        }else p1.x("=", p1.xi() + kx);
    }
}
```

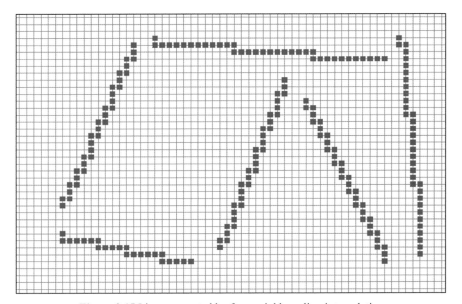

Figure 9.15 Lines generated by four-neighbour line-interpolation

The way this algorithm produces the same set of grid points interpolating from p1 to p2 as from p2 to p1 is illustrated in Figure 9.16 by setting up the line drawing procedure to give different sized "point" tiles, red for forwards and blue for backwards: the blue overwriting the red so they can both be seen,.

```
d.line(pa, pb, 1,Color.red);
d.line(pb, pa, 3,Color.blue)
```

In Figure 9.15 and 9.16 it can be seen that the vertical and horizontal lines appear lighter in weight than the oblique lines; and consequently, when using a higher resolution grid they look thinner than the diagonal lines.

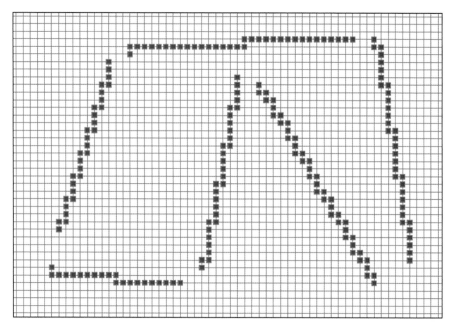

Figure 9.16 Four neighbour interpolation, forwards and backwards

Eight Point Neighbours

One approach to giving lines of different slope a more equal weight is to apply an eight neighbour interpolation scheme. Allowing movement to eight neighbouring positions means that diagonal grid-moves are permitted. This gives point sequences for lines matching those generated by the DDA interpolation algorithm, which in turn means that quadrants have to be divided into two to give an octant based framework

.

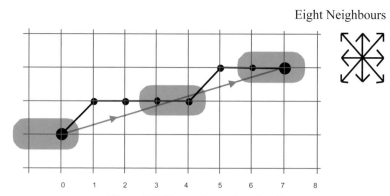

Figure 9.17 Forward stepping algorithm

An incrementing pattern for this method is illustrated in Figure 9.17 for a line in the first octant with a slope between $0°$ and $45°$ to the horizontal. In this case if the current point is above the line its distance to the line will be negative and a horizontal move towards the true line can be made using single steps in the X direction. However when the true line is crossed then a diagonal step is taken, stepping both X and Y together, to return to a position above the line. This diagonal step will only cross a line, if the line has a shallower slope than 45°. A line with a slope that is greater than 45° requires unit steps in the Y direction, when below the line, but a diagonal step as soon as the true line is crossed, to return to a position below it.

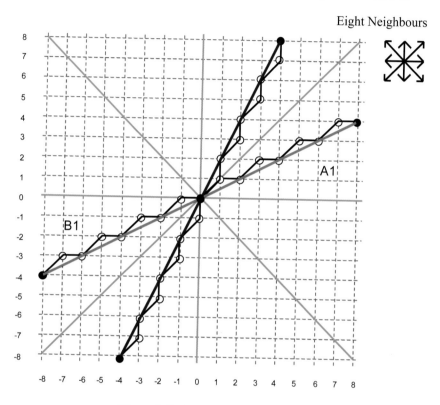

Figure 9.18 Eight neighbour interpolation

Figure 9.18 shows the way this allocates the line incrementing patterns to eight regions. The pattern for the A1 sequence is given in Figures 9.17, and Figure 9.19 shows the corresponding stepping sequence for the line drawn in the opposite direction, the B1 sequence.

As with the "four-neighbours" approach, using the same testing scheme to define the incrementing steps, in these two cases, does not give a matching set of points for the identical line drawn in opposite directions. In Figures 9.17 and 9.19 the critical line segments are highlighted for switching the direction of increments, and the same strategy that was applied before, can be applied again to this case. If an offset test is used for the second case: testing the point that, in the reverse direction would have led to the current point being chosen, then the same set of points whether moving forward or backwards can be selected. In the case of the *eight neighbours* scheme, however, this "offset" test has to diagonally cross two cells, in the way shown by the arrows in Figure 9.20.

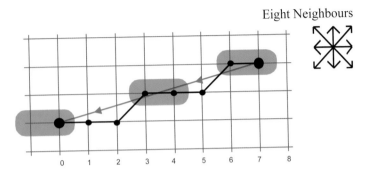

Figure 9.19 Backward stepping algorithm

The same strategy used in the "four neighbours" algorithm, reflecting incrementing patterns, can again be applied to this case, giving the stepping patterns in Figure 9.21.

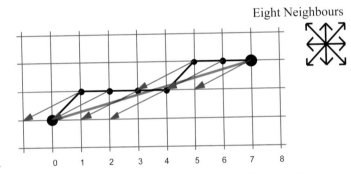

Figure 9.20 Backward stepping algorithm with look ahead test

Transferring the 1-4 labelling of the quadrants used in the "four-neighbours" examples to these octants makes comparisons easier. Once the stepping patterns for octants 1-4 have been generated by reflection across the x or y axes, they can be transferred to the remaining octants by reflecting the pattern across the 45° diagonal in the way illustrated in Figure 9.21.

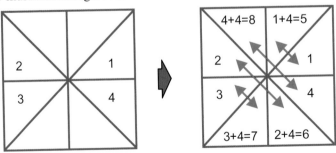

Figure 9.21 Octant labelling matching the quadrant labelling

Eight Neighbours

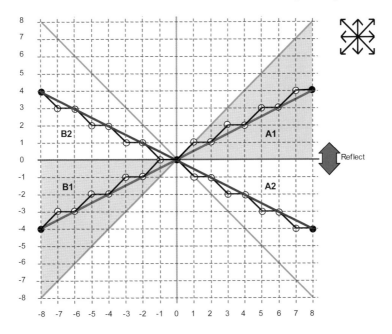

Eight Neighbours

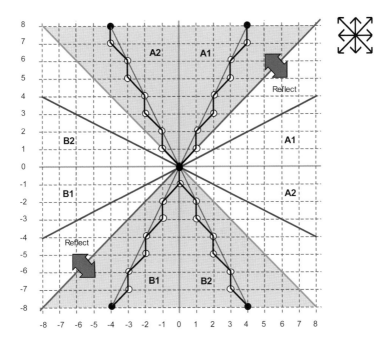

Figure 9.22 Reflecting line increment patterns vertically and diagonally

This diagonal reflection can be implemented, in the same way employed in the DDA algorithm, by swapping x and y ordinates. This solution means that only the stepping sequences labelled A1, A2, B1 and B2 in Figure 9.22 need to be analysed further.

Octant 2 $\qquad$ +ve dx < 0 dy > 0 $\qquad$ -ve d = -dx - 2dy (Initialised value)	**Octant 1** $\qquad$ +ve dx > 0 dy > 0 $\qquad$ +ve d = 0 (Initialised value)
d < 0: - ve : step in X & Y direction. d = 0: zero: step in X direction. d > 0: + ve: step in X direction. StepX: d= d-dy. StepX&Y: d = d-dy-dx	d < 0: - ve : step in X direction. d = 0: zero: step in X & Y direction. d > 0: + ve: step in X & Y direction. StepX: d= d+dy. StepX&Y: d = d+dy-dx
Octant 3 $\qquad$ -ve dx < 0 dy < 0 $\qquad$ -ve d = dx-2dy (Initialised value)	**Octant 4** $\qquad$ +ve dx > 0 dy < 0 $\qquad$ -ve d = 0 (Initialised value)
d < 0: - ve : step in X direction. d = 0: zero: step in X direction. d > 0: + ve: step in X & Y direction. StepX: d= d-dy. StepX&Y: d = d-dy+dx	d < 0: - ve : step in X & Y direction. d = 0: zero: step in X & Y direction. d > 0: + ve: step in X direction. StepX: d= d+dy. StepX&Y: d = d+dy+dx

Setting out these relationships in a table again makes it possible to see how the variables in each octant relate to each other. In this case it can be seen that the incrementing patterns for octants 1 and 2 are opposites, as are 3 and 4. Figure 9.22 shows how the pattern in octant 2 can be reflected from the line stepping scheme produced in octant 3. Similarly, how octant 1 can be reflected from octant 4: by treating the two *(dx < 0)* cases as the same but using opposite *y* stepping directions, and the *(dx > 0)* cases as the same again with opposite *y* stepping directions. For forward and backward drawing to be identical, the table's diagonal *(d = 0)* tests must still select opposite stepping actions.

Taking all these relationships into account led to the following implementation. As before, in order to modify the signs of *dx* and *dy* to implement mapping operations between octants, the stepping directions for the final line were stored in the variables *kx* and *ky*. These were assigned values of -1 or +1 to match the original corresponding signs of *dx* and *dy*. There appeared to be a variety of ways to map one octant to another, however the following approach provided a relatively direct scheme to satisfy the relationships laid out in the overall octant table. The distinguishing test to select either X or X&Y steps depends on *(d > 0)* for octants 3 and 4, and on *(d < 0)* for octants 2 and 1. Mapping octant 4 to 1 and 2 to 3, by inverting the sign of *dy*, still leaves two distinct cases. However the following operation offers a way to merge the two into one testing sequence which satisfies the original design target for the algorithm.

case A	(d < 0)	(d < 0)	R	case B	((-d) < 0)	(d >= 0)	R
case A	(d = 0)	(d >= 0)	S	case B	((-d) = 0)		R
case A	(d > 0)		S	case B	((-d) > 0)	(d < 0)	S

If the selection is made by the *(d < 0)* test then in '*case A*' action R will be taken if this test is true otherwise action S. If for '*case B*' the sign of *d* is reversed then the same test *(d < 0)* can be set up: to select action S if *(d < 0)* is true, otherwise action R. What this reversal achieves is to swap the action from R to S when *(d = 0)*, in the two cases. This makes it possible to map octant 3 to 1, using the code segment:

```
if(d < 0) j= kx; else j= -kx;
if(j < 0)  step X&Y   else  step X;
```

The sign of *kx* is used to select which stepping pair should be used. If *kx* is positive, then it selects octant 1, if negative it selects octant 3. This is the same code structure used to map from quadrant 2 to 1 with no sign change, in the four neighbours interpolation algorithm, and can be used here, in the same way, if so desired, as the following table shows.

Octant	dy	dx	d < 0	d = 0	d > 0
1	+ve	+ve	step X	step X & Y	step X & Y
2	+ve	-ve	step X & Y	step X	step X
3	-ve	-ve	step X	step X	step X & Y
4	-ve	+ve	step X & Y	step X & Y	step X

Thus, reversing the sign of *d* in octant 3 allows octant 3 to be mapped diagonally: to use the same selection test as octant 1, and vice versa. Applying these ideas led to the following program code, which will handle all eight octants.

```
DisplayGrid d = new DisplayGrid(f,T.tileColour,T.cols,T.rows);
d.paintGridArray(); d.drawDualGrid(Color.black,Color.gray);
f.writeString("Please enter number of line segments: ");
int number = f.readInteger();

for(int i = 0; i <number; i++){
   f.writeString("Use the mouse to enter the line vertices");  f.writeLine();
   Point pa = d.getCell();  Point pb = d.getCell();
   d.line(pa, pb, 1, Color.red);
   d.line(pa, pb, 3 ,Color.blue);
}
public void line(Point p1, Point p2, int wd, Color cc){
   int kx=1,ky=1;  int dx = p2.x-i()p1.xi();  int dy = p2.yi()-p1.yi();
   if ( dx < 0 ) { kx = -1;  dx = -dx;}
   if ( dy < 0 ) {ky = -1;  dy = -dy;}
   if(dx>dy) octant(p1.xi(), p1.yi(), p2.xi(), p2.yi(), wd, kx, ky, dx, dy, false, cc);
   else octant(p1.yi(), p1.xi(), p2.yi(), p2.xi(), wd, ky, kx, dy, dx, true, cc);
}
```

```
private void octant(int x, int y, int xend, int yend, int width,
                      int kx, int ky, int dx, int dy, boolean steep, Color cc){
    int  d= 0,  j = 0;
    if(kx<0){ d= 2*dy-dx;  dx = -dx;  dy = -dy;}
    for(;;){
        if(steep) paintInnerCell(y, x, width, cc);  else paintInnerCell(x, y, width, cc);
        if (x ==  xend) return;
        if(dy != 0){
            if(d > 0) j= kx; else j= -kx;
            if(j < 0){d= d+dx-dy; y= y+ky;} else d= d-dy;
        } x = x+ kx;
    }
}
```

As before, line variable initialisation maps octant 2, 3 and 4 to 1, by making the values of *dx* and *dy* both positive. The initialisation of d is then carried out with these values for octant 1. Lines from octants 3 and 2 are then mapped back to octant 3 by making both *dx* and *dy* negative. Examining the innermost loop of the line interpolation procedure, the increment to *d* for the X&Y step is made by $d = d+dx-dy$, and for the X step by $d = d-dy$. For lines from octant 1 *dx* and *dy* are positive and *d* is initialised to 0, which means this *'negative'* incrementing operation maps the line to octant 3. The offset value for the lines originally from octant 2 or 3, is initialised by $d = 2.dy-dx$ but with negated increments for *dx* and *dy*. This defines *d* correctly for octant 3. Applying the code given above effectively allows a mapping from octant 1 to octant 3, so giving a single unified selection test for an "eight-neighbour" line interpolation algorithm.

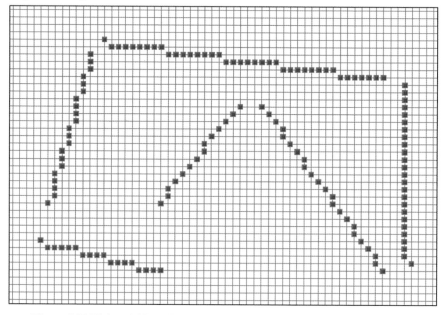

Figure 9.23 Eight neighbour line interpolation: forwards (red)& backwards (blue)

Figure 9.23 illustrates the output of this program for different line directions, which compared with Figure 9.15 goes some way to giving more balanced line weights. It also shows that the forward and backwards behaviour of this algorithm has been correctly implemented, by overlaying different sized "point" tiles for the two directions.

The diagram shown in Figure 9.24, where the algorithm is used to display nearly horizontal lines: prompted the final step in this program development. The stepping sequence is biased: selecting mostly points above the true line. The reason for this can be found in the interpolation patterns illustrated in Figures 9.10 and 9.17. The solution to this imbalance would be to move the resulting pattern of points downwards so that, on average, as many points lie below the line as above it. Clearly working on a grid this is not an option. However, an equivalent operation is to slide the pattern along the line still using grid-based points but using a different distribution along the line. The way this can be done is illustrated in Figure 9.24 for a "four-neighbours" scheme. The trick is to initialise the distance-measure d from the true line to be that for a point placed half a cell width above the line. However, the original incremental values of dx and dy, used to set up this distance, may not be *even* numbers, which can be halved as integers. The half value can be obtained by doubling both the initial values of dx and dy used as increments to d, but using half the result or the original dx or dy values to initialise the distance d.

Four Neighbours

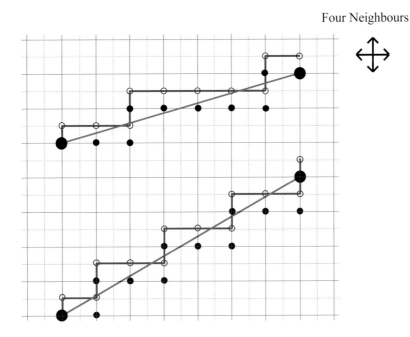

Figure 9.24 Offset line patterns improve near horizontal interpolation

The interesting development which arises from this adjustment, when the "eight-neighbour" interpolation scheme is examined, is that this offset testing can be set up symmetrically so the forward and backward initialisation and tests can be treated in virtually the identical way. The offset distance, moving forwards in the first octant, is half a step in the Y direction followed by a full step in the X direction.

Eight Neighbours

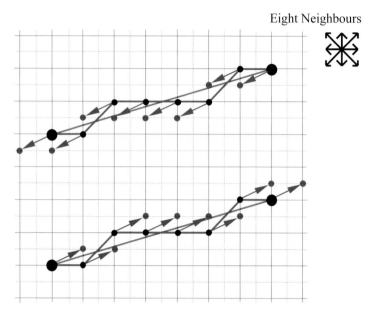

Figure 9.25 "Offset-testing" set up symmetrically

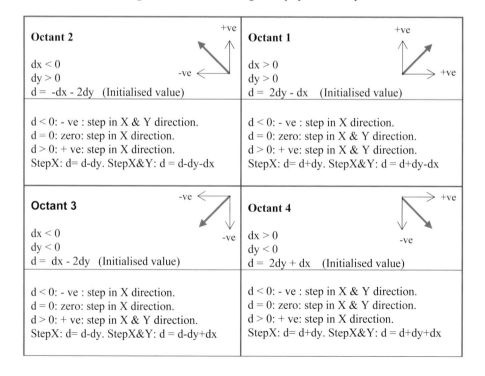

		+ve			+ve
Octant 2			**Octant 1**		
dx < 0		-ve	dx > 0		
dy > 0			dy > 0		+ve
d = -dx - 2dy (Initialised value)			d = 2dy - dx (Initialised value)		
d < 0: - ve : step in X & Y direction.			d < 0: - ve : step in X direction.		
d = 0: zero: step in X direction.			d = 0: zero: step in X & Y direction.		
d > 0: + ve: step in X direction.			d > 0: + ve: step in X & Y direction.		
StepX: d= d-dy. StepX&Y: d = d-dy-dx			StepX: d= d+dy. StepX&Y: d = d+dy-dx		
		-ve			+ve
Octant 3			**Octant 4**		
dx < 0		-ve	dx > 0		-ve
dy < 0			dy < 0		
d = dx - 2dy (Initialised value)			d = 2dy + dx (Initialised value)		
d < 0: - ve : step in X direction.			d < 0: - ve : step in X & Y direction.		
d = 0: zero: step in X direction.			d = 0: zero: step in X & Y direction.		
d > 0: + ve: step in X & Y direction.			d > 0: + ve: step in X direction.		
StepX: d= d-dy. StepX&Y: d = d-dy+dx			StepX: d= d+dy. StepX&Y: d = d+dy+dx		

The corresponding initialisation for the third octant moving in the opposite direction is again half a step in the Y direction, in this case down, followed by a full step in the X direction backwards, as shown in Figure 9.25. This will select the same set of points in both directions, and the points will be distributed in a balanced way on each side of the true line. As before, the only special case is where test points fall exactly on the line, when $d = 0$. These readjustments give the following revised table and the modified code:

octant	dy	dx	d: -ve	d: zero	d: +ve
1	+ve	+ve	step X	step X & Y	step X & Y
2	+ve	-ve	step X & Y	step X	step X
3	-ve	-ve	step X	step X	step X & Y
4	-ve	+ve	step X & Y	step X & Y	step X

```
private void octant(int x, int y, int xend, int yend, int width,
                              int kx, int ky, int dx, int dy, boolean dir, Color cc)
{
    int d = 2*dy-dx; dx = 2*dx; dy = 2*dy;
    if (ky < 0){ d = -d; dx = -dx; dy = -dy;}
    while(true){
        if(dir) paintInnerCell(y,x,width,cc); else paintInnerCell(x,y,width,cc);
        if (x == xend) return;
        if (d < 0) int j = -ky;  else j =ky;
        if (j > 0){d= d+dy-dx; y = y+ky;} else d= d+dy;
        x = x +kx;
    }
}

public void line(Point p1, Point p2, int wd, Color cc)
{
    int dx = p2.xi()-p1.xi(); int dy = p2.yi()-p1.yi(); int kx = 1; int ky = 1;
    if (dx < 0){ kx = -1; dx = -dx; }
    if (dy < 0){ ky = -1; dy = -dy; }
    if (dx < dy)octant(p1.yi(),p1.xi(),p2.yi(),p2.xi(),wd,ky,kx,dy,dx,true,cc);
    else octant(p1.xi(),p1.yi(),p2.xi(),p2.yi(),wd,kx,ky,dx,dy,false,cc);
}
```

This implementation of the program follows the same strategy as the previous one. Octant 2 is mapped to 3, and 4 to 1. However, in this case octant 3 and 1 are then processed as octant 1, to unify the code rather than as octant 3. There are in this case two offset values, which have to be initialised. The first is set up in octant 1; the second is set up for octant 3 by negating this initial value. In the inner loop of this interpolation procedure, the distance increments are carried out for octant 1, so d values defined using octant 3 values for dx and dy will be negative in this context. The outputs of this algorithm and the previous algorithm are illustrated in Figures 9.26 and 9.27 for two sets of line segments, to show the improved treatment of acute, internal and external angles.

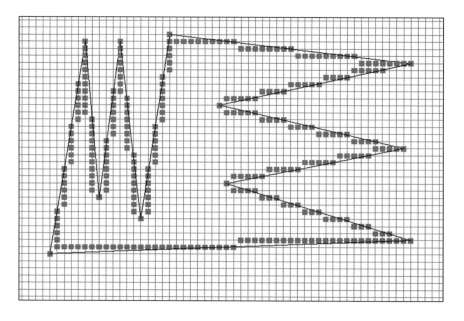

Figure 9.26 Biased line interpolation

Comparing the inner vertices and the outer vertices of the comb shapes using the "biased" interpolation algorithm in Figure 9.26 shows an unbalanced treatment, the outer angles are sharp the inner angles are rounded where they should be more or less the same.

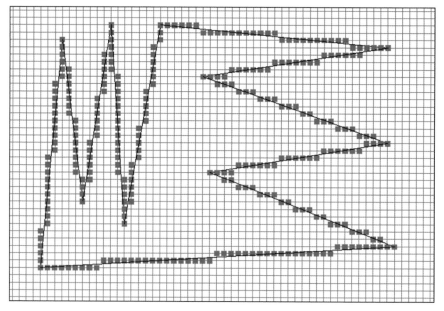

Figure 9.27 Balanced line interpolation

In contrast Figure 9.27 shows a substantial improvement resulting from using the balanced algorithm, giving a similar treatment for both inner and outer angles.

This final form of the algorithm gives the same sequence of points generated by the original *dda* algorithm, without using floating-point numbers and minimising the use of the multiplication operation. However, though this was critical in early systems, advances in hardware make the distinction less important in current systems, except where the line interpolation needs to be implemented in a minimum way, say on an integrated circuit for operations like polygon fill. On the other hand discussion in later chapters shows there are important advantages to using floating point numbers to represent line end-point co-ordinates, for example, when interpolating sections of a line represented by the same line equation where the interpolated point sequences from two line segments with different end points need to match up exactly if they overlap.

The step-by-step exploration of a line following algorithm on a grid, given above, by progressively refining the operation, generates a fairly complex analysis for what started out as a simple idea. The complexity arises mainly from the need to make the point selection consistent for lines drawn in any direction. However, it is also made particularly difficult because each step was being determined by a single distance test. If the *dda* algorithm is revisited it can be seen that the grid point choice implicitly considers two distances. For shallow-gradient lines, the exact y value is calculated as a floating-point number for a sequence of integer x values. These y values are then rounded to the nearest integer to give a y value on the grid. This is in effect comparing the exact y value with two integer grid values and selecting the nearer. An alternative approach to implementing the line following algorithm can be set up by adopting a similar selection process.

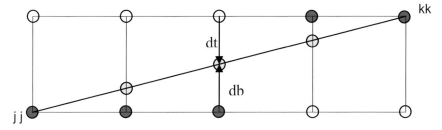

Figure 9.28 Line following comparing two distances

```
static void gridline(Point jj,Point kk,Color cc,Grid gd, int bdr){
    int x1 = jj.xi(),y1 = jj.yi(), x2 = kk.xi(),y2 = kk.yi();
    int x = x1, y = y1,  dx = x2-x1, dy = y2-y1, db=0, dt= dx;
    while(true){
        if(dt+db > 0){
            gd.paintInnerCell(x,y,bdr,cc);}else{ gd.paintInnerCell(x,y+ky,bdr,cc);}
        if(x==x2)return; else{  x=x+kx; db = db-dy;  dt=dt-dy;}
        if (dt < 0){ y=y+ky; db = db+dx;  dt=dt+dx;}
    }
}
```

However, it is still necessary to consider lines drawn in different directions. Considering shallow gradients first there are four quadrants to be catered for:

```
static void gridline(Point jj,Point kk,Color cc,Grid gd,int bdr){
    int x1 = jj.xi(),y1 = jj.yi(), x2 = kk.xi(),y2 = kk.yi();
    int x = x1,y = y1;
    int kx=1, ky=1,j=0,db=0, dt= 0;
    int dx = x2-x1, dy = y2-y1;
    if(dx<0){kx=-1; j=j+1;}
    if(dy<0){ky=-1; j=j+2;}

    switch(j){
    case 0:                                    //first quadrant
        while(true){
            if(dt+db > 0){ gd.paintInnerCell(x,y,bdr,cc);}
            else { gd.paintInnerCell(x,y+ky,bdr,cc);}
            if(x==x2)return;else{ x=x+kx; db = db-dy;  dt=dt-dy;}
            if (dt < 0){ y=y+ky; db = db+dx;  dt=dt+dx;}
        }
    case 1:                                    //second quadrant
        while(true){
            if(dt+db <= 0){gd.paintInnerCell(x,y,bdr,cc);}
            else {gd.paintInnerCell(x,y+ky,bdr,cc);}
            if(x==x2)return;else{ x=x+kx; db = db+dy;  dt=dt+dy;}
            if (dt >=0){ y=y+ky; db = db+dx; dt=dt+dx;}
        }
    case 3:                                    //third quadrant
        while(true){
            if ( dt+db <= 0){ gd.paintInnerCell(x,y,bdr,cc);}
            else { gd.paintInnerCell(x,y+ky,bdr,cc);}
            if(x==x2)return;else{ x=x+kx; db = db-dy; dt=dt-dy;}
            if (dt >=0){ y=y+ky; db = db+dx;  dt=dt+dx;}
        }
    case 2:                                    //fourth quadrant
        while(true){
            if (dt+db > 0){gd.paintInnerCell(x,y,bdr,cc);}
            else {gd.paintInnerCell(x,y+ky,bdr,cc);}
            if(x==x2)return;else{ x=x+kx; db = db+dy;  dt=dt+dy;}
            if (dt < 0){y=y+ky; db = db+dx; dt=dt+dx;}
        }
    }
}
```

Steep gradients can be handled by swapping the x and y ordinates, reflecting the pattern of points about the main diagonal, in the way outlined in Figure 9.21. Adopting the same approach as before it is possible to map these eight cases into a single parameterised procedure in the following way.

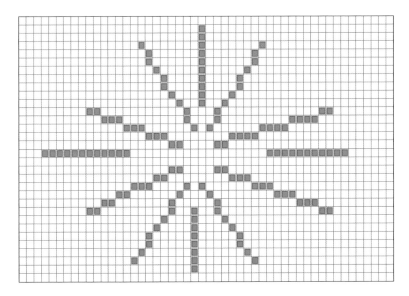

Figure 9.29 Radial line test

```
public static void main(String[] args){
    double xmin,xmax,ymin,ymax;
    int cols = 50,rows = 32;
    Color[][] cc = new Color[cols][rows];
    for(int i=0; i<cols;i++){ for(int j=0; j<rows; j++){ cc[i][j] = Color.white; }}
    Grid gd = new Grid(IO, dW , cc , cols, rows);
    gd.paintGridArray();  gd.drawGridLines(Color.black,Color.gray);
    Point kk = gd.getCell();Point jj = gd.getCell();
    while(gd.contains(jj)){                                // for a series of lines
        gridline(kk, jj, Color.red, gd,1);  gridline(jj, kk, Color.green, gd,3); //forward & backwards
        kk = gd.getCell();  jj = gd.getCell();             // get next line
    }
}
static void gridline(Point jj, Point kk, Color cc, Grid gd, int bdr){
    int x1=jj.xi(),y1=jj.yi(),x2=kk.xi(),y2=kk.yi(),x=x1,y=y1,kx=1,ky=1,j=0;
    int db=0,dt=0,dx=x2-x1,dy=y2-y1,xp=x,yp=y,fy=0; boolean up=false,t1=false;
    if(Math.abs(dy)>Math.abs(dx)){up=true;int tmp=x1; xp=x=x1=y1;
    yp=y=y1=tmp; tmp=x2; x2=y2; y2=tmp; tmp=dx; dx = dy; dy = tmp;}
    if(dx<0){kx=-1; j=j+1;} if(dy<0){ky=-1; j=j+2;}db=0;dt = dx; fy=dy;
    if((j==0)||(j==3))fy= -dy;      if((j==0)||(j==2))t1=true;
    while(true){
        if ((t1&&(dt+db>0))||(!t1&&(dt+db<=0))){yp=y;}else{yp=y+ky;}
        if(up)gd.paintInnerCell(yp,xp,bdr,cc);else gd.paintInnerCell(xp,yp,bdr,cc);
        if(x==x2)return; else {xp = x = x+kx; db=db+fy;dt=dt+fy;}
        if((t1&&(dt<0))||(!t1&&(dt>=0))){y=y+ky;db=db+dx;dt=dt+dx;}
    }
}
```

Anti-Aliased Lines

At first sight this gives a no better result than before, though perhaps it is slightly easier to understand. However, this approach opens up an alternative to using simple black or white pixels. The values, for the two pixels on each side of the line used to select which pixel to shade, are proportional to the pixel distances from the true line. These values can be used to give the same effect as a line captured by an analogue camera and displayed on a black and white TV screen, by shading each pixel in proportion to its distance from the true line, in the way illustrated in Figures 9.31 and 9.32:

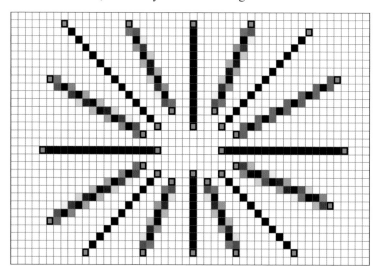

Figure 9.30 Anti-aliased lines on a grid

```
static void antiAliasedGridLine(Point jj,Point kk,Color cc,Grid gd ){
    int x1=jj.xi(),y1=jj.yi(),x2=kk.xi(),y2=kk.yi(),x=x1,y=y1,kx=1,ky=1,j=0;
    int db=0,dt=0,dx=x2-x1,dy=y2-y1,xp=x,yp=y,yq=y,fy=0;
    boolean up=false,t1=false;  float h=1, s =0,  b1=0, b2= 0;
    if(Math.abs(dy)>Math.abs(dx)){
        up=true;int tmp=x1; xp=x=x1=y1;yp= y=y1=tmp;
        tmp=x2; x2=y2; y2=tmp; tmp=dx; dx = dy; dy = tmp;}
    if(dx<0){kx=-1; j=j+1;}   if(dy<0){ ky=-1; j=j+2;}
    db=0;dt = dx; fy=dy;
    if((j==0)||(j==3))fy= -dy; if((j==0)||(j==2))t1=true;
    while(true){
        yp=y; yq=y+ky; b1= 1-dt/(float)(dt-db); b2=1+db/(float)(dt-db);
        Color c1 = Color.getHSBColor(h,s,b1), c2 = Color.getHSBColor(h,s,b2);
        if(up){gd.paintInnerCell(yp,xp,0,c1);gd.paintInnerCell(yq,xp,0,c2);}
        else{ gd.paintInnerCell(xp,yp,0,c1);gd.paintInnerCell(xp,yq,0,c2);}
        if(x==x2)return; else {xp = x = x+kx; db=db+fy;dt=dt+fy;}
        if((t1&&(dt<0))||(!t1&&(dt>=0))){y=y+ky;db=db+dx;dt=dt+dx;}
    }
}
```

A justification for this construction can be made in the following way. The value for each pixel is being treated as a point sampled value in the simple line interpolation algorithm whereas in the case of the analogue TV camera the value of each pixel is an integration of the brightness over each pixel area. Consider a raster line through two objects with a uniform shading value against a uniform back ground with the pixel values sampled on a regular grid placed in the way shown in Figure 9.31a.

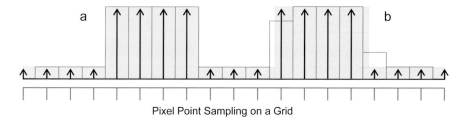

Pixel Point Sampling on a Grid

Figure 9.31 Pixel point sampling model

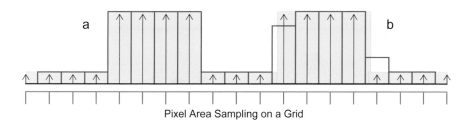

Pixel Area Sampling on a Grid

Figure 9.32 Pixel area sampling model

If the pixel values are considered to be measures from pixel area sampling then a block diagram of the form shown in Figure 9.32 is probably the best way of representing them. Notice that the point sampling technique will only reflect the *true position* of the object if the boundary of the object lines up with the edges of the pixel blocks and if the object is an integral number of pixel widths wide in the way shown in Figure 9.31a otherwise information is lost as in the case of Figure 9.31b. In contrast if the values are block area values, which are the best estimate of the area covered then information about the true size and location of the object is not totally lost, shown in Figure 9.32b.

If an edge is considered as a point value but it is placed on the sampling grid then it will be represented by a block centred on the edge as in Figure 9.31a. The area of this block can be converted into a triangle with its base points at the bottom of the two neighbouring pixel point values in the way shown in Figure 9.33a. If two point values are placed next to each other then by adding together their triangular representations: a standard form of linear interpolation between the initial point values is obtained shown in Figure 9.33b. Given a continuous distribution it is possible to break it down into a series of equivalent blocks.

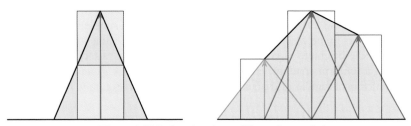

Figure 9.32 Point triangles and linear interpolation

Where the linear sections are unevenly spaced point samples need to be taken at the vertex points of the boundary shape. If a regular grid is applied to the same boundary errors will be generated at the changes in direction: shown in red in Figure 9.33b.

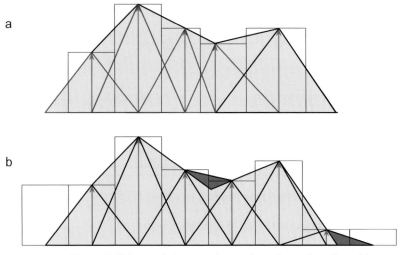

Figure 9.33 Interpolating a surface on irregular and regular grids

Applying the same approach to line interpolation helps to identify the position of the line. This can be done by placing a triangular distribution centred on the true position of the line then sampling it using the regular grid positions. This gives a centre of gravity for the resulting pixel values lying on the position of the true line.

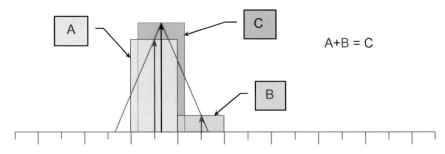

Figure 9.34 Interpolating pixel values for a line not centred on a pixel

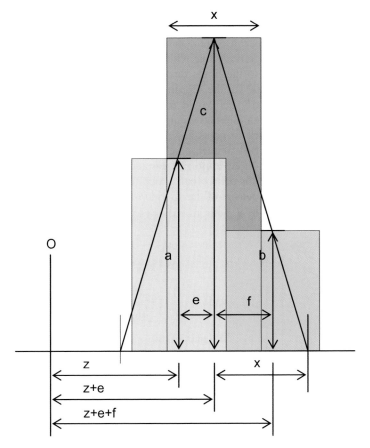

Figure 9.35 The relationship between pixel values used for line anti-aliasing

By similar triangles :

$$\frac{a}{c} = \frac{x-e}{x} \; ; \qquad \frac{b}{c} = \frac{x-f}{x} \; ;$$

$$\frac{a+b}{c} = \frac{x-e+e-f}{x} = \frac{2x-(e+f)}{x} = \frac{x}{x}$$

$$a+b = c$$

Taking Moments about O and assuming that $a + b$ balances c

$$z.a.x + (z+e+f).b.x = (z+e).c.x$$

$$z.a + z.b + z.e + z.f = z.c + c.e$$

$$z.(a+b) + e.b + f.b = z.(a+b) + (a+b).e$$

$$e.b + f.b = e.a + b.e$$

$$f.b = e.a$$

Hence the centre of gravity of the two blocks $a + b$ lies on the centreline of block c.

Although this extension distributes the weight of the pixel values correctly about the position of the true line, Figure 9.30 shows that the distribution along the length of the lines varies with direction. This is because the distances used to evaluate the pixel weightings are taken parallel to the axes. If the line width and the matching triangular distribution were taken perpendicular to the line's direction this variation would be reduced. The difficulty doing this is that for the $45°$ line pixels above and below the central pixel would have to be visited. This would require a major change in the program structure. However, there is an alternative approach, which can be applied within the existing program framework to adjusting the pixel value density along lines of different orientation. If the density of each pixel is modified by a scale factor dependent on the line's slope then a more uniform treatment of lines in different directions should result. This scale factor can be calculated for straight lines before the interpolation loop is entered. If the density of the $45°$ line is taken as unity then the density of other lines must be adjusted by a factor of cosine $(45-\theta)$.

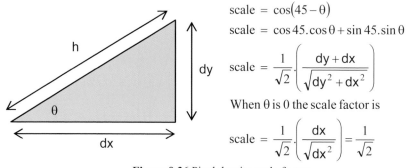

$$scale = \cos(45 - \theta)$$

$$scale = \cos 45.\cos\theta + \sin 45.\sin\theta$$

$$scale = \frac{1}{\sqrt{2}}\cdot\left(\frac{dy + dx}{\sqrt{dy^2 + dx^2}}\right)$$

When θ is 0 the scale factor is

$$scale = \frac{1}{\sqrt{2}}\cdot\left(\frac{dx}{\sqrt{dx^2}}\right) = \frac{1}{\sqrt{2}}$$

Figure 9.36 Pixel density scale factor

```
static void antiAliasedGridLine(Point jj,Point kk,Color cc,Grid gd){
    int x1=jj.xi(),y1=jj.yi(),x2=kk.xi(),y2=kk.yi(),x=x1,y=y1,kx=1,ky=1,j=0;
    int dx=x2-x1,dy=y2-y1, db=0, dt=0, xp=x,yp=y,yq=y,fy=0;
    boolean up=false,t1=false; float h=1, s =0;
    double DX = Math.abs(dx), DY= Math.abs(dy), b1=0, b2= 0;
    if(DY>DX)){up=true;int tmp=x1; xp=x=x1=y1;yp= y=y1=tmp;
        tmp=x2; x2=y2; y2=tmp; tmp=dx; dx = dy; dy = tmp;}
    if(dx<0){ kx = -1; j=j+1;}  if(dy<0){ky = -1; j=j+2;} db=0;dt = dx; fy=dy;
    fy=dy; if((j==0)||(j==3))fy= -dy; if((j==0)||(j==2))t1=true;
    double scale = (DX+DY)/Math.sqrt(2.0*(DX*DX+DY*DY))/(double)dx;
    while(true){
        yp=y;  yq=y+ky; b1 = 1-dt*scale; b2= 1+ db*scale;
        Color c1 = Color.getHSBColor(h,s,(float)b1);
        Color c2 = Color.getHSBColor(h,s,(float)b2);
        if(up){gd.paintInnerCell(yp,xp,0,c1);gd.paintInnerCell(yq,xp,0,c2);}
        else{ gd.paintInnerCell(xp,yp,0,c1);gd.paintInnerCell(xp,yq,0,c2);}
        if(x==x2)return; else {xp = x = x+kx; db=db+fy;dt=dt+fy;}
        if((t1&&(dt<0))||(!t1&&(dt>=0))){y=y+ky;db=db+dx;dt=dt+dx;}
    }
}
```

Applying this adjustment to the gridline code gives the result shown in Figure 9.37.

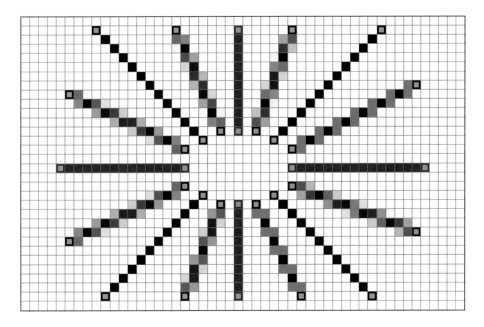

Figure 9.37 Anti-aliased lines adjusted for direction

As the resolution is increased so the effect of this refinement improves

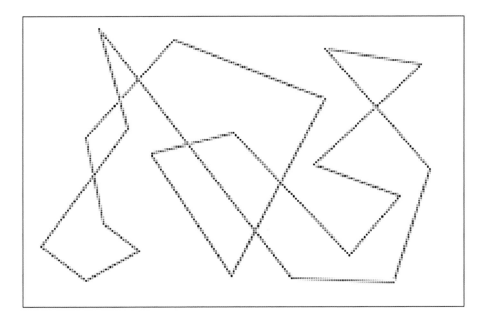

Figure 9.38 Anti-aliased lines adjusted for direction

Triangle and Polygon Fill

If the boundaries of a triangle are interpolated using the procedures from the previous section the result will be that shown in Figure 10.1. At first sight shading or filling in the points within the triangle with a different colour would seem to require drawing lines from each left boundary line point to the corresponding right boundary line point on the grid, for all the boundary points generated by the boundary line interpolation procedure.

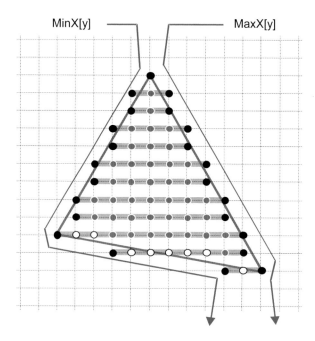

Figure 10.1 Selecting left and right edge points for shading

However, closer scrutiny reveals a problem. Shallow lines such as the base of the triangle in Figure 10.1 generate more than one potential edge point for such a '*scan-line*' or grid-row shading line. Since it is necessary to store these edge points while the boundary is being generated one of these multiple points must be selected for each row of the grid, one for the left edge and one for the right edge.

The simplest approach, (which will be extended later), is to select the outermost points in each case. This can be done for the left edge by selecting the co-ordinate with the minimum x value for a given y value, and for the right edge the co-ordinate with the maximum x value for a given y value. As the code given below shows it is not necessary in this initial solution to this problem to distinguish between left edge and right edge points, A minimum and maximum selection operation applied to all the boundary points will select the appropriate points for the final shading task. In Figure 10.1 the resulting edge points are shown as solid black circles, the "inner" unused boundary points by black circles with white interiors and the other shading points by dark grey circles.

```
public void polygonFill( Polygon p, Color color){
   int ymin = Integer.MAX_VALUE;
   int ymax = Integer.MIN_VALUE;
   for (int i= 0; i < p.length-1; i++){
      if(ymin > p.p[i].yi()) ymin = p.p[i].yi();
      if(ymax < p.p[i].yi()) ymax = p.p[i].yi();
   }
   int  len = ymax - ymin + 1;
   Shading S = new Shading(len, ymin);
   for(int i= 0; i< p.length-1; i++){
      line(p.p[i].xi(),  p.p[i].yi(), p.p[i+1].xi(),  p.p[i+1].yi(),  color,  S);
   }
   for(int i= 0; i < len; i++){
      line(S.leftedge[i], i+ymin, S.rightedge[i], i+ymin, color, null);
   }
}


public void defineEdges(Point p,Shading s){
   if (s.leftedge[p.yi() - s.miny]  > p.xi() ){s.leftedge[p.i()y-s.miny] = p.xi();}
   if (s.rightedge[p.yi() - s.miny]< p.xi()){s.rightedge[p.yi()-s.miny]= p.xi();}
}


private void line(int x1,int y1,int x2, int y2,Color color,Shading s){
   int kx, ky, dx, dy;
   dx = x2-x1; dy = y2-y1; kx = 1; ky = 1;
   if (dx < 0){ kx = -1; dx = -dx; }
   if (dy < 0){ ky = -1; dy = -dy; }
   if (dx < dy) {this.octant(y1, x1, y2, x2, ky, kx, dy, dx, 2, color, s);}
   else {this.octant(x1, y1, x2, y2, kx, ky, dx, dy, 1, color, s);}
}


private void octant(int x, int y, int xend, int yend, int kx, int ky, int dx, int dy,
                                   int dir, Color color, Shading s){
   int d,j;  Point p = new Point(2);   d = 2*dy-dx; dx = 2*dx; dy = 2*dy;
   if (ky < 0){ d = -d;  dx = -dx;  dy = -dy;}
   while(true){
      if (dir == 1){p.x("=", x); p.y("=", y);}
      else {p.y("=", x); p.x("=", y);}
      if (s == null) plotPoint(p,color);
      else defineEdges(p, s);
      if (x == xend) return;
      if (d < 0)  j = -ky;
      else j =ky;
      if (j > 0) {d =  d+dy-dx;  y = y+ky;}
      else d= d+dy;
      x = x +kx;
   }
}
```

```
class Shading{
    public int leftedge[],rightedge[],length,miny;
    Shading(int len,int minimumy){
        this.leftedge = new int[len];      this.rightedge= new int [len];
        for(int i=0;i<len;i++)
            {leftedge[i]= Integer.MAX_VALUE; rightedge[i]=Integer.MIN_VALUE;}
        this.length = len;    this.miny = minimumy;
    }
    public void defineEdges(Point p){
        if (this.leftedge[ p.yi()-this.miny] > p.xi())
            { this.leftedge[ p.yi()-this.miny] = p.xi();}
        if (this.rightedge[p.yi()-this.miny] < p.xi())
            { this.rightedge[p.yi()-this.miny] = p.xi();}
    }
}
class Polygon{
    private DisplayWindow dW = null;
    private TextWindow IO = null;
    public Point p[];
    public int length=0;
    public List ply = null;
    Polygon(){}
    Polygon(int len){  p = new Point[len];length = len; }
}
```

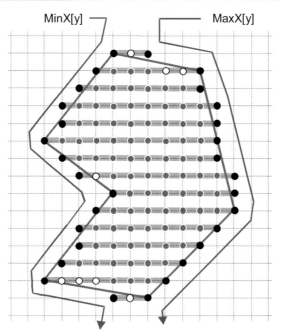

Figure 10.2 Non-convex polygon

The triangle in Figure 10.1 is a convex area. This algorithm clearly will only work with polygons that can be represented by single left-edge point-sequences and single right-edge point-sequences. This will include polygons of the form shown in Figure 10.2, but not, for example, a spiral polygon. To handle more complex figures of this type, it is necessary to decompose them into simpler shapes, usually convex shapes, which can then be shaded in sequence; or to develop a more complex shading algorithm. However it is also possible to shade polygons of the form shown in Figure 10.3 if the shading lines are drawn in vertically. Extending the *Shading class* to give the *Shadings class* in the following way opens up a series of useful applications in cartography

```
public static void main(String[ ] args){
    String str =""; int len=0;
    IO.writeString("do you wish to enter another polyon? y/n: ");
    str = IO.readString();IO.readLine();
    while(str.equals("y")){
        IO.writeString("please enter the number of vertices: ");
        int pnts = IO.readInteger();IO.readLine();
        Point poly[] = new Point[pnts+1];
        for(int i = 0; i<pnts;i++){
            poly[i] = dW.getCoord();
            if(i>0)dW.plotLine(poly[i-1],poly[i],Color.blue);
        } poly[0].c("->",poly[pnts]= new Point(2));
        dW.plotLine(poly[pnts-1],poly[pnts],Color.blue);
        int xmin=poly[pnts].xi(),xmax=xmin;  int ymin=poly[pnts].yi(),ymax=ymin;;
        for(int i = 0; i<pnts;i++){
            if(xmin > poly[i].xi())xmin = poly[i].xi(); if(xmax < poly[i].xi())xmax = poly[i].xi();
            if(ymin > poly[i].yi())ymin = poly[i].yi(); if(ymax < poly[i].yi())ymax = poly[i].yi();
        } IO.writeString("Is shading vertical y/n : ");
        str = IO.readString();IO.readLine();
        if(str.equals("y")){
            len = xmax-xmin+1;
            Shadings s= new Shadings(len,xmin,true);
            for(int i=0;i<pnts;i++){
                s.shadingEdge(poly[i].xi(),poly[i].yi(),poly[i+1].xi(),poly[i+1].yi());
            } s.scanfill(dW,len, Color.green);
        }else{
            len = ymax-ymin+1;
            Shadings s= new Shadings(len,ymin,false);
            for(int i=0;i<pnts;i++){
                s.shadingEdge(poly[i].xi(),poly[i].yi(),poly[i+1].xi(),poly[i+1].yi());
            } s.scanfill(dW,len, Color.red);
        }
        IO.writeString("do you wish to enter another polyon? y/n: ");
        str = IO.readString();IO.readLine();
    }
}
```

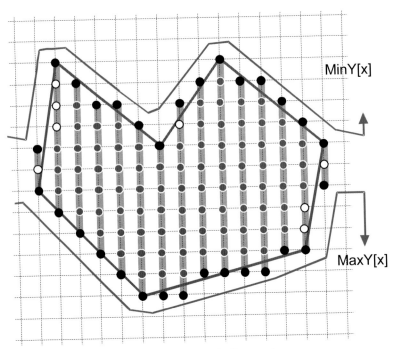

Figure 10.3 Vertically shading a polygon

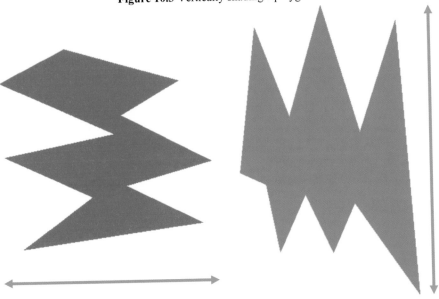

Figure 10.4 Horizontally and vertically shaded polygons

One application where this vertical shading algorithm can be applied directly is to produce profile drawings of three-dimensional surfaces.

```
class Shadings{
    public boolean vertical = false;
    public int leftedge[], rightedge[], length,min;
    Shadings(int len, int minimum,boolean vertical){
        this.IO=IO; this.vertical= vertical; this.length = len;   this.min = minimum;
        this.leftedge = new int[len];     this.rightedge= new int [len];
        for(int i=0;i<len;i++)
        {leftedge[i]= Integer.MAX_VALUE; rightedge[i]=Integer.MIN_VALUE;}
    }
    public void defineEdgePoint(Point pp){
        Point p = new Point(2);
        if(vertical){p.n[1]=pp.n[2];p.n[2]=pp.n[1];}else{p.n[1]=pp.n[1];p.n[2]=pp.n[2];}
        int i = p.yi()- this.min;
        if (this.leftedge[i] > p.xi()){ this.leftedge[i] = p.xi();}
        if (this.rightedge[i] < p.xi()){ this.rightedge[i] = p.xi();}
    }
    public void shadingEdge(int x1,int y1,int x2, int y2){
        int kx,ky,dx,dy;boolean notSteep=true;
        dx = x2-x1; dy = y2-y1; kx = 1; ky = 1;
        if (dx < 0){ kx = -1; dx = -dx; }
        if (dy < 0){ ky = -1; dy = -dy; }
        if (dx < dy) {this.octant(y1,x1,y2,x2,ky,kx,dy,dx,false);}
        else {this.octant(x1,y1,x2,y2,kx,ky,dx,dy,true);}
    }
    private void octant(int x,int y,int xend,int yend,int kx,int ky,
                                                  int dx, int dy,boolean notSteep){
        int d,j;Point p = new Point(2);
        d = 2*dy-dx; dx = 2*dx; dy = 2*dy;
        if (ky < 0){ d = -d; dx = -dx; dy = -dy;}
        while(true){
            if (notSteep){p.n[1] = x; p.n[2] = y;}else { p.n[1] = y; p.n[2] = x;}
            this.defineEdgePoint(p);
            if (x == xend) return;
            if (d < 0)j = -ky; else j =ky;
            if (j > 0){d = d+dy-dx; y = y+ky;}else d= d+dy;
            x = x +kx;
        }
    }
    public void scanfill(DisplayWindow dW, Color color){
        if (this.vertical) for(int i= 0; i < this.length; i++){
            if(this.leftedge[i] <= this.rightedge[i])
            dW.line(i+this.min, this.leftedge[i], i+this.min, this.rightedge[i], color, null);
        }else  for(int i= 0; i < this.length; i++){
            if(this.leftedge[i] <= this.rightedge[i])
            dW.line(this.leftedge[i], i+this.min, this.rightedge[i],i+this.min, color, null);
        }
    }
}
```

```
public void scanfill(DisplayWindow dW,int len, Color color){
    this.length =len;   scanfill(dW,color);
}
private void octantClip(DisplayWindow dW,  int x,  int y,  int xend,  int yend,
                        int kx,  int ky,  int dx,  int dy,  boolean steep,  Color color){
    int d, j;   Point p = new Point(2);   d = 2*dy-dx; dx = 2*dx; dy = 2*dy;
    if (ky < 0){ d = -d; dx = -dx; dy = -dy;}
    while(true){
        int jj=0; if(vertical)jj=jj+1; if(steep)jj=jj+2;
        switch(jj){
            case 0:if((y-min<0)||(y-min >length)||(x<leftedge[y-min])||(x>rightedge[y-min]))
                        dW.plotPoint(x,y,color);  break;
            case 1:if((y-min<0)||(y-min >length)||(y<leftedge[y-min])||(y>rightedge[y-min]))
                        dW.plotPoint(x,y,color);  break;
            case 2:if((x-min<0)||(x-min >length)||(y<leftedge[x-min])||(y>rightedge[x-min]))
                        dW.plotPoint(y,x,color);  break;
            case 3:if((x-min<0)||(x-min >length)||(x<leftedge[x-min])||(x>rightedge[x-min]))
                        dW.plotPoint(y,x,color);  break;
        } if (x == xend) return;
        if (d < 0)j = -ky; else j =ky;
        if (j > 0){d = d+dy-dx; y = y+ky;}  else d= d+dy;
        x = x +kx;
    }
}
```

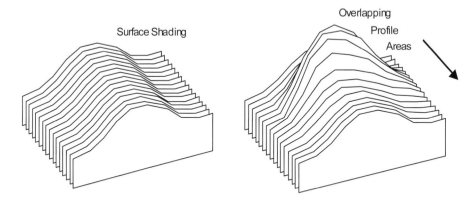

Figure 10.5 Profile block models

If contours are defined to be the set of lines which are created when a set of
parallel planes intersect a surface in three dimensions, then it is clear that grid data
lends itself to presentation in the form of contour drawings. If all the values along
each row or column are projected upwards, into the third dimension and linked
together the result is a "contour" or "profile" drawing of the surface. It is true: profile
lines are not in the conventional orientation used for cartographic contour lines, a
plan view of these profile lines cannot be used to show the shape of the surface.

However, geometrically they are the same thing, and if these profile lines are viewed from an oblique angle the result is a 'block model' presentation of the surface, which as a graphic device works well in portraying the shape of the surface. This is partly because the grid provides a regular frame of reference, and partly, because the lines of the profiles create a shading effect which, conforming to Lambert's reflection law, allows the viewer to perceive a surface representing the shape of the data distribution.

Simply drawing the oblique view of the profile lines however, produces the result shown in Figure 10.6, where nearer profiles overlay and interfere with, further profiles in a confusing way. If each polygon representing a profile is draw in from the back moving forwards filling in the new polygon, then the hidden sections of further away profiles will be removed as shown in Figure 10.7. Although these polygons are not convex they can employ the same polygon fill algorithm developed earlier with one simple modification. Switch the x and y values for each point, then instead of filling with horizontal scan lines, filling with vertical lines, creates the required objective, in the way shown in Figure 10.8.

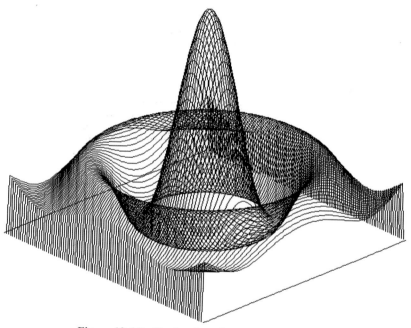

Figure 10.6 Profile drawing of a rotated sync function

Although this approach works, it serves to high light one of the inefficiencies of the painter's algorithm. Where there are many overlapping polygonal areas the same pixel-cells will be visited many times to white out or overwrite what has already been entered into them.

```
public class BlockModel{
    static TextWindow IO = new TextWindow(10,810,1000,100);
    static DisplayWindow dW = new DisplayWindow(IO,10, 10, 1000,800,Color.white);
```

```java
static double minVal = Double.MAX_VALUE,  maxVal= Double.MIN_VALUE;
static double[][] values = null;
static int minx = 0, maxx = 0, len = 0, rows = 0, cols = 0;
static CoordinateFrame frm = new CoordinateFrame();

public static void main(String[] args){
    Point p1=new Point(2); Point p2=new Point(2);
    IO.writeString("please enter the number of rows: ");
    rows = IO.readInteger();IO.readLine();
    IO.writeString("please enter the number of cols: ");
    cols = IO.readInteger();IO.readLine();
    displayFunction();
    IO.writeString("please enter boundary value: ");
    double val = IO.readLongReal(); IO.readLine();
    for(int i=0;i<rows;i++){ values[0][i]= val; values[cols-1][i]= val; }
    for(int i=0;i<cols;i++){ values[i][0]= val; values[i][rows-1]= val; }
    IO.writeString("please enter corners of the display space using the mouse\n");
    Point pa = dW.getCoord();   Point pb = dW.getCoord();
    minx= pa.xi(); maxx = pb.xi(); len= maxx-minx;
    p1.n[1]= 0 ; p1.n[2]= 0;  p2.n[1]= rows+cols ;p2.n[2]= rows*2+cols;
    frm.setScales(pa,pb,p1,p2);
    displaySurface();
}

static void displayFunction(){
    double PI = 3.1415962;
    values = new double[cols][rows];
    double c = ((double)cols)/2.0; double d = ((double)rows)/2;
    for(int j=0; j<rows;j++){
        for(int i=0;i<cols;i++){
            double x = (double)i; double y = (double)j;
            double r = Math.sqrt((c-x)*(c-x)+(d-y)*(d-y));
            double X = 3.0*PI*r/c;
            double sinx = Math.sin(X);
            if(X==0)values[i][j] = 1.0;  else values[i][j] = sinx/X;
        }
    }
}

static void displaySurface(){
    double x = (double)cols;         double y = (double)(cols+rows);
    double h = (double)rows*2.0;      double range = x+y;
    Point p1 = new Point(2);         Point p2 = new Point(2);        Point p3 = new Point(2);
    Point p4 = new Point(2);         Point pa = new Point(2);        Point pb = new Point(2);
    Point pc = new Point(2);         Point pd = new Point(2);
    Point pstart = new Point(2);     Point pend = new Point(2);
    Shadings s=null;
    for(int i = 0; i<rows-1; i++){
```

```
        len=maxx-minx;
        s=new Shadings(IO,maxx-minx,minx,true);
        double xx =x, yy=y;
        for (int j=0; j<cols-1; j++){
            p1.x("=",xx); p1.y("=",yy + h*values[cols-1-j][rows-1-i]);
            p2.x("=",xx+1.0); p2.y("=",yy - 1.0 + h*values[cols-1-j][rows-2-i]);
            p3.x("=",xx); p3.y("=",yy - 2.0 + h*values[cols-2-j][rows-2-i]);
            p4.x("=",xx-1.0); p4.y("=",yy -1.0 + h*values[cols-2-j][rows-1-i]);
            if(j==0)pstart = frm.scaleWtoS(p2);
            if(j==(cols-2))pend = frm.scaleWtoS(p3);
            pa = frm.scaleWtoS(p1);    pb = frm.scaleWtoS(p2);
            pc = frm.scaleWtoS(p3);    pd = frm.scaleWtoS(p4);
            s.shadingEdge(pb.xi(),pb.yi(),pc.xi(),pc.yi());
            dW.plotLine(pa,pd,Color.black);
            xx=xx-1; yy=yy-1;
        }
        s.shadingEdge(pstart.xi(),pstart.yi(),pend.xi(),pend.yi());
        s.scanfill(dW,Color.lightGray);
        dW.plotLine(pstart,pend,Color.black);
        x=x+1; y=y-1;
    }
  }
}
```

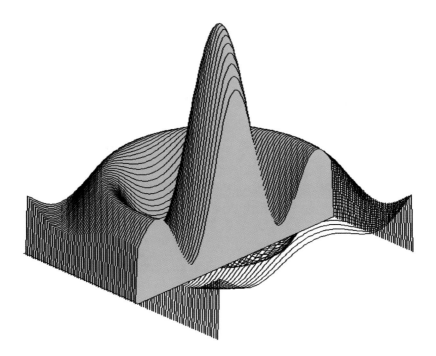

Figure 10.7 Overlaying profiles using the painter's algorithm

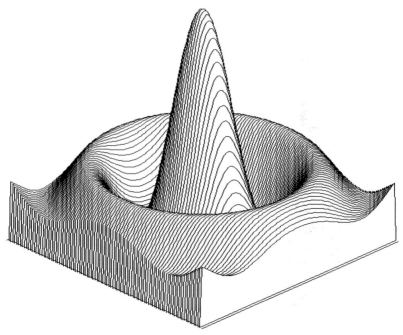

Figure 10.8 Block model with hidden lines overwritten

An early approach to this problem was to use a hidden line removal algorithm rather than the hidden area approach of the painter's algorithm. Frank Rens implemented this approach in the SYMVU program to create block model drawings of topographic surfaces, using the Calcomp plotter, in the way illustrated in Figure 10.10 by using line masking within the line interpolation procedure.

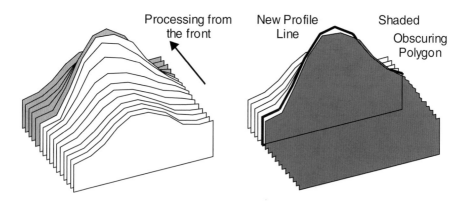

Figure 10.9 Profile block model drawing

Using the array of data-values output from both SYMAP and GRID cartography programs, SYMVU generated profile block model drawings of both abstract statistical surfaces and terrain models. Each profile was considered as a three-dimensional polygon, transformed as required by the viewing position. Profiles nearest to the observer were drawn first, but were only drawn in where previously drawn profile sections did not mask them in the way shown in Figure 10.9. To carry out this process it was necessary to implement a masking operation to remove new section-lines, which passed behind already drawn profiles. This mask was defined as the union of all previously drawn profile polygons. Each time that a profile line was completed this mask had to be updated, to include any new outstanding areas.

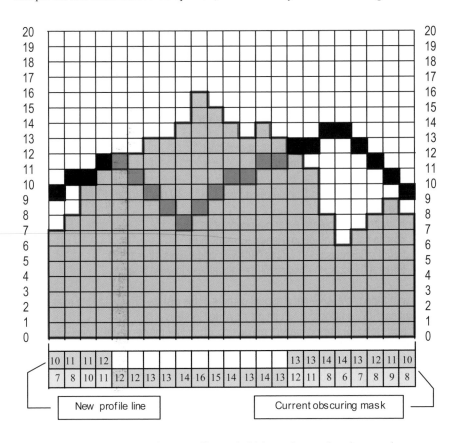

Figure 10.10 Updating a profile mask: higher values replace lower values

In SYMVU Frank Rens implemented this operation as part of the line interpolation procedure, which generated Calcomp plotting instructions. The mechanism is shown diagrammatically in Figure 10.10. The maximum y values for all profile line points so far drawn are stored in an array, one for each x position of the plotting grid. Only new line points which have y co-ordinates greater than these stored values are plotted, and when they are they replace the existing values. A 20-inch plotter with 200 steps per inch only requires a Mask array of 4000 y values to support the program code:

```
if (y > Mask[x]) { Mask[x] := y;  Plot(x,y) };
```

Including this statement in the plotting loop of Bresenham's line drawing algorithm gave a very fast hidden line removal procedure, implementing, perhaps, one of the first examples applying the depth buffer principle.

```
public class BlockModel{
    static TextWindow IO = new TextWindow(10,810,1000,100);
    static DisplayWindow dW = new DisplayWindow(IO,10, 10, 1000,800,Color.white);
    static double minVal = Double.MAX_VALUE, maxVal = Double.MIN_VALUE;;
    static double[][] values = null; static boolean vertical = false;
    static int minx = 0, maxx = 0, lenx = 0, miny = 0, maxy = 0, leny = 0;
    static int rows = 0, cols = 0; static String str ="";
    static CoordinateFrame frm = new CoordinateFrame();
    public static void main(String[] args){
        Point p1=new Point(2); Point p2=new Point(2);
        IO.writeString("please enter the number of rows: ");
        rows = IO.readInteger();IO.readLine();
        IO.writeString("please enter the number of cols: ");
        cols = IO.readInteger();IO.readLine();
        displayFunction();
        IO.writeString("set boundary elements y/n: ");  str = IO.readString();IO.readLine();
        if(str.equals("y")){
            IO.writeString("please enter boundary value: ");
            double val = IO.readLongReal(); IO.readLine();
            for(int i=0;i<rows;i++){  values[0][i]= val; values[cols-1][i]= val;  }
            for(int i=0;i<cols;i++){ values[i][0]= val; values[i][rows-1]= val; }
        }
        IO.writeString("please enter diagonal corners of the display area using the mouse\n");
        Point pa = dW.getCoord();   Point pb = dW.getCoord();
        if (pa.xi()<pb.xi()) { minx= pa.xi(); maxx = pb.xi(); }else{ maxx= pa.xi(); minx = pb.xi(); }
        if (pa.yi()<pb.yi()){ miny= pa.yi(); maxy = pb.yi(); }else{ maxy= pa.yi(); miny = pb.yi();  }
        lenx= maxx-minx+1; leny= maxy-miny+1;
        p1.n[1]= 0 ; p1.n[2]= 0; p2.n[1]= rows+cols ;  p2.n[2]= rows*3+cols;
        frm.setScales(pa,pb,p1,p2);
        displaySurface(true);
    }
    static void displayFunction(){
        double PI = 3.1415962;
        values = new double[cols][rows];
        double c = ((double)cols)/2.0; double d = ((double)rows)/2;
        for(int j=0; j<rows;j++){
            for(int i=0;i<cols;i++){
                double x = (double)I, y = (double)j,  r = Math.sqrt((c-x)*(c-x)+(d-y)*(d-y));
                double X = 3.0*PI*r/c,  sinx = Math.sin(X);
                if(X==0)values[i][j] = 1.0;  else values[i][j] = sinx/X;
            }
        }
    }
}
```

```
static void displaySurface(boolean vertical){
    double x = (double)rows,  y = (double)0,  h = (double)rows*1.0;
    Point p1 = new Point(2); Point p2 = new Point(2);  Point p3 = new Point(2);
    Point p4 = new Point(2); Point pa = new Point(2);  Point pb = new Point(2);
    Point pc = new Point(2); Point pd = new Point(2);
    Shadings s=null;
    if(vertical){ s=new Shadings(IO,maxx-minx+1,minx,true); }
    else{ s=new Shadings(IO,maxy-miny+1,miny,false); }
    s.shadingEdge(minx,800,maxx,800);
    Point ppp[] = new Point[rows];
    for(int i = 0; i<rows-1; i++){
        double xx =x, yy=y;
        for (int j=0; j<cols-1; j++){
            p1.x("=",xx); p1.y("=",yy + h*values[j][i]);        pa = frm.scaleWtoS(p1);
            p2.x("=",xx-1.0); p2.y("=",yy + 1.0 + h*values[j][i+1]);   pb = frm.scaleWtoS(p2);
            p3.x("=",xx); p3.y("=",yy  +2+ h*values[j+1][i+1]);   pc = frm.scaleWtoS(p3);
            p4.x("=",xx+1.0); p4.y("=",yy +1.0 + h*values[j+1][i]);   pd = frm.scaleWtoS(p4);
            if(i==50){ s.lineClip(dW,pa.xi(),pa.yi(),pd.xi(),pd.yi(),Color.red);
            }else if(i<50){ s.lineClip(dW,pa.xi(),pa.yi(),pd.xi(),pd.yi(),Color.black);
            }else{ s.lineClip(dW,pa.xi(),pa.yi(),pd.xi(),pd.yi(),Color.green);
            ppp[j]=pa;  xx=xx+1; yy=yy+1;
        } ppp[cols-1]=pd;
        for (int j=0; j<cols-1; j++){
            s.shadingEdge(ppp[j].xi(),ppp[j].yi(),ppp[j+1].xi(),ppp[j+1].yi());
        } x=x-1; y=y+1;
    }
}
```

b   c

a   d

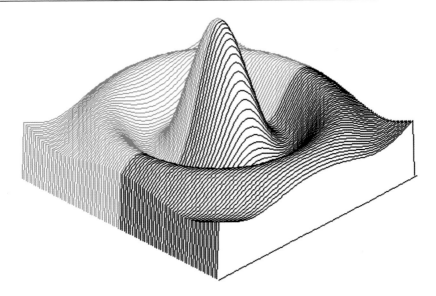

Figure 10.11 Central profile line drawn in red

The generation and use of the obscuring polygon as the union of previously processed profile lines is independent of the selection and drawing of the actual lines from each grid data cell that are used in the display. In Figure 10.12 the profile polygons used for hidden line removal are only used to render parts of the surface. In selected areas coloured, diagonal, red, green and blue lines are drawn in across the grid-data cells.

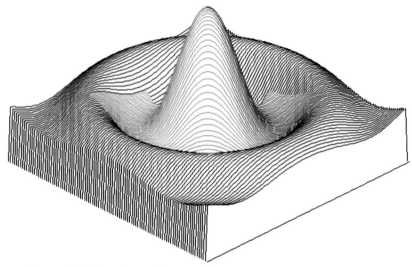

Figure 10.12 Changing cell based line segments in a profile drawing

```
s.shadingEdge(minx,800,maxx,800);  Point ppp[] = new Point[rows];
for(int i = 0; i<rows-1; i++){
    double xx =x, yy=y;
    for (int j=0; j<cols-1; j++){
        p1.x("=",xx); p1.y("=",yy + h*values[j][i]);           pa = frm.scaleWtoS(p1);
        p2.x("=",xx-1.0); p2.y("=",yy + 1.0 + h*values[j][i+1]);  pb = frm.scaleWtoS(p2);
        p3.x("=",xx); p3.y("=",yy +2+ h*values[j+1][i+1]);       pc = frm.scaleWtoS(p3);
        p4.x("=",xx+1.0); p4.y("=",yy +1.0 + h*values[j+1][i]);   pd = frm.scaleWtoS(p4);
        if ((i>=50)&&(i<80)&&(j>20)&&(j<50))
            s.lineClip(dW,pb.xi(),pb.yi(),pd.xi(),pd.yi(),Color.green);
        else if ((i>20)&&(i<50)&&(j>20)&&(j<50))
            s.lineClip(dW,pb.xi(),pb.yi(),pd.xi(),pd.yi(),Color.red);
        else if ((i>20)&&(i<50)&&(j>=50)&&(j<80))
            s.lineClip(dW,pb.xi(),pb.yi(),pd.xi(),pd.yi(),Color.blue);
        else  s.lineClip(dW,pa.xi(),pa.yi(),pd.xi(),pd.yi(),Color.black);
        } ppp[j]=pa;  xx=xx+1; yy=yy+1;
    }
    ppp[cols-1]=pd;
    for (int j=0; j<cols-1; j++){
        s.shadingEdge(ppp[j].xi(),ppp[j].yi(),ppp[j+1].xi(),ppp[j+1].yi());
    } x=x-1; y=y+1;
}
```

The line clipping operation can be implemented in the *Shadings* class in the following way. This allows simple profile drawings to be constructed but also allows other surface lines to be drawn on the surface of the block model. Employing profile polygons for hidden-line or hidden-area removal however, has limitations using either the painter's algorithm, or an obscuring polygon. Both these methods can be improved by processing the surface, in grid-cell steps rather than in profile strips.

```
private void octantClip( DisplayWindow dW,int x,int y,int xend,int yend,int kx,int ky,
                         int dx, int dy,boolean steep,Color color){
    int xx=0,yy=0, d, j;   Point p = new Point(2);
    d = 2*dy-dx; dx = 2*dx; dy = 2*dy;
    if (ky < 0){ d = -d; dx = -dx; dy = -dy;}
    while(true){
        if(steep){xx=y;yy=x;}else{xx=x;yy=y;}
        if(vertical){
            if((leftedge[xx-min]>yy)||(rightedge[xx-min]<yy))dW.plotPoint(xx,yy,color);
        }else{
            if((leftedge[yy-min]>xx)||(rightedge[yy-min]<xx))dW.plotPoint(xx,yy,color);
        } if (x == xend) return;
        if (d < 0)j = -ky; else j =ky;
        if (j > 0){d = d+dy-dx; y = y+ky;}else d= d+dy;
        x = x +kx;
    }
}
public void lineClip(DisplayWindow dW, int x1,int y1,int x2, int y2,Color color){
    int kx, ky, dx, dy, x, y, j=0,k=0;
    dx = x2-x1; dy = y2-y1; kx = 1; ky = 1;
    if (dx < 0){ kx = -1; dx = -dx; }
    if (dy < 0){ ky = -1; dy = -dy; }
    if (dx < dy){this.octantClip(dW,y1,x1,y2,x2,ky,kx,dy,dx,true,color);}
    else {this.octantClip(dW,x1,y1,x2,y2,kx,ky,dx,dy,false,color);}
}
```

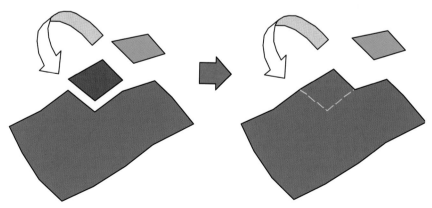

Figure 10.13 Incremental cell by cell update of the obscuring polygon

This cell-based approach has already been used in the painter's algorithm, in chapter 4 in Figure 4.30. However it is possible to obtain the same improvement for the obscuring polygon technique, by adding the boundary of each grid cell tile to the obscuring polygon as it is processed as shown in Figure 10.13.

```
s.shadingEdge(minx,800,maxx,800);
for(int i = 0; i<rows-1; i++){
    double xx =x, yy=y;
    for (int j=0; j<cols-1; j++){

        p1.x("=",xx); p1.y("=",yy + h*values[j][i]);            pa = frm.scaleWtoS(p1);
        p2.x("=",xx-1.0); p2.y("=",yy + 1.0 + h*values[j][i+1]);  pb = frm.scaleWtoS(p2);
        p3.x("=",xx); p3.y("=",yy +2+ h*values[j+1][i+1]);       pc = frm.scaleWtoS(p3);
        p4.x("=",xx+1.0); p4.y("=",yy +1.0 + h*values[j+1][i]);   pd = frm.scaleWtoS(p4);

        if ((j>=50)&&(j<80)&&(i>20)&&(i<50))
            s.lineClip(dW,pb.xi(),pb.yi(),pc.xi(),pc.yi(),Color.green);
        else if ((j>20)&&(j<50)&&(i>20)&&(i<50))
            s.lineClip(dW,pb.xi(),pb.yi(),pd.xi(),pd.yi(),Color.red);
        else if ((j>20)&&(j<50)&&(i>=50)&&(i<80))
            s.lineClip(dW,pc.xi(),pc.yi(),pd.xi(),pd.yi(),Color.blue);
        else   s.lineClip(dW,pc.xi(),pc.yi(),pd.xi(),pd.yi(),Color.lightGray);

        if(j==0){s.lineClip(dW,pb.xi(),pb.yi(),pc.xi(),pc.yi(),Color.black);
                 s.lineClip(dW,pc.xi(),pc.yi(),pd.xi(),pd.yi(),Color.black);}
        if(j==97)  s.lineClip(dW,pc.xi(),pc.yi(),pd.xi(),pd.yi(),Color.black);
        if((j==49))s.lineClip(dW,pc.xi(),pc.yi(),pd.xi(),pd.yi(),Color.black);
        if(i==0)  s.lineClip(dW,pc.xi(),pc.yi(),pb.xi(),pb.yi(),Color.black);
        if(i==49)  s.lineClip(dW,pb.xi(),pb.yi(),pc.xi(),pc.yi(),Color.black);
        if(i==97)  s.lineClip(dW,pc.xi(),pc.yi(),pb.xi(),pb.yi(),Color.black);

        if(j==0) s.shadingEdge(pa.xi(),pa.yi(),pb.xi(),pb.yi());
        s.shadingEdge(pb.xi(),pb.yi(),pc.xi(),pc.yi());
        s.shadingEdge(pc.xi(),pc.yi(),pd.xi(),pd.yi());
        if(i==0) s.shadingEdge(pd.xi(),pd.yi(),pa.xi(),pa.yi());
        xx=xx+1; yy=yy+1;
    }
    x=x-1; y=y+1;
}
```

The code segment above produced the output shown in Figure 10.14. The use of cell-based updates to the obscuring polygon allows a better treatment of the right outer edge of the block model, and makes it easier to avoid line leakage between profiles at the peripheries of the drawing. One remaining problem with this scheme, which will be explored more fully in a later chapter, is that the left and lower edges of a new grid cell tile will be masked out by this immediate update of the obscuring polygon. This can still cause drawing errors if care in ordering instructions is not taken.

```
static void displayFunction(){
    double PI = 3.1415962, valx=0,valy=0;  values = new double[cols][rows];
    double c = ((double)cols)/2.0;   double d = ((double)rows)/2;
    for(int j=0; j<rows;j++){
        for(int i=0;i<cols;i++){
            double x = 4*PI*(double)(i-c)/cols; double y = 4*PI*(double)(j-c)/cols;
            if (x==0) valx = 1.0;  else valx = Math.sin(x)/x;
            if (y==0) valy = 1.0;  else valy = Math.sin(y)/y;
            values[i][j]= valx*valy;
        }
    }
}
```

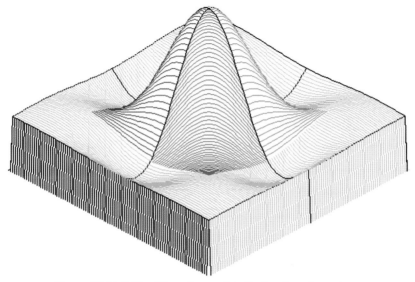

Figure 10.14 Grid cell boundary update for the obscuring polygon

The algorithms outline in this chapter originated in work carried out on two cartography programs. The first was OBLIX an experimental program set up to extend the capabilities of SYMVU in the laboratory for Computer Graphics and Spatial Analysis in Harvard University in 1968-69, the second was GIMMS started in the same place in 1970 as a Geographic Information Management and Mapping System. The use of the obscuring polygon was initially a hidden-line removal strategy. The hidden area approach of the painter's algorithm evolved as the simplest way to create three dimensional images, using the framestore developed during the seventies to support raster display systems, such as the "Bugstore" in Cambridge University. The area fill algorithms were a simplification of line shading developed for GIMMS, and the surface image or texture mapping onto block model surfaces was developed for OBLIX. OBLIX was also used to explore scan-line shading again based on line fill algorithms. What is interesting now is that as the resolution of displays has increased much line drawing has become an area-fill operation.

Circles and Thick Lines

The scan line fill of polygons based on interpolating their boundary points onto a grid can be extended to fill convex areas initially represented in other ways. The simplest of these is the circle represented by an algebraic equation. If the boundary points of the circle can be interpolated, then filling in the scan lines from the left edge to the right edge gives an efficient way of shading the circular area of a disk.

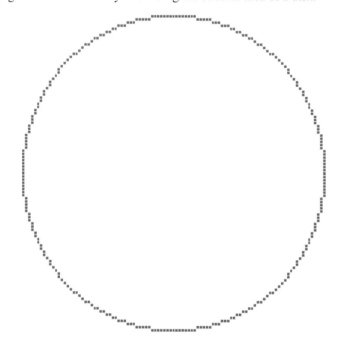

Figure 10.15 Circle interpolation

In chapter 3 a tile pattern for a circular area was obtained by testing each grid point for being inside or outside a circle using the sign of k in the equation:

$$k = x^2 + y^2 - r^2$$

to determine whether the point was inside the circle or not. By generating the boundary points, then using scan lines: an order $O(n)$ complex operation dependent on the circle equation is employed followed by a simpler and smaller $O(n^2)$ fill operation for the circle, compared with a more complex $O(n^2)$ operation applied to the whole area.

Circle interpolation can be based on the same principles employed for straight-line interpolation by dividing the circle into eight octants and then filling in the point sequence for each octant. What is useful in the case of the circle is its symmetry, only one octant stepping sequence needs to be determined and the same sequence by the appropriate reflection and translation operations can be applied to the other seven octants. Applying this approach generated the display shown in Figure 10.15.

If the equation of a circle centred on the origin, is written in the following way:

$$0 = x^2 + y^2 - r^2$$

It is possible to define the distance of a point (x, y) from this curve in a similar way to that employed in the case of the straight line by the value of k given by the following equation:

$$k = x^2 + y^2 - r^2 \qquad\qquad 1$$

If a step in the x direction of Δx is taken then this distance k becomes:

$$k + \Delta k = (x + \Delta x)^2 + y^2 - r^2$$
$$k + \Delta k = x^2 + 2.x.\Delta x + \Delta x^2 - r^2 \qquad\qquad 2$$

Subtracting 1 from 2 gives:

$$\Delta k = 2.x.\Delta x + \Delta x^2 \qquad\qquad 3$$

If the increment is a unit step then, for both steps in x, and steps in y, the values of Δk become:

$$\Delta k_x = 2.x + 1$$
$$\Delta k_y = 2.y + 1 \qquad\qquad 4$$

These values will themselves change with position:

$$\Delta k_x = 2.x + 1$$
$$\Delta k_x + \Delta^2 k_x = 2.(x + \Delta x) + 1 \qquad\qquad 5$$
$$\Delta^2 k_x = 2.\Delta x = 2$$

Similarly: $\qquad\qquad \Delta^2 k_y = 2.\Delta y = 2 \qquad\qquad 6$

If the circle is divided up into octants in the way shown in Figure 10.16 then starting in octant 1, at the top and moving to the right, the slope of the curved line will vary from a horizontal tangent line to one at 45° to the horizontal. This can be approximated by a point sequence on a grid using the simple eight-neighbour, line-incrementing scheme used for straight-lines, selecting either a move in a horizontal direction or in a downward diagonal direction for the next point depending on the position on the circle. The curve following algorithm simply has to select the point that lies closest to the true circle for each step taken.

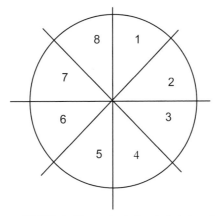

Figure 10.16 Labelling octants for circle interpolation

Starting with octant 1, the incrementing sequence can be defined in the following way.

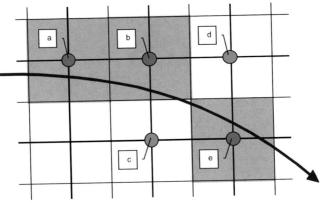

Figure 10.17 Select the nearest point using minimum k

Starting from position 'a' in Figure 10.17 the value of k is incremented using the relationship to define the new distance from position 'b':

$$\text{ktop} = k_b = k_a + \Delta k_x$$

and by the relationship to define the distance of the curve from position 'c':

$$\text{kbot} = k_c = k_a + \Delta k_x + \Delta k_y$$

In this case, moving from point 'a', 'ktop' is less than 'kbot' so point b is selected and cell b painted, in the way shown. The process is repeated for the next point, however, before incrementing the value of k, the two increments Δk_x and Δk_y must themselves be incremented to be applicable to the new position. The x increment is automatically assumed for each step in this octant, the choice is therefore between the two points 'd' and 'e'. In the example shown point 'e' is the nearer of the two and is shaded in accordingly.

```
int x=0, y = radius,k=0, kx = 1, ky = -2*radius+2,ktop,kbot;
while(x<=y){
    //  Paint pixel ( x, y) red.
    ktop = k+kx;  kbot = k+ky;
    if(ktop <= -kbot){ k = ktop;  ky = ky + 2; kx = kx + 2; x = x+1; }
    else{ k = kbot;  ky = ky + 4; kx = kx + 2;  x = x+1;   y = y-1; }
}
```

This gives a very simple incrementing algorithm for one octant. The pattern of increments defined by this repeating loop can then be transferred to the other octants by a process of reflection and translation, in a similar way to that employed interpolating straight lines in different octant directions.

```
f.writeString("please enter radius ");
int radius = f.readInteger(); int range = 3* radius;
int Cls = range, Rws = range;
Tiles T = new Tiles(Cls,Rws,Color.white);
Grid d = new Grid(f,T.tileColour,T.cols,T.rows);
d.paintGridArray(); d.drawDualGrid(Color.black,Color.gray);
int x=0, y = radius,k=0, kx = 1, ky = -2*radius+2,ktop,kbot;
while(x<=y){
    d.paintInnerCell( x+range/2, y+range/2,0,Color.red);
    d.paintInnerCell(-x+range/2, y+range/2,0,Color.red);
    d.paintInnerCell( x+range/2,-y+range/2,0,Color.red);
    d.paintInnerCell(-x+range/2,-y+range/2,0,Color.red);
    d.paintInnerCell( y+range/2, x+range/2,0,Color.red);
    d.paintInnerCell(-y+range/2, x+range/2,0,Color.red);
    d.paintInnerCell( y+range/2,-x+range/2,0,Color.red);
    d.paintInnerCell(-y+range/2,-x+range/2,0,Color.red);
    ktop = k+kx;   kbot = k+ky;
    if(ktop <= -kbot){  k = ktop;  ky = ky + 2; x = x+1; }
    else{ k = kbot;  ky = ky + 4;  x = x+1;   y = y-1; }
    kx = kx + 2;
}
```

The circle and sectors of the circle are necessary elements in many drawings. However, the full circle also makes an ideal link between line segments for drawing *thick* "poly-line" arcs in the way illustrated diagrammatically in Figure 10.18. A simple way of implementing thick line drawing is to pass the boundary points for the two endpoint-circles and the rectangle representing the main body of each line segment to the edge generating method used for polygon shading, then filling in the resulting border with scan line segments. The selection of the minimum x values for the left- hand edge and the maximum x values for the right hand edge removes the redundant lines. An alternative is to generate the endpoint circles but only create the top and bottom lines tangent to the two circles.

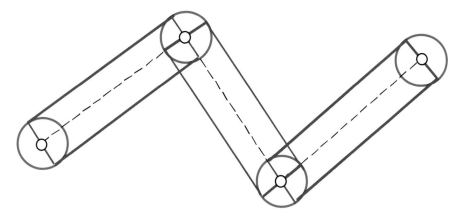

Figure 10.18 Thick poly-line arcs

```java
class ThickLine{
    public Point pa,pb,pc,p1,p2;      public Shadings s = null;
    public int ymin,len,r,incx,incy,dx,dy,x,y,kx,ky,k,ktop,kbot;
    public int dk1,dk2,dk3,dk4,dk5,dk6,dk7,dk8, mink=10000000;
    private DisplayWindow d=null;  private TextWindow IO=null;
    ThickLine(Point pp1,Point pp2,int w,DisplayWindow d,TextWindow f){
        this.d = d;  this.IO = f;   r = w/2;
        pa = new Point(2);  pb = new Point(2);  pc = new Point(2);
        p1 = new Point(2);  p2 = new Point(2);  p1.c("<-", pp1);  p2.c("<-", pp2);
        if (p2.yi() > p1.yi()){ len = p2.yi()-p1.yi()+1+w;   ymin = p1.yi()-r;}
        else { len = p1.yi()-p2.yi()+1+w;  ymin = p2.yi()-r;}
        s = new Shadings(IO,len,ymin, false);
        incx = (p2.xi()-p1.xi());      incy = (p2.yi()-p1.yi());
        dk6= dk7 =  -(dk2= dk3 = (p2.xi()-p1.xi())*r);
        dk4 = dk5 =-(dk1=dk8=(p2.yi()-p1.yi())*r);
        int x1=0;  int y1 = r;  kx=1;  ky = -2*r+2; k=0;
        while(x<=y){
            x=x1; y=y1;  ktop = k+kx;  kbot = k+ky;
            if(ktop <= -kbot) { k = ktop;  ky = ky + 2;  x1 = x+1;  dy = 0;}
            else { k = kbot;  ky = ky+4;  x1 = x+1;  y1 = y-1;  dy = -1;}
            kx = kx+2;  dx = 1;
            dk1 = edgePnt(dk1, x,  y,  dx, dy ); dk2 = edgePnt(dk2, y,  x,  dy, dx );
            dk3 = edgePnt(dk3,-y,  x,  dy,-dx ); dk4 = edgePnt(dk4, x, -y,  dx,-dy );
            dk5 = edgePnt(dk5,-x, -y, -dx,-dy );dk6 = edgePnt(dk6,-y, -x, -dy,-dx );
            dk7 = edgePnt(dk7, y, -x, -dy, dx ); dk8 = edgePnt(dk8,-x,  y, -dx, dy );
        }
        pa.x("=",p1.xi()+pc.xi());pa.y("=",p1.yi()+pc.yi());
        pb.x("=",p2.xi()+pc.xi());pb.y("=",p2.yi()+pc.yi());
        d.line(pa.xi(), pa.yi(), pb.xi(), pb.yi(), null, s);
        pa.x("=",p1.xi()-pc.xi());pa.y("=",p1.yi()-pc.yi());
        pb.x("=",p2.xi()-pc.xi());pb.y("=",p2.yi()-pc.yi());
        d.line(pa.xi(), pa.yi(), pb.xi() ,pb.yi(), null, s);
    }
    public int edgePnt(int dk,int x,int y,int dx,int dy){
        if(dk<0){if(-dk<mink){mink = -dk; pc.x("=",x); pc.y("=",y);}
        }else if(dk<mink) { mink = dk;  pc.x("=",x); pc.y("=",y);}
        pa.x("=", p1.xi()+x); pa.y ("=", p1.yi()+y);
        pb.x("=", p2.xi()+x); pb.y( "=", p2.yi()+y);
        s.defineEdgePoint(pa);s.defineEdgePoint(pb);
        return dk+incx*dx+incy*dy;
    }
    public void draw(Color cc){
        if (r<1){ d.plotLine(p1,p2,cc);return;}
        for(int i= 0; i<len; i++) { if(s.leftedge[i]<s.rightedge[i])
            d.line(s.leftedge[i],i+ymin,s.rightedge[i],i+ymin,cc,null);}
    }
}
```

This is done by finding the point on the boundary of an endpoint circle that lies closest to the line, through the centre of the circle, which is normal to the axis of the "thick" line. The end points of the tangent lines linking these circles are then generated as symmetrical offsets from this point, before linking them by linear interpolations. The code to implement this approach is given above; and the output it produces for a thick poly-line, drawn line segment by line segment, is shown in Figure 10.19.

Figure 10.19 Thick shaded poly-line

Shading more Complex Polygons

The algorithms decribed above depend either on having convex shapes or polygons that can be treated as only having one left and one right side. In the general case it is necessary to shade polygons with more complex shapes and even to handle polygons with boundaries that self intersect.

One possibility is to take the more complex polygon and subdivide it into a set of simpler pieces that can be shaded separately. Again there are several ways in which this can be done. One is to triangulate the interior region of the polygon, another is to subdivide the area into regions that have single left and right side sequences of edge points that can be processed by the existing algorithms.

There are two test polygons that pose contrasting problems for these subdivision shading algorithms. Shown in Figure 10.20 one is a spiral polygon and the other is a zigzag shape where concave sections of the boundary interleave. There are also polygons that contain holes which also create their own problems. Triangulation is an important process that is returned to in a later chapter. At first sight it does not appear to be a difficult task, certainly as a manual drawing exercise it is relatively easy. If starting points such as those labelled A in Figure 10.20 that are stationary points are identified -- where the boundary changes from a downwards direction to an upwards

This process generates all the new vertex points needed to define the trapezium strips shown in Figure 10.21a which can then be triangulated by inserting diagonals. However there are many more triangles than are necessary. If the new point generated by each vertex, as it is encountered, is compared with any previous point generated for the same vertex, and the nearer point selected in the appropriate direction (depending on the orientation of the current line segment), then only those points needed to create the trapezium strips shown in Figure 10.21b will be retained: thus giving a simpler triangulation.

This processing strategy using a threaded list to access vertices in coordinate order will be discussed in a later chapter to calculate the intersection points in overlapping polyline arcs. The structure of this triangulation of a polygon depends on the coordinate axes chosen, again, in a late chapter triangulation techniques that give the same result for a set of points whatever coordinate framework is used to represent them, will be examined.

One reason for triangulating polygons is that many display processor boards include triangle shading as a hardware or firmware primitive. Where the display system processes triangles then further work is unnecessary however if the shading has to be carried out in the program then if larger areas than triangles can be used it should simplify the subdivision task. A first step in this direction would be to subdivide complex polygons into regions that have single left and right-side edgepoint sequences. A similar approach to the definition of trapezium strips can be employed to this approach in the following way:

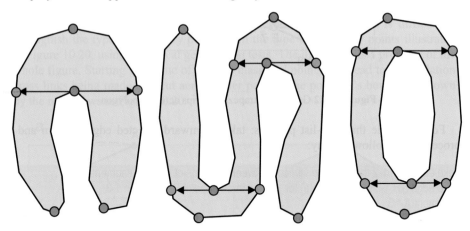

Figure 10.23 Subdivision based on horizontal lines through stationary points

Again the vertices will have to be sorted into (y,x) coordinate order, but only subdivisions through stationary points are necessary to give the required regions. When these subregions are shaded-in they will still have to have their boundary points sorted from boundary order to coordinate order to define the shading scan lines, albeit using a fast bucket sort into the left and right edge arrays. A simpification to this algorithm can therefore be obtained by unifying the two sorting stages into one. Using a "natural merge sort" will also take advantage of highly ordered boundary sequences to give a fast and efficient process.

There are several ways the polygon fill algorithm can be extended in this way. One is illustrated in Figure 10.23. In this case the shading scan-line end points are calculated directly rather than selecting them from the boundary line's pixel grid points. The boundary line is still interpolated but only where it cuts the scan lines. This is done by calculating boundary line x values for unit steps in y irrespective of the boundary line segment's slope. Each line segment is interpolated into a new array, in polygon boundary order, but this is then merge sorted with a list of the previous boundary points arranged in an array in y-x order. This final array is then used to fill the complete polygon by taking pairs of points in turn as the ends of shading scan lines. The principle is the same as before the points are created in boundary order but are then sorted, not by the "bucket sort" provided by the left and right edge arrays but by a "natural merge sort" to give points in the raster scan order needed for subsequent scanline fill.

Figure 10.24 A general polygon network fill algorithm with overlaid boundaries

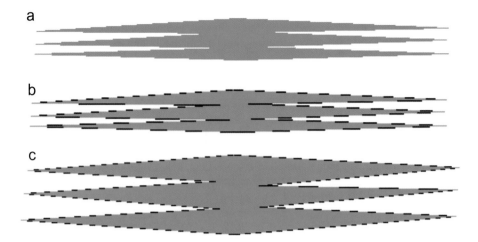

Figure 10.25 Shading algorithm with (a) & without (b,c) redrawn edges

```
public class PolygonShading{   //  general polygon fill
    static TextWindow IO = new TextWindow(0,800,800,100 );
    static DisplayWindow dW=new DisplayWindow(IO,1,1,800,800,Color.white);

    public static void main(String[] args){
        IO.writeString("please enter the number of vertices ");
        int num= IO.readInteger(); IO.readLine();
        Polygon poly = new Polygon();
        Point[] p = new Point[num+1];
        p[0]= dW.getCoord();
        for(int i=1;i<num;i++){
        p[i]= dW.getCoord();
        dW.plotLine(p[i-1],p[i],Color.black);
        }
        p[num]= p[0];
        dW.plotLine(p[num-1],p[num],Color.black);
        poly.p=p;
        shade(poly,Color.green);
        for(int i=0;i<num;i++){ dW.plotLine(p[i],p[i+1],Color.green);}
    }
    static void shade(Polygon p, Color cc){
        Point[] scanArray = null,s=null;
        s = scanArray = scan(p);
        if(s!=null){
            for(int i=0; i<s.length; i=i+2){
                dW.plotLine(s[i],s[i+1],cc);
            }
        }else IO.writeString("no scan array");
    }
```

```
static Point [ ] scan(Polygon p){
    int i = 0, j=0, starty=0, endy=0, starti=0, endi=0, len=0;
    if(p == null)return null;
    Point[] changeArray = null , newArray = null, oldArray = null;
    while(i+1 < p.p.length){
        if(p.p[i].yi()!=p.p[i+1].yi()){
            if(p.p[i].yi()<p.p[i+1].yi()){
                starty = p.p[i].yi(); starti = j = i;
                while((j+1<p.p.length)&&(p.p[j].yi()<p.p[j+1].yi())){j=j+1;}
                endy = p.p[j].yi();endi = j;
            }else if(p.p[i].yi()>p.p[i+1].yi()){
                endy = p.p[i].yi();starti =j=i;
                while((j+1<p.p.length)&&(p.p[j].yi()>p.p[j+1].yi())){j=j+1;}
                starty = p.p[j].yi();endi = j;
            }
            changeArray = new Point[endy-starty];
            for(int k = starti; k<endi; k++){
                lineSegment(p.p[k], p.p[k+1], changeArray, starty);
            } oldArray = newArray;
            newArray = merge(oldArray,changeArray);
            i= endi;
        }
        else i++;
    } return newArray;
}
static Point[] merge(Point[] oA, Point[] cA){
    if (oA==null)return cA;
    if (cA==null)return oA;
    Point[] nA = new Point[oA.length+cA.length];
    int i=0,j=0,k = 0;
    while(((i<cA.length)||(j<oA.length))){
        int t = 0;
        if((i<cA.length)&&(j<oA.length))t=1;
        if((i>=cA.length)&&(j<oA.length)||((t==1)&&(compare(oA[j],"<=",cA[i])))){
            nA[k] = new Point(2);  nA[k].x("=",oA[j].xd());
            nA[k].y("=",oA[j].yd()); nA[k].tag = oA[j].tag;
            k=k+1;  j=j+1;
        }
        t=0;
        if((i<cA.length)&&(j<oA.length))t=1;
        if((j>=oA.length)&&(i<cA.length) ||((t==1)&&(compare(cA[i],"<=",oA[j])))){
            nA[k] = new Point(2); nA[k].x("=", cA[i].xd());
            nA[k].y("=", cA[i].yd()); nA[k].tag = cA[i].tag;
            k=k+1;i=i+1;
        }
    } return nA;
}
```

```
static void lineSegment(Point p1,Point p2, Point[] array,int start){
    double xx = 0.0; int x,y,x1,y1,x2,y2;
    if(p1.yi()==p2.yi())return;
    double incr = ((double)p1.xi()-p2.xi())/(p1.yi()-p2.yi());
    if(p1.yi() < p2.yi()){
        y1=p1.yi();  y2=p2.yi();  xx = (double)(x = p1.xi());
    }else{
        y1=p2.yi();  y2=p1.yi();  xx = (double)(x = p2.xi());
    }
    for(y = y1-start; y < y2-start; y++){
        array[y] = new Point(2);  array[y].x("=",x); array[y].y("=",y+start);
        x=(int)(xx = xx + incr);
    }
}
}
public static boolean compare(Point a,String str, Point b){
    if(str == "<="){
        if(a.yi()<b.yi())return true;
        else if(a.yi()==b.yi()){
            if(a.xi()>b.xi())return false;
            else if(a.xi()==b.xi()){ if(a.tag>=b.tag)return true;else return false;}
            else return true;
        }return false;
    }
    if(str == ">="){
        if(a.yi()>b.yi())return true;
        else if(a.yi()==b.yi()){
            if(a.xi()<b.xi())return false;
            else if(a.xi()==b.xi()){ if(a.tag <= b.tag) return true;else return false;}
            else return true;
        }return false;
    }return false;
}
}
```

In the case of the simple convex polygons, the left edge of the polygon and the outer pixel, in a left edge line sequence, is determined by a single test selecting the minimum x value, similarly for the right hand edges by selecting the maximum x value for each scan line. As long as there is only one point for each boundary line to scan line intersection, when multiple left and right edge sequences have to be processed, then selecting left point and right point pairs along scan lines generates the shading shown in Figures 10.24 and 25. In Figure 10.26 the same approach is shown for a self-intersecting boundary line. This approach is suitable for shading a network of polygon areas where no boundary line is drawn in to define edges. Where a complete boundary line is needed or an existing boundary line needs to be fully shaded over, then it has to be redrawn after shading. This is because scan lines defined in this way start and end on points midway along any horizontal sequence of pixels which occur in a boundary line sequence.

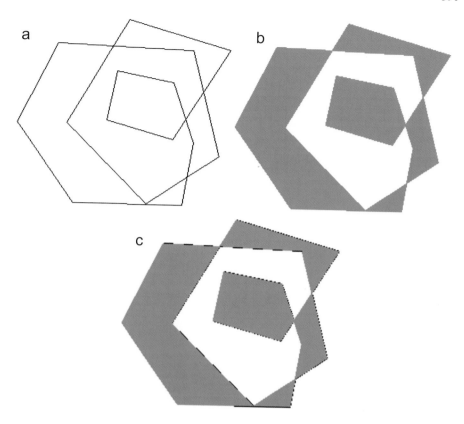

Figure 10.26 Self- intersecting boundaries with (b) & without (c) redrawn edges

If a shading algorithm is required that will automatically cover existing boundary lines, then identifying the outer pixels in a line sequence and identifying left and right edges has to be carried out separately. This is done in the following program by dividing up the boundary line between stationary points to give sequences of line segments in upwards or downwards directions. The left edge for the upward sequences can then be identified and similarly the right edge of the downward sequences, by selecting the minimum and maximum x values for each y value, as before. These sequences of points can then be merge sorted as they are created to give a final set of points in raster scan order. Pairing up these points gives the result shown in Figure 10.27.

Figure 10.27 A polygon shading algorithm to cover boundary edges

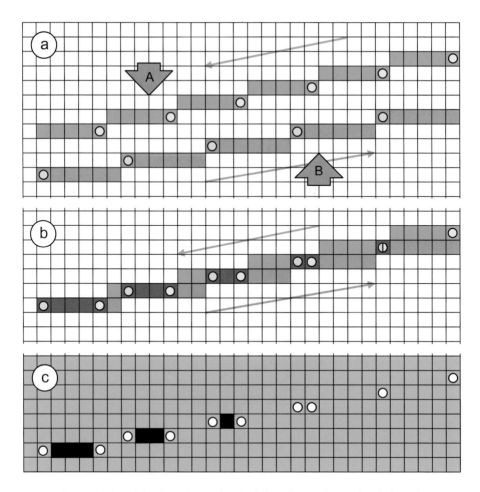

Figure 10.28 Pairing boundary points for inlets allows edges to break through

Unfortunately, there are problems with this algorithm when hanging-edges, spurs or narrow inlets are encountered. Although hanging edges can be detected and removed, narrow spurs and inlets cannot, and they are easy to generate automatically for example, if a polygon is scaled to a smaller size relative to the pixel grid. In Figure 10.28 two lines A and B are shown that form the right edge and the left edge respectively, of an inlet. In 11.28a the two lines are shown separately. In 10.28b they are shown superimposed with their common points shown in purple. The outer edge points for A as a right edge are shown with a yellow circle, the similar outer points of B which is a left edge are shown by a green circle. When these are ordered and paired up, it can be seen that these points can swap over where the lines overlap allowing the boundary line points to break through.

A different approach is needed to cope with this problem, using a different mechanism to identify the inside region of the polygon that is to be shaded in.

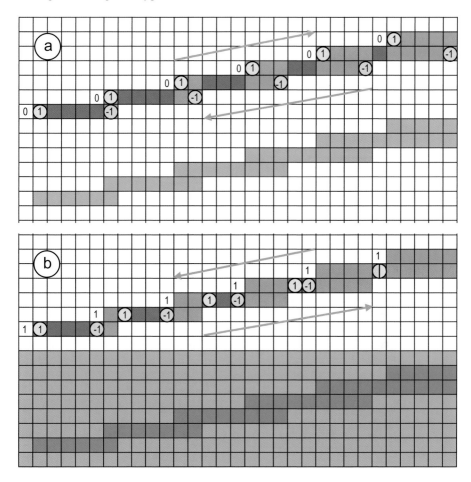

Figure 10.29 A spur and inlet fully covered using the winding number algorithm

The new approach is based on a value called the "winding number", which will be explored more fully in a later chapter. The method consists of allocating a number to each scanline intersection point with the boundary, and then controlling the selection of the points that will be used as the endpoints of shading scanlines, by the value of a running total adding these allocated values together as the scan line is traversed from left to right in ascending values of x.

For the simple case: pairing intersection points, it can be seen that by giving the left edge intersection points the value of 1 and the right edge intersection points the value of –1, the shading scan lines will exist where the running total is 1. The benefit of this approach is illustrated in Figure 10.29, where this technique is applied to a narrow spur and inlet. It becomes possible to implement a shading algorithm that will cover both the polygon interior and the boundary line without edge line points breaking through. What is more the same approach will allow the points inside a boundary line to be selected for shading without overwriting the boundary.

```
static void shade(Polygon p, Color cc){
        Point[] scanArray = null, s=null;
        Point p1=null,p2=null;
        s = scanArray = scan(p);
        if(s!=null){
            int count =0;  int i=0;
            while(i<s.length){
                count = 0;  int y = s[i].yi();
                do{
                    if((count==0)&&(s[i].tag >0)){p1= s[i];}
                    else if((count+s[i].tag ==0)&&(p1!=null)){
                        p2=s[i];  dW.plotLine(p1,p2,cc);  p1=null;
                    }count=count+s[i].tag;  i++;
                }
                while((i<s.length)&&(s[i].yi()== y));
            }
        }else IO.writeString("no scan array");
}
static Point [ ] scan(Polygon p){
    int i = 0, j=0, starty=0, endy=0, starti=0, endi=0, len=0;boolean up = true;
    if(p == null)return null;
    Point[] changeArray = null , newArray = null, oldArray = null;
    int n=1, m=0;
    while(p.p[n].yi()==p.p[m].yi()){m = n; n = incr(n,p.p.length-1);}
    if(p.p[n].yi()<=p.p[m].yi())
        while(p.p[n].yi()<=p.p[m].yi()){ m = n; n = incr(n,p.p.length-1);}
    else while(p.p[n].yi()>p.p[m].yi()){ m = n; n = incr(n,p.p.length-1);}
    int start = m;
    do{
        if(p.p[n].yi()>p.p[m].yi()){
            up=false; starty = p.p[m].yi(); starti = m;
            while((p.p[n].yi()>=p.p[m].yi())){ m=n; n=incr(n, p.p.length-1);}
            endy = p.p[m].yi();endi = m;
        }else if(p.p[n].yi()<p.p[m].yi()){
            up=true; endy = p.p[m].yi(); starti = m;
            while(p.p[n].yi()<=p.p[m].yi()){ m=n; n=incr(n, p.p.length-1);}
            starty = p.p[m].yi(); endi = m;
        } changeArray = new Point[endy-starty+1];
        int k = starti; j=incr(k,p.p.length-1);
        do{
            lineSegment(p.p[k], p.p[j], changeArray, starty,up);
            k=j;  j=incr(j,p.p.length-1);
        }while(k!=endi);
        oldArray = newArray;   newArray = merge(oldArray,changeArray);
    }while(m!=start);
    return newArray;
}
```

```
static void lineSegment(Point p1,Point p2, Point[] array, int start,boolean up){
    int len=0, miny=0;  Shadings s = null;
    if(! up){
        len = p2.yi()-p1.yi();  miny = p1.yi();
        s = new Shadings(IO,len+1, miny,false);
        s.shadingEdge(p2.xi(), p2.yi(), p1.xi(), p1.yi());
        for(int i=0;i<s.length;i++){
            int j= miny+i-start;
            if (array[j]!=null){
                if((array[j].xi()!=0)&&(array[j].xi()<s.rightedge[i]))array[j].x("=",s.rightedge[i]);
            }else{ array[j] = new Point(2); array[j].x("=",s.rightedge[i]); }
            array[j].y("=",i+miny);  array[j].tag= -1;
        }
    }else{
        len = p1.yi()-p2.yi();  miny = p2.yi();
        s = new Shadings(IO,len + 1, miny,false);
        s.shadingEdge(p1.xi(), p1.yi(), p2.xi(), p2.yi());
        for(int i=0;i<s.length;i++){
            int j=miny+i-start;
            if (array[j]!=null){
                if((array[j].xi()!=0)&&(array[j].xi()>s.leftedge[i]))array[j].x("=",s.leftedge[i]);
            }else{ array[j]= new Point(2); array[j].x("=",s.leftedge[i]); }
            array[j].y("=",i+miny);  array[j].tag= 1;
        }
    }
}
static int incr(int index, int bound) {return (index+1)%bound; }
```

The modified procedures to implement this algorithm are given above. Again the first step is to separate the boundary into left-hand sections and right-hand sections by subdividing the boundary at stationary points. Notice that this is done by including any horizontal steps into the current sequence whether it is going up or down. Each line interpolation sequence for boundary line segments is then treated as though it were a convex area so that its left and right edge points can be extracted into two arrays. In this case where the line is part of a left hand edge only the leftmost points are kept and they are merge sorted into a larger array that will ultimately be used to shade the whole polygon. Similarly where a line is part of a right-hand sequence only the rightmost points are kept and merge sorted into the final array.

The left edges are tagged with a value of 1, and the right edges tagged with the value of −1. The final array of points is in raster scan order so points can be accessed in x ordered sequences for each value of y. Starting with a running total of 0, as each point is processed its tag value can be added to this total. When it becomes 1 then a new shading line can be started and when it returns to 0 then the end point of the shading line is defined. It is the way these values distribute themselves along a scan line for the overlapping edges that make up a spur or an inlet shown in Figure 10.29 in a way that ensures boundary-points do not break through the shading.

leftMinx[y] -1   rightMaxx[y] +1

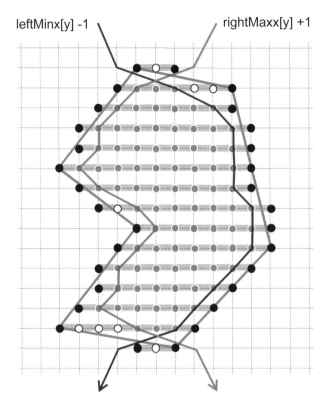

Figure 10.30 Polygon fill: inside edges

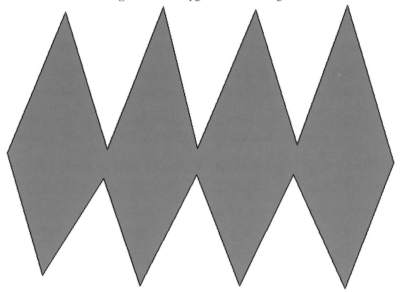

Figure 10.31 A polygon shading algorithm to fill between existing edges

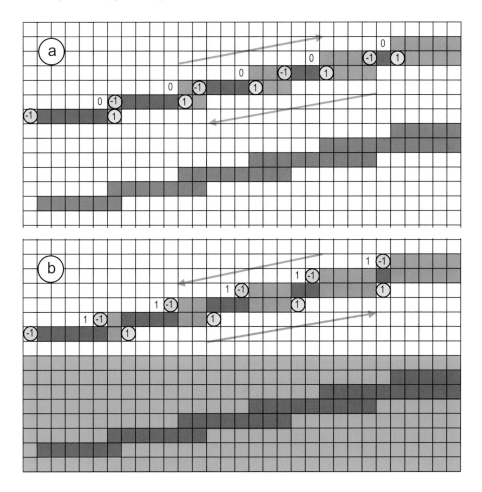

Figure 10.32 Polygon fill inside boundaries for spurs and inlets

If the right most point of the left edge of a polygon is incremented by one and the left most point of a right edge of a polygon is decremented by one then the resulting x values, for the same y value, define the interior scan line for shading a polygon without over writing its boundary, in the way illustrated in Figure 10.30. This approach is a simple extension of the program described above. If instead of taking the leftmost edge points for a left edge, the rightmost points are taken rather than discarding them, and their x values are incremented by 1, similarly if the leftmost edge points from right edges are taken and their x values decremented by 1, then the same scan fill algorithm will produce the result shown in Figure 10.31. What is more the same arrangements of tags: 1 for the left edges and –1 for right edges gives the correct behaviour for narrow spurs and inlets in the way illustrated in Figure 10.32. This algorithm works for non self-intersecting polygon boundaries with spurs and inlets demonstrated in Figure 10.33. It also works for polygons with self-intersecting boundaries but only if they contain no spurs or inlets.

```
static void lineSegment(Point p1,Point p2, Point[] array, int start,boolean up){
    int len=0, miny=0;  Shadings s = null;
    if(! up){
        len = p2.yi()-p1.yi();  miny = p1.yi();
        s = new Shadings(IO,len+1, miny,false);
        s.shadingEdge(p2.xi(), p2.yi(), p1.xi(), p1.yi());
        for(int i=0;i<s.length;i++){
            int j= miny+i-start;
            if (array[j]!=null){ if((array[j].xi()!=0)&&(array[j].xi()>s.leftedge[i]-1))
                array[j].x("=",s.leftedge[i]-1);
            }else{ array[j] = new Point(2); array[j].x("=",s.leftedge[i]-1); }
            array[j].y("=",i+miny);  array[j].tag= -1;
        }
    }else{
        len = p1.yi()-p2.yi(); miny = p2.yi();
        s = new Shadings(IO,len + 1, miny,false);
        s.shadingEdge(p1.xi(), p1.yi(), p2.xi(), p2.yi());
        for(int i=0;i<s.length;i++){
            int j=miny+i-start;
            if (array[j]!=null){ if((array[j].xi()!=0)&&(array[j].xi()<s.rightedge[i]+1))
                array[j].x("=",s.rightedge[i]+1);
            }else{ array[j]= new Point(2); array[j].x("=",s.rightedge[i]+1); }
            array[j].y("=",i+miny);  array[j].tag= 1;
        }
    }
}
```

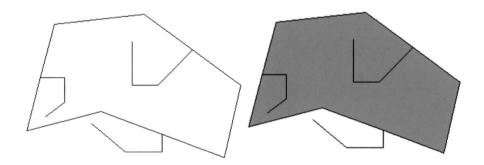

Figure 10.33 Polygon with a spur and two inlets shaded without covering edge lines

Flood Fill Algorithms

An alternative approach to shading the interior points of a polygon is to start from a point inside a polygon and then search the interior for non edge points. One implementation of this approach has already been developed in chapter 4 in the maze solving program. A small modification of the finite state maze solving code can be used to fill a closed region in a polygon in the way illustrated in Figures 10.34 to 36.

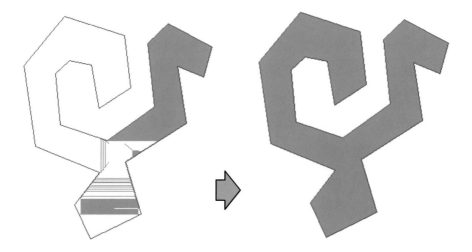

Figure 10.34 Polygon flood fill using a modified space search algorithm

The advantage of this approach appears in interactive work on images. Identifying a cell's colour using the mouse can then be used to search out all contiguous cells containing the same colour and modifying them to another colour. Like the maze solving algorithm on which it is based it depends on having a boundary of colour values that form a closed area, otherwise the fill operation escapes and can fill in more than is required!

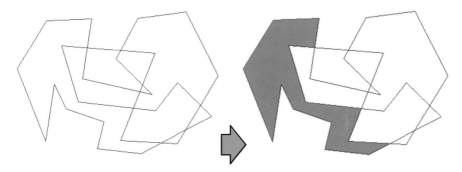

Figure 10.35 Polygon flood fill used to selectively shade a sub-area

Where a complex boundary intersects itself and the different areas generated by the line need to be coloured separately then this approach is easier to apply than the geometry based fill operations explored in the first part of this chapter.

Floodfill also has the added property that it can be used to fill in background colours without overwriting foreground features – as long as they are a different colour from the existing background.

This algorithm will also work on areas defined using other boundary interpolation schemes as the following chapter will demonstrate.

Figure 10.36 Overlayed polygons interactively flood filled

```
public class PolygonFill {
    static TextWindow IO = null;  static DisplayWindow dW = null;
    static int rows = 0;  static int cols = 0;  static Shadings s = null;
    public static void main(String[] args){
        boolean newPolygon=true,newColour=true, finished = false;
        IO = new TextWindow(10,810,1000,100);
        dW = new DisplayWindow(IO,10, 10, 1000,800,Color.white);
        String str ="",strr = "black";
        Color cc = Color.cyan, lineColour = getColour(strr);
        while(! finished){
            if(newPolygon){
                IO.writeString("  please enter the number of vertices: ");
                int pnts = IO.readInteger();IO.readLine();
                Point poly[] = new Point[pnts+1];
                IO.writeString("  please enter vertices using the mouse \n ");
                for(int i = 0; i<pnts;i++){
                    poly[i] = dW.getCoord();
                    if(i>0)dW.plotLine(poly[i-1],poly[i],lineColour);
                }
                poly[0].c("->",poly[pnts]= new Point(2));
                dW.plotLine(poly[pnts-1],poly[pnts],Color.black);
            }
            if(newColour){
                do{
                    IO.writeString("  please enter a colour: ");
                    str= IO.readString();IO.readLine();
                    if(str.equals(strr))IO.writeString("cannot use boundary colour:  ");
                }while (str.equals(strr));
                cc = getColour(str);                 // convert string to Color
            }
            IO.writeString("  please identify the region using the mouse \n ");
            Point seed= dW.getCoord();
            s = new Shadings();
            s.setText(IO);
            s.floodFill(dW,seed.xi(),seed.yi(), cc,lineColour);
            IO.writeString("  do you wish to finish? y/n: ");
            str = IO.readString();IO.readLine();
            if(str.equals("y"))break;  else finished=false;
            IO.writeString("  do you wish to change the colour? y/n: ");
            str = IO.readString();IO.readLine();
            if(str.equals("y")) newColour = true;else newColour=false;
            IO.writeString("  do you wish to enter another polyon? y/n: ");
            str = IO.readString();IO.readLine();
            if(str.equals("y"))newPolygon = true;else newPolygon = false;
        }
    }
}
```

```
public void floodFill(DisplayWindow dW,int xx, int yy, Color cc,Color edge){
    pixel = new int [dW.c.b.width][dW.c.b.height];
    for(int i=0;i<dW.c.b.width;i++){ for(int j=0;j<dW.c.b.height;j++) pixel[i][j]=0; }
    state=0; savevalue=0; savedir=0; dir=0; cellcount=1;
    int col = edge.getRGB(); int count=0; int alpha = 255;  alpha = alpha << 24;
    boundary = alpha | col;
    x1 = x = xx;  y1 = y = yy;
    while(true){
        switch(setstate(dW,cc)){
            case 0: move(dW,null); pixel[x][y]=cellcount++; break;
            case 1:case 2:case 3:case 4:  checknext();  dir=(dir+1)%4; break;
            case 5: dir=savedir;  move(dW,null);
                    if((xx==x)&&(yy==y)){ if (++count==4)return; } break;
        }nx = x+xincr[dir]; ny = y+yincr[dir];
    }
}
private void move(DisplayWindow dW,Color cc){
    int xx=x, yy=y;  x=x+xincr[dir];  y= y+yincr[dir];
    if (cc!=null) dW.c.setPixel(xx,yy,cc);
}
private int setstate(DisplayWindow dW,Color cc){
    if((pixel[nx][ny]==0)&& !(dW.c.getPixel(nx,ny)==boundary)){
        if(state != 0){x1=nx;y1=ny;savedirection = dir;}
        state=0;
    } else if(state<5){
        if ((state==0)){ x2=x; y2=y;
            switch(savedirection){
                case 0:  dW.plotRectangle(x1,y1,x2-x1+1,1,cc); break;
                case 1:  dW.plotRectangle(x1,y1,1,y2-y1+1,cc); break;
                case 2:  dW.plotRectangle(x2,y2,x1-x2+1,1,cc); break;
                case 3:  dW.plotRectangle(x2,y2,1,y1-y2+1,cc); break;
            }
        } state++;  return state;
    } else state =1;
    savevalue=0; savedir=dir; return state;
}
private void checknext(){
    if((pixel[nx][ny]<pixel[x][y])&&(pixel[nx][ny]>savevalue)) {
        savevalue=pixel[nx][ny];  savedir = dir;
    }
}
```

This approach also has difficulties with very sharp spurs: being unable to get to pixels that are isolated from the main body of interior points by the angle at which edge lines meet. In the next chapter a different interpolation sheme is introduced that allows curved boundaries to be created, and this requires the shading process to be revisited to give a more powerful general purpose approach.

11

Parametric Line Interpolation & Keyframe Infill "Inbetweening" Film Animation

Introduction

Figure 11.1 A linear inbetweening sequence transforming a rabbit to a fox

In this chapter an alternative form of interpolation is investigated. In the previous chapters lines and areas were filled in on a grid: the number of new points depending on the length of each line or the extent of each area. In this sequence of infill operations the process is controlled by a parameter independent of the length of any line. The application where this approach is appropriate is the inbetweening operation used to generate cartoon animation sequences for film and video work. In Figure 11.1

A. Thomas, *Integrated Graphic and Computer Modelling*,
DOI: 10.1007/978-1-84800-179-4_11, © Springer-Verlag London Limited 2008

an example of this process is shown not only creating multiple images between two given key-frames, but also illustrating a morphing operation changing one object into another in smoothly changing steps.

Creating the effect of movement in an image requires objects to be moved in the image frame in a time sequence. If the movement is to be seen as a smooth coherent action this movement has to be incremental along a trajectory or path. If the frames in an animation sequence showing a moving object are overlaid then each point of the object will follow a linear path in the resulting multi-image. A simple example using a moving triangle is shown in Figure 11.2.

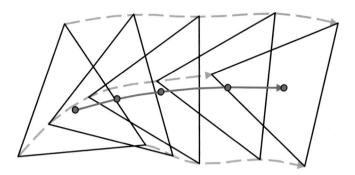

Figure 11.2 An overlaid animation sequence for a moving triangle

The path followed by each triangle vertex is different and relatively complex, even though the centre of gravity of the triangle follows a simple smooth path because the object is being translated and rotated at the same time. The idea behind inbetweening is that if such a movement can be broken down into steps, each step defined by a key-frame drawing or image, then straight-line segments can approximate segments of the trajectories lying between matching points without distorting the resulting motion too much. New triangles can then be drawn in linking the new vertices located on these straight-line segments. In this process however each trajectory line segment will have to have the same number of subdivisions to produce the correct sequence of new triangles. This form of interpolation can be provided by a parametric line equation based on the coordinates of the end points of each line segment.

$$p_\lambda = (\lambda - 1)p_1 + (\lambda)p_2$$

Whatever the length of the line, the parameter λ will place a point in the same relative position along the line. If λ is $\frac{1}{2}$ then the mid-point of each line will be calculated.

Figure 11.2 shows a series of overlaid key frames for a moving triangle. Because the triangle is a rigid body, in other words an object that is the same shape and size in each frame, it is easy to identify the matching points. Since straight edges stay straight, following the vertices of the triangle from frame to frame can capture the motion sequence and then linearly interpolating the triangle edges in each frame, once the correct vertex positions have been identified completes the sequence. In this case the simplest way to do this is to link matching points with straight lines and then select points along these trajectory lines for the triangle vertices: evenly spaced for uniform motion and unevenly spaced to give acceleration or deceleration effects.

The use of straight lines will distort the "in-between" triangles to some extent, as the correct trajectories shown in Figure 11.2 indicate. However a close spacing of the key-frames can reduce the error to what ever level is necessary. If this approach is used to animate three dimensional shapes and non-rigid objects this distortion becomes less apparent because the shape-boundary in each frame is different and cannot be compared directly in the way a simple shape like a triangle can. However in these cases the identification of matching points becomes more difficult or in many cases arbitrary.

A variety of semi-automatic schemes have been developed to cope with specific subjects. Although these many alternatives can be considered to be a vocabulary of special effects provided by an interactive cartoon generating systems. Animators might expect to construct many of these for themselves, from more general operations, tailoring them to their particular "content generation" tasks. Consequently they lie outside the remit for this book, where the objective is to introduce the operations needed to construct these effects. Two general aspects however, are of interest, the first is the use and extension of the parametric interpolation process, the second is the way the key-frame inbetweening approach can be used to morph from one object to another, as illustrated in Figure 11.1.

The starting point for the illustration in Figure 11.1 is inbetweening from one poly-line to another. This was then extended to give an inbetweening operation between one closed single polygon-boundary to another. Closing the polyline to give a polygon loop introduces an arbitrary choice, which is where in the boundary of each polygon to place the starting point for vertex matching. There is also the choice of which way the loop should be made to rotate about its contained area. Both of which affect the resulting sequence of in between polygons. The final stage was to include an area shading stage that could cope with a self-crossing or fragmenting polygon boundary.

Inbetweening Two Polylines

If two polylines contain the same number of vertices then one approach is to match up corresponding vertices and interpolate the in-between positions for these points in the way outlined in Figure 11.2. However if there are a different number of vertices then a different approach is called for. The approach implemented in these examples is to scale the lines to a common length and then map the vertices of one polyline onto the other. This gave two lines containing the same number of vertices that could be matched with each other and used to give in between points in the same way used in Figure 11.2. Unfold the boundary lines and scale them to match each other, map the missing vertex points across, then transfer the new vertex points back to the other boundary in the way shown in Figure 11.3

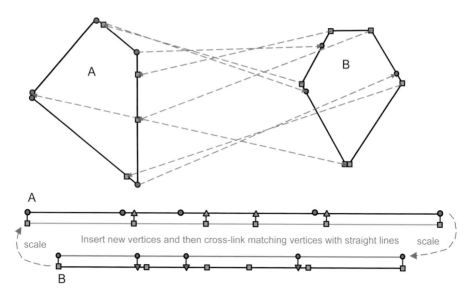

Figure 11.3 Mapping vertices from one polyline boundary to another

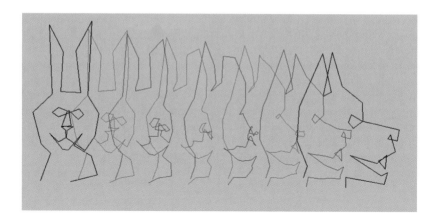

Figure 11.4 Polyline inbetweening

```
public class Animation{   // Polyline Inbetweening
    static TextWindow IO = new TextWindow(0,500,1000,100 );
    static DisplayWindow dW = new DisplayWindow(IO,1,1,1000,500,Color.white);
    public static void main(String[] args){
        Color cc = Color.black;
        Polyline a = new Polyline(IO,dW);
        a.getPolyloop();
        double alength = a.lengths();
        IO.writeLongReal(alength,10,5);   IO.writeLine();
        Polyline b = new Polyline(IO,dW);
```

```
        b.getPolyloop();
        double blength = b.lengths();
        IO.writeLongReal(blength,10,5);              IO.writeLine();
        double[] c = new double[a.p.length];
        for(int i=0; i<a.p.length; i++) c[i]= a.p[i].zd()*blength/alength;
        double[] d = new double[b.p.length];
        for(int i=0; i < b.p.length; i++) d[i] = b.p[i].zd()*alength/blength;
        Polyline aa =          match(a,d);
        Polyline bb =          match(b,c);
        double q = 1.0/29.0, L = 0;               //Polyline 29 spaces 30 vertices
        Point np = new Point(2);Point lp = new Point(2);
        Polygon op = null;
        for(int n=0; n<30; n++){
            Polygon p = new Polygon( aa.length);
            np.x("=", aa.p[0].xd()*(1-L) + bb.p[0].xd()*L);
            np.y("=", aa.p[0].yd()*(1-L) + bb.p[0].yd()*L);
            p.p[0] = new Point(2);
            p.p[0].x("=",np.xd()); p.p[0].y("=",np.yd());
            for(int i=1;i<aa.length;i++){
                p.p[i] = new Point(2);
                lp.x("=", aa.p[i].xd()*(1-L) + bb.p[i].xd()*L);
                lp.y("=", aa.p[i].yd()*(1-L) + bb.p[i].yd()*L);
                p.p[i].x("=",lp.xd()); p.p[i].y("=",lp.yd());
                dW.plotLine(np,lp,Color.white);
                np.x("=",lp.xd()); np.y("=",lp.yd());
            }
            if(op!=null)shade (op,Color.white);
            if(n%2==0)cc = Color.cyan; else cc = Color.red;
            shade (p,cc);
            op = p;
            L=L+q;
        }
    }
    static Polyline match(Polyline a,double[] d){
        int i = 0; int j=0;int k=0;double ll=0.0;
        Polyline aa = new Polyline(IO,dW);
        aa.p = new Point[a.length+d.length];
        while((i < a.length)||(j < d.length)){
            int t=0;
            if((i<a.length)&& (j<d.length)&& (((ll= a.p[i].zd()-d[j])*ll)<0.001)){
                aa.p[k]= a.p[i];i++; j++; k++;
            }else if((j>= d.length)|| ((i<a.length)&&(a.p[i].zd()< d[j]))){
                aa.p[k]= a.p[i];i++; k++;
            }else if((i>= a.length)|| ((j<d.length)&&(a.p[i].zd()> d[j]))){
                double less = a.p[i-1].zd(); double more = a.p[i].zd();
                double scale = (d[j]-less)/(more-less);
                aa.p[k]= new Point(3);
```

```
public static void main(String[] args){  //animated morphing sequence
    IO.writeString("please enter start direction: two points  \n");
    Point p1 = dW.getCoord();  Point p2 = dW.getCoord();
    IO.writeString("please enter finish direction: two points  \n");
    Point p3 = dW.getCoord();  Point p4 = dW.getCoord();
    double a1 = p1.yd()-p2.yd(),  b1 = p2.xd()-p1.xd();
    double a2 = p3.yd()-p4.yd(),  b2 = p4.xd()-p3.xd();
    int frames = 50;
    Color cc = Color.black;
    Polyline a = new Polyline(IO,dW);
    a.getPolyloop();                          // get object 1
    double alength = a.lengths();
    Polyline b = new Polyline(IO,dW);
    b.getPolyloop();                          // get object 2
    double blength = b.lengths();
    double[] c = new double[a.p.length];
    for(int i=0; i<a.p.length; i++)   c[i]= a.p[i].zd()*blength/alength;
    double[] d = new double[b.p.length];
    for(int i=0; i < b.p.length; i++) d[i] = b.p[i].zd()*alength/blength;
    Polyline aa = match(a,d), bb = match(b,c);
    double q = 1.0/(frames-1), L = 0;
    Point np = new Point(2), lp = new Point(2);
    Polygon op = null;
    for(int n=0; n < frames; n++){
        double x1 = aa.p[0].xd(), y1 = aa.p[0].yd();
        double x2 = bb.p[0].xd(), y2 = bb.p[0].yd();
        double c1 = -(a1*x1+b1*y1), c2 = -(a2*x2+b2*y2);
        Polygon p = new Polygon( aa.length);
        double x= b1*c2-b2*c1, y= a2*c1-a1*c2, w= a1*b2-a2*b1;
        double cx=0,cy=0;
        if(w!=0){ cx=x/w; cy=y/w;}
        else  IO.writeString("error division by zero  \n");
        dW.plotLine((int)x1,(int)y1,(int)cx,(int)cy,Color.green);
        dW.plotLine((int)cx,(int)cy,(int)x2,(int)y2,Color.green);
        double ax = x1*(1-L)+cx*L, ay = y1*(1-L)+cy*L;
        double bx = cx*(1-L)+x2*L, by = cy*(1-L)+y2*L;
        np.x("=", ax*(1-L) + bx*L); np.y("=", ay*(1-L) + by*L);
        p.p[0] = new Point(2);  p.p[0].x("=",np.xd());  p.p[0].y("=",np.yd());
        if(n%2==0)cc = Color.cyan;   else cc = Color.red;
        for(int i=1;i<aa.length;i++){
            p.p[i] = new Point(2);
            x1 = aa.p[i].xd();  y1 = aa.p[i].yd();
            x2 = bb.p[i].xd();  y2 = bb.p[i].yd();
            c1 = -(a1*x1+b1*y1);  c2 = -(a2*x2+b2*y2);
            x= b1*c2-b2*c1;  y= a2*c1-a1*c2;  w= a1*b2-a2*b1;
            if(w!=0){ cx=x/w; cy=y/w;}
            else  IO.writeString("error division by zero  \n");
```

```
            ax = x1*(1-L)+cx*L;  ay = y1*(1-L)+cy*L;
            bx = cx*(1-L)+x2*L;  by = cy*(1-L)+y2*L;
            lp.x("=", ax*(1-L) + bx*L);  lp.y("=", ay*(1-L) + by*L);
            p.p[i].x("=", lp.xd());   p.p[i].y("=", lp.yd());
            dW.plotLine(np, lp,cc);
            np.x("=",lp.xd()); np.y("=",lp.yd());
        }shade (p,cc);                               // render object p
        np.x("=", aa.p[0].xd()*(1-L) + bb.p[0].xd()*L);
        np.y("=", aa.p[0].yd()*(1-L) + bb.p[0].yd()*L);
        op = p;
        L=L+q;
    }
}
```

Figure 11.8 Changing shape along a curved trajectory

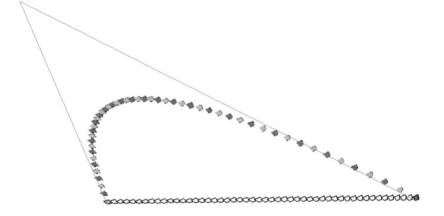

Figure 11.9 Skewed parabolic interpolation paths showing non-linear spacing

Even where the value of λ changes uniformly, in a curve it is applied to line segments of different and changing lengths: the greater the difference in the lengths of the control triangle the larger the effect. The task is to select the λ values that give the required motion.

Equal Subdivisions of a Parametric Line

If equal steps along a curved path are needed then it is necessary to calculate the length of the path. This is a relatively complex calculation if it needs to be repeated many times.

$$Pa = (1-\lambda)P1 + (\lambda)P3$$

$$Pb = (1-\lambda)P3 + (\lambda)P2$$

$$P = (1-\lambda)[(1-\lambda)P1 + (\lambda)P3] + (\lambda)[(1-\lambda)P3 + (\lambda)P2]$$

$$x = \left(1 - 2\lambda + \lambda^2\right)x1 + 2.\left(\lambda - \lambda^2\right)x3 + \lambda^2 x2$$

$$y = \left(1 - 2\lambda + \lambda^2\right)y1 + 2.\left(\lambda - \lambda^2\right)y3 + \lambda^2 y2$$

$$x = (x1 - 2.x3 + x2)\lambda^2 + 2(x3 - x1)\lambda + x1$$

$$y = (y1 - 2.y3 + y2)\lambda^2 + 2(y3 - y1)\lambda + y1$$

$$\frac{dx}{d\lambda} = 2\lambda.(x1 - 2.x3 + x2) + 2(x3 - x1)$$

$$\frac{dy}{d\lambda} = 2\lambda.(y1 - 2.y3 + y2) + 2(y3 - y1)$$

let $A \equiv 2.(x1 - 2.x3 + x2)$, $B \equiv 2.(x3 - x1)$

let $C \equiv 2.(y1 - 2.y3 + y2)$, $D \equiv 2.(y3 - y1)$

$$dx = (\lambda A + B)d\lambda$$

$$dy = (\lambda C + D)d\lambda$$

arc length $\quad ds = \sqrt{dx^2 + dy^2}$

$$ds = \sqrt[2]{(\lambda A + B)^2 + (\lambda C + D)^2}\,d\lambda$$

$$ds = \sqrt[2]{\left(\lambda^2 A^2 + 2\lambda.AB + B^2\right) + \left(\lambda^2 C^2 + 2\lambda.CD + D^2\right)}\,d\lambda$$

$$ds = \sqrt[2]{\lambda^2.\left(A^2 + C^2\right) + 2\lambda.(AB + CD) + B^2 + D^2}\,d\lambda$$

match to $\int ds = \int \sqrt{u^2 + a^2}\, du = \dfrac{u}{2}\sqrt{a^2 + u^2} + \dfrac{a^2}{2}.\ln\left|u + \sqrt{a^2 + u^2}\right| + c$

$$ds = \sqrt{\left[\lambda.\sqrt{A^2 + C^2} + \dfrac{AB + CD}{\sqrt{A^2 + C^2}}\right]^2 + \dfrac{\left(B^2 + D^2\right)\left(A^2 + C^2\right) - (AB + CD)^2}{A^2 + C^2}}\,.d\lambda$$

$u = \lambda.\sqrt{A^2 + C^2} + \dfrac{AB + CD}{\sqrt{A^2 + C^2}}\quad$ therefore $\quad \dfrac{du}{d\lambda} = \sqrt{A^2 + C^2}$

$$\int ds = \int \sqrt{a^2 + u^2}\,.du = \dfrac{u}{2}\sqrt{a^2 + u^2} + \dfrac{a}{2}.\ln\left|u + \sqrt{a^2 + u^2}\right| + c$$

$$ds = \dfrac{1}{\sqrt{A^2 + C^2}}\sqrt{\left[\lambda.\sqrt{A^2 + C^2} + \dfrac{AB + CD}{\sqrt{A^2 + C^2}}\right]^2 + \dfrac{\left(B^2 + D^2\right)\left(A^2 + C^2\right) - (AB + CD)^2}{A^2 + C^2}}\,.du$$

$$\int ds = \int \dfrac{1}{\sqrt{A^2 + C^2}}\sqrt{\left[\lambda.\sqrt{A^2 + C^2} + \dfrac{AB + CD}{\sqrt{A^2 + C^2}}\right]^2 + \dfrac{\left(B^2 + D^2\right)\left(A^2 + C^2\right) - (AB + CD)^2}{A^2 + C^2}}\,.du$$

$$\int_{\lambda1}^{\lambda2} ds = \left(\begin{aligned} &\dfrac{1}{2.\sqrt{A^2 + C^2}}\left[\lambda.\sqrt{A^2 + C^2} + \dfrac{AB + CD}{\sqrt{A^2 + C^2}}\right].\sqrt{\lambda^2.\left(A^2 + C^2\right) + 2\lambda(AB + CD) + B^2 + D^2}\\ &+ \dfrac{1}{2.\sqrt{A^2 + C^2}}\left[\dfrac{\left(B^2 + D^2\right)\left(A^2 + C^2\right) - (AB + CD)^2}{A^2 + C^2}\right]\\ &\times \ln\left|\lambda.\sqrt{A^2 + C^2} + \dfrac{AB + CD}{\sqrt{A^2 + C^2}} + \sqrt{\lambda^2.\left(A^2 + C^2\right) + 2\lambda(AB + CD) + B^2 + D^2}\right| \end{aligned} \right)_{\lambda1}^{\lambda2}$$

$$\int_{\lambda1}^{\lambda2} ds = \left(\begin{aligned} &\dfrac{1}{2}\left[\lambda + \dfrac{AB + CD}{A^2 + C^2}\right].\sqrt{\lambda^2.\left(A^2 + C^2\right) + 2\lambda(AB + CD) + B^2 + D^2}\\ &+ \dfrac{1}{2}.\left[\dfrac{\left(A^2 + C^2\right)\left(B^2 + D^2\right) - (AB + CD)^2}{\left(A^2 + C^2\right)\sqrt{A^2 + C^2}}\right]\\ &\times \ln\left|\lambda.\sqrt{A^2 + C^2} + \dfrac{AB + CD}{\sqrt{A^2 + C^2}} + \sqrt{\lambda^2.\left(A^2 + C^2\right) + 2\lambda(AB + CD) + B^2 + D^2}\right| \end{aligned} \right)_{\lambda1}^{\lambda2}$$

This result is undefined where $\sqrt{A^2 + C^2} = 0$. If $x1 + x2 = 2.x3$ and $y1 + y2 = 2.y3$ where $P3$ is the mid point of a straight line the length of the path S is undefined by this integral, but can be seen to be $\sqrt{(x2 - x1)^2 + (y2 - y1)^2}$. Setting $x1 + x2 \neq 2.x3$ and $y1 = y2 = y3$ and substituting the values of A B C and D into this formula gives a correct result, and also gives one way in which an accelerating or decelerating movement can be created along a straight line path.

$$\int_{\lambda 1}^{\lambda 2} ds = \left(\begin{array}{l} \frac{1}{2}\left[\lambda + \frac{AB+CD}{A^2+C^2}\right].\sqrt{\lambda^2.(A^2+C^2)+2\lambda.(AB+CD)+B^2+D^2} \\ \\ + \frac{1}{2}.\left[\frac{(A^2+C^2)(B^2+D^2)-(AB+CD)^2}{(A^2+C^2)\sqrt{A^2+C^2}} \right] \\ \\ \times \ln\left|\lambda.\sqrt{A^2+C^2} + \frac{AB+CD}{\sqrt{A^2+C^2}} + \sqrt{\lambda^2.(A^2+C^2)+2\lambda.(AB+CD)+B^2+D^2}\right| \end{array} \right)_{\lambda 1}^{\lambda 2}$$

$A = 2.(x1 - 2x3 + x2)$

$B = 2.(x3 - x1)$

$C = 2.(y1 - 2y3 + y2) = 0$

$D = 2.(y3 - y1) = 0$

substitute the zero values

$$\int_{\lambda 1}^{\lambda 2} ds = \left(\frac{1}{2}.\left[\lambda + \frac{AB}{A^2}\right].\sqrt{\lambda^2.(A^2)+2\lambda.(AB)+B^2} \right)_{\lambda 1}^{\lambda 2}$$

$$\int_{\lambda 1}^{\lambda 2} ds = \left(\frac{1}{2}.\left[\lambda + \frac{B}{A}\right].\sqrt{(\lambda A + B)^2} \right)_{\lambda 1}^{\lambda 2} = \left(\frac{1}{2}.\left[\lambda + \frac{B}{A}\right].(\lambda A + B) \right)_{\lambda 1}^{\lambda 2}$$

$$= \left(\frac{1}{2}.\left[\lambda^2 A + \lambda B + \lambda B + \frac{B^2}{A}\right] \right)_{0}^{1} = \frac{1}{2}.\left(A + 2B + \frac{B^2}{A}\right) - \frac{1}{2}.\frac{B^2}{A} = \frac{1}{2}.(A + 2B)$$

$$= (x1 - 2x3 + x2) + 2.(x3 - x1) = x2 - x1 \quad \text{substituting for A and B}$$

Once the length S has been calculated then the size of the equal intervals can be calculated. The problem then becomes one of calculating the values of λ needed to give this step size. If $\lambda = 0$ and $\lambda = \lambda 1$ were used for the first interval to calculate the distance S/n, where there are n equal steps required, an equation could be produced, the solution of which would give the value of $\lambda 1$. This value could then be used as the starting value for the next interval of length S/n to give the next value $\lambda 2$. Solving these equations is a more complex task than finding the path length, however, a numerical alternative is to calculate the values for λ needed to give the required step lengths using a successive approximation technique.

```
static int num=0;
static double x1=0,x2=0,x3=0,y1=0, y2=0,y3=0;
static double ABCD = 0, A2C2 = 0, B2D2 = 0, DD = 0, t = 0.5, e = 0;
public static void main(String[] args){
    IO.writeString("please enter the number of control points  \n");  num = IO.readInteger();
    for(int i=0;i<num;i++){
        IO.writeString("please enter control point\n");  cP[0][i]= dW.getCoord();
        if(i>0)dW.plotLine(cP[0][i-1],cP[0][i],Color.green);
    }
    for(int i=0;i<num-1;i++){  dW.plotLine(cP[0][i],cP[0][i+1],Color.red); }
    if(num==3){
        Point P1= cP[0][0], P2= cP[0][1], P3= cP[0][2];
        x1=P1.xd(); x2=P3.xd(); x3=P2.xd();
        y1=P1.yd(); y2=P3.yd(); y3=P2.yd();
        double A = 2*(x1-2*x3+x2),  C = 2*(y1-2*y3+y2);
        double B = 2*(x3-x1), D = 2*(y3-y1);
        ABCD = A*B+C*D;  A2C2 = A*A+C*C;  B2D2 = B*B+D*D;
        double Ll=0,  Lr=1.0;
        double result = span(Lr)-span(Ll);
        double DD=result/9;
        double q= subdivide(0,0,1, Color.red );
    }
}
static double span(double L){
    double S1= Math.sqrt(L*L*A2C2+2*L*ABCD+B2D2);
    double S2= (A2C2*L+ABCD)/Math.sqrt(A2C2);
    double S3= B2D2-ABCD*ABCD/A2C2;
    double S4= Math.log(Math.abs(S2+S1));
    double S5= (S2*S1+S3*S4)/2.0/Math.sqrt(A2C2);
    return S5;
}
static Point node(double L){
    Point np = new Point(2);
    double ax = x1*(1-L)+x3*L,  ay = y1*(1-L)+y3*L;
    double bx = x3*(1-L)+x2*L,  by = y3*(1-L)+y2*L;
    np.x("=", ax*(1-L) + bx*L);  np.y("=", ay*(1-L) + by*L);
    return np;
}
static double subdivide(double LS,double LL,double LR ,Color cc){
    double dd= distance(LS,LR);
    if((D+t-e>=dd)&&(D-e<=dd)) { e = dd-DD; dW.plotRectangle(node(LR),4,cc);return LR;}
    if(dd < D-e)return LS;
    double L = (LL+LR)/2;
    LS=subdivide(LS,LL,L,cc);
    LS=subdivide(LS,L,LRcc);
    return LS;
}
```

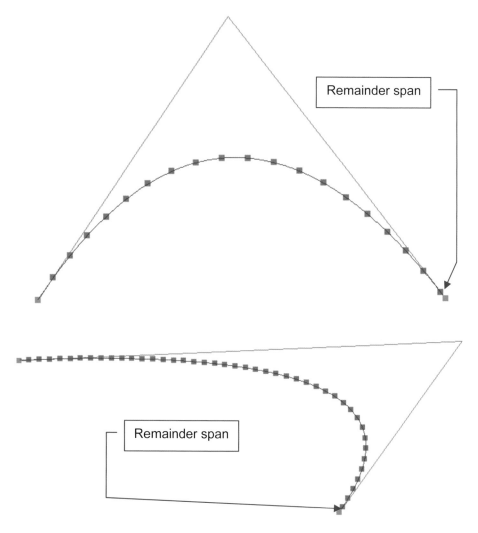

Figure 11.12 Equally space point symbols along curved paths

```
static Point subdivide(Point left,Point right,double LL,double LR,double D,double t){
    double dx = (left.xd()-right.xd()), dy= (left.yd()-right.yd()), d = dx*dx+dy*dy;
    if((D+t>=d)&&(D<=d)){dW.plotRectangle(right,4,Color.red);return right;}
    if(d < D)return left;
    double L = (LL+LR)/2;
    Point np = node(L);
    left=subdivide(left,np,LL,L,D,t);
    left=subdivide(left,right,L,LR,D,t);
    dx = (left.xd()-right.xd());  dy= (left.yd()-right.yd());  d = dx*dx+dy*dy;
    return left;
}
```

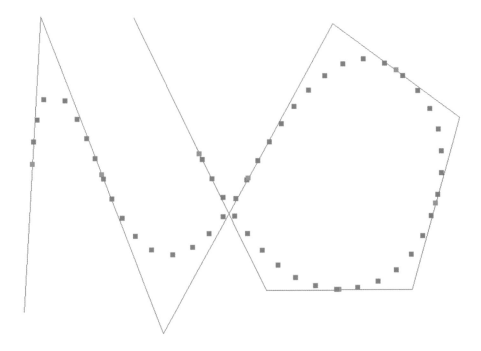

Figure 11.13 Linking curves with tangent continuity at their junctions

Curved paths can be constructed in a piecewise manner shown in Figure 11.13. This allows the remainder length from one "piece" of the path to be taken into the next "piece" of the curved path, to give a smooth motion along the total route. The two vectors used in the first example in Figures 11.9, are replaced by a sequence of control points, shown linked together as a control polyline drawn in grey. In Figure 11.13 the control triangles for each piece of the curve are linked together in this control polyline, so the edge of one triangle continues as a straight line into the next triangle. As a result the control points for this new arrangement are the vertices of this polyline and the midpoints of the polyline line-segments shown by green squares in Figure 11.13.

Because these parametric curves are tangent to the sides of their control triangles, this construction ensures that the resulting composite curve has tangent continuity at the positions where the separate parabolic curves join. This also allows the sequential subdivision into equal lengths to take the last span endpoint found in one curve segment to be the starting point for the next span linking into the next curve-piece, so keeping the spacing of points the same along the whole path.

Pixel Grid Parametric Line Interpolation

The uneven way that the parameter defined points occur along a curved line appeared to make a simple grid based interpolation impractical. However, the approximation process can be used in a slightly different way to interpolate curved lines onto a pixel grid. The algorithms which give fixed distances between points can be modified to give grid based interpolation by changing the tests used to terminate the recursive calls.

In this case the subdivision is continued until the point located, when converted to the integer coordinates of the pixel grid, gives a point one-pixel space away from the current point.

```
static Point subdivide(Point left,Point right,double LL,double LR ){
    int idx = (left.xi()-right.xi());
    int idy= (left.yi()-right.yi());
    int dix = idx*idx, diy=idy*idy;
    if(((dix<=1)&&(diy==1))||((dix==1)&&(diy<=1)))
        {dW.plotPoint(right,Color.black);  return right;}
    if((dix==0)&&(diy==0))return left;
    Point np = new Point(2);
    double L = (LL+LR)/2;
    double ax = x1*(1-L)+x3*L;
    double  ay = y1*(1-L)+y3*L;
    double bx = x3*(1-L)+x2*L;
    double  by = y3*(1-L)+y2*L;
    np.x("=", ax*(1-L) + bx*L);
    np.y("=", ay*(1-L) + by*L);
    left=subdivide(left,np,LL,L);
    left=subdivide(left,right,L,LR);
    return left;
}
```

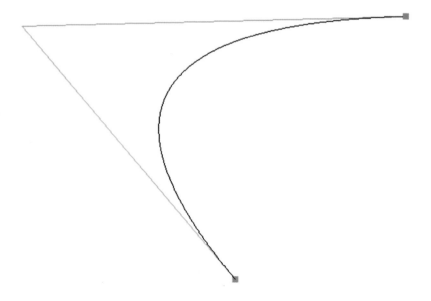

Figure 11.14 Parabolic curved line interpolation

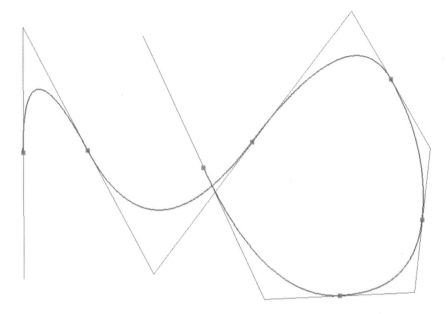

Figure 11.15 Piece-wise curve interpolation

Sharp corners in a piecewise curve, can be created in the way shown in Figure 11.16. by entering double points or control line segments with no length.

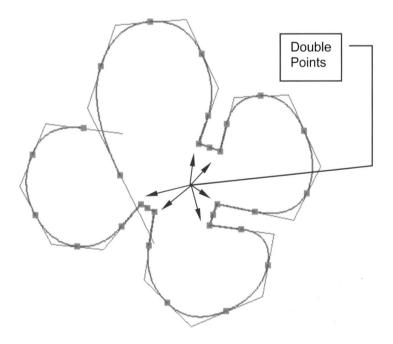

Figure 11.16 Piece-wise curve interpolation using double points

```java
static Point[][] cP = new Point[10][10];

static int num=0;
```

```java
public static void main(String[] args){
    IO.writeString("please enter the number of control points points  \n");
    num = IO.readInteger();
    for(int i=0;i<num;i++){
        IO.writeString("please enter control point\n");
        cP[0][i]= dW.getCoord();
        if(i>0) dW.plotLine(cP[0][i-1],cP[0][i],Color.green);
    }
    for(int i=0;i<num-1;i++){
        dW.plotLine(cP[0][i],cP[0][i+1],Color.red);
    }
    Point left= cP[0][0];
    Point right=cP[0][num-1];
    dW.plotRectangle(left,4,Color.green);
    Point pp= subdivide(left,right,0,1);
    dW.plotRectangle(right,4,Color.green);
}
```

```java
static Point subdivide(Point left,Point right,double LL,double LR){
    int idx = (left.xi()-right.xi());
    int  idy= (left.yi()-right.yi());
    int dix = idx*idx, diy=idy*idy;
    if(((dix<=1)&&(diy==1))||((dix==1)&&(diy<=1))){
        dW.plotPoint(right,Color.blue);
        return right;
    }
    if((dix==0)&&(diy==0))
        return left;
    Point np = new Point(2);
    double L = (LL+LR)/2;
    for(int j=1;j<num;j++){
        for(int i=0;i<num-j; i++){
            cP[j][i]= new Point(2);
            cP[j][i].x("=",(cP[j-1][i].xd()*(1-L) + cP[j-1][i+1].xd()*L));
            cP[j][i].y("=",(cP[j-1][i].yd()*(1-L) + cP[j-1][i+1].yd()*L));
        }
    }
    np= cP[num-1][0];
    left=subdivide(left,np,LL,L);
    left=subdivide(left,right,L,LR);
    return left;
}
```

Higher Order Parametric Line Interpolation

The algorithm to generate the quadratic curve in Figure 11.5 can be generalised to handle higher order curves in the way shown in Figure11.17 and illustrated in 11.18 to 20.

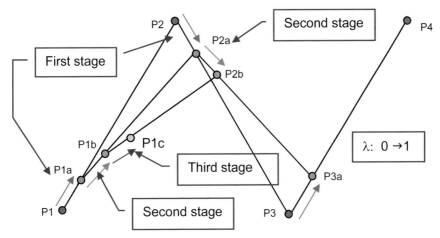

Figure 11.17 Third order curve interpolation for a cubic polynomial

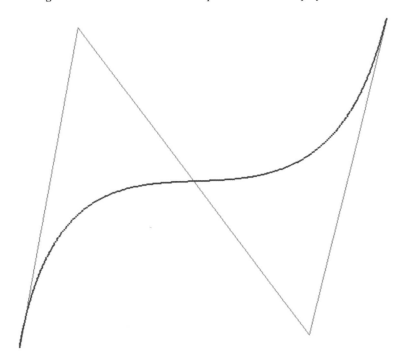

Figure 11.18 Cubic, order-three curve

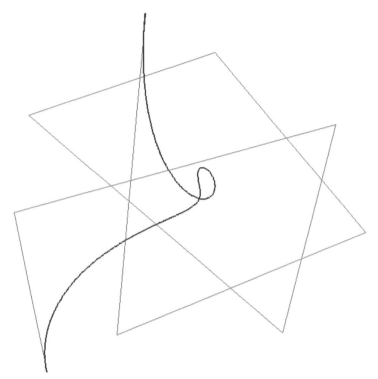

Figure 11.19 Order-eight curve

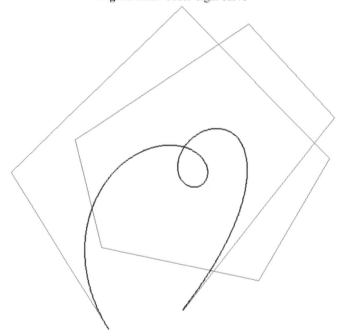

Figure 11.20 Order-nine curve

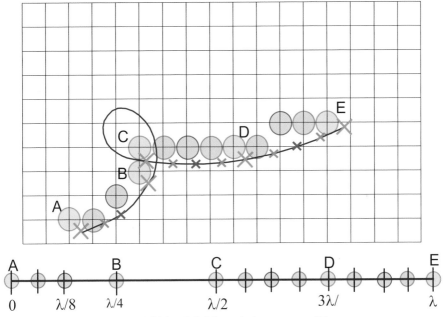

Figure 11.21 λ subdivision missing out a small loop

As curves get more complex and loops and cusps occur, further tests need to be added to cover special cases that the initial form of the algorithm could miss. Where sharp curves and loops are encountered, recursion returning after locating a new point could move to a next λ position far enough along a curved loop to come back to within a pixel width of the current point so missing out the whole loop in the way illustrated in Figure 11.21

```
static Point subdivide(Point left,Point right,double LL,double LR){
    boolean looptest = false;
    int idx = (left.xi()-right.xi()), idy= (left.yi()-right.yi());   int dix = idx*idx, diy=idy*idy;
    if((dix<=1)&&(diy<=1)) looptest = true;
    double L = (LL+LR)/2;
    Point np = node(L);
    int jdx = (left.xi()-right.xi()), jdy= (left.yi()-right.yi());
    int djx = jdx*jdx, djy=jdy*jdy;
    if(looptest&&(djx<=1)&&(djy<=1)){ looptest = false;
        if(((dix<=1)&&(diy==1))||((dix==1)&&(diy<=1))){
            dW.plotPoint(right,Color.blue); return right;
        }if((dix==0)&&(diy==0))return left;
    }
    left=subdivide(left,np,LL,L);
    left=subdivide(left,right,L,LR);
    return left;
}
```

```
static Point node(double L){
    for(int j=1;j<num;j++){
        for(int i=0;i<num-j;i++){
            cP[j][i]= new Point(2);
            cP[j][i].x("=",(cP[j-1][i].xd()*(1-L) + cP[j-1][i+1].xd()*L));
            cP[j][i].y("=",(cP[j-1][i].yd()*(1-L) + cP[j-1][i+1].yd()*L));
        }
    }return cP[num-1][0];
}
```

When a new point is found to be in a neighbouring pixel position, action is postponed until the next subdivision shows whether the next subdivision of λ also gives the same or a neighbouring pixel location. If it is a loop the point should lie further away. Though this is still not a watertight solution, the chances of multiple loops occurring, with exactly the correct spacing of pixel points with their corresponding λ values to cause loops to be missed out, becomes very much smaller. Adding more levels in the interpolation process gives smoother lines in a mathematical sense. This is important in some Computer Aided Design contexts, but a sequence of parabolic curved segments linked with tangent continuity is adequate for many interactive drawing applications.

Antialiased Curved Lines

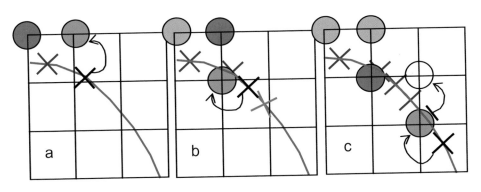

Figure 11.22 Locating curve-points on the pixel grid

The recursive subdivision of the values of λ place points on the curve in the positions indicated by crosses in Figure 11.22. These can be truncated to give integer indexes to the grid array shown by the red and green circles. Once the start point has been placed then subdivision of the rest of the curve will eventually place a new point in a pixel position (green circle) next door to the current pixel (red circle). This position can take over the current pixel-point role and the subdivision can be continued. As soon as a new point is located in the same grid cell as the current pixel: red cross in Figure 11.22b, then recursive subdivision can be stopped and control returned to the next level up in the recursive subdivision, shown in Figure 11.22c.

In this form this algorithm can miss out pixels containing small sections of the line shown by the hollow circle in Figure 11.22c. This should be compared with the same curve-grid relationship shown in Figure 11.22a.

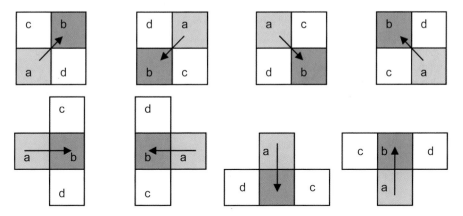

Figure 11.23 Adjacent pixel cell patterns that might contain elements of the line

The difference depends on the way the subdivision of λ occurs for a particular positioning of the curve and grid. Since the line will be represented by a sequence of contiguous pixels this is not a major problem as long as the display resolution is high, however further refinements are necessary to render antialiased lines.

The first problem is to identify all the pixel cells that the line passes through. The recursive search identifies neighbouring pixels. This information can be used to identify the different neighbourhood pixels that need to be tested to see if they contain segments of the line. Given the coordinates of the current point and a next point in the neighbouring pixel the integer indexes of the pixel array can be obtained by rounding or truncating these values. The changes in these x and y values can then be used to identify the patterns shown in Figure 11.23 where a is the current pixel, b is the next pixel and c and d are the adjacent pixels which may contain segments of the line.

```
//static int rows = 60, cols = 100;
//static Grid gd = null;
//gd= new Grid(IO,dW,a,cols,rows);
//gd.setGridBackground(Color.white);
//gd.paintGridArray();
//gd.drawGridLines(Color.black,Color.lightGray);
//Point left= subdivide(left,right,0,1,true);   ----
static Point subdivide(Point left,Point right,double LL,double LR,boolean antialiased){
    int[][] off = new int[][]{   //off sets from the next pixel b for the pixels c and d
    {0,0,0,0},{0,0,0,0},{0,0,0,0},{0,0,0,0},{0,0,0,0},{0,0,0,0},{0,1,0,-1},{0,-1,0,1},
    {0,0,0,0},{-1,0,1,0},{0,1,1,0},{-1,0,0,1},{0,0,0,0},{1,0,-1,0},{1,0,0,-1},{0,-1,-1,0}};
    boolean looptest = false;
    int ax=left.rx(), ay=left.ry(), bx=right.rx(), by=right.ry(); int cx=0, cy=0, dx=0, dy=0, j=0, k=0;
```

```
int idx = (right.rx()-left.rx()), idy= (right.ry()-left.ry()), dix = idx*idx, diy=idy*idy;
if((dix<=1)&&(diy<=1)) looptest = true;
double L = (LL+LR)/2;
Point np =node(L);
int jdx = (left.rx()-np.rx()), jdy= (left.ry()-np.ry()), djx = jdx*jdx, djy=jdy*jdy;
if(looptest&&(djx<=1)&&(djy<=1)){
    looptest = false; j=0;                                    //reset loop test
    if(idx== 0)j=j+1; if(idx== -1)j=j+2;  if(idx== 1)j=j+3;
    if(idy== 0)j=j+4;  if(idy== -1)j=j+8;  if(idy== 1)j=j+12;
    switch(j){                                    // select appropriate pixel pattern
        case 5:return left;
        case 6:case 7:case 9:case 13: case 10:case 11:case 14:case 15:
            cx= bx+off[j][0]; cy= by+off[j][1];
            dx= bx+off[j][2]; dy= by+off[j][3];
            double A= left.yd()-right.yd(),  B= right.xd()-left.xd();
            double C= left.xd()*right.yd()-right.xd()*left.yd();
            double db = Math.abs(A*bx+B*by+C),      // distance of cell b from the line
            double dc = Math.abs(A*cx+B*cy+C);      // distance of cell c from the line
            double dd = Math.abs(A*dx+B*dy+C);      // distance of cell d from the line
            double scale = (Math.abs(A)+Math.abs(B))/Math.sqrt(2)/(A*A+B*B);
            if(antialiased){          //set true for both grid based and standard displays
                double h= 1,s=0,b=1;
                dc= dc*scale;
                plotCell(dc, gd, cx, cy);
                dd=dd*scale;
                plotCell(dd, gd, dx, dy);
                db=db*scale;
                plotCell(db, gd, bx, by);
            }else{                        //set false for grid based displays only
                gd.paintInnerCell(cx,cy,1, Color.green );
                gd.paintInnerCell(dx,dy,1, Color.red );
                gd.paintInnerCell(bx,by,1, Color.black );
            }return right;
        }
    }
}
left=subdivide(left,np,LL,L,antialiased);
left=subdivide(left,right,L,LR,antialiased);
return left;
}
static void plotCell(double d,Grid gd,int x,int y){
    double h= 1,s=0;
    if(d<=1.0){
        Color cc= Color.getHSBColor((float)h,(float)s,(float)d);
        if(gd!=null)gd.paintInnerCell(x,y,1,cc ); else dW.plotPoint(x,y,cc);
    }else if(gd!=null) gd.paintInnerCell(x,y,1, Color.white);
    else dW.plotPoint(x,y,Color.white);
}
```

Parametric Area Boundary Line Interpolation

The natural extension to a control polyline is a control polygon. If the control polyline is closed in a loop then it is possible to generate a closed curve. The simplest scheme, shown in Figure 11.24, is provided by parabolic curves linked with tangent continuity. A smooth transition between curved segments is achieved by sharing each control polygon's first and last line segments between neighbouring curved arcs. This is done for quadratic curves by entering a polygon, in the way shown in Figure 11.24, then taking the mid points of its sides as the beginning and end points of a sequence of control triangles.

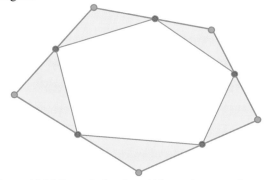

Figure 11.24 Control triangles within a polygon to give a closed loop

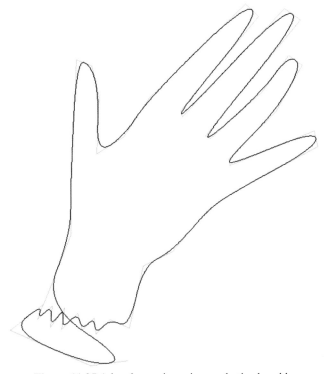

Figure 11.25 A hand as a piecewise quadratic closed loop

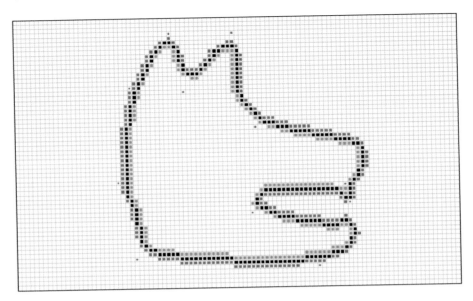

Figure 11.26 Painting in the cells that might contain line segments

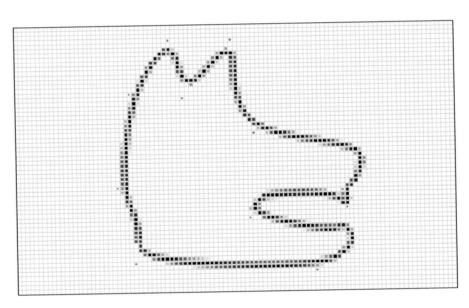

Figure 11.27 Antialiasing using the distance of pixel centres from the line

Using the patterns in Figure 11.23, Figure 11.26 shows a closed loop where the b pixels are coloured black, and the c pixels coloured green and the d pixels coloured red. The distance of the centres of these pixels from a locally estimated line position is then used to calculate a greyscale value for the pixel to give the antialiased version of the same curve shown in Figure 11.27.

Figure 11.28 Closed loops using control polygons: with and without antialiasing

Figure 11.28 shows that this gives a reasonably good result. However, an alternative slightly more flexible approach to antialiasing these lines, is to calculate the area of each pixel occupied by pieces of the line. The neighbouring pixel patterns are again identified but then the proportion of the area of each pixel taken up by the line sets the greyscale used to plot them. This allows the lines of different widths to be drawn in Figure 11.29.

The area of the cell occupied by the line is calculated assuming the cell is a unit square, giving it an area of 1, which will correspond to a fully black pixel. The cell is then tested against the top and bottom edges of the line to calculate how much of the cell is outside the line area. Each edge is tested against the vertices of the pixel cell and this gives two generic patterns either a triangle or a trapezium. Sixteen relationships between the line and the cell exist depending on whether a cell vertex is inside or outside the line.

Figure 11.29 Changing line widths using the antialiasing algorithm

```
int j=0;
if((d[0])<0)j=j+1;  if((d[1])<0)j=j+2;  if((d[2])<0)j=j+4;  if((d[3])<0)j=j+8;
```

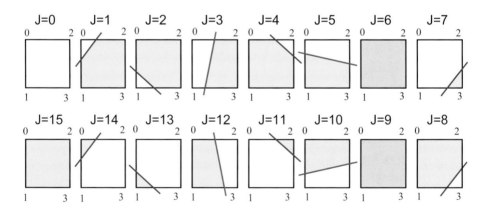

Figure 11.30 Classifying the relationship between a line edge and the pixel cell

Cells with a J value of 1, 2, 4, 7, 8, 11, 13 and 14 all correspond to a triangular pattern. The value of J is used as an index to a table giving the end-points of the cell's edge-lines that are cut by the line, which are then used to calculate the lengths of the sides of the triangles. Where the pattern is a triangle the area of the triangle is given by multiplying these lengths together and dividing by two. Cells with a J value of 3, 5, 10 and 12 all correspond to a trapezium pattern. In a similar way J is used to

```
static void shadeCell(NTuple ln,double w,Grid gd,int xx,int yy){
    double cell =1;
    double [] d= new double[4];
    double [] x= new double[]{xx-0.5,xx-0.5,xx+0.5,xx+0.5};
    double [] y= new double[]{yy-0.5,yy+0.5,yy-0.5,yy+0.5};
    double offset = w*Math.sqrt(ln.n[1]*ln.n[1]+ln.n[2]*ln.n[2]);
    for(int i=-1;i<2;i=i+2){                      //setting top and bottom edges of the line
        for(int j=0;j<4;j++) d[j]=((ln.n[1]*x[j]+ln.n[2]*y[j]+ln.n[0]))*i+offset;
        int j=0;
        if((d[0])<0)j=j+1;  if((d[1])<0)j=j+2;  if((d[2])<0)j=j+4;  if((d[3])<0)j=j+8;
        switch(j){
            case 0: break;
            case 1: case 2: case 4: case 8: cell=cell-area(d,x,y,j,1); break;
            case 7: case 11: case 13: case 14: cell=cell-1+area(d,x,y,j,1); break;
            case 3: case 5: cell=cell-area(d,x,y,j,2); break;
            case 10: case 12: cell=cell-1+area(d,x,y,j,2); break;
            case 15: cell=cell-1;break;
        }
    }
    if(cell>0)plotCell(1-cell,gd,xx,yy);
}
static double area(double[] d,double[] x,double[] y,int j,int n){
    int [][]k=new int[][]
        {{0,0,0,0},{0,2,0,1},{1,3,1,0},{0,2,1,3},{2,0,2,3},{0,1,2,3},{0,0,0,0},{3,1,3,2}};
    if(j>7)j=15-j;
    int [] i = k[j];
    double d1=0,d2=0;
    if(n==1){
        d1= d[i[0]]*(x[i[1]]-x[i[0]])/(d[i[0]]-d[i[1]]);
        d2= d[i[2]]*(y[i[3]]-y[i[2]])/(d[i[2]]-d[i[3]]);
        return Math.abs(d1*d2/2);
    }else if(j==3){
        d1= d[i[0]]*(x[i[1]]-x[i[0]])/(d[i[0]]-d[i[1]]);
        d2= d[i[2]]*(x[i[3]]-x[i[2]])/(d[i[2]]-d[i[3]]);
    }else{
        d1= d[i[0]]*(y[i[1]]-y[i[0]])/(d[i[0]]-d[i[1]]);
        d2= d[i[2]]*(y[i[3]]-y[i[2]])/(d[i[2]]-d[i[3]]);
    }return Math.abs((d1+d2)/2);
}
```

select the pairs of end points of the edge lines of the cell that are cut by the line. This allows the length of the sides of the trapezium to be calculated, and adding these two lengths together and dividing the result by two provides its area. In Figure 11.30, the side of the line that is shaded yellow determines the area of the pixel that lies inside the line. In case J=1 the triangle area lies outside the line's edge, whereas the area outside the edge of the line in the matching pattern given by J=14, lies on the other side of the edge and is one minus the triangle area. The same applies to the matching pairs of patterns given by J and 15-J classification values. Subtracting the outside areas for the two edge lines of the curve determines the area of each pixel occupied by the curve and hence the pixel's greyscale.

Thick Parametric Lines

Where the resolution of the display is high then the need for this kind of antialiasing treatment is less severe. In this context, for most applications, thick lines will replace single pixel-wide lines. A simple approach to thick line rendering can be achieved by plotting a series of overlapping squares over the pixel positions. This involves duplicate pixel plotting but is simple to implement. There is however an upper limit to the size of the squares used before the line can be seen to change width with orientation. One solution to this difficulty is to change the square to a rectangle that is adjusted so that its diagonal is perpendicular to the line and is set to the line width required. Figures 11.31 and 11.32 show the differences between the two approaches.

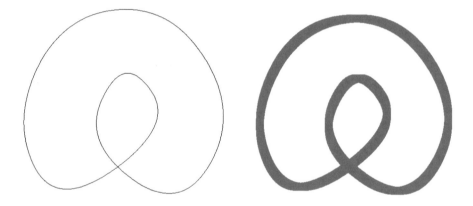

Figure 11.31 Replace pixel points by overlapping squares

Figure 11.32 Replace squares by overlapping scaled rectangles

```
static Point subdivide(Point left,Point right,double LL,double LR,int j,Color cc){
    boolean looptest = false;
    int idx = (left.rx()-right.rx()), idy= (left.ry()-right.ry()), dix = idx*idx, diy=idy*idy;
    if((dix<=1)&&(diy<=1)) looptest = true;
    double L = (LL+LR)/2;  Point np =node(L);
    int jdx = (left.rx()-np.rx()), jdy= (left.ry()-np.ry()), djx = jdx*jdx, djy=jdy*jdy;
    if(looptest&&(djx<=1)&&(djy<=1)){
        looptest = false;
        if(((dix<=1)&&(diy==1))||((dix==1)&&(diy<=1))){
            if (j==0)dW.plotPoint(right.rx(),right.ry(),cc);
            else if(j>0) {
                double A=left.yd()-right.yd(),  B=right.xd()-left.xd();
                double norm = Math.sqrt(A*A+B*B);
                double w = j*Math.abs(A)/norm;
                double h = j*Math.abs(B)/norm;
                int xx = (int)(right.rx()-w/2.0);
                int yy = (int)(right.ry()-h/2.0);
                dW.plotRectangle(xx,yy,(int) w,(int)h,cc);

            }
            else dW.plotRectangle((int)(right.rx()+j/2.0),(int)(right.ry()+j/2.0),-j,-j, cc);
            return right;}
        if((dix==0)&&(diy==0))return left;
    }
    left=subdivide(left,np,LL,L,j,cc);
    left=subdivide(left,right,L,LR,j,cc);
    return left;

}
```

Infilling Areas "Inside" Parametric Line Boundaries

Once a closed boundary has been created then it is possible to apply an area fill procedure to the shape. The same general approach explored in chapter 10 should be applicable. Merge sorting the pixel sequence should provide the ordered pairs needed to define scan lines. Again there are the options for filling in the areas that include or exclude the boundary line, and the intermediate scheme where the ends of scan lines are placed as close to the true boundary line as possible.

The most flexible and useful approach is to fill the area inside the boundary-line since this allows the boundary to be rendered in a different colour or the same colour as the interior. It also allows adjacent areas with shared edges to be filled in without overwriting an existing boundary line should this be necessary. At the least it allows the matter to be a choice depending on the application. Again it is the special cases that need to be studied, since a simple application of sorting and pairing will break down. Whether a particular pixel is the beginning or the end of an area-fill scan line depends on its neighbours in boundary drawing order. Taking a point's two neighbours would seem to provide the minimum local information needed to process a given pixel boundary point. However, area fill requires both global and local properties to be taken into account.

Line Segment Orientation Patterns

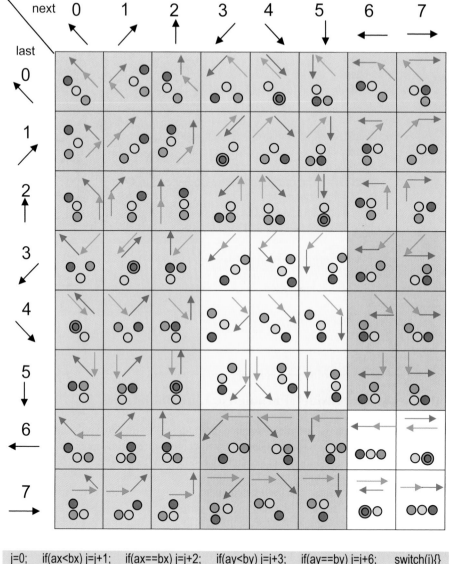

j=0;   if(ax<bx) j=j+1;   if(ax==bx) j=j+2;   if(ay<by) j=j+3;   if(ay==by) j=j+6;   switch(j){}

Figure 11.33 ⬤ last point ◯ current point ⬤ next point

Starting with the detail: the table in Figure 11.33 shows all the local relationships that can exist between any three neighbouring pixels in a boundary. If these three neighbouring points are treated as two line segments then there are eight orientations for each line segment, giving 64 relationships that need to be analysed.

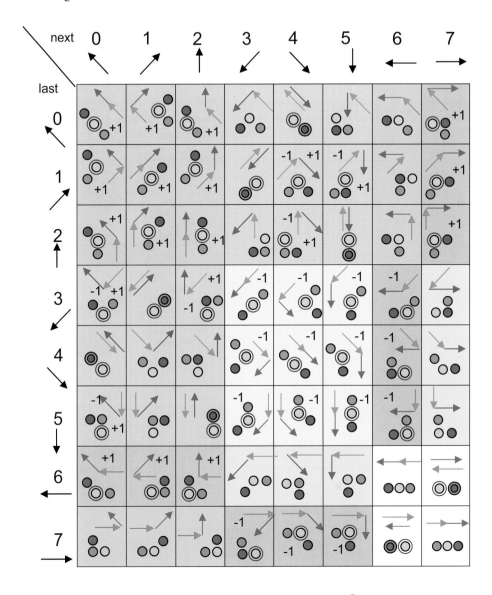

Figure 11.34 ⬤last point ◎current point ⬤next point ◯ scan-line point

The directions of the lines are shown by the colour-code for the pixels given below the table. The table is itself colour coded to show groups of relationships that demand the same or a similar treatment. Given a clockwise rotation round the "interior", the pink block in rows and columns 0, 1, 2 correspond to left-hand boundary pixels where the pale yellow block in rows and columns 3, 4, 5 contain the right-hand boundary line pixels. The green blocks identify the pixel sequences that will generate a single infill pixel on a scan line that will need special treatment. Columns and rows 6 and 7 contain horizontal pixel sequences that will occupy the same scan line. These also require different treatment to the standard infill operation. The table given in Figure 11.34 shows the selection of boundary points needed to define the infill scan lines that will provide full shading covering the interior of a simple area and its boundary. Assuming a clockwise boundary order round an area's interior, the beginning and end points of infill scan lines are shown circled in blue. When these points are sorted into coordinate or raster order from their boundary order then they can be paired up to define infill scan-line segments. This is done by the procedure called "*shade(..)*" in the code segment given below. This successfully fills shapes of the form shown in Figure 11.35a.

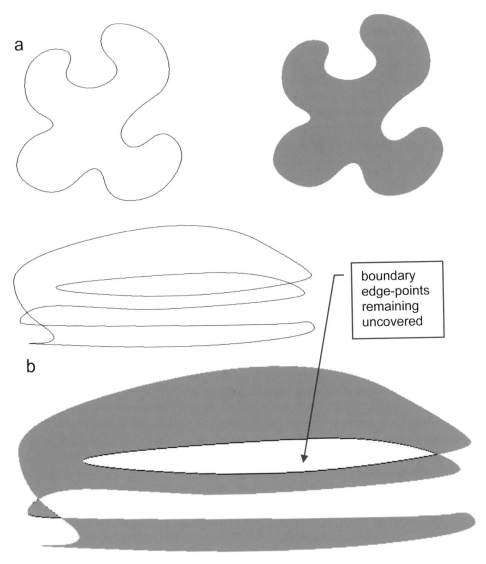

Figure 11.35 Area fill over non self-crossing and self-crossing boundaries

```
static void fill(Point[] pnt,int[] dir, int cpnt, Color c1, Color c2){
    int length = cpnt, jj=0;
    Point [] boundary = new Point[length+2];
    pnt[length]= pnt[0];   pnt[length+1]= pnt[1];
    length=length+2;
    for(int i=0;i<length;i++){ dW.plotPoint(pnt[i],Color.black); }
    for(int i=2;i<length;i++){                              // point order c--a--b
        int cx = pnt[i-2].xi(),ax = pnt[i-1].xi(),bx = pnt[i].xi();
        int cy = pnt[i-2].yi(),ay = pnt[i-1].yi(),by = pnt[i].yi();
        int k=0;if(cx<ax)k=k+1;if(cx==ax)k=k+2;if(cy<ay)k=k+3;if(cy==ay)k=k+6;
        int j=0;if(ax<bx)j=j+1;if(ax==bx)j=j+2;if(ay<by)j=j+3;if(ay==by)j=j+6;
        if(((j<3)||(j==7))&&((k<3)||(k==6))){                      // up
            Point pt = new Point(2);
            pt.n[1]=pnt[i-1].n[1]; pt.n[2]=pnt[i-1].n[2];
            boundary[jj]= pt; boundary[jj].tag= 1; jj++;
        } if((j>2)&&(j<7)&&(k>2)&&(k!=6)){                         // down
            Point pt = new Point(2);
            pt.n[1]=pnt[i-1].n[1]; pt.n[2]=pnt[i-1].n[2];
            boundary[jj]= pt; boundary[jj].tag= -1; jj++;
        } switch(k*8+j){
            case 24 : case 26 : case 40 :
                Point pt = new Point(2);
                pt.n[1]=pnt[i-1].n[1]; pt.n[2]=pnt[i-1].n[2];
                boundary[jj]= pt; boundary[jj].tag = -1; jj++;
                pt = new Point(2);
                pt.n[1]=pnt[i-1].n[1]; pt.n[2]=pnt[i-1].n[2];
                boundary[jj]= pt; boundary[jj].tag = +1; jj++; break;        //down up
            case 12 : case 13 : case 20 :
                pt = new Point(2);
                pt.n[1]=pnt[i-1].n[1]; pt.n[2]=pnt[i-1].n[2];
                boundary[jj]= pt; boundary[jj].tag = +1; jj++;
                pt = new Point(2);
                pt.n[1]=pnt[i-1].n[1]; pt.n[2]=pnt[i-1].n[2];
                boundary[jj]= pt; boundary[jj].tag = -1; jj++; break;        //up down
        }
    }Point[] s = reorderBoundaryPoints(boundary,jj);
    shade(s,Color.green); dW.getCoord();
    shade(s,Color.white);
    for(int i=0;i<length-2;i++){ dW.plotPoint(pnt[i],Color.white);}dW.getCoord();
    shade(s,0,Color.pink); dW.getCoord();
    for(int i=0;i<length-2;i++){dW.plotPoint(pnt[i],Color.black);}
}
static void shade(Point[] s,Color cc){
    if(s!=null){
        for(int i=0; i<s.length; i=i+2){ dW.plotLine(s[i],s[i+1],cc); }
    }else IO.writeString("no scan array");
}
```

```
static void shade(Point[] s,int dummy,Color cc){
    if(s!=null){
        int i=0;int cnt = 0; int y1=0,y2=0;  Point p1=null,p2=null;
        while(i<s.length){
            while((i<s.length)&&(cnt<=0)){
                cnt=cnt+s[i].tag; p1=s[i]; y1=s[i].yi();  i++;
                if(i<s.length) y2=s[i].yi();
            }while((i<s.length)&&(y1==y2)&&(cnt>0)){
                p2=s[i]; cnt=cnt+s[i].tag; i++;
                if(i<s.length)y2=s[i].yi();
            }if(p1.yi()==p2.yi())dW.plotLine(p1,p2,cc);
            if(y1!=y2) cnt=0;
        }
    }else IO.writeString("no scan array");
}
static Point[] reorderBoundaryPoints(Point[]b,int jj){
    int i = 0, j=0, starty=0, endy=0, starti=0, endi=0, len=0, num=0;
    if(b == null)return null;
    Point[] changeArray = null , newArray = null, oldArray = null;
    while(i < jj){
        if((i < jj)&&(b[i].tag<0)){
            endi = starti = i;
            while((i < jj)&&(b[i].tag<0)){dW.plotPoint(b[i++],Color.blue);}
            endi = i; num = endi-starti;
            changeArray = new Point[num];
            for(int k=0;k<num;k++){changeArray[k]=b[starti+k]; }
        }else if((i < jj)&&(b[i].tag>0)){
            starti= endi = i;
            while((i < jj)&&(b[i].tag>0)){dW.plotPoint(b[i++],Color.blue);}
            starti = i; num = starti-endi;
            changeArray = new Point[num];
            for(int k=0;k<num;k++){ changeArray[k]=b[starti-k-1]; }
        }
        oldArray = newArray;   newArray = merge(oldArray,changeArray);
    }
    return newArray;
}
```

If the aim is to cover all boundary as well as interior points of simple areas, the classification of scan line end points given in Figure 11.34 is sufficient. This selection of points is also adequate for an alternative infill strategy that gives the "silhouette" shading needed for three-dimensional surface boundaries when they are projected onto a display surface. An algorithm purely based on processing a pixel sequence in boundary order cannot provide the on-off area fill algorithm for self crossing boundary lines, without some of the boundary pixels from a previously drawn boundary line showing through, in the way shown in Figure 11.35b and 11.36a.

The On/Off strategy provides the simplest way to fill a non self-crossing area boundary. However, another approach is possible based on the orientation of the boundary segments that are cut by each infill scan line. Where a boundary does not cut itself the directions of these segments alternate along the scan line matching the On/Off cycle. Where a clockwise rotation round an area is employed to define its interior, the left edge cut by an infill scan-line will have an upward direction whereas the end or right hand edge will be cut by an edge with a downward direction. In contrast where the boundary intersects itself it is possible for an "off" step to align with cutting an "up" edge segment and for "on" steps to correspond with "down" segments, again assuming a clockwise order round the interior of the area.

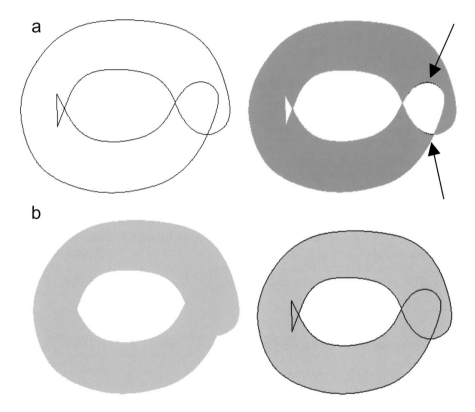

Figure 11.36 On-Off and silhouette or winding number area fill

The silhouette or winding number approach adds up the number of "up" crossing points and subtracts the number of "down" crossing points taken to reach each region within a self-crossing boundary. Where this count is less than or equal to zero the scan line is switched off where it is positive the infill line is switched on.

In Figure 11.34 the selected points have been given an "*up*" or "*down*" value of +1 or –1. If these tag values are carried with the points when they are sorted into raster order, then it is possible to write a new "*shade(..)*" procedure also given above, (overloading the function name), which by creating a running total of these tag values along each scan line is able to implement this alternative approach. Where the total is less than or equal to zero no line is drawn, when it is greater than zero the scan line is filled in. The result of this approach is shown in Figure 11.36b for a pink shaded shape. In Figure 11.36a the same shape with the same boundary is filled using the previous On/Off algorithm. This illustrates that, when the boundary cuts itself, boundary points can remain uncovered.

In order to get an On/Off algorithm that can cover the boundaries of a self-crossing boundary, in a consistent way, requires a further extension to this counting approach shown in the code segment given below. It is the treatment of horizontal sequences of boundary pixels along a scan line that creates the problem with the On/Off treatment of self-crossing boundary lines. The boundary order will consistently select the end of such a horizontal sequence based on local boundary order in a way that is suitable for the silhouette algorithm, where the selection in the case of the On/Off approach has to depend on the context set up by the overall arrangement of the boundary line. This global structure only emerges at the scan line level after raster sorting the points.

Consequently both ends of horizontal boundary point sequences need to be selected at the boundary order stage, as shown in Figure 11.37, and which end of such a sequence is used to define scan infill lines must be left to the "*shade (..)*" fill procedure. Selection in the "*shade(...)*" fill procedure can be done by extending the counting procedure employed in the silhouette fill algorithm. When a horizontal sequence of boundary points occurs following a simple sequence of up or down points, the next step can be either up or down. It is only when the two end relationships to the horizontal sequence are obtained that it is clear whether there is to be a cut in the infill line. Consequently if each point triple in Figure 11.37, which contains a horizontal pair is given a value of plus a half or minus a half, depending on the third point, then in a scan line sequence: two half ups will give a full up-intersection whereas an up and a down will indicate a horizontal tangent line.

```
static void shade(Point[] s,int dummy,Color cc){
    if(s!=null){
        int i=0;int cnt = 0; int y1=s[0].yi(),y2=0;
        Point p1=null,p2=null;
        while(i<s.length){
            while((i<s.length)&&(y1==s[i].yi())&&(cnt%4<=0)){cnt= cnt+s[i].tag;i++;}
            p1=s[i-1];
            while((i<s.length)&&(y1==s[i].yi())&&(cnt%4>0)){cnt=cnt+s[i].tag;i++;}
            p2=s[i-1];
            if(p1.yi()==p2.yi()) dW.plotLine(p1, p2,cc);
            if(i<s.length)y2=s[i].yi();
            if(y1!=y2){ cnt=0;y1=y2;}
        }
    }else IO.writeString("no scan array");
}
```

```
static void fill(Point[] pnt,int[] dir, int cpnt, Color c1, Color c2){
    int length = cpnt; int jj=0;
    Point [] boundary = new Point[length+2];
    pnt[length]= pnt[0];
    pnt[length+1]= pnt[1];
    length=length+2;
    for(int i=0;i<length;i++){dW.plotPoint(pnt[i],Color.black);}
    for(int i=2;i<length;i++){                              // point order c--a--b
        int cx = pnt[i-2].xi(),ax = pnt[i-1].xi(),bx = pnt[i].xi();
        int cy = pnt[i-2].yi(),ay = pnt[i-1].yi(),by = pnt[i].yi();
        int k=0;if(cx<ax)k=k+1;if(cx==ax)k=k+2;if(cy<ay)k=k+3;if(cy==ay)k=k+6;
        int j=0;if(ax<bx)j=j+1;if(ax==bx)j=j+2;if(ay<by)j=j+3;if(ay==by)j=j+6;
        if((j<3)&&(k<3)){                                   // up
            Point pt = new Point(2);  pt.n[1]=pnt[i-1].n[1];  pt.n[2]=pnt[i-1].n[2];
            boundary[jj]= pt;  boundary[jj].tag= 2;  jj++;
        }if(((j<3)&&(k>5))||((j>5)&&(k<3))){                 // up
            Point pt = new Point(2);  pt.n[1]=pnt[i-1].n[1];  pt.n[2]=pnt[i-1].n[2];
            boundary[jj]= pt;  boundary[jj].tag= 1;  jj++;
        }if(((j>2)&&(j<6)&&(k>5))||((k>2)&&(k<6)&&(j>5))){   // down
            Point pt = new Point(2); pt.n[1]=pnt[i-1].n[1]; pt.n[2]=pnt[i-1].n[2];
            boundary[jj]= pt; boundary[jj].tag= -1; jj++;
        }if((j>2)&&(j<6)&&(k>2)&&(k<6)){                     // down
            Point pt = new Point(2);  pt.n[1]=pnt[i-1].n[1];  pt.n[2]=pnt[i-1].n[2];
            boundary[jj]= pt;  boundary[jj].tag= -2;  jj++;
        }
        switch(k*8+j){
        case 24 : case 26 : case 40 :
            Point pt = new Point(2);  pt.n[1]=pnt[i-1].n[1];  pt.n[2]=pnt[i-1].n[2];
            boundary[jj]= pt;  boundary[jj].tag = -2; jj++;
            pt = new Point(2);  pt.n[1]=pnt[i-1].n[1];  pt.n[2]=pnt[i-1].n[2];
            boundary[jj]= pt;  boundary[jj].tag = +2;  jj++; break;       //down & up
        case 12 : case 13 : case 20 :
            pt = new Point(2);  pt.n[1]=pnt[i-1].n[1];  pt.n[2]=pnt[i-1].n[2];
            boundary[jj]= pt;  boundary[jj].tag = +2; jj++;
            pt = new Point(2); pt.n[1]=pnt[i-1].n[1]; pt.n[2]=pnt[i-1].n[2];
            boundary[jj]= pt;  boundary[jj].tag = -2;  jj++;  break;      //up & down
        }
    }
    Point[] s = reorderBoundaryPoints(boundary,jj);
    for(int i=0;i<s.length;i++){
        if((i>1)&&(s[i-1].tag==s[i].tag)&&(!compare(s[i-1],"<=",s[i]))){
            Point temp =s[i-1]; s[i-1]=s[i];s[i]=temp;
        }
    }
    shade(s,0,Color.pink);
    for(int i=0;i<length-2;i++){ dW.plotPoint(pnt[i],Color.black); }
}
```

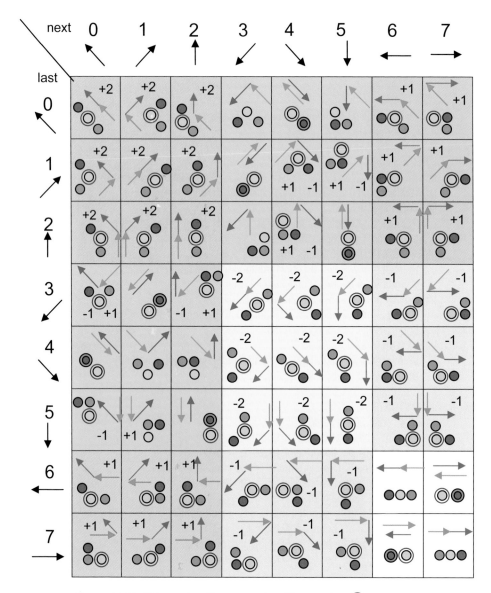

Figure 11.37 ⬤last point ◯current point ●next point ◯scan-line point

To keep the tag values as integers, if the half values are given as one then the standard cut values will have to be plus or minus two in the way shown in Figure 11.37. The "*shade(..)*" procedure by taking the running total *modulo four* can generate an On/Off infill line. Drawing a line where this value is not zero and leaving a space where it is zero gives the result shown in Figure 11.39a with the boundary points totally over written and in Figure 11.39b where a new boundary has been redrawn on top of the shading. Figure 11.38 illustrates the boundary order relationships that can be used for a shading algorithm that simply fills in the area

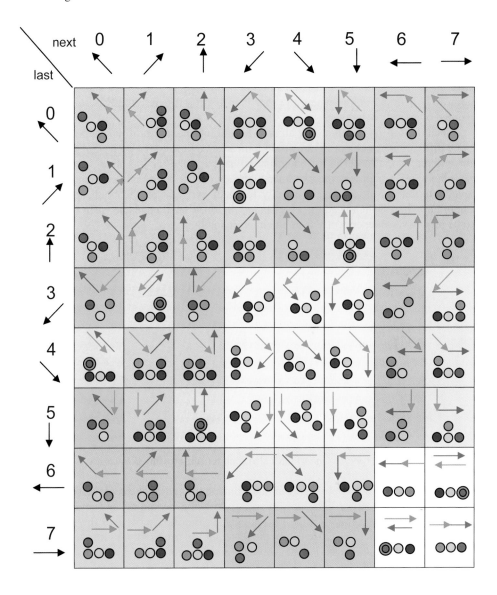

Figure 11.38 ◐ last point ○ current point ◑ next point ● scan-line end point

inside a boundary line, where there are no spurs or what are sometimes called hanging edges. These occur where forward and backward boundary line sequences coincide. The ends of spurs are indicated by the relationships colour coded yellow in Figure 11.38. If hanging edges occur one approach is to treating them as line features and remove them before shading the rest of the area, allowing them to be ignored in the table as only occurring by error. Two coincident boundary lines in opposite directions will not be apparent at the boundary order stage, except where they create end spurs. These end spur relationships can be used to match up coincident points leading up to and following from them, in a pre-processing step to remove them. End spurs occur in the same group of relationships containing singleton boundary points. When singleton boundary points are encountered two points have to be generated for the scan fill stage developed above to operate correctly. For infill giving total cover such an identical pair when it is outside the main area as a spur will cause a single point to be painted in, but where the spur is inside it will cause the two points to be painted on top of each other.

When the infill algorithm only aims to fill the internal areas of the shape this gets more difficult. The cells in Figure 11.38 indicate two blue coded points are generated for one directional ordering of boundary points but none for the other, but it is not possible to tell which way the two lines in the spur relate to each other from the pixel level information alone. The capability to shade only inside the boundary line is important where, when shading a tessellated region, shared edges between tiles need to be processed independently from the interior shading.

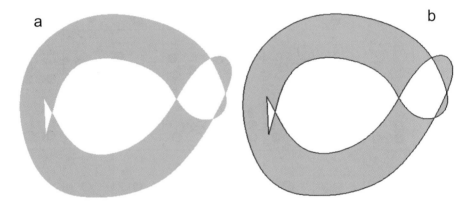

Figure 11.39 An On/Off infill algorithm for self-crossing boundaries

There is a different approach. The algorithms explored above remove sequences of horizontal boundary points, only working with the beginning and end of such sequences. If all boundary points are included in the data passed to the scan-line *shade()* procedure, it is possible to use the different ordering of the points after raster ordering to simplify the task. Because all the boundary points are passed to this stage of the process it is easy to plot or not plot these points as needed, and to generate the infill points between these points using the previous classification obtained from the boundary order stage. In fact by changing the coding at this stage it is possible to simplify the first stage in the way shown in Figure 11.40. The main blocks stay the same. The change is to the blocks containing singleton corner points lying on a scan line. This can be done because they are by definition boundary points and all boundary points are treated in the same way in the new scan line *shade()* procedure. With a tag value of zero where before they had two values which cancelled each other out, they do not affect the running total along the scan line that is used to switch the fill-line on or off. The major change to the program is that all the boundary points have to be sorted into raster scan order not just change points.

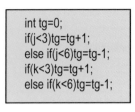

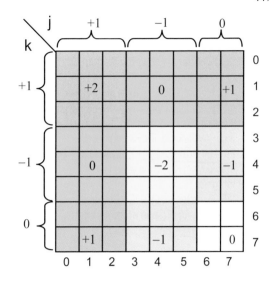

Figure 11.40 Alternative classification of boundary point sequences

```
static void shade(Point[] s,int dummy,Color cc,Color cb){
    if(s!=null){
        int i=0,j=0,cnt = s[i].tag;
        Point sb=s[i],eb=s[i]; Point ss= new Point(2); Point es= new Point(2);
        int liney = s[i].yi();i++;
        while(i<s.length){
            j=0;
            if(s[i-1].xi()==s[i].xi())j=j+1; if(s[i-1].xi()==s[i].xi()+1)j=j+2;
            if(s[i-1].xi()==s[i].xi()-1)j=j+3; if(s[i].yi()!=liney)j=j+4;
            switch(j){
                case 0:
                    dW.plotLine(sb,eb,cb);
                    if(cnt/2%2!=0){                              // On/Off test
                        ss.n[1] = s[i-1].n[1]+1; ss.n[2] = s[i-1].n[2];
                        es.n[1] = s[i].n[1]-1; es.n[2] = s[i].n[2];
                        dW.plotLine(ss,es,cc);
                    }sb = eb = s[i]; cnt=cnt+s[i].tag;  break;
                case 1: case 2: case 3:
                    eb=s[i];  cnt = cnt+s[i].tag; break;
                case 4: case 5: case 6: case 7:
                    dW.plotLine(sb,eb,cb);
                    cnt =s[i].tag; sb = eb = s[i]; liney = s[i].yi(); break;
            }i++;
        }
        dW.plotLine(sb,eb,cb);
    }
    else IO.writeString("no scan array");
}
```

```
static void fill(Point[] pnt,  int length, Color c1, Color c2){
    int  jj=0;
    Point [] boundary = new Point[length+2];
    for(int i=2;i<length;i++){                          // point order c--a--b
        int cx = pnt[i-2].xi(),  ax = pnt[i-1].xi(),  bx = pnt[i].xi();
        int cy = pnt[i-2].yi(),  ay = pnt[i-1].yi(),  by = pnt[i].yi();
        int k=0;  if(cx<ax)k=k+1;  if(cx==ax)k=k+2;
        if(cy<ay)k=k+3;  if(cy==ay)k=k+6;
        int j=0;  if(ax<bx) j=j+1; if(ax==bx) j=j+2;
        if(ay<by)j=j+3;  if(ay==by) j=j+6;
        int tg=0;
        if(j<3) tg=tg+1;  else if(j<6) tg=tg-1;
        if(k<3) tg=tg+1; else if(k<6)tg=tg-1;
        boundary[jj]= pnt[i-1];  boundary[jj].tag=tg;
        jj++;
    }
    boundary[jj]=boundary[0];jj++;
    Point[] s = reorderBoundaryPoints(boundary,jj);
    for(int i=0;i<length-2;i++){ dW.plotPoint(pnt[i],Color.gray);}
    shade(s,0,c1,c2);
}
```

```
static Point[ ] reorderBoundaryPoints(Point[ ] b, int  jj ){
    int i = 0, j=0, starty=0, endy=0, starti=0, endi=0, len=0;
    if(b == null)return null;
    Point[] changeArray = null , newArray = null, oldArray = null;
    while(i < jj-1){
        if((i < jj-1)&&  compare(b[i],"<=", b[i+1])){
            starti = i;
            endi = i+1;
            while((i < jj-1)&&  compare(b[i],"<=", b[i+1])){ i++;  endi = i; }
            int num = endi-starti;
            changeArray = new Point[num];
            for(int k=0;k<num;k++){changeArray[k]=b[starti+k]; }
        }else{
            starti= i+1;
            endi = i;
            while((i < jj-1)&&  compare(b[i],">=", b[i+1])){i++; starti = i; }
            int num = starti-endi;
            changeArray = new Point[num];
            for(int k=0;k<num;k++){ changeArray[k]=b[starti-k-1]; }
        }
        oldArray = newArray;
        newArray = merge(oldArray,changeArray);
    }
    return newArray;
}
```

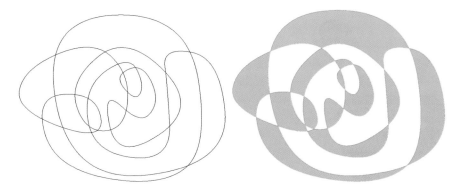

Figure 11.41 Shaded self intersecting, curved, boundary loop

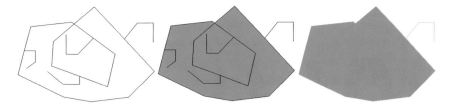

Figure 11.42 Winding number shaded self-intersecting polygon with spurs and inlets

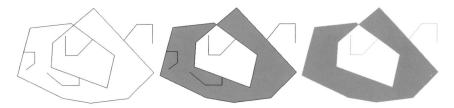

Figure 11.43 On/off shaded self-intersecting polygon with spurs and inlets

The final version of this infill algorithm will process self-intersecting curved boundary loops in the way shown in Figure 11.41. It will also handle all the variations of polygon shading discussed in chapter 10. In Figure 11.42 a self-intersecting polygon boundary is shaded using the winding number approach, showing boundaries left uncovered and also covered. In Figure 11.43 the same polygon is shown but filled in using the on/off approach, again with the boundaries uncovered and covered.

Figure 11.44 Shaded, curvilinear, morphed, sequence

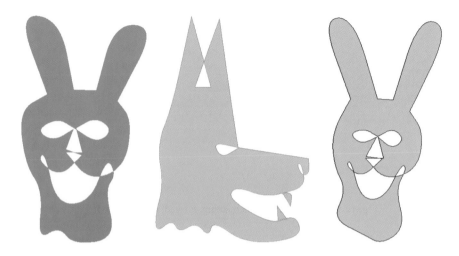

Figure 11.45 The rabbit and the fox

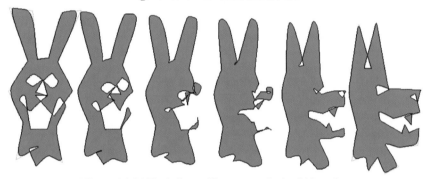

Figure 11.46 Shaded, curvilinear, morphed, rabbit to fox

Inbetweening control polygons, and then interpolating curved boundaries allows the sequences shown in Figures 11.44 and 11.46 to be created. Comparing the images in Figure 11.45 and 11.46 it can be seen that the mapping of polygon control points from the rabbit to the fox and from the fox to the rabbit generates many more line segments in the control polygons. This results in the curved sequences being shorter and more straight-line sections appearing compared with the independently interpolated shapes in Figure 11.45. However, it is possible to implement this morphing operation in a variety of other ways using the elements developed in this chapter. In many cases inbetweening images matching separately linked boundary pieces can give a better control of a resulting morphed animation sequence.

The elements explored in this chapter make it possible to construct an animated sequence from curved-boundary, shaded areas. However, the motion between keyframes is limited to that determined by the interpolation process. To get a more flexible range of movements more powerful mathematical structures need to be introduced to provide the way the necessary spatial or geometrical operations can be expressed in computer program statements.

Algebra: Formal Structures and Operations

The processes in this chapter depend on mathematical operations, which involve structural changes to the data, obeying rules that ensure the resulting data still represent correct information. A simple example, for an arithmetic expression, is 2×4: which by using a multiplication look-up table can be replaced by 8. In algebraic expressions various swapping rules can be used depending on the context.

Where the order of operands does not affect the outcome an operation is said to commute:

$$a . b = b . a$$

This is clearly true for addition and multiplication but is not true for subtraction and division.

$$4 + 5 = 5 + 4 = 9 \quad \text{but} \quad 4 - 5 \ \neq \ 5 - 4$$

In sequences of binary operators, where an operator is both left and right associative the order of execution does not change the result:

$$(a . b) . c = a . (b . c).$$

This is true for addition and multiplication

$$a + b + c : \quad (3+5)+2 = \ 8+2 = 11 = \ 3+(5+2) = 3+7$$

but is not true for subtraction nor is it true for division

$$a - b - c : \quad (5\text{-}3)\text{-}2 = 2 - 2 = 0, \quad \text{but} \quad 5 - (3\text{-}2) = 5 - 1 = \ 4$$

$$a / b / c : \quad (12 / 6) / 2 \ = 2 / 2 = \ 1 \quad \text{but} \quad 12 / (6 / 2) = 12 / 3 \ = 4.$$

Using Equations

The relationship between variables where they are interdependent can often be expressed by an equation. For example the relationship between the x value and the y value of the coordinates of points lying on a line can be expressed by the line's equation. However, in order to use equations for programming it is necessary that unknown variables are moved to one side of the equation and known variables and values are moved to the other. This is a standard manipulation used to solve simple equations.

A variable or value being used to multiply the right side of an equation can be moved to the left side but the left side has to be divided by the variable rather than being multiplied by it.

$$(x - 5) = (y - 8) . x \quad \rightarrow \quad \frac{(x - 5)}{x} = (y - 8)$$

When processing an equation, by definition, the same operation must to be carried out to each side of the equation for the relationship it represents to remain true.

Dividing both sides by the variable implements this rule. However a variable multiplied by its reciprocal, in other words divided by itself, gives unity, which multiplies to leave the right side of the equation with the variable removed.

$$(x-5)=(y-8).x \quad \rightarrow \quad (x-5).\frac{1}{x}=(y-8).x.\frac{1}{x} \quad \rightarrow \quad \frac{(x-5)}{x}=(y-8)$$

The reciprocal of a variable is its inverse relative to the multiply operator. The same ideas apply to addition. There are two ways of viewing the relationship between both multiplication and division, and addition and subtraction. In this case the addition of the inverse value involves adding the negated variable. This gives zero when combined with the original variable, and so adding zero for addition is equivalent to multiplying by one in the case of multiplication.

Consider the equation:

$$9x-3=6.(2x-4) \quad \rightarrow \quad 9x-3=12x-24 \quad \rightarrow \quad 21=3x \quad \rightarrow \quad 7=x$$

This apparently complicated way of presenting these operations allows the process to be generalised to manipulate equations that contain matrix elements.

Matrix Operations

A matrix is an array of values that can be operated on in a similar way to simple variables that only have a single value. Two matrixes that have the same structure and size can be added together:

$$\begin{bmatrix} a & b \\ c & d \end{bmatrix} + \begin{bmatrix} e & f \\ g & h \end{bmatrix} = \begin{bmatrix} a+e & b+f \\ c+g & d+h \end{bmatrix}$$

They can also be multiplied together:

$$\begin{bmatrix} a & b \\ c & d \end{bmatrix} \times \begin{bmatrix} e & f \\ g & h \end{bmatrix} = \begin{bmatrix} a.e+b.g & a.f+b.h \\ c.e+d.g & c.f+d.h \end{bmatrix}$$

Each row in the first matrix is multiplied by each column in the second matrix, where the row-column multiplication follows the following rule:

$$\begin{bmatrix} a & b \end{bmatrix} \times \begin{bmatrix} c \\ d \end{bmatrix} = a.c+b.d$$

Notice that the row column order is important:

$$\begin{bmatrix} a \\ b \end{bmatrix} \times \begin{bmatrix} c & d \end{bmatrix} = \begin{bmatrix} a.c & a.d \\ b.c & b.d \end{bmatrix}$$

Matrix multiplication does not commute. This is illustrated in the next chapter by the way the order in which rotation matrices are multiplied cannot be reversed, and still obtain the same outcome

$$[A].[B] \neq [B].[A]$$

This directional convention in carrying out the internal multiplication operations obeys the following pattern:

Matrix Equations

The equation of a line through the origin of the coordinate system can be expressed by the matrix equation:

$$\begin{bmatrix} x & y \end{bmatrix}.\begin{bmatrix} a \\ b \end{bmatrix} = a.x + b.y = 0$$

The equation of a line that does not pass through the origin can be handled in a similar way by introducing the homogenous coordinate. This will be discussed more fully in the next chapter, however, in this context it consists of converting the two dimensional coordinate column matrix into a three dimensional matrix by adding a third element of 1 in the way shown in the equation:

$$\begin{bmatrix} a & b & c \end{bmatrix} \times \begin{bmatrix} x \\ y \\ 1 \end{bmatrix} = a.x + b.y + c = 0$$

Formally this merely ensures the two matrixes match so they can be multiplied together.

Manipulating Matrix Equations

Given an equation of the form:

$$\begin{bmatrix} x & y \end{bmatrix}.\begin{bmatrix} a & b \\ c & d \end{bmatrix}_A = \begin{bmatrix} X & Y \end{bmatrix}$$

Where the values of a, b, c, d, X and Y are known how can x and y be evaluated? If [A] could be treated as a simple variable, then dividing both sides of the equation by [A] would transfer the unknowns to one side of the equation. Matrix division does not exist, but an equivalent operation can be expressed as the multiplication by an inverse matrix to obtain the same reordering of the equation.

$$[x \quad y].\begin{bmatrix} a & b \\ c & d \end{bmatrix}.\begin{bmatrix} a & b \\ c & d \end{bmatrix}^{-1} = [X \quad Y].\begin{bmatrix} a & b \\ c & d \end{bmatrix}^{-1} \rightarrow [x \quad y].\begin{bmatrix} 1 & 0 \\ 0 & 1 \end{bmatrix} = [X \quad Y].\begin{bmatrix} a & b \\ c & d \end{bmatrix}^{-1}$$

$$\rightarrow [x \quad y] = [X \quad Y].\begin{bmatrix} a & b \\ c & d \end{bmatrix}^{-1}$$

Rearranging matrix equations requires the idea of an inverse of a variable relative to an operator. Expressing operations in terms of a standard set of algebraic operations allows a series of simple procedures to be written to spatially rearrange objects represented by sets of points.

The algebraic properties of the matrix-structure, allows it to be used to manipulate other entities than point coordinates. This is a consequence of a duality of form in the representation of points and other spatial objects. For example, in two dimensions using homogeneous coordinates the point and the line, as data structures are 3-tuples. The three numbers that define the coefficients of a line's equation not only can be used to represent a line but support similar operations to those applied to point coordinates. This can be illustrated by the relationship between the linear combination of the 3-tuples representing these two spatial objects.

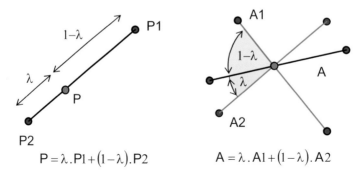

$$P = \lambda.P1 + (1-\lambda).P2 \qquad\qquad A = \lambda.A1 + (1-\lambda).A2$$

Figure 11.47 Dual interpolation of points and lines

In the case of two points this operation, linearly combining their coordinates, creates new points interpolated on the straight line between the original points, in the case of two lines, a linear combination of their coefficients creates a new line passing through the intersection point where the two original lines cross. A similar dual relationship occurs in three dimensions, only in this case, the duality of structure and operations link points and planes: both represented by 4-tuples. The application of this form of interpolation or inbetweening has already been introduced in this chapter. It is also the basis for constructing more complex shapes by sweeping lines and areas through space to create new areas and volumes respectively. The application of this approach to volume building will be explored in a later chapter.

The Determinant of a Matrix

Another important operation that can be carried out on a square matrix is to calculate its determinant. This pattern of calculations has a variety of applications again explored in the next chapter. A two dimensional matrix when treated as a determinant expresses the following expansion:

$$\begin{bmatrix} a & b \\ c & d \end{bmatrix} \xrightarrow{det} \begin{vmatrix} a & b \\ c & d \end{vmatrix} = a.d - b.c$$

This pattern is extended for a three dimensional matrix in the following recursive or nested way:

$$\begin{bmatrix} a & b & c \\ d & e & f \\ g & h & k \end{bmatrix} \xrightarrow{det} \begin{vmatrix} a & b & c \\ d & e & f \\ g & h & k \end{vmatrix} = a.\begin{vmatrix} e & f \\ h & k \end{vmatrix} - b.\begin{vmatrix} d & f \\ g & k \end{vmatrix} + c.\begin{vmatrix} d & e \\ g & h \end{vmatrix}$$

where

$$a.\begin{bmatrix} e & f \\ h & k \end{bmatrix} = a.e.k - a.f.h \quad b.\begin{bmatrix} d & f \\ g & k \end{bmatrix} = b.d.k - b.f.g \quad c.\begin{bmatrix} d & e \\ g & h \end{bmatrix} = c.d.h - c.e.g$$

An important property of this operation is that the determinant of the product of two matrices is the product of the determinants of the individual matrices.

$$\|[A].[B]\| = \|[A]\|.\|[B]\|$$

in two dimensions

$$\left\|\begin{bmatrix} a & b \\ c & d \end{bmatrix} . \begin{bmatrix} e & f \\ g & h \end{bmatrix}\right\| = \left\|\begin{bmatrix} a & b \\ c & d \end{bmatrix}\right\| . \left\|\begin{bmatrix} e & f \\ g & h \end{bmatrix}\right\|$$

$$\left\|\begin{bmatrix} a & b \\ c & d \end{bmatrix} . \begin{bmatrix} e & f \\ g & h \end{bmatrix}\right\| = \left\|\begin{bmatrix} a.e+b.g & a.f+b.h \\ c.e+d.g & c.f+d.h \end{bmatrix}\right\| = (a.e+b.g)(c.f+d.h) - (a.f+b.h)(c.e+d.g)$$

$$= a.e.c.f + a.e.d.h + b.g.c.f + b.g.d.h - [a.f.c.e + a.f.d.g + b.h.c.e + b.h.d.g]$$

$$= a.d.e.h + c.b.g.f - a.d.g.f - c.b.e.h$$

$$\left\|\begin{bmatrix} a & b \\ c & d \end{bmatrix}\right\| . \left\|\begin{bmatrix} e & f \\ g & h \end{bmatrix}\right\| = (ad - bc)(eh - fg) = a.d.e.h + c.b.g.f - a.d.g.f - c.b.e.h$$

The order in which rows and columns are entered into a matrix affect the value of its determinant. Swapping two rows or two columns negates the value of the determinant.

$$\begin{vmatrix} a & b & c \\ d & e & f \\ g & h & i \end{vmatrix} = a.\begin{vmatrix} e & f \\ h & i \end{vmatrix} - b.\begin{vmatrix} d & f \\ g & i \end{vmatrix} + c.\begin{vmatrix} d & e \\ g & h \end{vmatrix} = a.e.i - a.f.h - b.d.i + b.f.g + c.d.h - c.e.g = Dl$$

$$\begin{vmatrix} d & e & f \\ a & b & c \\ g & h & i \end{vmatrix} = d.\begin{vmatrix} b & c \\ h & i \end{vmatrix} - e.\begin{vmatrix} a & c \\ g & i \end{vmatrix} + f.\begin{vmatrix} a & b \\ g & h \end{vmatrix} = d.b.i - d.c.h - e.a.i + e.c.g + f.a.h - f.b.g = D2$$

$\therefore \quad Dl = -D2$

Where a point is interpolated between two other points on a straight line its elements are the linear combination of the two original point coordinates. If these two points and their interpolated value are placed in a matrix then the determinant of the matrix is zero. This property can be used to show that three points are not linearly independent but lie on a line.

$$\begin{vmatrix} p1 \\ p2 \\ p \end{vmatrix} \equiv \begin{vmatrix} x1 & y1 & 1 \\ x2 & y2 & 1 \\ \lambda.x1 + (1-\lambda)x2 & \lambda.y1 + (1-\lambda)y2 & 1 \end{vmatrix}$$

$$= x1.\begin{vmatrix} y2 & 1 \\ \lambda.y1 + (1-\lambda).y2 & 1 \end{vmatrix} - y1\begin{vmatrix} x2 & 1 \\ \lambda.x1 + (1-\lambda).x2 & 1 \end{vmatrix} + \begin{vmatrix} x2 & y2 \\ \lambda.x1 + (1-\lambda).x2 & \lambda.y1 + (1-\lambda).y2 \end{vmatrix}$$

$$= x1.y2 - x1.(\lambda.y1 + (1-\lambda).y2) - y1.x2 + y1.(\lambda.x1 + (1-\lambda).x2)$$
$$\qquad + x2.(\lambda.y1 + (1-\lambda).y2) - y2.(\lambda.x1 + (1-\lambda).x2)$$

$$= x1.y2 - x1.\lambda.y1 + x1.\lambda.y2 - x1.y2 - y1.x2 + y1.\lambda.x1 - y1.\lambda.x2 + y1.x2$$
$$\qquad + x2.\lambda.y1 - x2.\lambda.y2 + x2.y2 - y2.\lambda.x1 + y2.\lambda.x2 - y2.x2$$
$$= 0$$

In a similar way the determinant of the three sets of line equation coefficients from lines that pass through a common point is also equal to zero.

$$\begin{vmatrix} A1 \\ A2 \\ A \end{vmatrix} \equiv \begin{vmatrix} a & b & c \\ d & e & f \\ g = (\lambda.a + (1-\lambda)d) & h = (\lambda.b + (1-\lambda)e) & k = (\lambda.c + (1-\lambda)f) \end{vmatrix} = 0$$

In the next chapter further mathematical operations that can be used to move and manipulate the shape of object made up from collections of points and lines are examined further

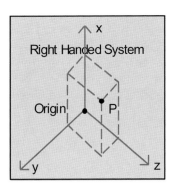

12

Geometry, Algebras, Co-ordinate Systems, and Transformations

Introduction

The task in this chapter is to explore the way the formal structure of a language, or mathematical algebra system can be used to represent geometry; in other words spatial properties, and support spatial operations on them. Attempts to do this can be traced at least as far back in history as the early Egyptian civilisation on the banks of the Nile, possibly even earlier. There appears to have been two motivating forces to this development in Egypt. Firstly, the desire to maintain coded records and laws, by an educated priestly civil service, and secondly the need to use these records to re-establish accurate property boundaries for land, that was periodically inundated by river flooding, often totally removed existing boundary markings. Among other things, this provided a continuing, sound basis for tax gathering. How could the correct spatial structure of land ownership be specified in non-ambiguous statements acceptable in a legal document that would allow it to be correctly reinstated whenever required? Even when limited to a primitive tool such as a measuring line, it was possible to define complex relationships between sets of points by defining each point using a collection of measurements from fixed landmarks in the environment that were not changed by the flooding.

Representing Point Locations

If a single point landmark is given then simple geometry indicates that one measurement can determine a circle of points. Two landmarks allow two measurements to define two circles. Where these intersect, if they do, it is possible to define two points. A third landmark, appropriately placed, gives three distances, and this permits a location to be unambiguously specified. This scheme is illustrated in Figure 12.1 and uniquely defines a point on a surface by three numbers. It is self

A. Thomas, *Integrated Graphic and Computer Modelling*,
DOI: 10.1007/978-1-84800-179-4_12, © Springer-Verlag London Limited 2008

evident that these measurements must be ordered and the order must be related to a defined sequence of landmarks in a reference list. This reference framework sets up a co-ordinate system. It can be considered as a form of abstract jig. Standard sequences of measurements relative to such a jig or framework can be use to represent a variety of geometric entities, such as circles, lines, and triangles and also to establish their spatial relationship to each other on a surface. The simplest of these geometric entities the point and sets of points will be discussed in the rest of this chapter. Different frameworks clearly require the list of measurements taken from them to be interpreted in different ways to give the same information. It is interesting to notice in the first example for the co-ordinate system given in Figure 12.1 that for a given point, the three values of its co-ordinate measurements are not independent.

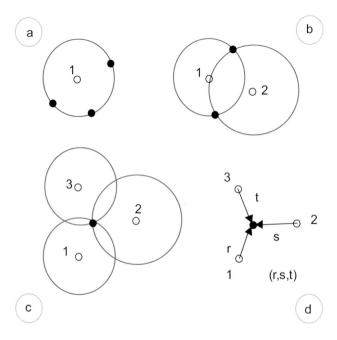

Figure 12.1 Locating a point relative to landmarks

Once the first two distances have been defined the third cannot be just any value. It is there to select between the two possible positions shown in Figure 12.1b. To uniquely define a point using this scheme all three values need to be present.

However, the three values in this scheme can be normalised by manipulating the identity:

$$r + s + t = r + s + t$$

$$\frac{r}{(r+s+t)} + \frac{s}{(r+s+t)} + \frac{t}{(r+s+t)} = 1.0$$

$$r' + s' + t' = 1.0$$

The values of r, s and t are clearly always positive, because they are lengths of measuring lines laid out in any direction, consequently the values of r', s', and t' in this normalised form each lie in the range 0.0 to 1.0 units. From any two of these normalised values the third can be calculated. Normalising these values permits a position to be recorded using only two numbers. However, comparing two location using this normalised representation can become difficult, generally requiring the original three measurements to be recalculated using the original landmark geometry in a process which makes the approach less useful.

Given a more sophisticated set of geometrical tools, an alternative measurement scheme base on a triangle of landmarks can be set up. If each side of the triangle is laid out as a straight line, then the measurements to an undefined point can be made perpendicular to these lines in the way shown in Figure 12.2. This approach also gives a *triple* of three measurements, but a triple that has slightly different properties. The first is that each measurement must have a direction relative to its associated triangle-edge. This can be expressed by making the measurement positive or negative: the direction *into* the triangle being conventionally taken as the positive direction.

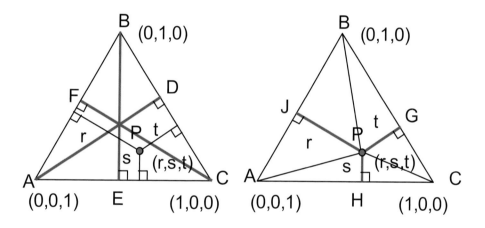

Figure 12.2 Triangular frame of reference

The three co-ordinate elements are still not independent of each other. The three values when appropriately scaled can also be made to add up to 1.0. If **r** is scaled by dividing JP by FC, and similarly **s** by dividing HP by EB and **t** by dividing GP by DA, then the new co-ordinate values (**r'**, **s'**, **t'**) sum to unity.

$$r'+s'+t'=1.0$$

This property can be demonstrated by observing that JP is the perpendicular height of the triangle APB, and FC is the perpendicular height of the triangle ABC. Both these triangles have the same base AB, so dividing the area of these triangles will give the same ratio defining the scaled value of **r**. However, if the areas of triangles APB, BPC and CPA are added together they give the area of ABC.

$$\Delta APB + \Delta BPC + \Delta CPA = \Delta ABC$$

$$\tfrac{1}{2}.AB.JP + \tfrac{1}{2}.BC.AD + \tfrac{1}{2}.CA.HP = \Delta ABC$$

$$\frac{\tfrac{1}{2}AB.JP}{\Delta ABC} + \frac{\tfrac{1}{2}BC.GP}{\Delta ABC} + \frac{\tfrac{1}{2}CAHP}{\Delta ABC} = \frac{\Delta ABC}{\Delta ABC} = 1.0$$

$$\frac{\tfrac{1}{2}AB.JP}{\tfrac{1}{2}AB.FC} + \frac{\tfrac{1}{2}BC.GP}{\tfrac{1}{2}BC.DA} + \frac{\tfrac{1}{2}CA.HP}{\tfrac{1}{2}CA.EB} = 1.0$$

$$\frac{JP}{FC} + \frac{GP}{DA} + \frac{HP}{EB} = 1.0$$

This establishes a system for defining the location of a point called baricentric co-ordinates. These co-ordinates have a variety of valuable properties.

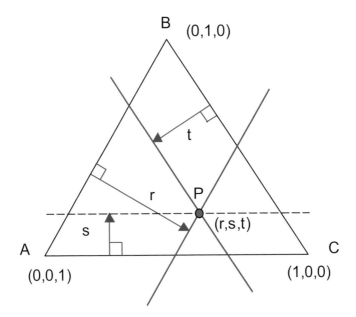

Figure 12.3 Independent co-ordinates

The simplest is again the fact that the third number of the three used to define a point in a plane, in this scheme, is unnecessary. Each co-ordinate measure defines a line in space parallel to the corresponding baseline of the reference triangle and where these lines intersect defines the required point. Only two lines, which are not co-linear or parallel, are needed for this task.

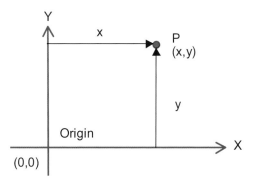

Figure 12.4 Two-dimensional Cartesian system

The minimum number of measurement values necessary to determine the position of a point in space is considered to be a property of the space itself. On a surface it is two, in 'real' space it is three. This property is called the dimension of the space.

If the reference triangle used to define baricentric co-ordinates is set up as a right angle triangle then a Cartesian co-ordinate system is generated. The two dimensional form of this co-ordinate system is shown in Figure12.4. Again in this case only two ordered numbers are needed to define a point position in a plane. The advantages of this system are that each co-ordinate pair represents a unique point, and the two numbers can be modified independently of each other in useful ways that are easy to visualise and are therefore effective for defining points in a formal language environment.

The ability to measure distances, and even to set up parallel lines a specific distance apart was established early in geometric work. Flexible ways of measuring and using accurately defined angles proved more difficult, however, once these became possible an alternative co-ordinate system emerged, the polar co-ordinate system.

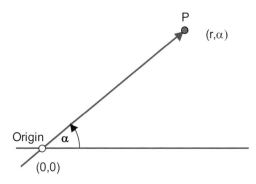

Figure 12.5 Two-dimensional polar co-ordinate system

In a two dimensional space, given a base line and a pivot point on this line, any other point can be defined in the following way. Join the pivot point to the new point and measure the length of the line, then measure the angle by which this line is rotated from the original baseline. The co-ordinate for the new point is the number pair giving the distance and the angle, as illustrated in Figure 12.5.

Three Dimensional Spaces

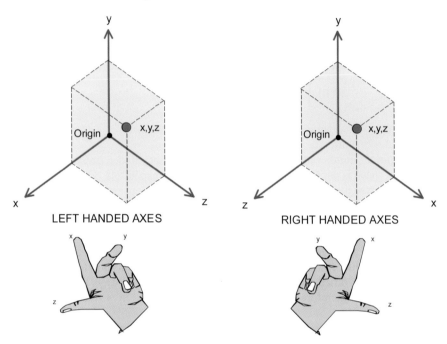

Figure 12.6 Three-dimensional Cartesian co-ordinate systems

Each of these schemes can be extended to define points in three-dimensional space. The main difference being that, to define a point position, a minimum of three numbers is needed.   Figure 12.6 introduces a new problem encountered in the higher dimensional space. There are two spatial relationships possible between three axes of measurement. The first giving a right-handed system and the second giving a left-handed system.   This naming arises from the following convention. If co-ordinate elements are taken in the order in which they are written down, the first one corresponding to the index finger of the hand then the other two will take on the rotational order given by the second finger and thumb, of either the left or the right hand. The two alternative arrangements are shown in Figure 12.6. Polar co-ordinates in three dimensions can also be implemented in several ways. The simplest form is the spherical polar system shown in Figure 12.7a. However, there is also an in-between, hybrid form shown in Figure 12.7b, called a cylindrical, polar co-ordinate system.

Coordinates and Vectors

Coordinates can be used to represent points in space. They can also be used to represent a different but related mathematical object the vector. A vector has direction and magnitude. A point can be located by moving a defined distance in a given direction from a starting point as specified in a polar coordinate or two distances in directions at right angles in a Cartesian coordinate. The movement is distinct from the result. The data structure is the coordinate, but two types immerge, a

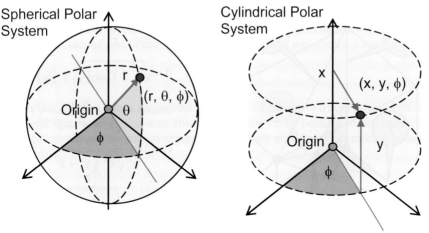

Figure 12.7 Three-dimensional Polar co-ordinate systems

point and a vector, the difference depending on the operations that are permitted on the structure. A similar distinction was made between absolute coordinates and relative coordinates in Chapter 4. The absolute coordinate gave a point-position the relative coordinates defined movements from this base to their final point locations defining a symbol or an image. The objective was a means for replicating an image in a display space. Adding two points together makes no sense however adding two movements together is totally reasonable and was implemented by adding the absolute and relative coordinate sequence to relocate an image in multiple positions.

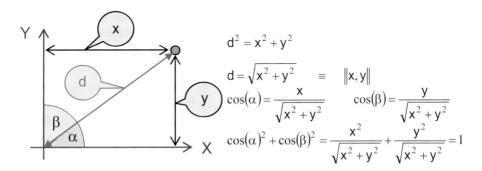

Figure 12.8 Distance of a point from the origin

Converting a vector in Cartesian coordinate form to Polar coordinate form involves calculating the distance of a point from the origin of the axes system being employed using Pythagoras' theorem in the way shown in Figure 12.8 for the two dimensional case.

If a vector is factored into components parallel to the coordinate axes, these vectors can in turn be converted into unit vectors multiplied by a scalar representing the magnitude of the vector. This creates a new notation for representing vectors. If the unit vectors are represented by the letters **i**, **j** and **k** in the x, y and z directions, then the vector (x, y, z) can be represented by (x.**i** + y.**j** + z.**k**). This provides a way of linking vector multiplication operations to standard arithmetic, algebraic operations.

Vector Multiplication: Dot and Cross Products

There are two standard forms of multiplication operation that can be applied to vectors, the first called the dot product, the second called the cross product.

The dot product is defined as: $\mathbf{a.b} = (x1, y1).(x2, y2) = (x1.x2 + y1.y2)$

If two vectors **a** and **b** represented by $(x1.\mathbf{i} + y1.\mathbf{j},)$ and $(x2.\mathbf{i} + y2.\mathbf{j},)$ are multiplied together treating **i** and **j** as simple variables the result will be:

$$(x1.\mathbf{i} + y1.\mathbf{j})(x2.\mathbf{i} + y2.\mathbf{j}) = x1.x2.\mathbf{i.i} + y1.y2.\mathbf{j.j} + x1.y2.\mathbf{j.i} + x2.y1.\mathbf{i.j}$$

Replacing the products of **i** and **j**: **i.i** = 1, **j.j** = 1, and **i.j** = **j.i** = 0 gives:

$$x1.x2.\mathbf{i}^2 + y1.y2.\mathbf{j}^2 + x1.y2.\mathbf{i.j} + x2.y1.\mathbf{i.j} = x1.x2. + y1.y2$$

The dot product of unit vectors provides a way of calculating the angle between the two vectors in the way shown in Figure 12.13. This demonstrates i.i = $\cos(0)$ = 1, and i.j = $\cos(90)$ =0.

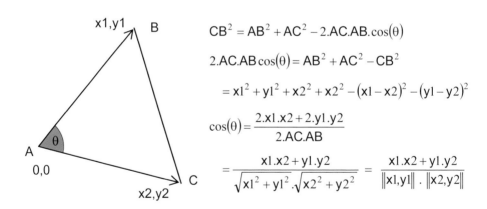

$$CB^2 = AB^2 + AC^2 - 2.AC.AB.\cos(\theta)$$

$$2.AC.AB\cos(\theta) = AB^2 + AC^2 - CB^2$$

$$= x1^2 + y1^2 + x2^2 + x2^2 - (x1 - x2)^2 - (y1 - y2)^2$$

$$\cos(\theta) = \frac{2.x1.x2 + 2.y1.y2}{2.AC.AB}$$

$$= \frac{x1.x2 + y1.y2}{\sqrt{x1^2 + y1^2}.\sqrt{x2^2 + y2^2}} = \frac{x1.x2 + y1.y2}{\|x1,y1\| . \|x2,y2\|}$$

Figure 12.13 The dot product gives the cosine of the angle between two unit vectors

The dot product can be expressed in matrix format as the product of a row and column matrix:

$$\begin{bmatrix} x1 & y1 \end{bmatrix}\begin{bmatrix} x2 \\ y2 \end{bmatrix} = \begin{bmatrix} x1.x2 + y1.y2 \end{bmatrix}$$

Similarly for the cross product: if two vectors **a** and **b** represented by $(x1.\mathbf{i} + y1.\mathbf{j},)$ and $(x2.\mathbf{i} + y2.\mathbf{j},)$ are multiplied together treating **i** and **j** like simple numerical variables where $x1.\mathbf{i} \times x2.\mathbf{j} = x1.x2.\mathbf{i} \times \mathbf{j}$, the result will be:

$$\mathbf{a} \times \mathbf{b} = (x1.\mathbf{i} + y1.\mathbf{j}) \times (x2.\mathbf{i} + y2.\mathbf{j}) = x1.x2.\mathbf{i} \times \mathbf{i} + y1.y2.\mathbf{j} \times \mathbf{j} + x1.y2.\mathbf{j} \times \mathbf{i} + x2.y1.\mathbf{i} \times \mathbf{j}$$

Replacing the products of **i** and **j**: $\mathbf{i} \times \mathbf{i} = 0$, $\mathbf{j} \times \mathbf{j} = 0$, $\mathbf{i} \times \mathbf{j} = -\mathbf{k}$, $\mathbf{j} \times \mathbf{i} = \mathbf{k}$

$$x1.x2.\mathbf{i} \times \mathbf{i} + y1.y2.\mathbf{j} \times \mathbf{j} + x1.y2.\mathbf{j} \times \mathbf{i} + x2.y1.\mathbf{i} \times \mathbf{j} = (x1.y2 - x2.y1)\mathbf{k}$$

where **k** is a unit vector orthogonal to the plane containing **i** and **j**.

The cross product gives a way of calculating the area of the parallelogram formed by adding two vectors in the way shown in Figure 12.14.

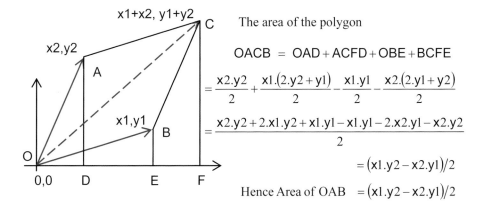

$$\text{OACB} = \text{OAD} + \text{ACFD} + \text{OBE} + \text{BCFE}$$

$$= \frac{x2.y2}{2} + \frac{x1.(2.y2 + y1)}{2} - \frac{x1.y1}{2} - \frac{x2.(2.y1 + y2)}{2}$$

$$= \frac{x2.y2 + 2.x1.y2 + x1.y1 - x1.y1 - 2.x2.y1 - x2.y2}{2}$$

$$= (x1.y2 - x2.y1)/2$$

Hence Area of OAB $= (x1.y2 - x2.y1)/2$

Figure 12.14 Area of the triangle OAB defined by two vectors OA and OB

This demonstrates that the cross products of unit vectors has the magnitude of 0 when the vector is multiplied with itself, and 1 or -1 when multiplied by a unit vector at right angles to itself. The magnitude of the cross product can be calculated in matrix format, by the determinant of the matrix containing the two vectors.

$$\begin{vmatrix} x1 & y1 \\ x2 & y2 \end{vmatrix} = [x1.y2 - x2.y1]$$

Changing the order of the rows in the determinant changes the sign of the result, this matches the relationships $\mathbf{i} \times \mathbf{j} = -\mathbf{k}$, $\mathbf{j} \times \mathbf{i} = \mathbf{k}$, and a corresponding change in the related geometric orientation and rotation.

Rotating Points and Vectors

The best choice of co-ordinate system will be determined by the application, and the spatial properties it needs to represent. Consider the problem of rotating a vector or a point or a set of points about the origin of the co-ordinate system. In the case of polar co-ordinates the problem is simple. The representation of the rotation can be achieved simply be adding on the angle of rotation to each coordinate's directional angle. However, if the same task is to be undertaken using Cartesian coordinates, then the task becomes more difficult.

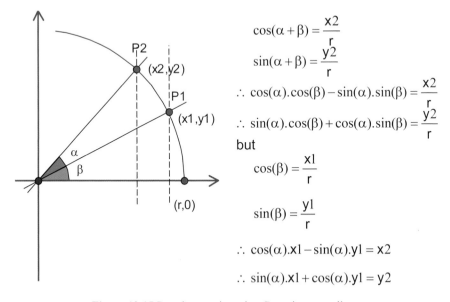

$$\cos(\alpha + \beta) = \frac{x2}{r}$$

$$\sin(\alpha + \beta) = \frac{y2}{r}$$

$$\therefore \cos(\alpha).\cos(\beta) - \sin(\alpha).\sin(\beta) = \frac{x2}{r}$$

$$\therefore \sin(\alpha).\cos(\beta) + \cos(\alpha).\sin(\beta) = \frac{y2}{r}$$

but

$$\cos(\beta) = \frac{x1}{r}$$

$$\sin(\beta) = \frac{y1}{r}$$

$$\therefore \cos(\alpha).x1 - \sin(\alpha).y1 = x2$$

$$\therefore \sin(\alpha).x1 + \cos(\alpha).y1 = y2$$

Figure 12.15 Rotating a point using Cartesian co-ordinates

Given a point defined in a rectangular Cartesian Co-ordinate system as (x, y) how can this point be moved to another position? In general by changing the values of the coordinate's (x, y) elements. The question is how. The answer must depend on the nature of the movement. Since the coordinate is a position specified relative to an origin and a set of axes, the simplest operations will be defined relative to this framework. One example is moving a point round the origin, in other words rotate the line linking it to the origin by an angle α. In other words rotate the vector.

In Figure 12.15 if the initial position of the point is given as P1, and its rotated position by P2. Then from this diagram the trigonometric relationships shown on the right can be set out between the co-ordinates (x1, y1) and (x2, y2). The final pair of these relationships can be rewritten in matrix form to give:

$$\begin{bmatrix} x1 & y1 \end{bmatrix} \begin{bmatrix} \cos(\alpha) & \sin(\alpha) \\ -\sin(\alpha) & \cos(\alpha) \end{bmatrix} = \begin{bmatrix} x2 & y2 \end{bmatrix} \qquad \text{or} \qquad \begin{bmatrix} \cos(\alpha) & -\sin(\alpha) \\ \sin(\alpha) & \cos(\alpha) \end{bmatrix} \begin{bmatrix} x1 \\ y1 \end{bmatrix} = \begin{bmatrix} x2 \\ y2 \end{bmatrix}$$

This provides a framework for specifying a series of similar operations defined relative to the origin in the following way. If α is set to be 90° then the point can be rotated relative to the origin by 90°, by substituting into the matrix the sines and cosines of 90°, 180°, 270°, 360°: either 0.0, 1.0 or −1.0, in the following way.

A	$[x1 \quad y1]\begin{bmatrix} 0 & 1 \\ -1 & 0 \end{bmatrix} = [x2 \quad y2] = [-y1 \quad x1]$		Rotate 90°
B	$[x1 \quad y1]\begin{bmatrix} -1 & 0 \\ 0 & -1 \end{bmatrix} = [x2 \quad y2] = [-x1 \quad -y1]$		Rotate 180°
C	$[x1 \quad y1]\begin{bmatrix} 0 & -1 \\ 1 & 0 \end{bmatrix} = [x2 \quad y2] = [y1 \quad -x1]$		Rotate 270°
D	$[x1 \quad y1]\begin{bmatrix} 1 & 0 \\ 0 & 1 \end{bmatrix} = [x2 \quad y2] = [x1 \quad y1]$		Rotate 360°

These rotation-based operations must be distinguished from reflection operations. Reflections require matrix entries that cannot be obtained by substituting the sine or cosine of a single angle into the matrix. Reflections about the x-axis, the y-axis and diagonally can be obtained by using matrix operations of the form:

1	$[x1 \quad y1]\begin{bmatrix} -1 & 0 \\ 0 & 1 \end{bmatrix} = [x2 \quad y2] = [-x1 \quad y1]$		Reflect x → -x
2	$[x1 \quad y1]\begin{bmatrix} 1 & 0 \\ 0 & -1 \end{bmatrix} = [x2 \quad y2] = [x1 \quad -y1]$		Reflect y → -y
3	$[x1 \quad y1]\begin{bmatrix} 0 & 1 \\ 1 & 0 \end{bmatrix} = [x2 \quad y2] = [y1 \quad x1]$		Reflect x → y Reflect y → x
4	$[x1 \quad y1]\begin{bmatrix} 0 & -1 \\ -1 & 0 \end{bmatrix} = [x2 \quad y2] = [-y1 \quad -x1]$		Reflect x → -y Reflect y → -x

The effect of these matrix, reflection operations is to change the signs of x and y values or to swap them over. Reflection transformations were employed in the line interpolation algorithms in Chapter 9 though not implemented as matrix operations. These examples show that rotations can be obtained from pairs of reflection operations, but reflections cannot be obtained using matrix rotations.

Rotations: A, B, C, D	First reflection	A: second reflection	B: second reflection	C: second reflection	D: second reflection
Sequence of Equivalent Reflections	1	4	2	3	1
	2	3	1	4	2
	3	1	4	2	3
	4	2	3	1	4

Multiple Rotations

$$[x1 \ \ y1].\begin{bmatrix} \cos(\alpha) & \sin(\alpha) \\ -\sin(\alpha) & \cos(\alpha) \end{bmatrix} = [x2 \ \ y2] \qquad [x1 \ \ y1].\begin{bmatrix} 1 & 0 \\ 0 & 1 \end{bmatrix} = [x1 \ \ y1]$$

A rotation of 0 radians is the same as a multiplication by the identity matrix.

Two successive rotations about the same axis are given by multiplying the two matrices together

$$[p1] = [p0]. \ [R(\alpha)]$$

$$[p2] = [p1].[R(\beta)] = \ [p0].[R(\alpha)].[R(\beta)]$$

alternatively in expanded form:

$$[x1 \ \ y1]\begin{bmatrix} \cos(\alpha) & \sin(\alpha) \\ -\sin(\alpha) & \cos(\alpha) \end{bmatrix}\begin{bmatrix} \cos(\beta) & \sin(\beta) \\ -\sin(\beta) & \cos(\beta) \end{bmatrix} = [x2 \ \ y2]$$

where:

$$\begin{bmatrix} \cos(\alpha) & \sin(\alpha) \\ -\sin(\alpha) & \cos(\alpha) \end{bmatrix}.\begin{bmatrix} \cos(\beta) & \sin(\beta) \\ -\sin(\beta) & \cos(\beta) \end{bmatrix}$$

$$= \begin{bmatrix} \cos(\alpha).\cos(\beta) - \sin(\alpha).\sin(\beta) & \sin(\alpha).\cos(\beta) + \cos(\alpha).\sin(\beta) \\ -(\sin(\alpha).\cos(\beta) + \cos(\alpha).\sin(\beta)) & \cos(\alpha).\cos(\beta) - \sin(\alpha).\sin(\beta) \end{bmatrix}$$

$$= \begin{bmatrix} \cos(\alpha+\beta) & \sin(\alpha+\beta) \\ -\sin(\alpha+\beta) & \cos(\alpha+\beta) \end{bmatrix}$$

giving: $\qquad\qquad\qquad [R(\alpha)].[R(\beta)] = \ [R(\alpha+\beta)]$

or, if the coordinate is given in column vector form:

$$\begin{bmatrix} \cos(\beta) & -\sin(\beta) \\ \sin(\beta) & \cos(\beta) \end{bmatrix}.\begin{bmatrix} \cos(\alpha) & -\sin(\alpha) \\ \sin(\alpha) & \cos(\alpha) \end{bmatrix}.\begin{bmatrix} x1 \\ y1 \end{bmatrix} = \begin{bmatrix} x2 \\ y2 \end{bmatrix}$$

$$\begin{bmatrix} \cos(\alpha+\beta) & \sin(\alpha+\beta) \\ -\sin(\alpha+\beta) & \cos(\alpha+\beta) \end{bmatrix}.\begin{bmatrix} x1 \\ y1 \end{bmatrix} = \begin{bmatrix} x2 \\ y2 \end{bmatrix}$$

Thus $$[P].[R(\alpha)].[R(\beta)]=\left[[R(\beta)]^T.[R(\alpha)]^T.[P]^T\right]^T$$

$$[P].[R(\alpha)].[R(\beta)]=[P].[R(\alpha+\beta)]=\left[[R(\alpha+\beta)]^T.[P]^T\right]^T$$

Rigid Body Movement: Rotation and Internal Angles

If a rigid object is represented by a set of points and the object is moved relative to the coordinate framework then all internal distances between pairs of points have to stay the same, as do the angles and areas defined by all point triples.

For simple rotation by an angle α about the origin the angles between pairs of points can be shown to be the same by the following algebraic sequence:

The points v1 and v2 treated as vectors will be rotated to $R(\alpha)$v1 and $R(\alpha)$v2

Calculating the angle between these two vectors using their dot product in a matrix format gives:

$$R(\alpha)v1.\ R(\alpha)v2 \rightarrow [R(\alpha)v1]^T.[R(\alpha)v2]$$

$$\rightarrow [v1]^T.[R(\alpha)]^T.[R(\alpha)].[v2]$$

$$\rightarrow [v1]^T.[v2]$$

$$\rightarrow v1.v2\ =\ \cos(\alpha).\|v1\|.\|v2\|$$

This depends on the inverse rotation in other words the reverse rotation being produced by inserting a negative angle into the forward rotation matrix, which in turn is equivalently to transposing the forward rotation matrix since:

$$[R(-\alpha)]=[R(\alpha)]^T\ =[R(\alpha)]^{-1}$$

$$\begin{bmatrix}\cos(-\alpha) & -\sin(-\alpha)\\ \sin(-\alpha) & \cos(-\alpha)\end{bmatrix}=\begin{bmatrix}\cos(\alpha) & \sin(\alpha)\\ -\sin(\alpha) & \cos(\alpha)\end{bmatrix}$$

Multiplying the rotation matrix by its inverse or transpose gives the unit matrix, which is equivalent to no rotation.

$$[R(\alpha)][R(\alpha)]^T\ =[R(\alpha-\alpha)]=[R(0)]=[1]$$

$$\begin{bmatrix}\cos(\alpha) & -\sin(\alpha)\\ \sin(\alpha) & \cos(\alpha)\end{bmatrix}.\begin{bmatrix}\cos(\alpha) & \sin(\alpha)\\ -\sin(\alpha) & \cos(\alpha)\end{bmatrix}\ =$$

$$\begin{bmatrix}\cos^2(\alpha)+\sin^2(\alpha) & \cos(\alpha).\sin(\alpha)-\cos(\alpha).\sin(\alpha)\\ \cos(\alpha).\sin(\alpha)-\cos(\alpha).\sin(\alpha) & \cos^2(\alpha)+\sin^2(\alpha)\end{bmatrix}=\begin{bmatrix}1 & 0\\ 0 & 1\end{bmatrix}$$

Rigid Body Movement: Rotation and Internal Distances

Distances between rotated points must also remain the same for a rigid body. This can be demonstrated in the following way

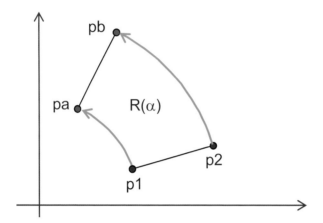

Figure 12.16 Distances between rotated sets of points

The distances d(.) between the pairs of points (p1, p2) and (pa, pb) are calculated using Pythagoras in the way described above.

$$d(p1,p2)^2 \rightarrow (x2-x1)^2 + (y2-y1)^2 = \left(x1^2 + x2^2 + y1^2 + y2^2\right) - 2(x1.x2 + y1.y2)$$

$$d(pa,pb)^2 \rightarrow (xb-xa)^2 + (yb-ya)^2 = \left(xa^2 + xb^2 + ya^2 + yb^2\right) - 2(xa.xb + ya.yb)$$

Each point is rotated to keep the same distance from the origin : r

$$d(p1,p2)^2 \rightarrow r1^2 + r2^2 - 2.(p1.p2) = k1 - 2.(p1.p2)$$

$$d(pa,pb)^2 \rightarrow ra^2 + rb^2 - 2.(pa.pb) = k2 - 2.(pa.pb)$$

but $r1^2 = ra^2$ and $r2^2 = rb^2$ so $k1 = k2$

Rotating p1 gives pa, and rotating p2 by the same angle gives pb

$$R(\alpha)p1 = pa \quad \text{and} \quad R(\alpha)p2 = pb$$

The angle between p1 and p2 will be the same as the angle between pa and pb.

$$R(\alpha)p1.R(\alpha)p2 = p1.p2 = pa.pb$$

Consequently : $d(p1,p2) = d(pa,pb)$

  Rotating the pair of points does not change the distance between the points. If angles and distance are maintained then areas should also be maintained.

Rigid Body Movement: Rotation and Internal Areas

If distances and angles are conserved by the rotation operation then it is natural to expect triangulated areas formed by any three points, to stay the same. The way the algebra supports this intuition can be shown in the following way.

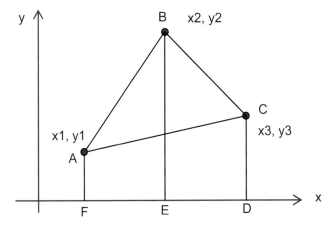

Figure 12.17 Calculating the area of a triangle using trapeziums

Area $ABC = ABEF + BCDE - ACDF$

$$\rightarrow (x2+x1).\frac{(y2+y1)}{2}+(x3+x2).\frac{(y3+y2)}{2}+(x1+x3).\frac{(y1+y3)}{2}$$

$$\rightarrow \frac{1}{2}.\left[(x2.y1-x1.y2)+(y2.x3-x2.y3)+(x1.y3-x3.y1)\right]$$

$$\rightarrow \frac{1}{2}.\left[p2\times p1+p3\times p2+p1\times p3\right] \qquad \text{vector cross products}$$

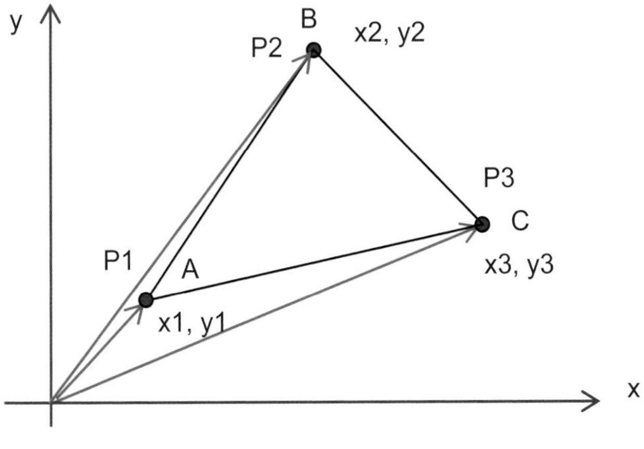

Figure 12.18 Calculating the area of a triangle using vectors

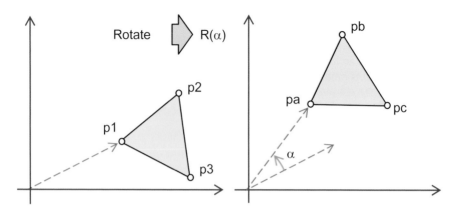

Figure 12.19 Rotating a triangle

$$\text{area } A \to \frac{1}{2}\left[x2.y1 - x1.y2 + y2.x3 - x2.y3 + x1.y3 - x3.y1\right]$$

$$\to \frac{1}{2}\cdot\left[\begin{vmatrix} x2 & y2 \\ x1 & y1 \end{vmatrix} + \begin{vmatrix} x3 & y3 \\ x2 & y2 \end{vmatrix} + \begin{vmatrix} x1 & y1 \\ x3 & y3 \end{vmatrix}\right] \qquad \text{determinants}$$

$$\to \frac{1}{2}\cdot\left[\begin{vmatrix} p2 \\ p1 \end{vmatrix} + \begin{vmatrix} p3 \\ p2 \end{vmatrix} + \begin{vmatrix} p1 \\ p3 \end{vmatrix}\right] \qquad \text{short hand}$$

$$\text{area } B \to \frac{1}{2}\cdot\left[xb.ya - xa.yb + yb.xc - xb.yc + xa.yc - xc.ya\right]$$

$$\to \frac{1}{2}\cdot\left[\begin{vmatrix} xb & yb \\ xa & ya \end{vmatrix} + \begin{vmatrix} xc & yc \\ xb & yb \end{vmatrix} + \begin{vmatrix} xa & ya \\ xc & yc \end{vmatrix}\right]$$

$$\to \frac{1}{2}\cdot\left[\begin{vmatrix} pb \\ pa \end{vmatrix} + \begin{vmatrix} pc \\ pb \end{vmatrix} + \begin{vmatrix} pa \\ pc \end{vmatrix}\right]$$

However:

$$\begin{vmatrix} pb \\ pa \end{vmatrix} = \left|\begin{bmatrix} p2 \\ p1 \end{bmatrix}\cdot[R(\alpha)]\right| = \left|\begin{bmatrix} p2 \\ p1 \end{bmatrix}\right|\cdot\left|[R(\alpha)]\right| \qquad \text{product of determinants}$$

$$\left|[R(\alpha)]\right| = \begin{vmatrix} \cos(\alpha) & -\sin(\alpha) \\ \sin(\alpha) & \cos(\alpha) \end{vmatrix} = \cos^2(\alpha) + \sin^2(\alpha) = 1$$

Hence $$\frac{1}{2}\left[\begin{vmatrix} pb \\ pa \end{vmatrix} \begin{vmatrix} pc \\ pb \end{vmatrix} \begin{vmatrix} pa \\ pc \end{vmatrix}\right] = \frac{1}{2}\left[\begin{vmatrix} p2 \\ p1 \end{vmatrix} \begin{vmatrix} p3 \\ p2 \end{vmatrix} \begin{vmatrix} p1 \\ p3 \end{vmatrix}\right]$$

Consequently the area of the triangle is unaffected by rotation.

Rigid Body Movement: Translation and Internal Areas

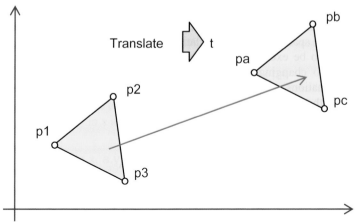

Figure 12.20 Translating a triangle

$$\text{area A} \to \frac{1}{2}.\left[\begin{vmatrix} p2 \\ p1 \end{vmatrix} + \begin{vmatrix} p3 \\ p2 \end{vmatrix} + \begin{vmatrix} p1 \\ p3 \end{vmatrix}\right] \to \frac{1}{2}.\left[\begin{vmatrix} x2 & y2 \\ x1 & y1 \end{vmatrix} + \begin{vmatrix} x3 & y3 \\ x2 & y2 \end{vmatrix} + \begin{vmatrix} x1 & y1 \\ x3 & y3 \end{vmatrix}\right]$$

$$\to \frac{1}{2}.[x2.y1 - x1.y2 + x3.y2 - x2.y3 + x1.y3 - x3.y1]$$

$$\text{area B} \to \frac{1}{2}.\left[\begin{vmatrix} pb \\ pa \end{vmatrix} + \begin{vmatrix} pc \\ pb \end{vmatrix} + \begin{vmatrix} pa \\ pc \end{vmatrix}\right] \to \frac{1}{2}.\left[\begin{vmatrix} xb & yb \\ xa & ya \end{vmatrix} + \begin{vmatrix} xc & yc \\ xb & yb \end{vmatrix} + \begin{vmatrix} xa & ya \\ xc & yc \end{vmatrix}\right]$$

$$\to \frac{1}{2}.\left[\begin{vmatrix} x2+xt & y2+yt \\ x1+xt & y1+yt \end{vmatrix} + \begin{vmatrix} x3+xt & y3+yt \\ x2+xt & y2+yt \end{vmatrix} + \begin{vmatrix} x1+xt & y1+yt \\ x3+xt & y3+yt \end{vmatrix}\right]$$

$$\to \frac{1}{2}.\left[\begin{array}{l} (x2+xt).(y1+yt) - (x1+xt).(y2+yt) \\ +(x3+xt).(y2+yt) - (x2+xt).(y3+yt) \\ +(x1+xt).(y3+yt) - (x3+xt).(y1+yt) \end{array}\right]$$

$$\to \frac{1}{2}.\left[\begin{array}{l} x2.y1 - x1.y2 + xt.y1 + yt.x2 + xt.yt - xt.yt - xt.y2 - yt.x1 \\ +x3.y2 - x2.y3 + xt.y2 + yt.x3 + xt.yt - xt.yt - xt.y3 - yt.x2 \\ +x1.y3 - x3.y1 + xt.y3 + yt.x1 + xt.yt - xt.yt - xt.y1 - yt.x3 \end{array}\right]$$

$$\to \frac{1}{2}.[x2.y1 - x1.y2 + x3.y2 - x2.y3 + x1.y3 - x3.y1]$$

Consequently the area of the triangle is also unaffected by translation.

since $i \times j = k, \quad j \times i = -k, \quad j \times k = i, \quad k \times j = -i \quad k \times i = j, \quad i \times k = -j$

$$a \times b = \begin{vmatrix} i & j & k \\ a_1 & a_2 & a_3 \\ b_1 & b_2 & b_3 \end{vmatrix}$$

$$\|a \times b\| = \|a\| . \|b\| . \sin(\theta) \qquad \text{therefore} \qquad \sin(\theta) = \frac{\|a \times b\|}{\|a\| . \|b\|}$$

The volume of a tetrahedron defined by three vectors, in other words with one vertex at the origin, can be calculated by the determinant of the three vectors representing the remaining points in a similar way that the area of a triangle is defined by the determinant of two, two-dimensional vectors.

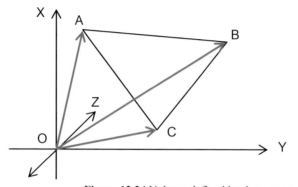

Figure 12.24 Volume defined by three vectors

If the vectors are added together then a parallelepiped is generated.

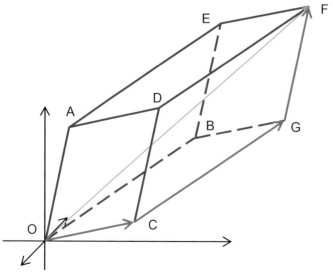

Figure 12.25 Adding three vectors

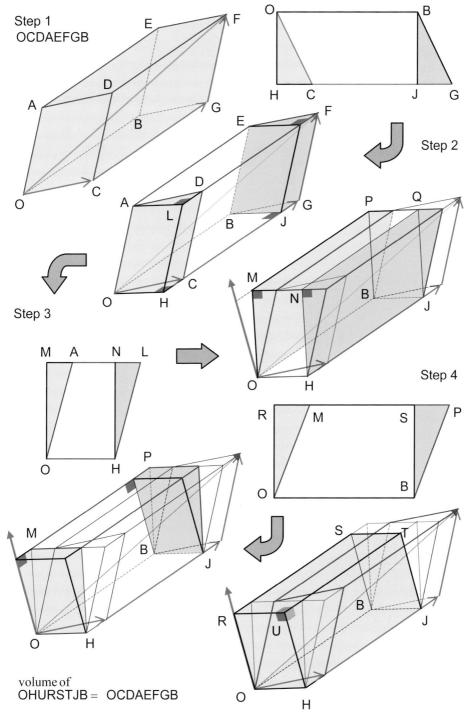

Step 1
OCDAEFGB

Step 2

Step 3

Step 4

volume of
OHURSTJB = OCDAEFGB

volume of OHURSTJB = OR×OH×HJ

Figure 12.26 Parallelepiped volume

Transforming and subdividing the volume in the way shown in Figure 12.26 can be used to calculate the volume of the parallelepiped generated when three vectors are added together. Figure 12.26 shows that the volume of a parallelepiped can be calculated by the area of one face, times the perpendicular distance to its opposite matching face. This is demonstrated be cutting wedges from one side of the figure and adding them to the other side to change the parallelepiped into a cuboid. The matching calculation can be carried out as follows:

The area of the base facet, which is a parallelogram, is given by $OB \times OH$

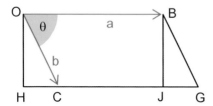

Figure 12.27 Area $a \times b$

In vector notation $OB \times OH \equiv \|a\|.\|b\|.\sin(\theta)$

$$\|a\|.\|b\|.\sin(\theta) = \|a\|.\|b\|.\sqrt{1 - \cos^2(\theta)} = \|a\|.\|b\|.\sqrt{1 - \left(\frac{a.b}{\|a\|.\|b\|}\right)^2}$$

if $a = (x_1, y_1, z_1)$, and $b = (x_2, y_2, z_2)$

$$\|a\|.\|b\|.\sin(\theta) = \sqrt{(\|a\|.\|b\|)^2 - (a.b)^2}$$

$$= \sqrt{(x_1^2 + y_1^2 + z_1^2).(x_2^2 + y_2^2 + z_2^2) - (x_1.x_2 + y_1.y_2 + z_1.z_2)^2} \qquad [1]$$

Alternatively the area can be given by $\|a \times b\|$

$$\|a \times b\| = \begin{Vmatrix} i & j & k \\ x_1 & y_1 & z_1 \\ x_2 & y_2 & z_2 \end{Vmatrix} = \left\| i.\begin{vmatrix} y_1 & z_1 \\ y_2 & z_2 \end{vmatrix} - j.\begin{vmatrix} x_1 & z_1 \\ x_2 & z_2 \end{vmatrix} + k.\begin{vmatrix} x_1 & y_1 \\ x_2 & y_2 \end{vmatrix} \right\|$$

$$= \sqrt{(y_1.z_2 - z_1.y_2)^2 + (z_1.x_2 - x_1.z_2)^2 + (x_1.y_2 - y_1.x_2)^2} \qquad [2]$$

Expanding either $[1]$ or $[2]$ gives

$$\sqrt{\begin{aligned} & x_1^2.y_2^2 + y_1^2.z_2^2 + z_1^2.x_2^2 + x_1^2.z_2^2 + y_1^2.z_2^2 + z_1^2.y_2^2 - \\ & 2.(y_1.y_2.z_1.z_2 + x_1.x_2.z_1.z_2 + x_1.x_2.y_1.y_2) \end{aligned}}$$

The vector $a \times b$ is perpendicular to the base so the perpendicular height is given by:

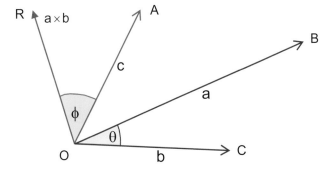

Figure 12.28 Vector $a \times b$

Perpendicular height $OR = \|c\| . \cos(\phi) = \|c\| . \left(\dfrac{(a \times b).c}{\|a \times b\| . \|c\|} \right)$

Volume $OADCGBEH$ $= OR \times OH \times OB$

$\qquad = $ Perpendicular height $\times$ Base area

$\qquad = \|c\| . \left(\dfrac{(a \times b).c}{\|a \times b\| . \|c\|} \right) . \|a \times b\|$

$\qquad = (a \times b).c$

$\qquad = \begin{bmatrix} i & j & k \\ x_1 & y_1 & z_1 \\ x_2 & y_2 & z_2 \end{bmatrix} . \begin{bmatrix} x_3 \\ y_3 \\ z_3 \end{bmatrix} = \begin{bmatrix} i. \begin{vmatrix} y_1 & z_1 \\ y_2 & z_2 \end{vmatrix} & -j. \begin{vmatrix} x_1 & z_1 \\ x_2 & z_2 \end{vmatrix} & +k. \begin{vmatrix} x_1 & y_1 \\ x_2 & y_2 \end{vmatrix} \end{bmatrix} . \begin{bmatrix} x_3 \\ y_3 \\ z_3 \end{bmatrix}$

$\qquad = \begin{bmatrix} \begin{vmatrix} y_1 & z_1 \\ y_2 & z_2 \end{vmatrix}, & -\begin{vmatrix} x_1 & z_1 \\ x_2 & z_2 \end{vmatrix}, & \begin{vmatrix} x_1 & y_1 \\ x_2 & y_2 \end{vmatrix} \end{bmatrix} . \begin{bmatrix} x_3 \\ y_3 \\ z_3 \end{bmatrix}$

$\qquad = \begin{vmatrix} x_3 & y_3 & z_3 \\ x_1 & y_1 & z_1 \\ x_2 & y_2 & z_2 \end{vmatrix} = -\begin{vmatrix} x_1 & y_1 & z_1 \\ x_3 & y_3 & z_3 \\ x_2 & y_2 & z_2 \end{vmatrix} = \begin{vmatrix} x_1 & y_1 & z_1 \\ x_2 & y_2 & z_2 \\ x_3 & y_3 & z_3 \end{vmatrix}$

The volume changes its sign depending on the rotational order in which the three vectors are combined.

$$a \times b.c = a.b \times c \quad \text{but} \quad a \times b.c = -a \times c.b$$

This is matched by the way a determinant changes sign if two adjacent rows or two adjacent columns are swapped over.

The area of a triangle is half the area of the parallelogram generated by adding two vectors together. The volume of a tetrahedron is one sixth of the volume of the parallelepiped generated by adding three vectors, and this can be demonstrated by the following partition of the parallelepiped into six tetrahedra of equal volumes

Volume OADCGBEF = OABC + DEFG + ADCGBE

Volume ADCGBE = CADE + CAEB + CDEG + CEGB

Volume CADE = CAEB = CDEG = CEGB   same base and vertical heights

Volume OABC = CAEB   same base and vertical heights

Volume OABC = DEFG   congruent volumes

Therefore

Volume OADCGBEF = 6. OABC

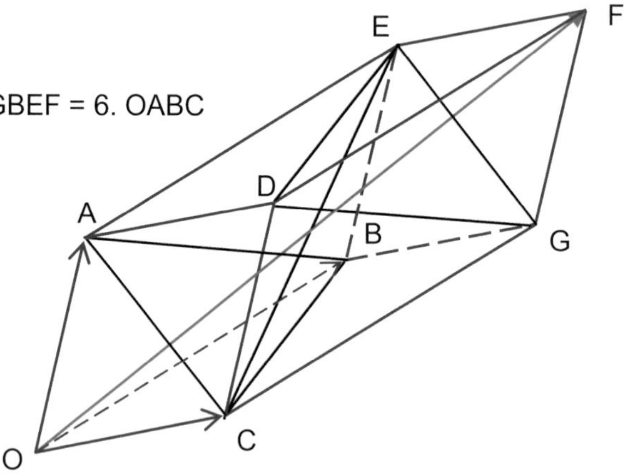

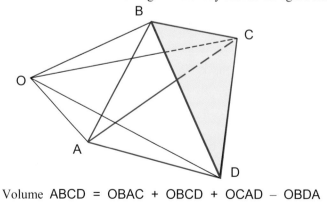

Figure 12.29 Subdividing the parallelepiped into six equal tetrahedral volumes

This result can then be extended in the same way used for the free standing triangle to give the volume of a tetrahedron with none of its vertices at the origin, by adding up four volumes linked to the origin in the way shown in Figure 12.30.

Volume ABCD = OBAC + OBCD + OCAD − OBDA

Figure 12.30 The volume of a tetrahedron as the sum of four tetrahedral volumes

The sign of the tetrahedral sub-volumes is determined by the orientation of each triangular facet relative to the origin. Where the origin lies inside the facet the vertices will be in anticlockwise order viewed from the origin and the volume will be positive. Where the origin lies outside the facet the vertices will be in a clockwise order viewed from the origin and the volume will be negative. The algebraic sum of these values leaves a positive value for the tetrahedron ABCD

Given four vertices for a tetrahedron P_1, P_2, P_3 and P_4 the volume is given by the sum :

$$\text{tetrahedron volume} = \frac{1}{6} \cdot \left(\begin{vmatrix} P_1 \\ P_2 \\ P_3 \end{vmatrix} + \begin{vmatrix} P_2 \\ P_4 \\ P_3 \end{vmatrix} + \begin{vmatrix} P_1 \\ P_3 \\ P_4 \end{vmatrix} + \begin{vmatrix} P_1 \\ P_4 \\ P_2 \end{vmatrix} \right)$$

$$= \frac{1}{6} \cdot \left(\begin{vmatrix} x_1 & y_1 & z_1 \\ x_2 & y_2 & z_2 \\ x_3 & y_3 & z_3 \end{vmatrix} + \begin{vmatrix} x_2 & y_2 & z_2 \\ x_4 & y_4 & z_4 \\ x_3 & y_3 & z_3 \end{vmatrix} + \begin{vmatrix} x_1 & y_1 & z_1 \\ x_3 & y_3 & z_3 \\ x_4 & y_4 & z_4 \end{vmatrix} + \begin{vmatrix} x_1 & y_1 & z_1 \\ x_4 & y_4 & z_4 \\ x_2 & y_2 & z_2 \end{vmatrix} \right)$$

Extending this approach, gives an algorithm for calculating the volume of any polyhedron in any location, in a similar way to that used to calculate the area of a polygon from adding up a sequence of triangular areas.

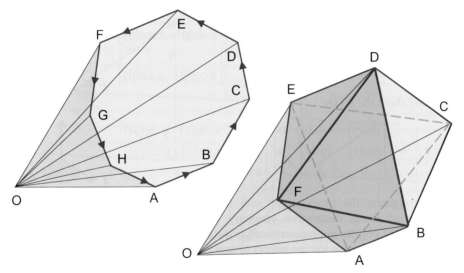

Figure 12.31 The area of a polygon and the volume of a polyhedron

In figure 12.31 if each of the boundary line segments for the polygon ABCDEFGH is taken in an anticlockwise order and linked in sequence to the origin then five anticlockwise triangles are created and three clockwise triangles. The sum of the anticlockwise triangles will be positive and give the area of OABCDEF, and the sum of the clockwise triangles will give the green area of OFGHA as a negative value. Adding them together will give the yellow area of ABCDEFGH. Similarly for

$$[x \ y \ z]_P \cdot \begin{bmatrix} \dfrac{x}{\sqrt{x^2+y^2}} & \dfrac{-y}{\sqrt{x^2+y^2}} & 0 \\ \dfrac{y}{\sqrt{x^2+y^2}} & \dfrac{x}{\sqrt{x^2+y^2}} & 0 \\ 0 & 0 & 1 \end{bmatrix}_A \rightarrow \begin{bmatrix} \sqrt{x^2+y^2} & 0 & z \end{bmatrix}_Q$$

The second step is to rotate the target axis about the Y-axis to line it up with the X-axis:

$$\text{rotate about the Y axis}: \quad \cos(\theta) = \frac{z}{r} \qquad \sin(\theta) = \frac{\sqrt{x^2+y^2}}{r}$$

$$\begin{bmatrix} \sqrt{x^2+y^2} & 0 & z \end{bmatrix}_Q \cdot \begin{bmatrix} \dfrac{\sqrt{x^2+y^2}}{r} & 0 & -\dfrac{z}{r} \\ 0 & 1 & 0 \\ \dfrac{z}{r} & 0 & \dfrac{\sqrt{x^2+y^2}}{r} \end{bmatrix}_B \rightarrow \begin{bmatrix} \dfrac{x^2+y^2+z^2}{r} & 0 & 0 \end{bmatrix}$$

$$\rightarrow \begin{bmatrix} r & 0 & 0 \end{bmatrix}_R$$

The third step is to calculate the required rotation about the target axis (now lined up with the X-axis). Once this has been done the two rotations to move this axis to line up with the X-axis have to be reversed to give the final orientation of the object.

$$[P'] \cdot [A] \cdot [B] \cdot \begin{bmatrix} 1 & 0 & 0 \\ 0 & \cos(\gamma) & -\sin(\gamma) \\ 0 & \sin(\gamma) & \cos(\gamma) \end{bmatrix}_D \cdot [B]^{-1} \cdot [A]^{-1} \rightarrow [P'']$$

The five concatenated matrices $[A] \cdot [B] \cdot [D] \cdot [B]^{-1} \cdot [A]^{-1}$ give the rotation about the axis through the origin defined by the vector [x, y, z], for each point in the object.

Evaluating a Reflection Ray

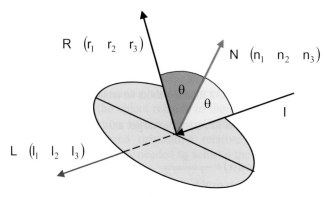

Figure 12.33 Reflection ray

A similar operation which also requires an axis to be rotated in order to line it up with one of the coordinate axes is the calculation of the reflection ray from a mirror surface given the incident ray. If the incident-ray, the normal to the surface, and the reflection ray are represented by unit vectors: L, N and R. This allows the calculation to be made treating the point where the ray strikes the surface to be the origin.

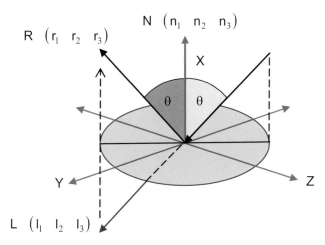

Figure 12.34 Rotate the reflection axis to be along the X axis

The angle of incidence equals the angle of reflection, and the incident ray the normal and the reflection ray lie in a plane perpendicular to the surface. Rotating the normal to lie along the X-axis can be done using the forward and backward transformation.

Rotations to align the normal with the X axis and return to the original orientation:

$$
[C] = \begin{bmatrix}
n_1 & \dfrac{-n_1}{\sqrt{n_1^2+n_2^2}} & \dfrac{-n_3 \cdot n_1}{\sqrt{n_1^2+n_2^2}} \\[3mm]
n_2 & \dfrac{n_1}{\sqrt{n_1^2+n_2^2}} & \dfrac{-n_3 \cdot n_2}{\sqrt{n_1^2+n_2^2}} \\[3mm]
n_3 & 0 & \sqrt{n_1^2+n_2^2}
\end{bmatrix}
\qquad
[C]^{-1} = \begin{bmatrix}
n_1 & n_2 & n_3 \\[3mm]
\dfrac{-n_1}{\sqrt{n_1^2+n_2^2}} & \dfrac{n_1}{\sqrt{n_1^2+n_2^2}} & 0 \\[3mm]
\dfrac{-n_3 \cdot n_1}{\sqrt{n_1^2+n_2^2}} & \dfrac{-n_3 \cdot n_2}{\sqrt{n_1^2+n_2^2}} & \sqrt{n_1^2+n_2^2}
\end{bmatrix}
$$

Reflection of the incident ray vector L in the YZ axes plane gives the reflection ray R :

$$
[R] = [L].[C].\begin{bmatrix} -1 & 0 & 0 \\ 0 & 1 & 0 \\ 0 & 0 & 1 \end{bmatrix}.[C]^{-1} \equiv [L].[A].[B].[E].[B]^{-1}.[A]^{-1}
$$

If these axes were not through the origin, for example, if they were defined by a line segment (p1, p2) then an initial translation of p1, p2 and any associated set of points, to place p1 at the origin, followed by the rotation or reflection, followed by a translation returning the new origin back to p1, would be necessary.

Homogeneous Coordinates

Complex mixtures of translation and rotation can create calculation sequences that make it impossible to reduce the matrix operations to a single matrix. However, it is possible to express translation as a matrix operation by representing two-dimensional operations in three dimensions and similarly three-dimensional operations in four. In order to express a 2D translation as a matrix operation it is represented by a movement in 3D space for a point (x, y, z). This movement is projected back onto a plane parallel to the x-y axes plane through the $z = 1$ point on the z-axis, in the way shown in Figure 12.34.

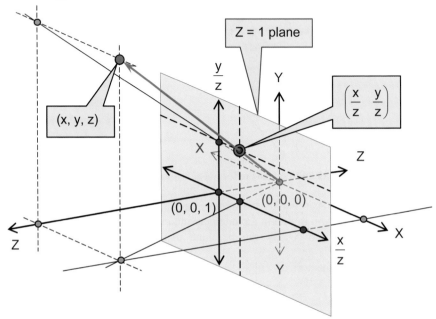

Figure 12.35 Homogeneous coordinates

The way this works for translation using three-dimensional matrices is as follows:

$$\begin{bmatrix} x_1 & y_1 & 1 \end{bmatrix} . \begin{bmatrix} 1 & 0 & 0 \\ 0 & 1 & 0 \\ \Delta x & \Delta y & 1 \end{bmatrix} \rightarrow \begin{bmatrix} x_1 + \Delta x & y_1 + \Delta y & 1 \end{bmatrix}$$

or

$$\begin{bmatrix} x_1 & y_1 & z_1 \end{bmatrix} . \begin{bmatrix} 1 & 0 & 0 \\ 0 & 1 & 0 \\ \Delta x & \Delta y & 1 \end{bmatrix} \rightarrow \begin{bmatrix} x_1 + z_1.\Delta x & y_1 + z_1.\Delta y & z_1 \end{bmatrix}$$

$$\rightarrow \begin{bmatrix} \dfrac{x_1}{z_1} + \Delta x & \dfrac{y_1}{z_1} + \Delta y & 1 \end{bmatrix}$$

The convention is to represent the extra dimension by the symbol w. This gives two-dimensional coordinates (w, x, y) or (x, y, w) and for three-dimensional coordinates (w, x, y, z) or (x, y, z, w).

The other matrix transformations can be extended to work within this framework. This includes reflections, various forms of scaling, shear and rotation. The extension provides translation, projection, and uniform scaling as new operations. Each of these basic matrices can be inverted by a simple change in the matrix entries, in the ways given below. This allows the composition of more complex transformations by concatenating the appropriate sequence of simpler transformation matrices into a single matrix in the way already discussed.

Homogeneous Reflections

Reflection about the line y=x;

$$
\begin{bmatrix} 0 & 1 & 0 \\ 1 & 0 & 0 \\ 0 & 0 & 1 \end{bmatrix} . \begin{bmatrix} x \\ y \\ 1 \end{bmatrix} = \begin{bmatrix} y \\ x \\ 1 \end{bmatrix} \quad \text{where} \quad \begin{bmatrix} 0 & 1 & 0 \\ 1 & 0 & 0 \\ 0 & 0 & 1 \end{bmatrix} . \begin{bmatrix} 0 & 1 & 0 \\ 1 & 0 & 0 \\ 0 & 0 & 1 \end{bmatrix} = \begin{bmatrix} 1 & 0 & 0 \\ 0 & 1 & 0 \\ 0 & 0 & 1 \end{bmatrix}
$$

Reflection about the line y= −x;

$$
\begin{bmatrix} 0 & -1 & 0 \\ -1 & 0 & 0 \\ 0 & 0 & 1 \end{bmatrix} . \begin{bmatrix} x \\ y \\ 1 \end{bmatrix} = \begin{bmatrix} -y \\ -x \\ 1 \end{bmatrix} \quad \text{where} \quad \begin{bmatrix} 0 & -1 & 0 \\ -1 & 0 & 0 \\ 0 & 0 & 1 \end{bmatrix} . \begin{bmatrix} 0 & -1 & 0 \\ -1 & 0 & 0 \\ 0 & 0 & 1 \end{bmatrix} = \begin{bmatrix} 1 & 0 & 0 \\ 0 & 1 & 0 \\ 0 & 0 & 1 \end{bmatrix}
$$

or

$$
\begin{bmatrix} 0 & 1 & 0 \\ 1 & 0 & 0 \\ 0 & 0 & -1 \end{bmatrix} . \begin{bmatrix} x \\ y \\ 1 \end{bmatrix} = \begin{bmatrix} y \\ x \\ -1 \end{bmatrix} = \begin{bmatrix} -y \\ -x \\ 1 \end{bmatrix} \quad \text{where} \quad \begin{bmatrix} 0 & 1 & 0 \\ 1 & 0 & 0 \\ 0 & 0 & -1 \end{bmatrix} . \begin{bmatrix} 0 & 1 & 0 \\ 1 & 0 & 0 \\ 0 & 0 & -1 \end{bmatrix} = \begin{bmatrix} 1 & 0 & 0 \\ 0 & 1 & 0 \\ 0 & 0 & 1 \end{bmatrix}
$$

Reflection about the y axis.

$$
\begin{bmatrix} -1 & 0 & 0 \\ 0 & 1 & 0 \\ 0 & 0 & 1 \end{bmatrix} . \begin{bmatrix} x \\ y \\ 1 \end{bmatrix} = \begin{bmatrix} -x \\ y \\ 1 \end{bmatrix} \quad \text{where} \quad \begin{bmatrix} -1 & 0 & 0 \\ 0 & 1 & 0 \\ 0 & 0 & 1 \end{bmatrix} . \begin{bmatrix} -1 & 0 & 0 \\ 0 & 1 & 0 \\ 0 & 0 & 1 \end{bmatrix} = \begin{bmatrix} 1 & 0 & 0 \\ 0 & 1 & 0 \\ 0 & 0 & 1 \end{bmatrix}
$$

Reflection about the x axis

$$
\begin{bmatrix} 1 & 0 & 0 \\ 0 & -1 & 0 \\ 0 & 0 & 1 \end{bmatrix} . \begin{bmatrix} x \\ y \\ 1 \end{bmatrix} = \begin{bmatrix} x \\ -y \\ 1 \end{bmatrix} \quad \text{where} \quad \begin{bmatrix} 1 & 0 & 0 \\ 0 & -1 & 0 \\ 0 & 0 & 1 \end{bmatrix} . \begin{bmatrix} 1 & 0 & 0 \\ 0 & -1 & 0 \\ 0 & 0 & 1 \end{bmatrix} = \begin{bmatrix} 1 & 0 & 0 \\ 0 & 1 & 0 \\ 0 & 0 & 1 \end{bmatrix}
$$

or

$$
\begin{bmatrix} -1 & 0 & 0 \\ 0 & 1 & 0 \\ 0 & 0 & -1 \end{bmatrix} . \begin{bmatrix} x \\ y \\ 1 \end{bmatrix} = \begin{bmatrix} -x \\ y \\ -1 \end{bmatrix} = \begin{bmatrix} x \\ -y \\ 1 \end{bmatrix} \quad \text{where} \quad \begin{bmatrix} -1 & 0 & 0 \\ 0 & 1 & 0 \\ 0 & 0 & -1 \end{bmatrix} . \begin{bmatrix} -1 & 0 & 0 \\ 0 & 1 & 0 \\ 0 & 0 & -1 \end{bmatrix} = \begin{bmatrix} 1 & 0 & 0 \\ 0 & 1 & 0 \\ 0 & 0 & 1 \end{bmatrix}
$$

Scaling Along Each Axis

$$\begin{bmatrix} a & 0 & 0 \\ 0 & b & 0 \\ 0 & 0 & 1 \end{bmatrix} . \begin{bmatrix} x \\ y \\ 1 \end{bmatrix} = \begin{bmatrix} a.x \\ b.y \\ 1 \end{bmatrix} \text{ where } \begin{bmatrix} a & 0 & 0 \\ 0 & b & 0 \\ 0 & 0 & 1 \end{bmatrix} . \begin{bmatrix} \frac{1}{a} & 0 & 0 \\ 0 & \frac{1}{b} & 0 \\ 0 & 0 & 1 \end{bmatrix} = \begin{bmatrix} 1 & 0 & 0 \\ 0 & 1 & 0 \\ 0 & 0 & 1 \end{bmatrix}$$

Uniform Scaling

$$\begin{bmatrix} 1 & 0 & 0 \\ 0 & 1 & 0 \\ 0 & 0 & \frac{1}{s} \end{bmatrix} . \begin{bmatrix} x \\ y \\ 1 \end{bmatrix} = \begin{bmatrix} x \\ y \\ \frac{1}{s} \end{bmatrix} = \begin{bmatrix} s.x \\ s.y \\ 1 \end{bmatrix} \text{ where } \begin{bmatrix} 1 & 0 & 0 \\ 0 & 1 & 0 \\ 0 & 0 & \frac{1}{s} \end{bmatrix} . \begin{bmatrix} 1 & 0 & 0 \\ 0 & 1 & 0 \\ 0 & 0 & s \end{bmatrix} = \begin{bmatrix} 1 & 0 & 0 \\ 0 & 1 & 0 \\ 0 & 0 & 1 \end{bmatrix}$$

Shear

In the x direction

$$\begin{bmatrix} 1 & a & 0 \\ 0 & 1 & 0 \\ 0 & 0 & 1 \end{bmatrix} . \begin{bmatrix} x \\ y \\ 1 \end{bmatrix} = \begin{bmatrix} x+ay \\ y \\ 1 \end{bmatrix} \text{ where } \begin{bmatrix} 1 & -a & 0 \\ 0 & 1 & 0 \\ 0 & 0 & 1 \end{bmatrix} \begin{bmatrix} 1 & a & 0 \\ 0 & 1 & 0 \\ 0 & 0 & 1 \end{bmatrix} = \begin{bmatrix} 1 & 0 & 0 \\ 0 & 1 & 0 \\ 0 & 0 & 1 \end{bmatrix}$$

in the y direction

$$\begin{bmatrix} 1 & 0 & 0 \\ b & 1 & 0 \\ 0 & 0 & 1 \end{bmatrix} . \begin{bmatrix} x \\ y \\ 1 \end{bmatrix} = \begin{bmatrix} x \\ b.x+y \\ 1 \end{bmatrix} \text{ where } \begin{bmatrix} 1 & 0 & 0 \\ -b & 1 & 0 \\ 0 & 0 & 1 \end{bmatrix} . \begin{bmatrix} 1 & 0 & 0 \\ b & 1 & 0 \\ 0 & 0 & 1 \end{bmatrix} = \begin{bmatrix} 1 & 0 & 0 \\ 0 & 1 & 0 \\ 0 & 0 & 1 \end{bmatrix}$$

Combining the two

$$\begin{bmatrix} 1 & a & 0 \\ 0 & 1 & 0 \\ 0 & 0 & 1 \end{bmatrix} . \begin{bmatrix} 1-ab & 0 & 0 \\ b & 1 & 0 \\ 0 & 0 & 1 \end{bmatrix} = \begin{bmatrix} 1 & a & 0 \\ b & 1 & 0 \\ 0 & 0 & 1 \end{bmatrix}$$

or

$$\begin{bmatrix} 1 & 0 & 0 \\ b & 1 & 0 \\ 0 & 0 & 1 \end{bmatrix} . \begin{bmatrix} 1 & a & 0 \\ 0 & 1-ab & 0 \\ 0 & 0 & 1 \end{bmatrix} = \begin{bmatrix} 1 & a & 0 \\ b & 1 & 0 \\ 0 & 0 & 1 \end{bmatrix}$$

shear in both directions

$$\begin{bmatrix} 1 & a & 0 \\ b & 1 & 0 \\ 0 & 0 & 1 \end{bmatrix} . \begin{bmatrix} x \\ y \\ 1 \end{bmatrix} = \begin{bmatrix} x+ay \\ b.x+y \\ 1 \end{bmatrix} \text{ where } \begin{bmatrix} \frac{-1}{a.b-1} & \frac{a}{a.b-1} & 0 \\ \frac{b}{a.b-1} & \frac{-1}{a.b-1} & 0 \\ 0 & 0 & 1 \end{bmatrix} . \begin{bmatrix} 1 & a & 0 \\ b & 1 & 0 \\ 0 & 0 & 1 \end{bmatrix} = \begin{bmatrix} 1 & 0 & 0 \\ 0 & 1 & 0 \\ 0 & 0 & 1 \end{bmatrix}$$

combining shear and scaling in the appropriate amounts gives a rotation

Rotation

$$\begin{bmatrix} a & b & 0 \\ c & d & 0 \\ 0 & 0 & 1 \end{bmatrix} . \begin{bmatrix} x \\ y \\ 1 \end{bmatrix} = \begin{bmatrix} a.x + b.y \\ c.x + d.y \\ 1 \end{bmatrix}$$

$$\text{where} \quad \begin{bmatrix} \dfrac{d}{d.a - c.b} & \dfrac{-b}{d.a - c.b} & 0 \\ \dfrac{-c}{d.a - c.b} & \dfrac{a}{d.a - c.b} & 0 \\ 0 & 0 & 1 \end{bmatrix} . \begin{bmatrix} a & b & 0 \\ c & d & 0 \\ 0 & 0 & 1 \end{bmatrix} = \begin{bmatrix} 1 & 0 & 0 \\ 0 & 1 & 0 \\ 0 & 0 & 1 \end{bmatrix}$$

If $a = d = \cos(\alpha)$ and $b = -c = \sin(\alpha)$ then $d.a - c.b = \cos^2(\alpha) + \sin^2(\alpha) = 1$;

The following matrix will carry out a rotation for two-dimensional points, by an angle α, and the inverse operation can again be obtained by using $-\alpha$ instead of α.

$$\begin{bmatrix} \cos(\alpha) & -\sin(\alpha) & 0 \\ \sin(\alpha) & \cos(\alpha) & 0 \\ 0 & 0 & 1 \end{bmatrix} . \begin{bmatrix} x \\ y \\ 1 \end{bmatrix} = \begin{bmatrix} x.\cos(\alpha) - y.\sin(\alpha) \\ x.\sin(\alpha) + y.\cos(\alpha) \\ 1 \end{bmatrix}$$

$$\begin{bmatrix} \cos(\alpha) & -\sin(\alpha) & 0 \\ \sin(\alpha) & \cos(\alpha) & 0 \\ 0 & 0 & 1 \end{bmatrix} . \begin{bmatrix} \cos(\alpha) & \sin(\alpha) & 0 \\ -\sin(\alpha) & \cos(\alpha) & 0 \\ 0 & 0 & 1 \end{bmatrix}$$

$$= \begin{bmatrix} \cos^2(\alpha) + \sin^2(\alpha) & \cos(\alpha).\sin(\alpha) - \cos(\alpha).\sin(\alpha) & 0 \\ \cos(\alpha).\sin(\alpha) - \cos(\alpha).\sin(\alpha) & \cos^2(\alpha) + \sin^2(\alpha) & 0 \\ 0 & 0 & 1 \end{bmatrix} = \begin{bmatrix} 1 & 0 & 0 \\ 0 & 1 & 0 \\ 0 & 0 & 1 \end{bmatrix}$$

Parallel Projections onto Axes Planes

Projections along with translations are the main extensions provided by using homogeneous coordinates. Simple projections give two-dimensional images from three dimensional scene models in the following way as projections on the axes planes. In practice this can be implemented merely by ignoring one of the coordinate values.

$$\begin{bmatrix} 1 & 0 & 0 & 0 \\ 0 & 1 & 0 & 0 \\ 0 & 0 & 0 & 0 \\ 0 & 0 & 0 & 1 \end{bmatrix} . \begin{bmatrix} x \\ y \\ z \\ 1 \end{bmatrix} = \begin{bmatrix} x \\ y \\ 0 \\ 1 \end{bmatrix} \quad \begin{bmatrix} 1 & 0 & 0 & 0 \\ 0 & 0 & 0 & 0 \\ 0 & 0 & 1 & 0 \\ 0 & 0 & 0 & 1 \end{bmatrix} . \begin{bmatrix} x \\ y \\ z \\ 1 \end{bmatrix} = \begin{bmatrix} x \\ 0 \\ z \\ 1 \end{bmatrix} \quad \begin{bmatrix} 0 & 0 & 0 & 0 \\ 0 & 1 & 0 & 0 \\ 0 & 0 & 1 & 0 \\ 0 & 0 & 0 & 1 \end{bmatrix} . \begin{bmatrix} x \\ y \\ z \\ 1 \end{bmatrix} = \begin{bmatrix} 0 \\ y \\ z \\ 1 \end{bmatrix}$$

More complex three-dimensional projections such as perspective will be the subject of a later chapter where the problem of rendering three-dimensional scenes is explored in greater depth, as well as the task of constructing sequences of transformations to give smooth three-dimensional trajectories.

Example Using Matrix Operations to Animate a Sequence

Once a curved shape is input then it can be rotated about a point on the screen by first setting the point to be the origin then rotating the object's control polygon about this point then returning the object to its original coordinate framework in the way illustrated in Figure 12.36.

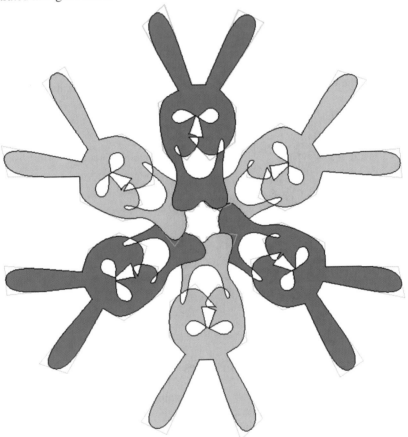

Figure 12.36 Rigid body rotation about an arbitrary point

Where the object needs to be kept upright then further steps in building the transformation matrix are necessary shown in Figure 12.37. Firstly the object is rotated forwards by the angle alpha about its centre then the whole figure is rotated back about the main centre of rotation by the same angle, so keeping the figure upright.

```
Transf2D TT = Tb.mult(Rf.mult(Tf.mult(Sb.mult(Rb.mult(Sf)))));
```

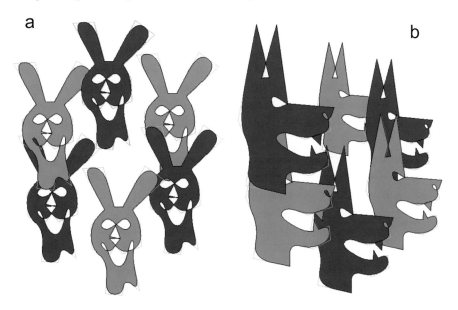

Figure 12.37 Rotation translation and scaling in combinations

```
class Transf2D{
    double [][] t = new double[3][3];
    Transf2D(){
        for(int i=0;i<3;i++){ for(int j=0;j<3;j++){ if(i==j)t[i][j]=1.0;else t[i][j]=0.0;} }
    }
    public Transf2D setScales(double x,double y,double s,Transf2D m){
        if (m==null) m=new Transf2D();
        t[0][0]= x;  t[1][1]= y;  t[2][2]= s;   m.t[0][0]=1/x;  m.t[1][1]=1/y;  m.t[2][2]=1/s;
        return m;
    }
    public Transf2D setShear(double a,double b,double c,double d,Transf2D m){
        if (m==null)m=new Transf2D();
        t[0][0]= a;  t[0][1]= b;  t[1][0]= c;  t[1][1]= d;  double D= c*b-a+d;
        m.t[0][0]=-d/D;  m.t[0][1]=b/D;  m.t[1][0]=c/D;  m.t[1][1]=-a/D;
        return m;
    }
    public Transf2D setTranslate(double x,double y,Transf2D m){
        if (m==null) m=new Transf2D();
        t[0][2]= x;t[1][2]= y;  m.t[0][2]= -x; m.t[1][2]= -y; return m;
    }
    public Transf2D setRotate(double alpha,Transf2D m){
        if (m==null)  m=new Transf2D();double a,b,c,d;
        a = Math.cos(alpha); b = Math.sin(alpha);
        t[0][0]= a;t[0][1]= -b; t[1][0]= b; t[1][1]= a;
        m.t[0][0]= a; m.t[0][1]= b ; m.t[1][0]= -b; m.t[1][1]= a; return m;
    }
```

```
public Transf2D mult(Transf2D m){
    Transf2D n = new Transf2D();
    for (int i=0;i<3;i++) for (int j=0;j<3;j++){
        n.t[i][j] = 0;
        for(int k=0;k<3;k++){ n.t[i][j] = n.t[i][j]+this.t[i][k]*m.t[k][j]; }
    }return n;
}
public Point premult(Point p){
    Point m = new Point(2);Point tmp = new Point(2);
    tmp.n[0]=p.n[1];tmp.n[1]=p.n[2];tmp.n[2]=p.n[0];
    for (int i=0;i<3;i++){
        m.n[i] = 0;  for(int k=0;k<3;k++){  m.n[i] = m.n[i]+this.t[i][k]*tmp.n[k]; }
    }
    Point s = new Point(2);  s.n[0]=m.n[2];  s.n[1]=m.n[0];  s.n[2]=m.n[1];  return s;
}
}

public static void main(String[] args){
    Polyline aa = new Polyline(IO,dW);  aa.getPolyloop();
    Polygon a = new Polygon();  a.p = aa.p;  a.length=aa.length;
    Polygon b = new Polygon(a.length);
    IO.writeString("please enter the centre of rotation \n");  Point p = dW.getCoord();
    IO.writeString("please enter the centre of object \n");    Point q = dW.getCoord();
    Transf2D Tf = new Transf2D();
    Transf2D Tb = Tf.setTranslate(-p.xd(),-p.yd(),null);
    Transf2D Sf = new Transf2D();
    Transf2D Sb = Sf.setTranslate(-q.xd(),-q.yd(),null);
    IO.writeString("please enter the nuber of rotations \n");
    int n= IO.readInteger(); IO.readLine();
    double apha =0; double scale =1, dscale = 0.1; double dangle = 2* Math.PI/n;
    for(int i= 0; i<n; i++){
        Transf2D Scf = new Transf2D();
        Transf2D Scb = Scf.setScales(scale,scale,1,null);
        Transf2D Rf = new Transf2D();
        Transf2D Rb = Rf.setRotate(alpha,null);
        Transf2D TT = Tb.mult(Scf.mult(Rf.mult(Tf.mult(Sb.mult(Rb.mult(Sf))))));
        for(int k=0; k<a.length;k++){
            b.p[k]= TT.premult(a.p[k]);
            if(k>0)dW.plotLine(b.p[k-1],b.p[k],Color.lightGray);
        }
        int nn= b.length*2-1;
        Point boundary [][] = new Point[nn][];
        Point[] bnd = boundaryFill(b,boundary,Color.black);
        if(i%2==0)cc=Color.green;else cc=Color.blue;
        fill(bnd, cc, Color.black);
        alpha = alpha + dangle;  scale = scale + dscale;
    }
}
```

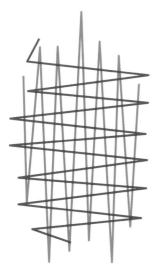

13
Spatial Relationships Overlap & Adjacency: Point to Line to Area

Introduction

To write intelligent programs that are more than a series of simple commands it is necessary to make commands conditional on data values and relationships between data values. For graphics programming and many spatial modelling tasks, it is necessary to make actions depend on spatial relationships, and to provide tests for them. In this chapter the relationships of overlap and adjacency or neighbourliness between spatial objects are examined. Since spatial objects are often represented by complex data structures, such relationship tests usually have to be implemented as sub-programs. Clearly such programs will depend on the nature of the data structure representing the objects even where the abstract objects, for example polygons, are the same. For this reason the main spatial relationships reflecting the primary differences in the structures involved, are set out in the table given below. The more detailed study of algorithms that follow, are based on the specific data structures commonly used to represent these objects in practice.

Table 13.1 Basic spatial relationships

Objects	Point	Line	Area	Volume
Point	Point on Point	Point in Line	Point in Area	Point in Volume
Line		Line on Line	Line in Area	Line in Volume
Area			Area on Area	Surface in Volume
Volume				Volume on Volume

A. Thomas, *Integrated Graphic and Computer Modelling*,
DOI: 10.1007/978-1-84800-179-4_13, © Springer-Verlag London Limited 2008

Point on Point

There are two essential relationship tests that are important when working with points. The first is *"the same as"*, the second is *"near to"*. Where co-ordinates represent points from the same co-ordinate space, then they are the same points if their co-ordinates are equal. For Cartesian co-ordinates this is the relatively simple test:

$$if((x1==x2)\&\&(y1==y2)\&\&(z1==z2))\{ \}.$$

The exception to this rule occurs where homogeneous co-ordinates are used, in which case the comparisons $(x1/w1 == x2/w2)$, $(y1/w1 == y2/w2)$ etc. must be used.

Using a Mouse to Select a Point on the Screen

The point-to-point test occurs repeatedly in geometry programs. However, perhaps its most common occurrence is where a mouse or pointing device is used to select a point on the display screen. A pointing device is a piece of hardware that generates a co-ordinate, which is used to display a cursor, either an arrow or cross hairs on the display screen. Moving or adjusting the pointing device changes the co-ordinate and the system then moves the cursor to reflect the change. This allows the user to interactively define a point on the display surface. However if the pointer is to be used to select an object on the display screen this point co-ordinate has to be matched with an element of an object displayed at that point. To do this the co-ordinate of the pointing device needs to be compared with the co-ordinates of points in the object's display file. When a match is found then the object in question can be selected.

With modern high-resolution displays the task of locating a particular point object on the screen becomes very difficult. In such cases a *"near to"* test is more appropriate. This test measures the distance between a potential target point and the input, test point, and when the distance is below some threshold size, the target point is selected.

There are various measures and approximations used for the distance between a test and target point. The natural choice for true distance uses Pythagoras' Theorem.

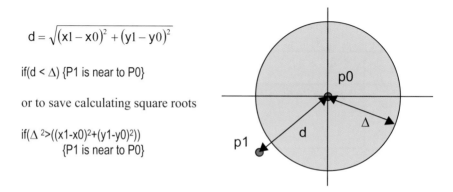

$$d = \sqrt{(x1 - x0)^2 + (y1 - y0)^2}$$

if(d < Δ) {P1 is near to P0}

or to save calculating square roots

if(Δ²>((x1-x0)²+(y1-y0)²))
       {P1 is near to P0}

Figure 13.1 Pythagorean distance between target and test points

However, Manhattan distances can also be used for this test to save calculating a square root when checking large display files.

$$d = |(x1-x0)|+|(y1-y0)|$$

In this case the x and y distance components can be positive or negative, consequently the final distance that is compared to the threshold size has to be the sum of two absolute values. This approach essentially involves testing the target point to see if it lies inside a small square box surrounding the test point. If a box of sides 2Δ is set up round a test point p0 the left and right edges of the box will be xa = x0-Δ, xb = x0+Δ, and the top and bottom edges will be ya = y0-Δ and yb = y0+Δ. This allows a single test to check whether the target point lies in the box surrounding the test point in the way shown in Figure 13.2

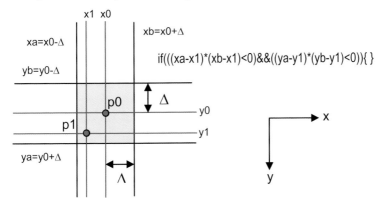

Figure 13.2 Manhattan "point near to point" test

Matching Multiple Points

Matching single points is a relatively straightforward task. However when two large files or lists of points need to be matched up then the pair by pair checking is not the most efficient way of executing the task. If the two lists of points are joined together and then the collection sorted, matching points will end up as neighbours in the sorted list. Also instead of an $O(n^2)$ task as a worst case, an $O(n.\log(n))$ operation results. If independent points were being processed this could be done using a standard sorting algorithm. However, where the points occur in a data structure representing a more complex object then the sorting operation has to be carried out using a linked-list sorting algorithm so the points can be kept in their original positions in the initial data structure.

The simplest example of this operation is the chaining of line segments into polyline arcs. Many processes generate output in the form of unordered straight-line segments represented by the pair of end points for each line segment. Where these line segments join up in the final drawing, it is efficient to chain them together in a polyline representation. This however, means matching corresponding line segment end points so that the duplicate co-ordinates can be removed.

```
public class ListOfTuples extends List {
    public ListOfTuples( ){  super.comparable = true;  }
    public NTuple mid(TextWindow tW){
        ListElement ref = this.start;  if (ref == null) return null;
        NTuple v = (NTuple)ref.object;
        NTuple min=new NTuple(v.dimension); NTuple max=new NTuple(v.dimension);
        NTuple midTuple = new NTuple(v.dimension);
        for(int i=0;i<v.dimension; i++)
            {min.n[i] = Double.MAX_VALUE;  max.n[i] = Double.MIN_VALUE; }
        while(ref != null){
            v = (NTuple)ref.object;
            if(v.dimension!= 0){
                for(int i=0;i<v.dimension; i++){
                    if(v.n[i]<min.n[i])min.n[i]=v.n[i]; if(v.n[i]>max.n[i])max.n[i]=v.n[i];
                }ref = ref.right;
            }
        }boolean test = true;
        for(int i=0;i<v.dimension;i++){
            if(min.n[i]!=max.n[i]) test = false;
            midTuple.n[i]= (min.n[i]+max.n[i])/2.0;
        }if (test)midTuple=null;
        return (NTuple)midTuple;//(Comparable)midTuple;
    }
    public ListOfTuples joinTo(ListOfTuples b){
        if(this.start==null)return b;  if(b.start==null)return this;
        ListOfTuples a = new ListOfTuples(); a.start = this.start; a.finish = b.finish;
        this.finish.right= b.start;  b.start.left = this.finish; return a;
    }
    public ListOfTuples sortn(TextWindow tW){
        this.setLength();   NTuple test = this.mid(tW);
        if(test== null)return this; if (this.start == this.finish) return this;
        ListElement ref = this.start;   ListOfTuples leftList = new ListOfTuples();
        ListOfTuples rightList= new ListOfTuples();
        while(ref!=null){
            if(((NTuple)ref.object).compareTo(test)<0){
                leftList.push(ref.object);
                leftList.start.link1 = ref.link1; leftList.start.link2 = ref.link2;
            }else{
                rightList.push(ref.object);
                rightList.start.link1 = ref.link1; rightList.start.link2 = ref.link2;
            }ref = ref.right;
        }if (leftList.start==null) return rightList = rightList.sortn(tW);
        if (rightList.start==null)return leftList  = leftList.sortn(tW);
        leftList = leftList.sortn(tW);   rightList = rightList.sortn(tW);
        return leftList.joinTo(rightList);
    }
}
```

```
public class NTuple extends ComparableObject {
    public double[] n = null;
    public int dimension= 0;
    public NTuple(){}
    public NTuple(int dim){
        this.dimension = dim;      this.n = new double[dim];
    }
    public int compareTo(Object b){
        for(int i=0; i<this.dimension; i++){
            if(this.n[i]> ((NTuple)b).n[i])return +1;
            if(this.n[i]< ((NTuple)b).n[i])return -1;
        }return 0;
    }
}
```

In the code given above a class called *ListOfTuples* is defined as an extension of the class *List*. This class allows a double linked list to be built from *NTuple* objects. Point objects are themselves extensions of *NTuple* objects: an *NTuple* simply being an array of *double* values such as a homogeneous coordinate or the coefficients of a line or plane equation. These elements are made "*comparable objects*" so they can be ordered using standard Java methods. However, to provide the linked, ordered list requires the methods defined in the class *ListOfTuples*, given above. The extreme *NTuple* values are established for a collection of *NTuples* then a central pivot "*mid*" element is set up and used to divide the collection into two collections: smaller or larger than the mid element. These in turn are recursively processed in the same way until singleton elements form each subset, at which point the recursive return links the sub-lists back into an ordered list.

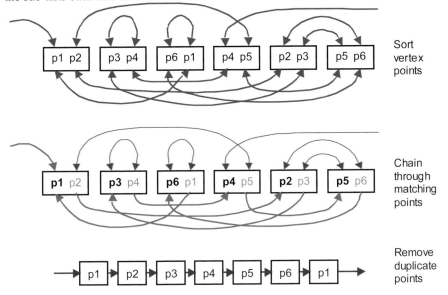

Figure 13.3 Chaining line segments together using co-ordinate sorting

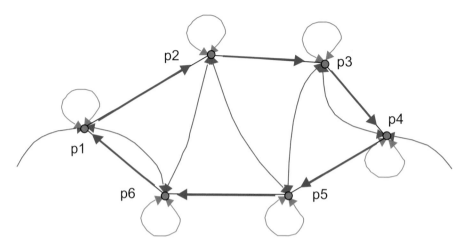

Figure 13.4 Ordered threaded list through the vertices of a polygon boundary

If the points in a set of line segment pairs are ordered in the way shown in Figure 13.3. Then by starting with the line segment containing the smallest point and linking to the line segment containing the point matching its other end, and by repeating the process for the subsequent line segments, a polyline arc can be threaded together, in the way illustrated in Figure 13.4, to give a closed polygon boundary.

Point on Line

The nature of the point on line test, as with all these relationship tests, depends on how the point and line are represented. Where the line is represented by a line-equation and the point by a co-ordinate then the line is in effect infinite in extent. If in contrast the line is only a line segment represented by its two end points, then linear interpolation between its two end points is the implicit representation of only part of an infinite line so the test has to be different.

The basic idea underlying the point to line tests is calculating the distance of the point from the line. If the distance is zero then the point is on the line, if the distance is less than some threshold value then the point is near the line. However because points and lines are approximated in raster display systems, the second approach usually has to be adopted with a threshold related to pixel dimensions. Where this is too fine a measure, for example when pointing to a line in a high-resolution display, then the threshold can be increased to a multiple pixel dimension.

If the test is for an infinite line then there is one step to the test, however for the line segment there are two steps. The common step for both is measuring the perpendicular distance of the point from the infinite line. The extra step for the line segment is locating the point within the neighbourhood of the segment. This second step applies the same box test used for testing the nearness of two points with a Manhattan metric, but uses a different definition of the box. The test box is set up as the min-max box for the line extended in each direction by the threshold distance.

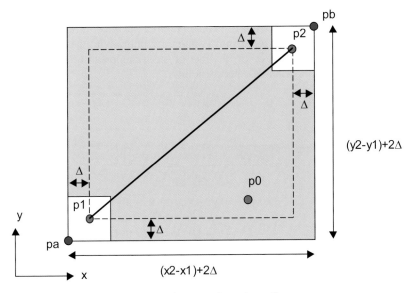

Figure 13.5 Box envelope test for point to line segment tests

The test becomes: *if (((xa-x0)*(xb-x0)<0)&&((ya-y0)*(yb-y0)<0)) { }* to test whether the point p0 is within the extended box around the line segment. Where xa = x1-Δ and xb = x2+Δ, and ya = y1-Δ and yb = y2+Δ.

Where the test point lies inside the box envelope round the line segment the next step is to calculate the perpendicular distance of the point from the line. Where the line segment is given as a pair of end points then the properties of determinants provide a way to obtain the distance of the test point from the line.

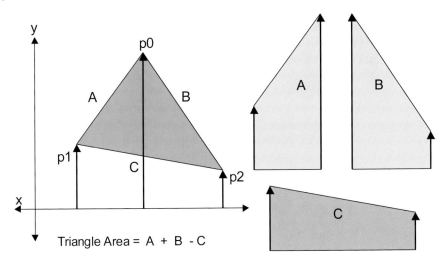

Figure 13.6 Calculating the area of a triangle

In Figure 13.6 the area of the triangle can be calculated by adding the area of trapezium A to the area of trapezium B then subtracting the area of trapezium C from the result. Where the area of a trapezium is given by multiplying its base times the average of its height, the formula for the area of the triangle can be expressed as a determinant containing the triangle's vertex points.

$$A = (x0 - x1)(y0 + y1)/2; \quad B = (x2 - x0)(y2 + y0)/2; \quad C = (x2 - x1)(y2 + y1)/2$$

$$\text{TriangleArea}(p0, p1, p2) = A + B - C$$

$$= \left(\frac{x0.y0 + x0.y1 - x1.y0 - x1.y1 + x2.y2 + x2.y0}{-x0.y2 - x0.y0 - x2.y2 - x2.y1 + x1.y2 + x1.y1} \right) / 2$$

$$= (x0.y1 - x0.y2 - x1.y0 + x2.y0 + x1.y2 - x2.y1)/2$$

$$= (x0.(y1 - y2) - y0.(x1 - x2) + x1.y2 - x2.y1)/2$$

$$= \frac{1}{2} \begin{vmatrix} x0 & y0 & 1 \\ x1 & y1 & 1 \\ x2 & y2 & 1 \end{vmatrix}$$

If the test point and the two line-segment end-points are entered into a 3×3 matrix as homogeneous co-ordinates the determinant of the matrix gives twice the area of the triangle defined by the points. Where the determinant value is zero then the triangle area is zero, in other words the test point p0 lies on the line through p1 and p2. If the determinant is zero it indicates that its rows or columns are not linearly independent. If the determinant is partially expanded then its relationship to the equation of the line through p1 and p2 can be demonstrated.

$$0 = \begin{vmatrix} x0 & y0 & 1 \\ x1 & y1 & 1 \\ x2 & y2 & 1 \end{vmatrix} = x0.\begin{vmatrix} y1 & 1 \\ y2 & 1 \end{vmatrix} - y0.\begin{vmatrix} x1 & 1 \\ x2 & 1 \end{vmatrix} + \begin{vmatrix} x1 & y1 \\ x2 & y2 \end{vmatrix}$$

$$0 = a.x0 + b.y0 + c$$

$$\text{where}: \quad a = \begin{vmatrix} y1 & 1 \\ y2 & 1 \end{vmatrix} = (y1\text{-}y2)$$

$$b = -\begin{vmatrix} x1 & 1 \\ x2 & 1 \end{vmatrix} = -(x1 - x2)$$

$$c = \begin{vmatrix} x1 & y1 \\ x2 & y2 \end{vmatrix} = (x1.y2\text{-}y1.x2)$$

If the line through points p1 and p2, is defined by the equation: a.x + b.y + c = 0 obtained in this way. The dot product of the equation coefficients (a, b, c) and the homogeneous co-ordinate (x0, y0, 1) of the test point will give twice the area of the triangle formed by the test point and the line segment. However, if the coefficients of the equation are normalised the result will be the perpendicular distance from the test point to the line. Dividing the area of the triangle by its base length can show the reason why the coefficients of the line equation need to be normalised to give this distance.

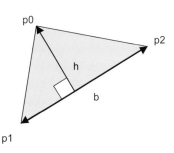

Area of Triangle:

$$= \big((y1 - y2)x0 - (x1 - x2)y0 + x1.y2 - x2.y1\big)/2$$
$$= (a.x0 + b.y0 + c)/2$$

Base Length:

$$b = \sqrt{(x1 - x2)^2 + (y1 - y2)^2} = \sqrt{a^2 + b^2}$$

Triangle area = (Triangle height × BaseLength)/2

Figure 13.7 Calculating the perpendicular height of a triangle

$$\text{Triangle height} \quad h = \frac{a.x0 + b.y0 + c}{\sqrt{a^2 + b^2}}$$

The combination of the two steps gives the test area shown in Figure 13.8, for a point p0 and line (p1, p2). The calculation of a square root can be avoided by squaring both sides of the equation.

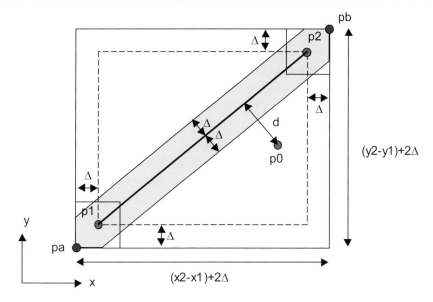

if(Δ*Δ > (a.x0+b.y0+c)*(a*x0+b.y0+c)/(a*a+b*b)) {point is near the line}

Figure 13.8 Nearness neighbourhood for a line segment to point test.

Point in Area Relationship Tests

There are various approaches to testing whether a point lies inside a polygon or not, partly dependent on the nature of the polygon and partly dependent on the way it is represented.

The simplest scheme, which can be used for polygons represented by polyline boundaries where the boundary does not cross itself, is the semi-infinite line test.

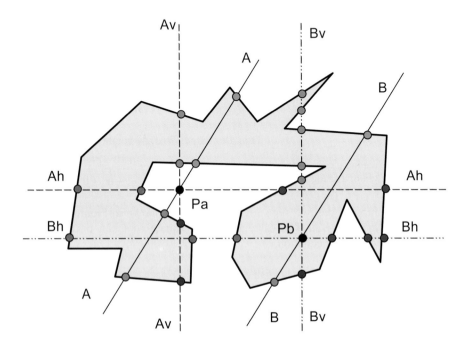

Figure 13.9 Semi-infinite line tests for point in polygon

If a line is taken out from the test point to infinity, or as far as is necessary to be clear of the polygon boundary then if the line cuts the boundary once or an odd number of times then the point is inside the polygon. Otherwise if the line does not cut the boundary or cuts it an even number of times then the point lies outside the polygon. In Figure 13.9 the simplest line to work with is the straight line. If oblique lines are used then it will be necessary to calculate the intersection points between the test lines such as A/A or B/B with each line making up the polygon boundary to see if it lies within its segment length. However if either vertical or horizontal test lines Av/Av or Bv/Bv or Ah/Ah or Bh/Bh are used then these tests can be simplified.

For vertical or horizontal test lines the only boundary segments which need to be considered are those which lie directly above or below the test point, or directly to the left or right sides of the point. These segments can be identified using a vertical line by a single conditional test of the form:

```
if ( (p1.x()-p0.x())*(p2.x()-p0.x()) < 0 )    // then the segment p1,p2 needs to be selected
```

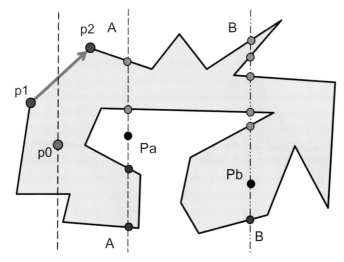

Figure 13.10 Vertical semi-infinite-line point-in-polygon testing

In figure 13.10 where the number of line segments (p1, p2) with both p1.y and p2.y greater than p0.y selected by this test, is odd, it classifies the point p0 as inside the polygon, and for an even number it indicates the point p0 lies outside the polygon. The actual co-ordinate of each crossing point, in these cases, does not need to be calculated. The equivalent approach can be implemented for horizontal test lines.

As usual, there are special cases that need to be handled to make this algorithm robust. The boundary segment test given above will miss cases where test lines pass through boundary segment end points. However, if these are included in the test:

if ((p1.x()-p0.x())*(p2.x()-p0.x()) <= 0) // then the segment p1,p2 needs to be selected

then double counting can occur in a way which will give false results for the point-in-polygon test. There are six ways in which this can occur for the vertical line testing procedure:

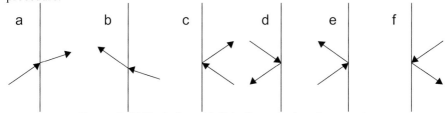

Figure 13.11 Vertical semi-infinite line point in polygon testing

The first two cases in Figure 13.11, cases a and b, are equivalent to a single crossing point, while the remaining cases must be treated as no crossing point or two crossing points. If a boundary line segment is co-linear with a test line then it will also be selected by the modified test given above, adding even further to the complexity of the counting task.

Considering the horizontal test-line example, shown in Figure 13.12, it is clear that all these awkward cases could occur in the same polygon.

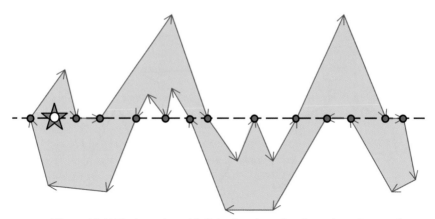

Figure 13.12 Horizontal semi-infinite line through polygon boundary vertices

A further complication occurs to spoil the initial simple implementation of this algorithm where a boundary line segment is not totally above or totally below the test point. When the test point lies inside a boundary-line-segment's min-max box-envelope, neither testing y values alone will be able to determine whether such line segments lie above the test point or below the test point, for vertical test lines, nor will testing x values determine whether line segments lie to the left or to the right of a test point for horizontal test lines.

In these cases it is necessary to test the boundary line segments in a different way to determine on which side a test point lies. This can be done using the orientation of the boundary line segment relative to the test point, employing a determinant to calculate the area of the triangle created by the three points, the test point and the two end points of the line segment. The sign of the determinant of the three points will give the required relationship.

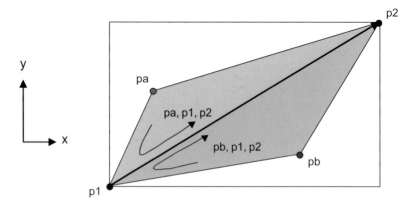

Figure 13.13 Points lying inside a boundary line segment box envelope

In Figure 13.13 it can be seen that changing a point p0 from one side of the line segment to the other causes the same order in the boundary list (p0, p1, p2), to change the geometry from a clockwise order to an anticlockwise order. This is matched by a change in the sign of the determinant from negative to positive.

$$\begin{vmatrix} x0 & y0 & 1 \\ x1 & y1 & 1 \\ x2 & y2 & 1 \end{vmatrix} = x0.(y1 - y2) - y0.(x1 - x2) + x1.y2 - y1.x2$$

$$= x0.(y1 - y2) - y0.(x1 - x2) + x1.y2 - x1.y1 - y1.x2 + x1.y1$$
$$= x0.(y1 - y2) - y0.(x1 - x2) - x1.(y1 - y2) + y1.(x1 - x2)$$
$$= (x0 - x1)(y1 - y2) - (y0 - y1)(x1 - x2)$$

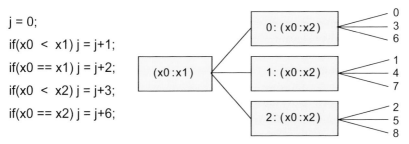

```
j = 0;

if(x0 < x1) j = j+1;

if(x0 == x1) j = j+2;

if(x0 < x2) j = j+3;

if(x0 == x2) j = j+6;
```

Table 13.2 Point line relationships:

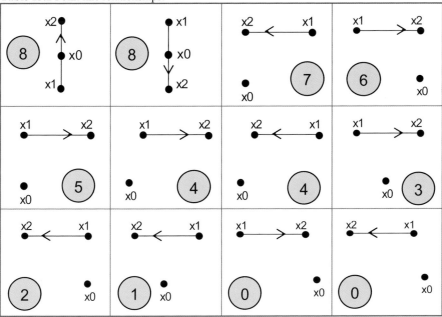

A program to implement these ideas can be written in the following way. The relationship between the test point and each boundary line segment is classified by a switch variable j, which is then used to handle the general and the special cases in an appropriate way. A similar classification can be set up for the relationship between the test point's y value and the boundary line-segment end-point y-values. However, it is simpler to handle the y relationships within each of the x-relationship cases given in table 13.2. These cases form three groups. For cases 2 and 5 the count increment has to be $+1$ or -1 as long as $(y1 > y0)$; also for cases 6 and 7: where $(y2 > y0)$. By incrementing the count in a way that depends on the direction in which the boundary meets the test line, the different cases shown in Table 13.2 can be handled in a self-consistent way. By making the increment 1 unit, where the test line passes through a boundary line segment's end point, and making its contribution -1 if the next segment changes direction and does not complete the crossing, its effect can be removed. For cases 1 and 3, as long as both y values are above the test point y value the count can be incremented or decremented by 2 depending on whether $(x1 < x2)$ or $(x2 < x1)$. Where the test y value lies between the y values of a segment's end points, then the determinant test given above has to be applied to decide the relationship, which is relevant. This means that the final count value has to be divided by 2 and then an odd-even test will determine whether the test point lies inside the polygon or not.

```
Public boolean contains(Point pp, boolean boundary){      // point in polygon test
    double x0, x1, x2, y0, y1, y2, k;   int inc = 1, val=0;
    if(this.length == 0) return false;
    x0 = pp.xd();   y0 = pp.yd();
    for(int i = 0; i < this.length –1; i++;){
        x1 = p[i].xd( );  y1 = p[i].yd( );  x2 = p[i+1].xd( );  y2 = p[i+1].yd( );
        if( x1 <  x2)    inc = 1;  else inc = -1;
        j=0;
        if( x0 <  x1)     j=j+1;
        if( x0 == x1)    j=j+2;
        if( x0 <  x2)    j=j+3;
        if( x0 == x2)    j=j+6;
        switch ( j ){
        case 2:case 5:
            if (y1 > y0) val = val+inc; else if (y1 == y0)  return (boundary); break;
        case 1: case 3:
            if ((y0<y1)&&(y0<y2)) val = val+inc+inc;
            else { k = ((x0-x1)*(y0-y2)-(x0-x2)*(y0-y1))*inc;
                   if(k<0.0) val = val+inc+inc; else if( k==0.0) return(boundary);
            } break;
        case 6: case 7:
            if (y2> y0)val = val+inc; else if (y2 == y0)  return boundary; break;
        case 8:  if ((y0 – y1)*(y0 – y2)<=0) return boundary;
        }
    }
    if(( val / 2) % 2 == 0) return false;       // AA
    return true;
}
```

The extra Boolean parameter *"boundary"* is included to allow any point lying on the boundary to be selectively classified as inside or outside the polygon depending on the nature of the application.

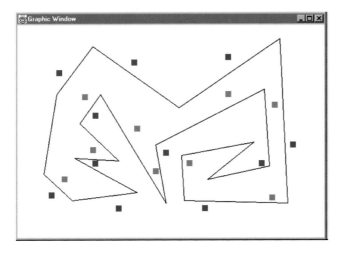

Figure 13.14 "Semi-infinite line" point in polygon with a non self-crossing boundary

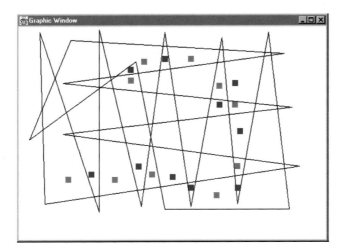

Figure 13.15 "Semi-infinite line" point in polygon with a self-crossing boundary

If this algorithm is applied to a polygon with a self-crossing boundary then it may not always give the required result. Where a polygon folded in three dimensions is projected onto a display surface it is possible for its boundary to become self-crossing. It would be convenient to be able to interactively select such a projected polygon using a point in polygon test. An alternative approach, which makes this task possible, is to use a winding number algorithm rather than the semi-infinite line algorithm.

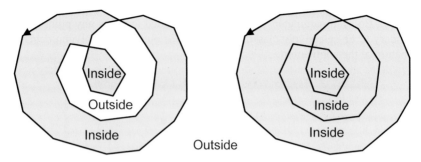

Figure 13.16 Points lying inside a self-crossing boundary line

The winding number can be measured by the angle a line following a point on the boundary sweeps out pivoting about the test point. In Figure 13.17 it can be seen that for a simple convex polygon where the point is inside it, the angle is 2π, but where the point lies outside the polygon some of sweep is clockwise some is anti clockwise giving positive and negative angles adding up to 0. Calculating angles is computationally expensive, however this approach can be simplified to a process, which counts the number of revolutions more directly.

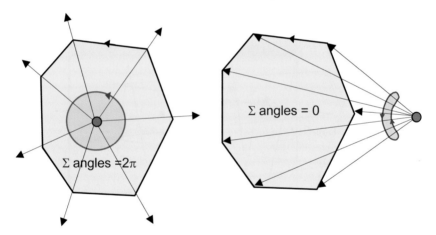

Figure 13.17 Boundary angle sum subtended at the test point

If all the vertices of the polygon are translated to make the test point lie at the origin of the co-ordinate system, each boundary point can be classified by the quadrant of the co-ordinate system, which it lies within. The simplest way to do this is to use the signs of the x and y values of the point's co-ordinate. The rotations can then be counted by the number of complete quadrant sequences that a point passes through, following the boundary from its beginning to its end.

If a boundary point starts in quadrant 1 then progresses to quadrant 2, through 3 and 4 back to 1 again then one rotation can be counted as shown in Figure 13.18. The difficulty with this process occurs when a boundary segment passes diagonally from quadrant 1 to quadrant 3, or from quadrant 2 to quadrant 4, or the reverse.

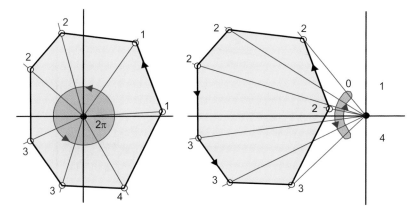

Figure 13.18 Winding number approach using quadrants

These cases correspond to the semi-infinite line algorithm where a test point lies within the min-max box envelope of a boundary line segment. It is not possible without an extra test to tell on which side of the line the test point lies. In the case of the winding number count a step from 1 to 3 has to be tested to determine whether the sequence is 1, 2, 3 or is the reverse rotation 1, 4, 3. Both of these cases can be resolved by the same determinant-based calculation.

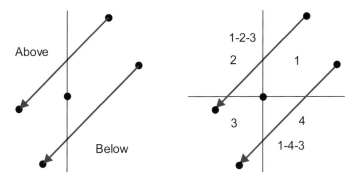

Figure 13.19 Tests needed to define which side of a test point a line segment lies

It is not necessary to write a different program to implement this approach. Treating the boundary line segments as moving in a positive or negative direction to solve the special cases shown in Figure 13.11 depends on the same information that is employed using the quadrant classification approach.

All that needs to be changed in the previous code is the final test. Changing code AA to BB will produce the required change of behaviour.

```
if(( val / 2) % 2 == 0) return false;      //AA
return true;

if( val == 0) return false;                //BB
return true;
```

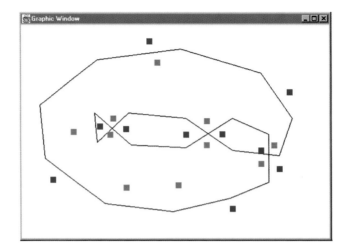

Figure 13.20 Point in polygon using the semi-infinite line algorithm

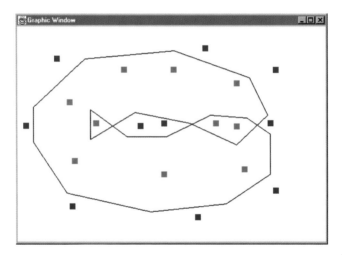

Figure 13.21 Point in polygon using the winding number algorithm

In figure 13.21 the result of a change to the winding number approach can be seen. This algorithm is useful for hidden line removal tests, the boundary given in this display represents the outer edge of a projection of a torque shaped ring. The area where the ends of the torque overlap is "inside" the projected image of the object, whereas points in the centre of the ring lie outside the object in its image.

Line on Line

The most common relationship test between two lines is probably to determine whether they intersect or not. This test is usually between two line segments and is executed in two steps. The first step tests the lines to see if they are in the same neighbourhood by testing their box envelopes to see if they overlap. The second step, if they do, then calculates the intersection point.

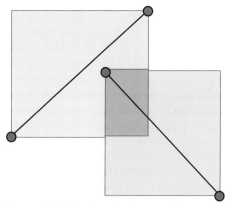

There are a variety of ways of programming the overlapping box test. One approach is to take the extreme points of one box and see if they lie between the extreme points of the other box. This allows the point-in-box test given above, to be used as a building block. The test has to be applied both in the x-axes and y-axes directions. Three separate point-in-box tests are needed in each direction to cover the four relationships shown in Figure 13.23.

Figure 13.22 Overlapping box envelopes

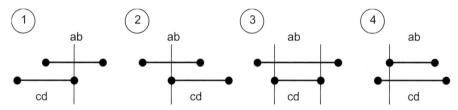

Figure 13.23 Extreme point relationships for overlap

if(((a-c)*(b-c)<=0)||((a-d)*(b-d)<=0)||((c-a)*(d-a)<=0)){overlap exists}

The first two point in box tests cover the first three relationships in Figure 13.23 however an extra test is needed to cover the final relationship shown in case 4.

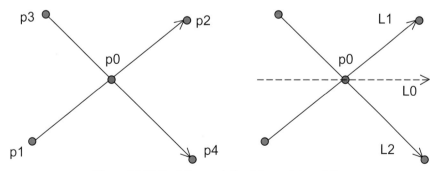

Figure 13.24 Dual linear relationships: points and lines

The second step, finding the intersection point of two line segments, involves calculating the equations of the two lines and then solving the resulting pair of simultaneous equations. Programming this task can becomes complicated by the special cases dealing with vertical lines, parallel lines and other conditions that can create divide by zero errors, if the steps of the algorithm are not carefully planned. However, a robust approach can be built up round the use of homogeneous co-ordinates, determinants, and the duality of points and lines in two dimensions. In Figure 13.24 p0 will be linearly dependent on p1 and p2 as well as p3 and p4, so the two determinants containing these line end points and p0 will be zero.

$$\begin{vmatrix} p0 \\ p1 \\ p2 \end{vmatrix} = \begin{vmatrix} x0 & y0 & 1 \\ x1 & y1 & 1 \\ x2 & y2 & 1 \end{vmatrix} = 0 \qquad\qquad \begin{vmatrix} p0 \\ p3 \\ p4 \end{vmatrix} = \begin{vmatrix} x0 & y0 & 1 \\ x3 & y3 & 1 \\ x4 & y4 & 1 \end{vmatrix} = 0$$

$$\begin{vmatrix} y1 & 1 \\ y2 & 1 \end{vmatrix}.x0 - \begin{vmatrix} x1 & 1 \\ x2 & 1 \end{vmatrix}.y0 + \begin{vmatrix} x1 & y1 \\ x2 & y2 \end{vmatrix} = 0 \quad \equiv \quad a1.x0 + b1.y0 + c1 = 0$$

$$\begin{vmatrix} y3 & 1 \\ y4 & 1 \end{vmatrix}.x0 - \begin{vmatrix} x3 & 1 \\ x4 & 1 \end{vmatrix}.y0 + \begin{vmatrix} x3 & y3 \\ x4 & y4 \end{vmatrix} = 0 \quad \equiv \quad a2.x0 + b2.y0 + c2 = 0$$

Similarly, in the dual arrangement, the arbitrary line L0 will linearly depend on the other two lines L1 and L2 passing through the same point p0.

$$\begin{vmatrix} L0 \\ L1 \\ L2 \end{vmatrix} = \begin{vmatrix} a0 & b0 & c0 \\ a1 & b1 & c1 \\ a2 & b2 & c2 \end{vmatrix} = 0$$

$$a0.\begin{vmatrix} b1 & c1 \\ b2 & c2 \end{vmatrix} - b0.\begin{vmatrix} a1 & c1 \\ a2 & c2 \end{vmatrix} + c0.\begin{vmatrix} a1 & b1 \\ a2 & b2 \end{vmatrix} = 0 \quad \equiv \quad a0.x0 + b0.y0 + c0.w0 = 0$$

$$x0 = \begin{vmatrix} b1 & c1 \\ b2 & c2 \end{vmatrix}$$

$$y0 = -\begin{vmatrix} a1 & c1 \\ a2 & c2 \end{vmatrix}$$

$$w0 = \begin{vmatrix} a1 & b1 \\ a2 & b2 \end{vmatrix}$$

If $w0 \neq 0$ $\quad x = x0/w0;$ $\qquad y = y0/w0;$

This means that the final intersection point p0 can be calculated by converting the homogeneous co-ordinate back to a conventional two-dimensional co-ordinate by dividing through by $w0$. The advantage of this approach is that this is the only division in the whole sequence. If $w0$ is zero, then it can be trapped, and it indicates that the point lies at infinity. This occurs when the original line segments are parallel or co-linear, or if either of the line segments is of zero length.

Once the intersection point has been calculated it still has to be tested for being within the length of each line segment. Though the envelopes of the lines overlap the lines themselves may not intersect, as Figure 13.22 illustrates.

An alternative approach which initially appears to involve more calculations, but which provides a more versatile sequence of steps, can be constructed as follows. Each end point is classified as inside or outside the other line segment -- treated as an infinite oriented line. Evaluating the determinant constructed from the test point and the end point co-ordinates of the other line segment will provide this classifying value "k". These values will be of opposite signs for pairs of test point on opposites sides of the line. Consequently if the product of the two determinant values for each line segment is negative for both line segments, then the two line segments must cross within their length.

If these determinant values are labelled k1, k2, ka and kb then the line relationships shown in Figure 13.25 can be discriminated by examining their relative k-values.

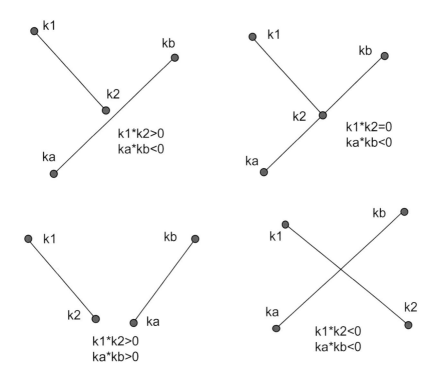

Figure 13.25 Classifying line segment end points

The advantage of this process is that if the line segments are found to cross within their length the values of k1, k2, ka and kb can be used to calculate the intersection point directly. Because the determinant value is proportional to the area of a triangle, relationships based on similar triangles can be used to derive the crossing point in the way illustrated in Figure 13.26.

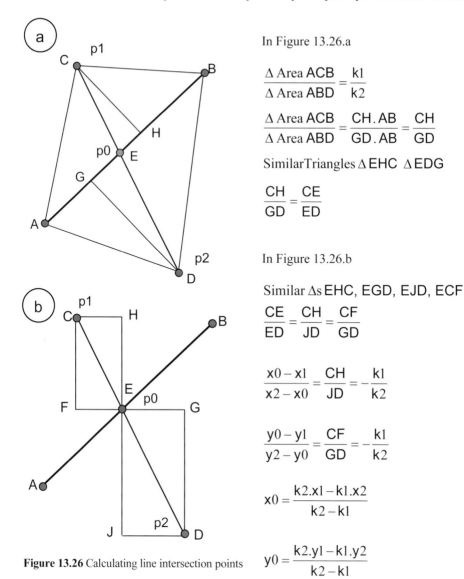

In Figure 13.26.a

$$\frac{\Delta \text{ Area ACB}}{\Delta \text{ Area ABD}} = \frac{k1}{k2}$$

$$\frac{\Delta \text{ Area ACB}}{\Delta \text{ Area ABD}} = \frac{CH.AB}{GD.AB} = \frac{CH}{GD}$$

Similar Triangles $\Delta\,EHC\;\;\Delta EDG$

$$\frac{CH}{GD} = \frac{CE}{ED}$$

In Figure 13.26.b

Similar $\Delta s\,EHC,\ EGD,\ EJD,\ ECF$

$$\frac{CE}{ED} = \frac{CH}{JD} = \frac{CF}{GD}$$

$$\frac{x0-x1}{x2-x0} = \frac{CH}{JD} = -\frac{k1}{k2}$$

$$\frac{y0-y1}{y2-y0} = \frac{CF}{GD} = -\frac{k1}{k2}$$

$$x0 = \frac{k2.x1-k1.x2}{k2-k1}$$

Figure 13.26 Calculating line intersection points

$$y0 = \frac{k2.y1-k1.y2}{k2-k1}$$

The advantage of this approach is that not only can these "k" values be used to calculate the crossing point of two lines but also when these values are zero they signal special cases that may need special handling. For example when one of these values is zero it means that the corresponding end point is the intersection point with the other line. The signs and values of these "k" values can be used to identify most of the relationships between line segments that require some form of special treatment. For example where the lines are collinear, parallel or where one of the lines has zero length, which can occur when some other process is automatically generating line segments, in a sequence of operations.

Polyline on Polyline

As with points, a better approach is needed, when handling large sets of potentially interacting line segments, than executing all the possible pair-wise tests between individual line segments. Sets of line segments can occur as separate line segments or as polyline arcs, where line segments are linked in chained vertex sequences. The same approach should be applicable to both these data structures. The first attempt to create an efficient algorithm for polylines was developed for the OBLIX program to draw surface lines and contour lines onto cartographic block models. The aim was to produce a drawing, in vector format rather than in the pixel grid format of chapter 9, so that it was easy to redraw at different scales. This is more efficient where output might be directed to different display devices. The raster size of the high-resolution printer being different from that employed in CRT or LCD display screens.

The task was to find the intersection points between surface based line segments and the obscuring polygon, in the way illustrated in Figure 13.27. All the surface based line-segments were linked together into a single polyline, alternating real and virtual segments. This allowed each line segment to be compared with only those segments of the obscuring polygon lying directly above or below it. (P1, P2) only needs to be tested against the sequence Pa, Pb, Pc. (P2, P3) in the opposite direction simply requires the same set to be tested in reverse order. By controlling the testing using the current line segment direction and end point relationships, many wasted comparisons were avoided. The data in this case were generated in a conveniently localised manner. To generalise this approach it is necessary to pre-process polyline data to gain the same advantage.

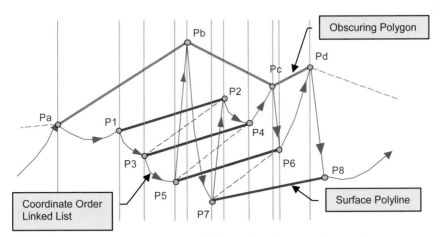

Figure 13.27 Sort the polyline vertices into co-ordinate order

The first step is to sort the vertices of the polylines into co-ordinate order. As before this sequence has to be set up as a linked list, to avoid destroying the data structure representing the lines. The second step is to take each line segment in order, as it is located following this ordered list. These line segments are then compared with all the line segments, whose leftmost points are encountered following the ordered list moving from the first point of the test line segment to its second point. In Figure 13.27 the testing sequence becomes:

(Pa,Pb) tested against (P1,P2), (P2,P3),(P3,P4),(P4,P5),(P5,P6)
(P1,P2) tested against (Pb,Pc)
(P3,P2) tested against (Pb,Pc)
(P3,P4) tested against (Pb,Pc)
(P5,P4) tested against (Pb,Pc)
(P5,P6) tested against (Pb,Pc), (Pc,Pd)
(Pb,Pc) tested against (P7,P6),(P7,P8)
(P7,P6) tested against (Pc,Pd)
(P7,P8) tested against (Pc,Pd)

This ordered processing is a general approach used for many geometric tasks called "plane sweep" processing, because it works as though a line or plane is swept through an image or object space, and every object it encounters is processed in the order in which it finds them. It localises comparisons, and consequently reduces the number of unnecessary tests. However, although it helps, each line segment will still have to be compared with all other lines that cross the vertical band above and below it. An alternative approach with an even better power to reduce unnecessary comparisons is given by binary sub-division of the image space firstly in the x-direction and then in the y-direction. This process continued recursively until there are only two line segments in each area, will very efficiently localise tests in the way shown diagrammatically in Figure 13.28. The coloured rectangles are where two line segments from different polylines are left in the quadrant. Where the colour is purple the lines cross where it is pale yellow the lines do not cross. The uncoloured quadrants contain line segments from only one polyline set, or no line segments.

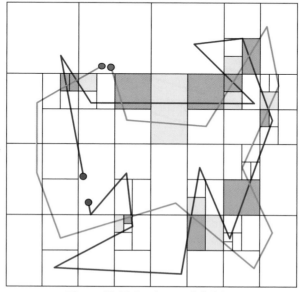

Figure 13.28 Quad tree subdivision

```
public class PolylineIntersectonExample{
    static TextWindow IO = new TextWindow(15,405,600,110);
    static DisplayWindow dW =new DisplayWindow(IO,15,5,600,400,Color.white);
    public static void main(String[] args){
        double xmin,xmax,ymin,ymax;
        Polyline p1 = new Polyline(IO,dW);   p1.getPolyline();
        Polyline p2 = new Polyline(IO,dW);   p2.getPolyline();
        List pointList = p1.intersect( p1, p2);
        if(pointList!= null){
            ListElement ref = pointList.start;
            while(ref!=null) {dW.plotRectangle((Point)ref.object,3,Color.red); ref= ref.right;}
        }
        for(int j=1;j<p1.length;j++) dW.plotLine(p1.p[j-1],p1.p[j],Color.black);
        for(int j=1;j<p2.length;j++) dW.plotLine(p2.p[j-1],p2.p[j],Color.black);
    }
    public List intersect(Polyline p1,Polyline p2){
        List pc = new List();  List pd = new List();
        for(int i=1;i<p1.length;i++){
            Line ln = new Line();ln.p1=new Point(2); ln.p2=new Point(2);
            p1.p[i-1].c("->",ln.p1);p1.p[i].c("->",ln.p2);  pc.append(ln);
        }for(int i=1;i<p2.length;i++)  {
            Line ln = new Line(); ln.p1=new Point(2);  ln.p2=new Point(2);
            p2.p[i-1].c("->",ln.p1); p2.p[i].c("->",ln.p2);  pd.append(ln);
        }Box bx1 = new Box(p1); Box bx2 = new Box(p2);
        Box bx  = new Box(bx1,bx2);  bx.draw(dW,Color.blue);
        List cP = new List();  return cP = divide(pc,pd,bx,cP,0,2);
    }
    private void partition(List pa,List pa1,List pa2,NTuple nt){
        if ((pa==null)||(pa.length==0)) return;
        ListElement ref = pa.start;
        while(ref!=null){
            Line a = (Line)ref.object;
            Line b = new Line(); b.p1=new Point(2); b.p2=new Point(2);
            a.p1.c("->",b.p1); a.p2.c("->",b.p2); Point p1= a.p1; Point p2= a.p2;
            double k1= nt.n[0]*p1.xd()+nt.n[1]*p1.yd()+nt.n[2];
            double k2= nt.n[0]*p2.xd()+nt.n[1]*p2.yd()+nt.n[2];
            if(k1*k2<0){
                Point p0 = new Point(2);
                p0.x("=",(p2.xd()*k1-p1.xd()*k2)/(k1-k2));
                p0.y("=",(p2.yd()*k1-p1.yd()*k2)/(k1-k2));
                p0.c("->",a.p2);  p0.c("->",b.p1);
                if(k2<0)pa1.append(b);else pa2.append(b);
            }
            if(k1<=0)pa1.append(a);else pa2.append(a);
            ref=ref.right;
        }
    }
}
```

```
public List divide(List pa,List pb,Box bx,List cP,int level,int alt){
    if((pa==null)||(pb==null)) return cP;  if((pa.length== 0)||(pb.length==0))  return cP;
    double x = (bx.minP.xd()+bx.maxP.xd())/2;  double y = (bx.minP.yd()+bx.maxP.yd())/2;
    bx.draw(dW,Color.green);   level = level+1; ListElement ref1= pa.start, ref2= pb.start;
    if((pa.length==1)&&(pb.length==1)){
        Point p = new Point(2);
        if(lineCross((Line)pa.start.object,(Line)pb.start.object,p)) cP.append(p);
        return cP;
    }if(level==40){
        while(ref1!=null){
            ref2= pb.start;
            while(ref2!=null){
                Point p = new Point(2);
                if(lineCross((Line)pa.start.object,(Line)pb.start.object,p))  cP.append(p);
                ref2=ref2.right;
            }ref1=ref1.right;
        }return cP;
    } Box bx1= new Box(bx); Box bx2= new Box(bx);  NTuple nt = new NTuple(3);
    switch(alt){
    case 1:  bx1.maxP.x("=",x); bx2.minP.x("=",x);
            nt.n[0]= 1; nt.n[1]= 0; nt.n[2]= -x; alt=2;    break;
    case 2: bx1.maxP.y("=",y); bx2.minP.y("=",y);
            nt.n[0]= 0; nt.n[1]= 1; nt.n[2]= -y; alt=1;    break;
    } List pa1=new List();List pa2=new List();  List pb1=new List();List pb2 = new List();
    partition(pa,pa1,pa2,nt);   partition(pb,pb1,pb2,nt);
    cP= divide(pa1,pb1,bx1,cP,level,alt);  cP= divide(pa2,pb2,bx2,cP,level,alt);
    return cP;
}
private boolean lineCross(Line a,Line b,Point p){
    Point p1= a.p1,  p2=a.p2, pa= b.p1,  pb=b.p2;
    NTuple nt1 = new NTuple(3); NTuple nt2 = new NTuple(3);
    nt1.n[0] =  p1.yd()- p2.yd();   nt1.n[1] =  p2.xd()- p1.xd();
    nt1.n[2] =  p1.xd()*p2.yd()-p1.yd()*p2.xd();
    double k1= pa.xd()*nt1.n[0] + pa.yd()*nt1.n[1] + nt1.n[2];
    double k2= pb.xd()*nt1.n[0] + pb.yd()*nt1.n[1] + nt1.n[2];
    nt2.n[0] =  pa.yd()- pb.yd();   nt2.n[1] =  pb.xd()- pa.xd();
    nt2.n[2] =  pa.xd()*pb.yd()-pa.yd()*pb.xd();
    double k3= p1.xd()*nt2.n[0]+p1.yd()*nt2.n[1]+nt2.n[2];
    double k4 =p2.xd()*nt2.n[0]+p2.yd()*nt2.n[1]+nt2.n[2];
    if(k1==0) k1=.00004; if(k2==0) k2=.00004; if(k3==0) k3=.00004; if(k4==0) k4=.00004;
    if((k1*k2<=0)&&(k3*k4<=0)){
        p.x("=",(p2.xd()*k3-p1.xd()*k4)/(k3-k4));    // temporary fix to
        p.y("=",(p2.yd()*k3-p1.yd()*k4)/(k3-k4));    // avoid k=0 cases
        return true;
    }return false;
}
}
```

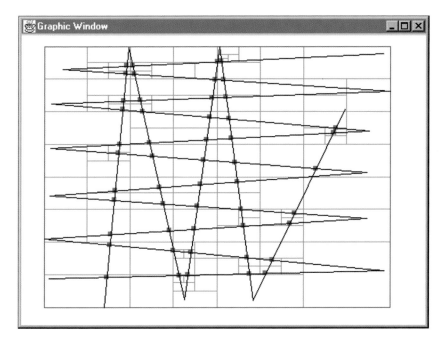

Figure 13.29 Binary subdivision calculating intersection points between polylines

Line and Line Sets on Polygons

In this section the interaction between lines and polygons is examined using two forms of line representation and two polygon representations. Firstly a single line is interacted with a single rectangle clipping it so that it is only visible in the rectangle's interior. Secondly a set of simple straight lines is interacted with a simple polygon to give a generalised form of shading. A naïve approach to this task is to repeatedly interact single lines with the polygon. However, where the lines are parallel this algorithm can be made more efficient. Finally, replacing the simple line by a polyline allows the subdivision algorithms described above to be used to calculate new boundary and line intersection points, which in turn links to the next chapter examining overlapping polygons and rectangles.

Line Rectangle Clipping

An elegant algorithm developed by Ivan Sutherland for the SKETCHPAD system in 1963 addressed the problem of clipping large line files, to fit into the refresh memory of an interactive display for a CAD system. The problem was that most of the lines in a large drawing would not be in the field of view, and only a few visible lines would need to be clipped. The majority of lines could be rejected, most of the remainder could be drawn as they were defined, and only a few would need modifying. The algorithm was therefore designed to reject lines lying outside the rectangle of the display space with minimum testing, and only spends resources on lines with a high chance of crossing the edges of the rectangle.

The first step was to divide the display space up into nine regions by projecting the four edges out to infinity. The end points of line segments were then classified by their relationship to these four lines. A four-bit value was set up each bit associated with an edge of the rectangle. If a point was outside a particular edge its bit position was set 1 otherwise 0. The two values generated for a line segment were then used together to totally reject the line, totally accept it or pass it on for further processing and testing. If these bit values when "*and-ed*" together gave other than zero, the line could be rejected because both points would be outside the same edge, and could not therefore appear in the window. If both of these values were zero then both points would be inside the rectangle, so the line could be drawn in directly, this could be established by "*or-ing*" the two values together, and a non zero result would again indicate further testing was necessary. Any line that failed these two tests would then have to be tested against each edge in turn to see if it needed to be clipped to that edge. The remaining segment after clipping could then be drawn in.

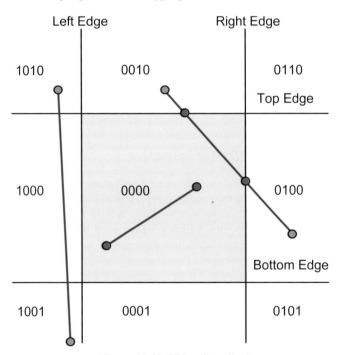

Figure 13.30 Widow line clipping

```
private int classify(int x, int y, Box r){
    int j=0;
    if(x < r.minP.xi())  j=j+1;
    if(x > r.maxP.xi())  j=j+2;
    if(y < r.minP.yi())  j=j+4;
    if(y > r.maxP.yi())  j=j+8;
    return j;
}
```

```
public boolean clipBox(Box cl,Point p1,Point p2){
    i = classify(p1.xi(),p1.yi(),cl);   j = classify(p2.xi(),p2.yi(),cl);
    if((i==0)&&(j==0))return true;
    if((i&j&15)!=0)return false;
    if(edgeClip(p1,p2,cl.minP.xi(), 1,false))     return false;
    if(edgeClip(p1,p2,cl.maxP.xi(),-1,false))     return false;
    if(edgeClip(p1,p2,cl.minP.yi(), 1, true))     return false;
    if(edgeClip(p1,p2,cl.maxP.yi(),-1, true))     return false;
    return true;
}
private boolean edgeClip(Point p1,Point p2,double e,double d,boolean up){
    double x1,x2,y1,y2,k1,k2;
    if(up){x1= p1.yd(); x2=p2.yd(); y1=p1.xd(); y2=p2.xd();}
    else{x1= p1.xd(); x2=p2.xd(); y1=p1.yd(); y2=p2.yd();}
    k1= x1-e;  k2=x2-e;
    if(k1*k2<=0){
        if(k1*d<0){
            y1=(k2*y1-k1*y2)/(k2-k1);
            if(up){p1.y("=",e);p1.x("=",y1);}else {p1.x("=",e); p1.y("=",y1);}
        }if(k2*d<0){
            y2=(k2*y1-k1*y2)/(k2-k1);
            if(up){p2.y("=",e);p2.x("=",y2);}else {p2.x("=",e); p2.y("=",y2);}
        }
    } else if(k1*d<0) return true;
    return false;
}
public static void main(String[] args){
    double xmin, xmax, ymin, ymax;
    IO.writeString("Please enter diagonal window corners\n");
    Point p1 = dW.getCoord();   Point p2 = dW.getCoord();
    if (p1.xd()<p2.xd()){xmin = p1.xd(); xmax= p2.xd();}
    else {xmin = p2.xd();xmax= p1.xd();}
    if (p1.yd()<p2.yd()){ymin = p1.yd(); ymax= p2.yd();}
    else { ymin = p2.yd(); ymax= p1.yd();}
    p1.x("=",xmin); p2.x("=",xmax);  p1.y("=",ymin); p2.y("=",ymax);
    Box bx = new Box(p1,p2); bx.draw(dW,Color.blue);
    IO.writeString("Please enter 10 lines for clipping\n");
    for(int j=0;j<10;j++){
        p1 = dW.getCoord();  p2 = dW.getCoord();
        Point p3 = new Point(2);  Point p4 = new Point(2);
        p3.c("<-",p1);  p4.c("<-",p2);
        if(dW.clipBox(bx,p1,p2)){
            dW.plotLine(p1,p3,Color.green); dW.plotLine(p4,p2,Color.green);
            dW.plotLine(p1,p2,Color.black);
        }else dW.plotLine(p3,p4,Color.magenta);
    }
}
```

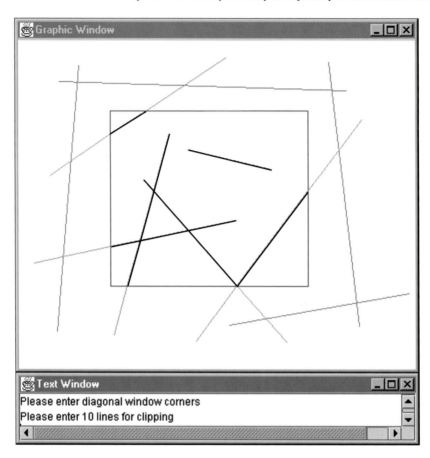

Figure 13.31 Widow line clipping: black inside, magenta rejected, green clipped

General Polygon Shading

An alternative to colour fill for shading polygons is the more traditional line drawing technique of hatch shading. This can be built up by overlaying grids of parallel lines. Clipping lines where they cut the boundary of a polygon will give the hatch shading. However, because the shading lines are parallel they can be generated by the same approach used for polygon fill. This algorithm depends on the lines being horizontal or vertical in order to reorder the end points of the shading lines, obtained by walking round the polygon boundary in grid-spaces steps, into a scan line sequences. The same sorting principle can be applied to give a more general line-shading scheme, for vertical or horizontal shading lines applied to complex polygon shapes. However hatching often requires oblique shading lines. Fortunately it is possible to allow the kernel of the shading algorithm to use the same approach by rotating the polygon in the way shown in Figure 13.32 so that the oblique lines become horizontal. These shading line segments will all have the correct size and interrelationship with the polygon boundary for oblique shading, so all that needs to be done is to rotate them back to give the required orientation and position in the final display.

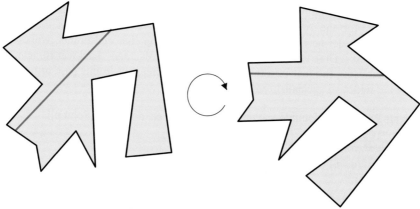

Figure 13.32 Oblique shading-lines

The bucket sorting approach was used to reorder the scan-line fill end points, as a fast linear-complexity, sorting algorithm. Its limitation was that it was only convenient to work with single left and right edge points. The general approach being studied in this section, required to handle any shape, will need to use a more standard co-ordinate sorting-algorithm. This at first sight will be rather less efficient. However, walking round the polygon boundary to generate the end points of the shading lines: most of these points will already be in co-ordinate order. If these points are placed into an ordered link list as they are generated then a form of merge sort will give very close to a linear complexity sorting performance. In the procedure *insertInOrder* given below each new point t is compared to the last point entered into the ordered list and placed before or after it depending on whether the current boundary line segment is orientated upwards or downwards.

```
public ListElement insertInOrder(ListElement l, NTuple t, NTuple i){
    if(l!= null){
        if (((NTuple)l.object).compareTo(t,i)>0){        // t smaller than l
            while((l.left!=null)&&(((NTuple)l.left.object).compareTo(t,i)>0)){
                l=l.left;}
            l = this.insertBefore(l,t);
        }else{                                            // t larger then l
            while((l.right!=null)&&(((NTuple)l.right.object).compareTo(t,i)<0)){
                l=l.right;}
            l = this.insertAfter(l,t);
        }
    }else l = this.append(t);
    return l;
}
```

The rotation and reverse rotation can be combined with a scaling transformation in the y direction to make the shading line spacing a unit step. It also is convenient to have an offset translation from the origin to adjust the relative placement of different grids. In practice, it is useful to add a large translation forwards and backwards, to maintain all the edge-point co-ordinates in the first quadrant, in other words so both x and y values for the rotated polygon remain positive. This ensures the integer clipping used to define the y ordinates of the shading line end-points, works in a consistent way. The two transformations, forwards and backwards, can be implemented using the matrix routines presented earlier, but in this case are simpler to carry out with two specialist routines.

```
private Point ftransf(double spc,double sa, double ca, double off,Point p){
    Point pp = new Point(2);
    pp.x("=",p.xd()*ca+p.yd()*sa+2000);        // +2000 all points to the first quadrant
    pp.y("=",(p.yd()*ca-p.xd()*sa)/spc+off+2000);
    return pp;
}       // forward transformation:  +2000 take all points to the first quadrant
private Point btransf(double spc,double sa, double ca, Point p){
    Point pp = new Point(2);
    pp.x("=",(p.xd()-2000)*ca-sa*(p.yd()f-2000)*spc);
    pp.y("=",(p.yd()-2000)*ca*spc+(p.xd()-2000)*sa);
    return pp;
}       // backward transformation:  -2000 all points returned to original space
```

In the transformed space used for calculating the shading lines the grid lines will fall on integer values of y. The only problem occurs when the end points of boundary line segments also lie on the same grid lines. Duplicate intersection points of the sort already discussed in the case of point-in-polygon testing make the pairing of beginning and end points for line shading more complicated.

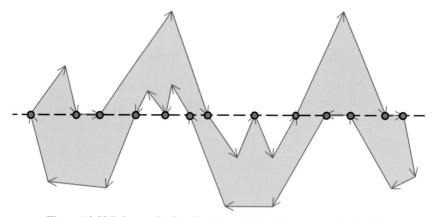

Figure 13.33 Polygon shading-line through boundary segment end points

If all the intersection points along the shading line are coded with the orientation of the boundary-line segment which produced them, then an accurate shading line can be calculated. For example a simple intersection point can be tagged with a $(+ -)$ or $(- +)$ property, whereas a compound intersection point can be made up from a sequence of points tagged $(+ 0) (0 0) (0 0) (0 -)$ for example or $(+ 0) (0 +)$. This labelling allows tangents to be separated from true crossing points.

For display purposes there is a simpler pragmatic solution to this task. If the end point of a boundary line segment is found to lie on a shading line, moving the boundary line end point a small Δ distance upwards, small enough not to visually affect the display will produce correct self-consistent results. This approach can be used to resolve similar problems in several corresponding cases however it is not always a good idea. Sometimes it merely moves the difficulty elsewhere in the processing. Figures 13.34 and 13.35 show the use of a single grid and two overlaid grids. The shading is generated by the method *hatch* included within the *polygon class*. Multiple calls to the same method with different parameters will give multiple grids for the same polygon boundary.

```
public ListOfTuples hatch(double sp, double angle, double off){
    ListOfTuples lnt= new ListOfTuples();
    Point [] pp = new Point[this.length]; NTuple index = new NTuple(3);
    index.n[0]=0; index.n[1]=2; index.n[2]=1;    // y, x co-ordinate order
    double a = 3.1415962*angle/180;
    double sa = Math.sin(a), ca= Math.cos(a);

    for(int i=0;i<this.length;i++){                //transform boundary forward
        pp[i]= this.ftransf(sp, sa, ca, off, this.p[i]);
        if(pp[i].yi()== pp[i].yd())pp[i].y("=", pp[i].yd()+0.001);
    }
    ListElement lst = null; int d;
    for(int i=1; i < this.length;  i++){
        Point pa= pp[i-1]; Point pb = pp[i];
        double y =0, ystep=0, x=0, xstep=0;
        if((pb.yi()-pa.yi()>=1) || (pb.yi()-pa.yi()<= -1)){
            if (pb.yd()>pa.yd()){y = pa.yi()+1; d=1;}
            else {y = pa.yi();d= -1;}
            x = ((pb.xd()-pa.xd())*y+(pa.xd()*pb.yd()-pa.yd()*pb.xd()))
            x = x/(pb.yd()-pa.yd());
            xstep = (pb.xd()-pa.xd())/(pb.yd()-pa.yd());
            while(d*y<d*pb.yd()){
                Point p = new Point(2);
                p.x ("=", x); p.y ("=", y);  y=y+d;  x=x+xstep*d;
                lst = lnt.insertInOrder(lst, p, index);  // y, x co-ordinate order
            }
        }
    }
    ListElement ref = lnt.start;       //transform back shading line points
    while(ref!=null){
        Point ppp = (Point)ref.object;
        ppp =this.btransf(sp,sa,ca,off,(Point)ref.object);
        ref.object= ppp;
        ref=ref.right;
    } return lnt;
}
```

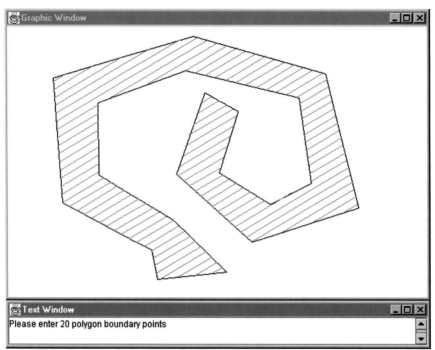

Figure 13.34 Shading using parallel lines

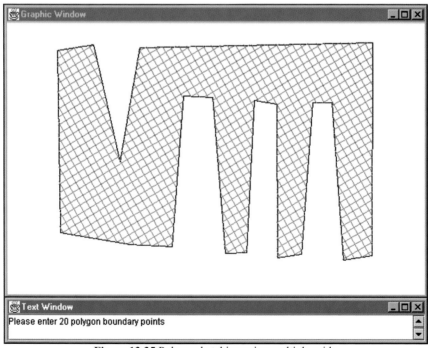

Figure 13.35 Polygon hatching using multiple grids

14

Spatial Relationships Overlap & Adjacency Polygon on Polygon

Introduction

In this chapter the general topic is an exploration of the relationships of overlap and adjacency between areas, highlighted yellow in Table 14.1.

Table 14.1 Basic Spatial Relationships

Objects	Point	Line	Area	Volume
Point	Point on Point	Point in Line	Point in Area	Point in Volume
Line		Line on Line	Line in Area	Line in Volume
Area			Area on Area	Surface in Volume
Volume				Volume on Volume

The algorithms analysed in this chapter developed from work in two application areas. The first was in a Geographic Information System, (GIS) cartographic subsystems. Using stored geographical information involved supporting many ways of combining and displaying spatial distributions in the form of maps. These ranged from presenting the complex spatial analysis provided by regional planning and urban simulation models, to drawing relatively traditional sieve maps, as precursors to more sophisticated spatial correlation studies.

A. Thomas, *Integrated Graphic and Computer Modelling*,
DOI: 10.1007/978-1-84800-179-4_14, © Springer-Verlag London Limited 2008

The second application was in computer aided design systems. A good example comes from the design and layout of printed circuit boards and integrated circuits. Where conducting paths were built up from rectangular and polygonal elements it was possible to test the layout to show that unintentional overlaps which would cause short circuits had not been created. Alternatively where a conducting path was made up from various materials, it was possible to create the union of their outlines and test for circuit continuity. In a different design area testing the sweep area of door and window swings against structural elements in an architectural plan could very quickly highlight errors that would be expensive to rectify, if left until construction, for discovery. By including envelopes around objects it is possible to check the nearness of objects to each other and include the necessary working space required to install or repair plant and pipe work, for example in narrow ducts.

Area on Area Relationships

Where areas are defined by boundary lines the *area-on-area* tests become extensions of the *polyline-on-polyline* tests already presented. Where one or more of the areas are rectangles with sides parallel to the coordinate axes, then different algorithms are needed. An interesting example of this case is the window clipping of standard polygons. Where its boundary line represents a polygon, this "line" can be clipped at the window edges. The polygon fill or the scan-line-fill extension of this approach can still be used to colour the portion of the polygon inside the window: each polygon-fill scan-line being clipped by the window boundaries. Where diagonal shading or hatching techniques are needed, then the "polygon" has to be clipped to have a complete boundary for the area that is visible within the window.

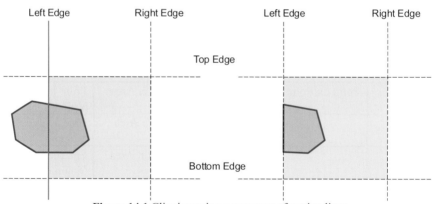

Figure 14.1 Clipping using a sequence of cutting lines

Polygon Window Clipping

In Figure 14.1 the first step of polygon window clipping is shown. A cutting edge is defined and the points of the polygon boundary are tested to see on which side of it they lie. Line segments with both end points inside the window can be plotted, or saved directly. Similarly line segments with both end points outside the window can be ignored. Any line segment with a point on each side of the cutting line can be clipped to give the section inside the window. Where the boundary is a closed polyline, processing its segments sequentially will allow a new boundary to be created as points from the original polygon are classified as inside or outside the window. This process will automatically enter the new points where the boundary is cut by the cutting edge to give a correct closed polyline for the new area. An interesting side effect of this ordered processing is shown in Figure 14.2 where all four edges of the window are processed in order, for a complex polygon that mostly lies outside the window.

Figure 14.2 Polygon clipping on window edges

Although the final area of the polygon becomes divided into two separate regions this algorithm still outputs a single boundary line. Sections of the new boundary coincide with each other as well as the edges of the window. It is quite possible to post process this new boundary to create independent loops for the resulting two areas. However, there is a useful bye-product to the original approach. In the case of some complex areas, retaining only the boundary sections visible inside a window, if they are representing holes in the original area, may create problems. In other words removing the extra information could result in the hole being shaded in as a solid area, instead of its surround. Figure 14.3 shows the output from this polygon-clipping algorithm followed by a polygon hatching operation to show the new area. The program used to create this display is given below the figure.

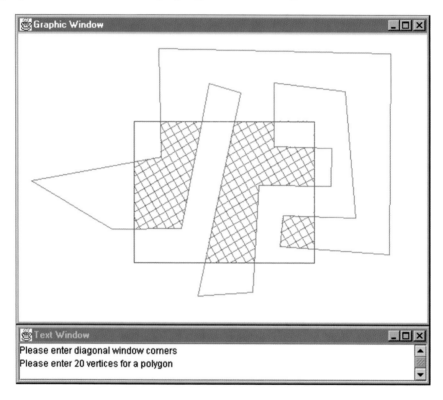

Figure 14.3 Polygon window clipping

```
public static void main( String[] args){
    double xmin,xmax,ymin,ymax;
    IO.writeString("Please enter diagonal window corners \n");
    Point p1 = dW.getCoord();    Point p2 = dW.getCoord();
    if (p1.xd()<p2.xd()){xmin = p1.xd();xmax= p2.xd();}
    else {xmin = p2.xd();xmax= p1.xd();}
    if (p1.yd()<p2.yd()){ymin = p1.yd();ymax= p2.yd();}
    else {ymin = p2.yd();ymax= p1.yd();}
    p1.x("=",xmin); p2.x("=",xmax); p1.y("=",ymin);  p2.y("=",ymax);
    Box bx = new Box(p1,p2); bx.draw(dW,Color.blue);
    IO.writeString("Please enter 20 vertices for a polygon \n");
    Polygon p = new Polygon(21);
    p.p[0] = dW.getCoord();
    for(int j=1;j<20;j++) {
```

```
                p.p[j] = dW.getCoord(); dW.plotLine(p.p[j-1],p.p[j],Color.red);
        }
        p.p[20] = new Point(2); p.p[20].c("<-",p.p[0]);
        dW.plotLine(p.p[19], p.p[20], Color.red);
        Polygon pp= p.clip(bx,p);
        double[][] htch = new double[][]{{10,-30,4},{10,60,4}};
        for(int i=0; i<2; i++){
            double space = htch[i][0], angle = htch[i][1], offset = htch[i][2];
            ListOfTuples ls = pp.hatch(space,angle,offset,IO);
            if(ls==null)IO.writeString("null list 1\n");
            else{
                ListElement ref = ls.start;
                if(ref==null)IO.writeString("null list 2\n");
                else while(ref != null){
                    Point pa = (Point)ref.object;  pa.y("=",pa.yd());pa.x("=",pa.xd());
                    ref= ref.right;
                    Point pb = (Point)ref.object; pb.y("=",pb.yd()); pb.x("=",pb.xd());
                    ref= ref.right;  dW.plotLine(pa,pb,Color.magenta);
                }
            }
        }
    }
    private Polygon polyClip(int edge,double ee,double dd ){
        boolean cut = false; int n=1;
        double[] x3 = new double[1]; double[] y3 = new double[1];
        double[] k1 = new double[1]; double[] k2 = new double[1];
        Polygon p1 = new Polygon(p.length+20);
        for(int i =1;i<this.length;i++){
            double x1 = this.p[i-1].xd();double y1 = this.p[i-1].yd();
            double x2 = this.p[i].xd();  double y2 = this.p[i].yd();
            switch(edge){
                case 1: cut = edgeCut(ee,x1,y1,x2,y2,x3,y3,dd,k1,k2); break;
                case 2: cut = edgeCut(ee,y1,x1,y2,x2,y3,x3,dd,k1,k2); break;
            }p1.p[n]= new Point(2);
            if(cut){
                p1.p[n].x("=",x3[0]);p1.p[n].y("=",y3[0]); n++;
                p1.p[n]= new Point(2); }
            if(k2[0]>=0){ p1.p[n].x("=",x2);p1.p[n].y("=",y2); n++; }
        }
        if(n==1) return null;
        else{
            p1.p[0]= new Point(2); n= n-1;
            p1.p[0].x("=",p1.p[n].xd());
            p1.p[0].y("=",p1.p[n].yd()); p1.length = n+1;
            return p1;
        }
    }
}
```

```
private boolean edgeCut(double ee,double x1,double y1,double x2,double y2,
                double[] x3,double[] y3,double dd,double[] k1,double[] k2){
    k1[0]= (x1-ee)*dd; k2[0]= (x2-ee)*dd;
    if(k1[0]*k2[0]<0){
        x3[0]=ee;  y3[0]=(k2[0]*y1-k1[0]*y2)/(k2[0]-k1[0]);   return true;
    } return false;
}
public Polygon clip( Box bx, Polygon p1){
    Polygon p2,p3,p4,p5;
    if ((p2= p1.polyClip(1, bx. minP.xd(),  1 )) == null) return null;
    if ((p3= p2.polyClip(1, bx. maxP.xd(), -1)) == null) return null;
    if ((p4= p3.polyClip(2, bx. minP.yd(),  1)) == null) return null;
    if ((p5= p4.polyClip(2, bx. maxP.yd(), -1)) == null) return null;
    return p5;
}
```

Polygon on Polygon Overlay

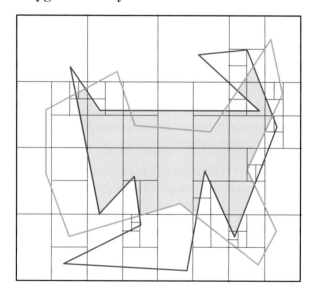

Figure 14.4 Selecting the intersection of two polygon regions

The first step, defining the overlap area of two polygons, represented by closed
polyline boundary lines is to intersect the polyline boundaries to locate the boundary
line crossing points. A small extension of the polyline intersection program given
above produces the output in Figure 14.5. Not only do the intersection points have to
be calculated, they also need to be inserted into the original boundaries if sub-regions
of the overlapping polygons are going to be made accessible. Figure 14.6 shows the
output of this insertion stage. Each new crossing point is linked into a linked list
representation of the boundary of each polygon. These new Point objects are
themselves then cross linked in a ring list so that boundary walks round the polygons

can transfer from one boundary to the other, depending on the sub area required. Large blue rectangles are plotted for the vertices of the first new polygon boundary line and smaller red rectangle for the second. This allows the new intersection points to show up as red rectangles with a blue boundary.

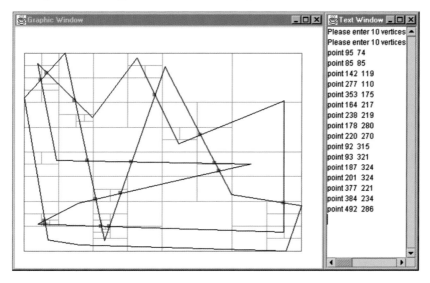

Figure 14.5 Binary subdivision calculating intersection points between polygons

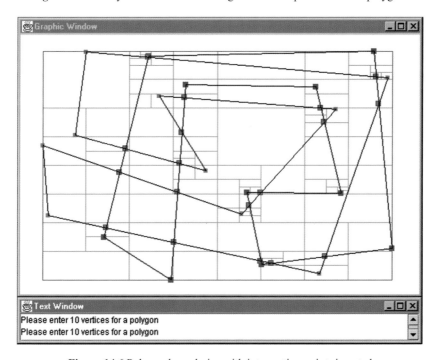

Figure 14.6 Polygon boundaries with intersection points inserted

Boolean Selectors

Given two polygons labelled A and B, there are four general regions that can be obtained by overlaying them. These are the Boolean combinations A.B, !A.B, A.!B and !A.!B. Each of these regions, because the original polygons are not restricted in their shapes, can be made up from sets of polygons. This is the reason why the operation is defined as the interaction of *Polygons* rather than two single polygon areas. In Figure 14.7 the region with the left most vertex is selected. This makes it easy to define this region as being labelled either A.!B or !A.B depending on which area has the left most point.

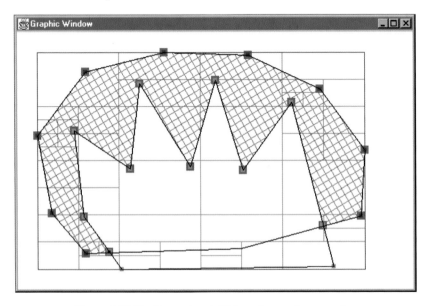

Figure 14.7 Shading region A.!B from the overlay operation

A more complete treatment requires the set of polygons for each of the regions to be generated and stored as polygon sets. This allows a collection of polygons to be shaded with different crosshatched patterns and colours, in the way shown in Figures 14.8 and 14.9.

```
public static void main(String[] args){
    double xmin,xmax,ymin,ymax;  int num1= 20;int num2= 20;
    IO.writeString("Please enter "+num1+" vertices for a polygon\n");
    Polygon p1 = new Polygon(num1+1);  p1.p[0] = dW.getCoord();
    for(int j=1;j<num1;j++){
        p1.p[j] = dW.getCoord(); dW.plotLine(p1.p[j-1],p1.p[j],Color.black);}
    p1.p[num1] = new Point(2);  p1.p[num1].c("<-",p1.p[0]);
    dW.plotLine(p1.p[num1-1],p1.p[num1],Color.black);
    p1.setTextWindow(IO);  p1.setDisplayWindow(dW);
    IO.writeString("Please enter "+num2+" vertices for a polygon \n");
    Polygon p2 = new Polygon(num2+1); p2.p[0] = dW.getCoord();
```

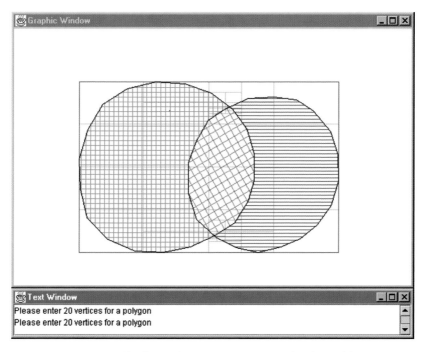

Figure 14.8 Shading regions A.!B green, A.B red and !A.B blue

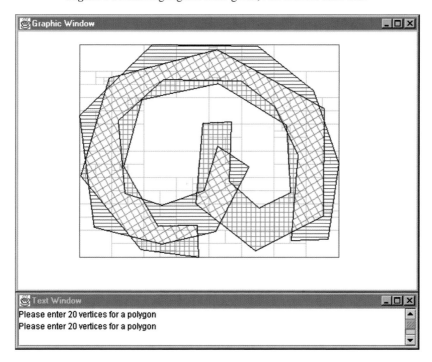

Figure 14.9 Shading regions A.B, A.!B and !A.B from the overlay operation

```
        for(int j=1;j<num2;j++){
            p2.p[j] = dW.getCoord();  dW.plotLine(p2.p[j-1],p2.p[j],Color.black);}
        p2.p[num2] = new Point(2);  p2.p[num2].c("<-",p2.p[0]);
        dW.plotLine(p2.p[num2-1],p2.p[num2],Color.black);
        p2.setTextWindow(IO); p2.setDisplayWindow(dW);
        Polygons p = new Polygons(p1,p2,IO);
        for(int j=1;j<num1;j++)dW.plotLine(p1.p[j-1],p1.p[j],Color.black);
        dW.plotLine(p1.p[num1-1],p1.p[num1],Color.black);
        for(int j=1;j<num2;j++)dW.plotLine(p2.p[j-1],p2.p[j],Color.black);
        dW.plotLine(p2.p[num2-1],p2.p[num2],Color.black);
    }
class Polygons{
    public Polygon bnd= null;
    public Point[] shading = null;
    public Color colour = null;
    public Polygon[] p = null;
    public List pL = null;
    private TextWindow IO = null;
    public Polygons(){ }
    public Polygons(TextWindow tW){  this.IO=tW;}
     public Polygons(int num,TextWindow tW)
        { this.p=new Polygon[num];  this.IO=tW;}
    public Polygons(Polygon p1,Polygon p2,TextWindow tW)
        { this.IO=tW;  p1.intersect(p1,p2,tW);}
    public void setPolygons(int num){  this.p = new Polygon[num];}
}
class Polygon extends java.lang.Object {
    private DisplayWindow dW = null;  private TextWindow IO = null;
    public Point p[];  public int length=0;  public List ply = null;
    Polygon(int len){ p = new Point[len]; length = len;}
    Polygon(int len, DisplayWindow dW){
        this.dW = dW;  p = new Point[len]; length = len;}
    Polygon(List poly,TextWindow tW){
        length= poly.length+1; ply = poly;
        p = new Point[poly.length+1];  ListElement pnt = poly.start;
        for(int i=0;i<poly.length+1;i++){ p[i]= (Point)pnt.object; pnt=pnt.right;}
    }
    public void intersect(Polygon p1,Polygon p2,TextWindow tW){
        List IP = new List(); ListElement le1=null,le2=null,lf=null;
        List pa = new List(); List pb = new List();
        List pc = new List(); List pd = new List();
        Point p = new Point(2);   p1.p[0].c("->",p); pa.append(p);le1 = pa.finish;
        for(int i=1;i<p1.length;i++){
          p = new Point(2); Line ln = new Line();
          ln.p1=new Point(2); ln.p2=new Point(2);
          p1.p[i-1].c("->",ln.p1); p1.p[i].c("->",ln.p2);  p1.p[i].c("->",p);
          pa.append(p); le2 = pa.finish;  pc.append(ln); lf = pc.finish;
```

```
        lf.link1= le1; lf.link2= le2;  le1 = le2;
    }
    p = new Point(2);p2.p[0].c("->",p);  pb.append(p);le1 = pb.finish;
    for(int i=1;i<p2.length;i++){
        p = new Point(2); Line ln = new Line();
        ln.p1=new Point(2); ln.p2=new Point(2);
        p2.p[i-1].c("->",ln.p1); p2.p[i].c("->",ln.p2);  p2.p[i].c("->",p);
        pb.append(p);le2 = pb.finish;  pd.append(ln); lf = pd.finish;
        lf.link1= le1;lf.link2= le2;  le1 = le2;
    }
    Box bx1 = new Box(p1);Box bx2 = new Box(p2);Box bx = new Box(bx1,bx2);
    IP=divide(pa,pb,pc,pd,bx,IP,0,2);
    pa.finish.left.right = pa.start;  pa.start.left = pa.finish.left;  //  close the loops

    if(pa.finish.link1!=null)
        { pa.start.link1=pa.finish.link1; pa.start.link2=pa.finish.link2;}
    pb.finish.left.right= pb.start; pb.start.left =pb.finish.left; //close the loops
    if(pb.finish.link1!=null)
        { pb.start.link1=pb.finish.link1; pb.start.link2=pb.finish.link2;}
    Polygons ov[] = new Polygons[4];
    ListElement refIP=null,refA=null,refB=null,ref=null,ref1=null,ref2=null;
    Point pT=null,pA=null,pB=null;Point pnt; boolean forward=true;
    ListElement refstart=null; refstart = ref = pa.start;
    do{
        ref.tag=-1;
        if(((NTuple)ref.object).b("==",(NTuple)ref.left.object)){
            if(ref.link1!= null){
                ref1 = ref.link1;  ref1.left.right = ref1.right;
                ref1.right.left = ref1.left; ref1.left=null;ref1.right=null;
            }
            ref.left.right = ref.right;  ref.right.left = ref.left;  ref.left=null;
            if(ref==refstart) {pa.start= refstart= ref.right;}
        }
        ref=ref.right;
    }while(ref!=pa.start);
    refstart = ref = pb.start;
    do{
        ref.tag= -1;
        if(((NTuple)ref.object).b("==",(NTuple)ref.left.object)){
            if(ref.link1!= null){
                ref1 = ref.link1; ref1.left.right = ref1.right;
                ref1.right.left = ref1.left; ref1.left=null; ref1.right=null;
            }
            ref.left.right = ref.right; ref.right.left = ref.left; ref.left=null;
            if(ref==refstart){pb.start=refstart=ref.right;}
        }ref=ref.right;
    }while(ref!=pb.start);
```

```
for(int j=0;j<4;j++){
    ov[j] = new Polygons();
    reflP=IP.start;
    while((reflP!=null)&&(((ListElement)reflP.object).left==null))
        reflP=reflP.right;
        do{
            boolean complete = false, first=true;
            List poly = new List();
            refA = (ListElement)reflP.object; refB = refA.link1;
            do{
                if(refA==null)tW.writeString("refA null \n");
                if(refA.link1==null)tW.writeString("refA.link1 null \n");
                int state=1;
                switch(j){
                case 0:                               // both forwards
                    if((refA.right!=null)&&
                                    ((Point)refA.right.object).inside(refB,dW)){
                        state=1;ref = refA; forward=true;
                        if(ref.tag==j)complete=true;ref.tag=j;
                    }else
                    if((refB.right!=null)&&
                                    ((Point)refB.right.object).inside(refA,dW)){
                        state=2;ref = refB;
                        forward=true;
                        if(ref.tag==j)complete=true;ref.tag=j;
                    }else complete = true;  break;
                case 1:                               // A forwards B backwards
                    if((refA.right!=null)&&
                                    !((Point)refA.right.object).inside(refB,dW)){
                        state=1;ref = refA;forward=true;
                        if(ref.tag==j)complete=true;ref.tag=j;
                    }else if((refB.left!=null)&&
                                    ((Point)refB.left.object).inside(refA,dW)){
                        state=2;ref = refB;
                        forward=false;
                        if(ref.tag==j)complete=true;ref.tag=j;
                    }else complete = true;  break;
                case 2:                               // B forwards A backwards
                    if((refA.left!=null)&&
                                    ((Point)refA.left.object).inside(refB,dW)){
                        state=1;ref = refA;forward=false;
                        if(ref.tag==j)complete=true;ref.tag=j;
                    }else  if((refB.right!=null)&&
                                    !((Point)refB.right.object).inside(refA,dW)){
                        state=2;ref = refB;forward=true;
                        if(ref.tag==j)complete=true;ref.tag=j;
                    } else complete = true;break;
```

```
                    case 3:                                              // both backwards
                        if((refA.left!=null)&& !((Point)refA.left.object).inside(refB,dW)){
                            state=1;ref = refA;
                            forward=false;
                            if(ref.tag==j)complete=true;
                            ref.tag=j;
                        } else if((refB.left!=null)&& !((Point)refB.left.object).inside(refA,dW)){
                            state=2;ref = refB;
                            forward=false;
                            if(ref.tag==j)complete=true;
                            ref.tag=j;
                        }else complete = true; break;
                    }
                    if(complete)break;
                    poly.append(ref.object);
                    if(forward){ref=ref.right;}
                    else {ref=ref.left;}
                    while(ref.link1==null){                               //follow polyline
                        poly.append(ref.object);
                        if(forward)ref= ref.right;
                        else ref= ref.left;
                    }
                    if(state==1){refA= ref;refB=ref.link1;}
                    else {refB=ref;refA=refB.link1;}
                }while(true);                    // complete polygon loop, assign to polygon list.
                if(poly.length!=0){
                    poly.start.left=poly.finish; poly.finish.right=poly.start;
                    Polygon pol = new Polygon(poly,tW);
                    if(ov[j]==null) tW.writeString(" ov[j] null \n");
                    if(ov[j].pL==null) ov[j].pL=new List();
                    ov[j].pL.append(pol);
                    if(j==0) hatching(pol,new double[][]{{10,-30,4},{10,60,4}}, 2,Color.red);
                    if(j==1)hatching(pol,new double[][]{{7,0,4},{7,90,4}}, 2,Color.green);
                    if(j==2)hatching(pol,new double[][]{{5,0,4}, {10,60,4}}, 1,Color.blue);
                }
                reflP = reflP.right;                // test each intersection point
            while((reflP!=null)&&(((ListElement)reflP.object).left==null))
                reflP=reflP.right;
        }while(reflP != null);                                // complete each polygon set
    }
    bx.draw(dW,Color.blue);
}..
}
```

These examples still depend on calculating the intersection points between a single pair of boundary lines. To complete this overlap operation it is necessary to cover the situation where one polygon boundary lies totally enclosed within a second, in the ways shown in Figure 14.10.

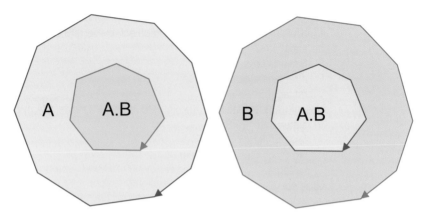

Figure 14.10 Nested overlapping polygons A and B

Nested boundary loops are necessary to define doughnut shaped areas such as those shown in Figure 14.11. To cope with these kinds of situation it is necessary to extend the overlap operation to handle "sets" of polygon loops. If each set is given a set label, the generalisation that is needed is to be able to process the overlap of "sets" of "sets" of polygon loops in one efficient operation.

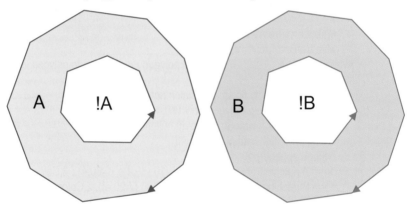

Figure 14.11 Nested doughnut polygon boundary loops for A and B

Where a nested boundary loop occurs there are no intersection points to relate it to other boundary loops. A simple solution is to run a point-in-polygon test for a starting point of such a loop on all the other polygons in the testing set. This, as well as extending the process which has so far been developed, requires a repeated series of pair wise tests. If large numbers of polygons are being processed a more efficient overall method is needed.

One approach to such a unified process can be set up in the following way. If all the polygon loops are represented by vertex list data structures, where each vertex is labelled by the area it belongs to, then all these loop-lists can be concatenated into a single list before calculating and cross-linking the new intersection points. This new structure will contain line links between loops that are "virtual edges" and have to be labelled as such. The advantage of doing this is that tracking the new intersection points generated by these virtual edges implicitly provides the information otherwise requiring multiple point-in-polygon tests. Each area-set of boundary loops can then be traversed by sequentially walking along the original concatenated list. As this is done the labelled boundary segments can be output, for the newly created minimum area cells, (or land parcels as they were called in early computer cartographic systems,) resulting from the intersection of overlapping polygons. The original areas can then be defined as collections of these new cells, the cells themselves being labelled by concatenating the names of the areas in which they fall in some standard order.

Polygon Sets on Polygon Sets and Polygon Networks

The first step is to concatenate the polygon loop lists into one list and treat it as a single polyline. This creates virtual edges in the way shown in Figure 14.12 between p0 and p1, p1 and p8, p8 and p13, and between p13 and p18.

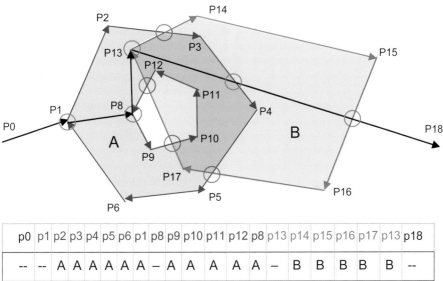

p0	p1	p2	p3	p4	p5	p6	p1	p8	p9	p10	p11	p12	p8	p13	p14	p15	p16	p17	p13	p18
--	--	A	A	A	A	A	–	A	A	A	A	A	–	B	B	B	B	B	--	

Figure 14.12 Concatenated polygon loop lists

Making the first vertex of each polygon-loop list have a special label and all other vertices have their polygon, area-labels, is one way these virtual edges can be identified and distinguished in subsequent processing. The second step is to test this extended polyline for self-crossing points. This will require the previous code: testing two polylines for crossing points, to be modified to process a single self crossing line. The major problem with modifying the previous code is to avoid re-linking all the line segment end points in the original boundary lists. If all the intersections at the end of line segments are filtered out, then the missing point problem shown in Figure 14.13 can result.

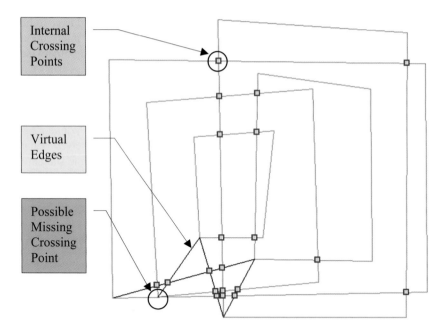

Figure 14.13 Internal line intersection points in three overlaid, polygon-area sets

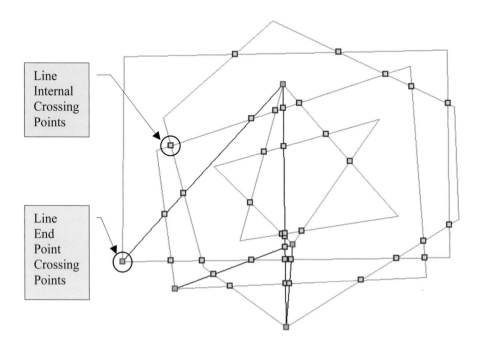

Figure 14.14 All intersection points in three overlaid, polygon-area sets

Displaying the link lines black and the boundary lines red gives Figure 14.14

```
ListElement  refp, refq;
refp = overlayList.start;   refq = refp.right;
while(refq!=null){
    if(refq.name.equals("linkedge"))
        dW.plotLine((Point)refp.object,(Point)refq.object,Color.black);
    else   dW.plotLine((Point)refp.object,(Point)refq.object,Color.red);
    refp= refp.right;   refq= refp.right;
}
```

The use of plane sweep processing makes it easier to generate the set of line intersection points from an extended polyline than the previous implementation of the spatial subdivision algorithm does. The first step is to create an array of the vertices for the new polyline from the overlaid polygon sets, and at the same time create a corresponding array of names.

```
Point[] vertices  = new Point[listLength];
String[] names  = new String[listLength];
int n=0;
for(i=0;i<polygonSets.length;i++){
    Polygons polx = polygonSets[i];
    String name = "xx"+i;                       // temporary name for the polgon set
    for(j=0;j<polx.p.length;j++){
        Polygon poly = polx.p[j];
        for(k=0;k<poly.p.length;k++){
            vertices[n]=poly.p[k];
            if (k!=0)names[n] = name;
            else names[n] = "**";               // name to indicate a virtual edge
            n++;
        }
    }
}
ListOfTuples overlayList = new ListOfTuples();
ListOfTuples order = new ListOfTuples();
ListElement ref1,ref2;
for(i=0; i<n;i++){
    overlayList.append(vertices[i]);
    order.append(vertices[i]);
    ref2= order.finish;
    ref1= overlayList.finish;
    ref1.name = names[i];
    ref1.link1 = ref2;  ref1.link2 = ref2;
    overlayList.finish.name = names[i];
    ref2.link1 = ref1;   ref2.link2 = ref1;
}
order = order.sort();
```

From this array a linked list of the vertices can be generated with each list element also including the name associated with its vertex. A secondary dependent link list for these vertex-points, cross-linked to the original link list can then be created, which can then be sorted giving a threaded list providing the coordinate order to control the plane sweep processing of line segments to follow. The program below implements the scheme shown in Figure 13.27.

```
//main
    ListElement outer1, outer2, ref1,ref2;
    ...
    ref1= order.start;
    while(ref1 != null){
        outer1 = ref1.link1;
        outer2 = outer1.right;
        if((outer2!=null)
                        &&(((Point)outer1.object).xd() < ((Point)outer2.object).xd())){
            ref2 = outer2.link1;
            innerLoop(outer1,outer2,ref1,ref2);
        }
        outer2 = outer1.left;
        if((outer2!=null)
                &&(((Point)outer1.object).xd() < ((Point)outer2.object).xd())){
            ref2 = outer2.link1;
            innerLoop(outer1,outer2,ref1,ref2);
        }
        ref1=ref1.right;
    }
    ...
static void innerLoop(ListElement outer1,ListElement outer2,
                                    ListElement ref1, ListElement ref2){
    ListElement inner1, inner2;
    int state = 1,count=1;
    while((ref1!=null)&&(ref1!=ref2)){
        inner1 = ref1.link1;
        inner2 = inner1.right;
        if( (inner2!=null) && (((Point)inner1.object).xd() < ((Point)inner2.object).xd())){
            crosstest(outer1,outer2,inner1,inner2);
        }

        inner2 = inner1.left;
        if((inner2!=null)
                &&(((Point)inner1.object).xd() < ((Point)inner2.object).xd())){
            crosstest(outer1,outer2,inner1,inner2);
        }
        ref1 = ref1.right;
    }
}
```

```
static void crosstest(ListElement o1,ListElement o2,ListElement i1,ListElement i2){
   Point p = new Point(2);
   if(lineCross((Point)o1.object,(Point)o2.object,
                             (Point)i1.object,(Point)i2.object,p)){
      dW.plotRectangle(p,5,Color.blue); dW.plotRectangle(p,3,Color.yellow);
}
static boolean lineCross(Point p1,Point p2,Point pa,Point pb,Point p){
   NTuple nt1 = new NTuple(3);
   nt1.n[0] =  p1.yd()- p2.yd();          nt1.n[1] =  p2.xd()- p1.xd();
   nt1.n[2] =  p1.xd()*p2.yd()-p1.yd()*p2.xd();
   double k1= pa.xd()*nt1.n[0] + pa.yd()*nt1.n[1] + nt1.n[2];
   double k2= pb.xd()*nt1.n[0] + pb.yd()*nt1.n[1] + nt1.n[2];
   NTuple nt2 = new NTuple(3);
   nt2.n[0] =  pa.yd()- pb.yd();          nt2.n[1] =  pb.xd()- pa.xd();
   nt2.n[2] =  pa.xd()*pb.yd()-pa.yd()*pb.xd();
   double k3= p1.xd()*nt2.n[0]+p1.yd()*nt2.n[1]+nt2.n[2];
   double k4 =p2.xd()*nt2.n[0]+p2.yd()*nt2.n[1]+nt2.n[2];

   int j=0;
   if(k1==0)j=j+1; if(k2==0)j=j+2; if(k3==0)j=j+4; if(k4==0)j=j+8;
   switch (j){
     case 0:
       if((k1*k2<0)&&(k3*k4<0)){
           p.x("=",(p2.xd()*k3-p1.xd()*k4)/(k3-k4));
           p.y("=",(p2.yd()*k3-p1.yd()*k4)/(k3-k4));
           return true;
       }break;
     case 1:  if(k3*k4<0){ pa.c("->",p); return true; }break;
     case 2:  if(k3*k4<0){ pb.c("->",p); return true;}break;
     case 4:  if(k1*k2<0){ p1.c("->",p); return true;}break;
     case 8:  if(k1*k2<0){ p2.c("->",p); return true;}break;
     default:  return false;
   }return false;
}
```

Including cases 1,2,4,8, handles end point to midpoint intersections shown in Figure 12.25, which is sufficient to cover the cases of collinear lines in closed boundaries.

Sorting the vertex list allows the endpoints, which match, illustrated in Figure 14.14, to be identified by a simple test.

```
ListElement ref1 = order.start;
ListElement ref2 = ref1.right;
Point p1=(Point)ref1.object;
while(ref2!=null){
    Point p2=(Point)ref2.object;
    if (p1.compareTo(p2)== 0){
```

```
            dW.plotRectangle(p2,5,Color.magenta);
            dW.plotRectangle(p2,3,Color.cyan);
      }
      ref2 = ref2.right;
      p1=p2;
}
```

On completion of this first stage the data structure consists of a polyline, double linked list of the sequence of points making up all the polygon area boundaries overlaid. Each area name is associated with each vertex in its boundaries –except for the first point in each loop, which has a special name identifying it as a virtual edge in the overall polyline sequence. The intersection tests will have inserted copies of crossing points into these boundary sequences and will have linked all identical points in separate two way ring lists.

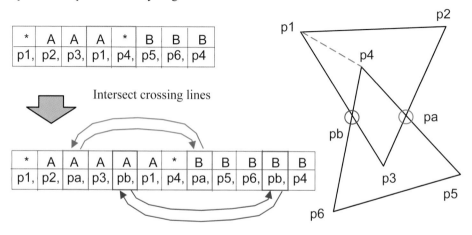

Figure 14.15 Intermediate data structure for polygon set overlay operations

This allows the second stage to be a single traversal of the polyline list outputting the boundary sequences of the minimum network cells resulting from the overlay. Linking the last point of the polyline back to its first point to give a single continuous loop allows the first point in the "ordered list" of these vertices used to identify the crossing points, to be taken as the starting point, to this traversal. The polygon area this point originated from can be used to label the first new cell region. Then as soon as a crossing point in the network is reached, indicated by the presence of the ring lists (*link1, link2* not null), then this label can be extended if the polyline traversal passes into the region of a second polygon.

This stage consists of constructing the names of the new minimum network cells. One way would be to construct a Boolean combination of all the original areas, as in earlier examples in this chapter, however a simpler approach is to merely collect the names of the regions each cell lies within and concatenate the names in some standard order for these new areas. The next stage, reconstructing the boundaries of these cells, can be done in a standard way chain sorting the boundary segments into order. If the beginning and end point pairs for boundary sequences that have the same name are collected in lists associated with the name in a name table then the third stage by chain sorting these endpoint pairs can generate the related cell boundaries as new polygon loops.

The immediate advantage of this approach is that a single operation is all that appears necessary as each cross-linked point in the polyline is traversed. When extracting the newly labelled boundary sequences for the minimum cells in the overlay network, every crossing point potentially changes the current cell's name and could therefore require new beginning or end points for boundary sequences to be defined and listed under the appropriate names in the new cell name table.

The crossing point operation for simple crossing line segments is not complicated and is illustrated in Figure 14. 16:

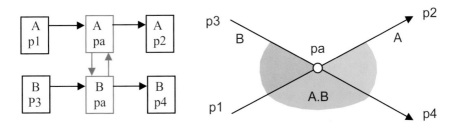

Figure 14.16 Simple crossing point

Following the boundary of area B, a point pa is reached which is cross-linked to the same point listed in the boundary of area A. If the point p4 is tested against line sequence p1-pa-p2, it will be classified as inside this boundary of A. This means that the line up to this intersection point will be labelled B, but after this point will be labelled AB.

Where intersections between boundary lines occur at existing line-segment end-points then though the principle remains the same, the coding becomes more complex. Similarly when virtual edges, linking polygon loops, are processed, the re-labelling process has to be modified. The various cases for end-point intersections are shown in Figure 14.17. Cases 1 and 2 can be treated in the same way as the simple case in Figure 14.16. Cases 3 and 4 represent point tangents, which can potentially be ignored, as they do not need to change the classification of boundary segments. However, if they are ignored then new regions with multiple lobes will result. If it proves possible to avoid this, subsequent tasks will be simpler if area boundaries can be constructed as sets of simple polygon loops.

In Figure 14.17, the current boundary being processed is shown by a black arrowed line. The input area it defines is labelled A, and is colour coded yellow. The boundary line it is intersecting is shown as a red line, for area B, which is shown

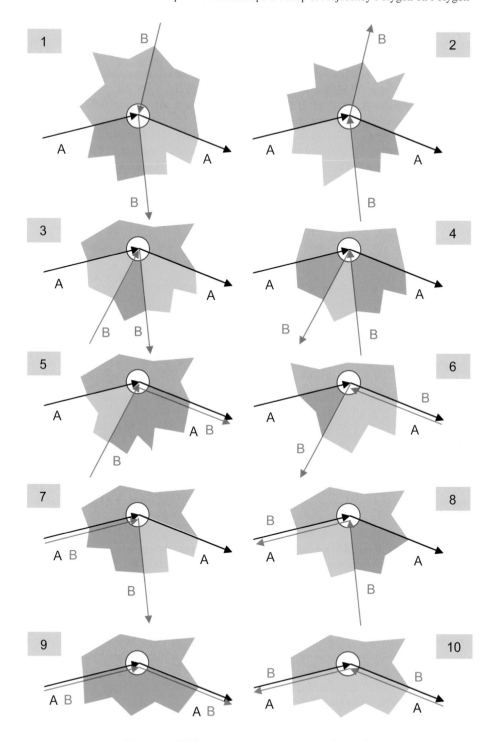

Figure 14.17 End point intersection point relationships

coloured blue. Where A overlaps B the colour is changed to green for the region A.B, and the area outside both A and B is shaded grey. This makes it possible to identify the new boundary labelling needed for the next line segment from A's boundary in the resulting network of edges in a reasonably simple way.

In cases 1, 3, 6, 7 and 10 this line segment will stay labelled A. In cases 2, 4, 5, 8 and 9 this line will be relabelled A.B. In order for complete boundary loops to be generated for the minimum cells in the resulting network, it is also necessary to generate boundaries for the region outside A. In cases 1, 3, 5, 7 and 9, where this is shown grey in Figure 14.17 this boundary can be missed out, however, in cases 2, 4, 6, 8 and 10, where it lies inside B, this boundary line is needed to complete the B region in the new network.

In the general case where many areas are overlaid, the label associated with the A boundary may have become A.X.Y before it reached the current vertex, in other words the line lies inside the X.Y region. In this case where the outer area is shown grey the outer boundary will have to be generated for the region X.Y, while where it is shown blue this boundary will be labelled B.X.Y. In summary the general process handling the re-labelling starting with A.X will generate either A.B.X or A.X inside the boundary and either X or B.X outside the boundary.

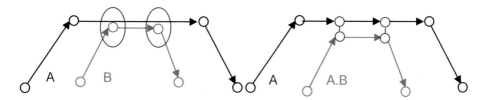

Figure 14.18 Tangent labelling problem

In cases 5 and 9 where the lines are co-linear there are potentially two ways in which the line A.B will be generated. The first following the original A boundary the second following the B boundary. In this case an extra test: to only generate the line for the "greater" label of the pair is needed to avoid duplication.

Another problem, which has to be addressed in the case of tangent boundaries, is the way vertices can be duplicated in incorrect ways. In the example shown in Figure 14.18 there are two ways in which the new vertices can be generated and inserted in the boundary line A. In the first case only the leading point in each line segment is processed for insertion into the other line. This approach however requires the co-linear tangent section of the line B to be processed. This can be done but it requires more complex testing to ensure the selected intersection point lies within the end points of the matching co-linear line from A. An apparently simpler scheme is to ignore the co-linear lines altogether and merely consider the endpoints of the lines linking into and out from the tangent segments. This however can create a double insertion problem in the situation shown in Figure 14.19, unless it is tested for explicitly.

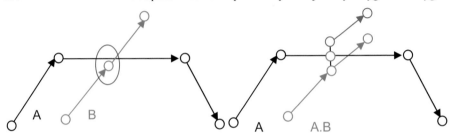

Figure 14.19 Potential point duplicating problem

Where intersections occur through the first point in a polygon loop within the polyline data structure, the label for the point will be for a virtual edge. The label for the area will be found on the next vertex in order along the boundary. Where this occurs it will be necessary to cycle through all the vertices linked to the current point to find the matching end point to the loop from the same area. These two line segments crossing the current line, can then be processed as a linked pair in the way outlined in Figure 14.17 to redefine the current line's labels. Where the current line is from a virtual edge, intersections with other boundary lines still need to be processed, in order to update the current area label sequence, though no new boundary line segment sequence will be generated. However where a virtual edge intersects a virtual edge no action is needed.

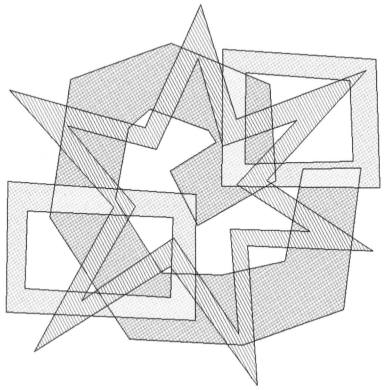

Figure 14.20 Multiple overlaid polygons

The relationships shown in Figure 14.17 can be transformed to local operations on each line segment in the polyline as it is reached in sequence.

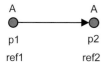

Check all boundary lines passing through point p1 to test whether p2 lies inside the areas with these boundaries or outside them. The boundary line segment generated should contain A in its inside name but not in its outside name.

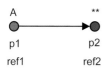

A linking "virtual" boundary line: checks need to be made to determine whether p2 is entering or leaving area A, as well as checking all other boundary line sequences passing through p1, to identify the region p2 lies within.

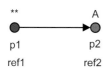

The first line segment of a new boundary line. This will require the label A to be added to the inside name sequence for this line, but not to its outside name sequence, along with the other names of regions that p2 lies within.

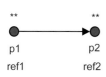

Boundary lines crossing through p1 will modify both inside and outside name sequences for later boundary lines from the polyline, either deleting the name if p2 is outside the boundary or adding the name if it is inside the boundary through p1.

The test whether these line segments lie inside or outside the boundary lines passing through p1 depend on the relative directions of the line segments. Three triangle area calculations provide the information needed to produce this inside/outside classification. If the triangle (pa, p1, pb) is positive the angle is acute; if it is negative the angle is reflexive, and if it is zero the lines lie in a straight line. If the angle is acute both triangles (pa, p1, p2) and (p1, pb, p2) need to be positive, if it is reflexive just one positive triangle classifies the point p2 as inside, otherwise the point p2 lies outside the boundary (pa, p1, p2).

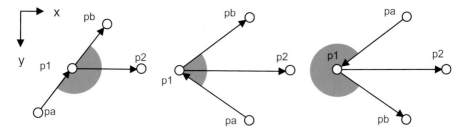

Figure 14.21 Inside outside testing

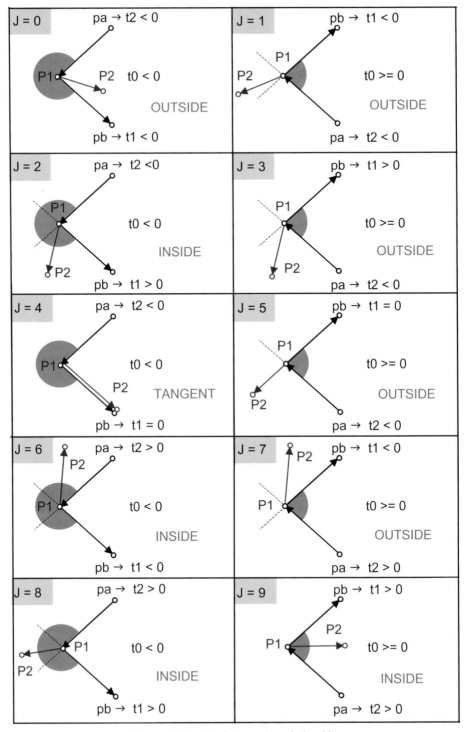

Figure 14.22a Line intersection relationships

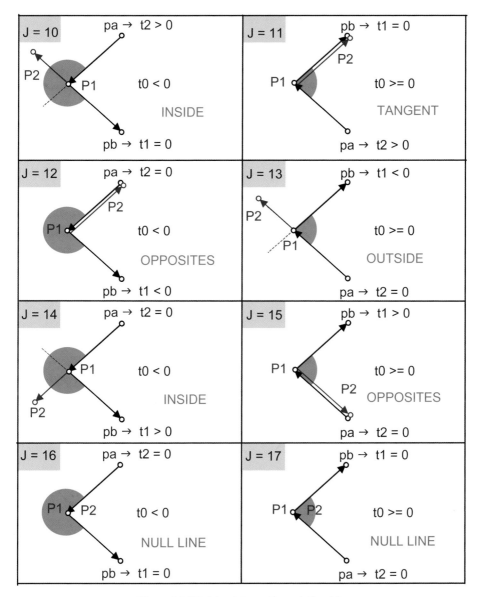

Figure 14.22b Line intersection relationships

The three triangles potentially return three values of interest: *greater than*, *less than* and *equal* to zero. This could give 27 different cases. However most of the cases where $t0 = 0$ can be mapped onto either the $t0<0$ or the $t0>0$ cases, giving 18 cases. The exceptions arise when "null line segments" or "spikes of zero width" occur in a polygon's boundary. The current approach is to make these structures illegal, and take care that the overlay process and other operations do not create boundary sequences containing such features and filter them out from input data structures.

```
TRAVERSE THE POLYLINE AND GENERATE LABELLED BOUNDARY
  SEGMENTS FROM THE OVERLAID POLYGON BOUNDARIES


  Table nameTable = new Table();
  NameSet outsideName = new NameSet(),  newOutside = new NameSet();
  NameSet insideName  = new NameSet(),  newInside  = new NameSet();
  String lineName = null,  bName=null;
  boolean begin = true, found=false; int count = 0;
  ListElement ref=null, refa=null, refb=null, refc=null;

  outsideName.add("X"); insideName.add("X");
  newOutside.add("X");  newInside.add("X");
  ref1 = order.start;   ref1 = ref1.link1;  ref1 = ref1.link2;
  newOutside = outsideName.copy();   newInside = insideName.copy();

  start = ref1;              // traverse the extended polyline boundary sequence
  do{
      ref2 = ref1.right;
      lineName = ref2.name;
      if((((Point)ref1.object).compareTo((Point)ref2.object)!=0)){
          lineName = ref2.name;
          if((begin)&&(!lineName.equals("**")))
              {newInside.add(lineName); begin=false;}
          ref=ref1.link1;
          if((lineName.equals("**"))&&(!ref1.name.equals("**"))){
              newOutside.delete(ref1.name);
              newInside.delete(ref1.name);
          }else if((!lineName.equals("**"))&&(ref1.name.equals("**"))){
              newInside.add(lineName);
              newOutside.delete(lineName);
          }
          while((ref!=null)&&(ref!=ref1)){
              refb= ref.right;
              if(((Point)refb.object).compareTo((Point)ref.object)!=0){
                  bName= refb.name;
                  if((!bName.equals("**"))&&(!bName.equals(lineName))){
                      found=false;
                      if(ref.name.equals("**")){
                          refc=ref.link1;
                          while(refc!=ref){
                              if(refc.name.equals(bName))
                                  { refa = refc.left;  found=true;  break;}
                              refc=refc.link1;
                          }
                      }else{ found = true;  refa = ref.left;}
                      if((found)&&
                          (((Point)refa.object).compareTo((Point)ref.object)!=0)){
```

```
                    double t0 = triangle((Point)refa.object,
                                        (Point)ref.object,(Point)refb.object);
                    double t2 = triangle((Point)refa.object,
                                        (Point)ref.object,(Point)ref2.object);
                    double t1 = triangle((Point)ref.object,
                                        (Point)refb.object,(Point)ref2.object);
                    j=0;
                    if(t0>=0)    j=j+1;
                    if(t1>0)     j=j+2;
                    if(t1==0)    j=j+4;
                    if(t2>0)     j=j+6;
                    if(t2==0)    j=j+12;
                    switch(j){

                    case 0:case 1:case 3:case 5:case 7: case 13:
                        newOutside.delete(lineName);
                        newOutside.delete(bName);
                        newInside.delete(bName);
                        if(!lineName.equals("**"))newInside.add(lineName);
                        else newInside = newOutside.copy();
                        break;                                    // outside

                    case 2:case 6:case 8:case 9: case 10: case 14:
                        newOutside.delete(lineName);
                        if(!lineName.equals("**"))newInside.add(lineName);
                        if(!bName.equals("**"))
                            {newOutside.add(bName);newInside.add(bName);};
                        break;                                    // inside

                    case 4:case 11:
                        if(lineName.compareTo(bName)<0){
                            newOutside.delete(lineName);
                            newOutside.delete(bName);
                            newInside = outsideName.copy();
                            if(!bName.equals("**")) newInside.add(bName);
                            if(!lineName.equals("**"))
                                newInside.add(lineName);
                        }break;                        // tangent lines A.B once

                    case 12:case 15:
                        if(lineName.compareTo(bName)<0){
                            newOutside.delete(lineName);
                            newOutside.delete(bName);
                            if(!lineName.equals("**"))newInside.add(lineName);
                            if(!bName.equals("**"))newOutside.add(bName);
                        } break;        // tangent lines A and B separately once
```

```
                        case 16: case 17:
                            newOutside.delete(lineName);
                            newOutside.delete(bName);
                            newInside = newOutside.copy();
                            if(lineName.compareTo(bName)<0){
                                if(!lineName.equals("**"))newInside.add(lineName);
                                if(!bName.equals("**")){newInside.add(bName);}
                            }
                        }                                              //switch
                    }
                }
            }
            ref = ref.link1;
        }
}
```

The composite names for the inside and outside regions adjacent to each line segment are constructed and stored in a name table. This allows a collection of line segments to be accumulated for each different, name combination, as the polyline linking all the original area boundaries is sequentially traversed.

```
if(newInside.compareTo(newOutside)!=0){
    if(outsideName.compareTo(insideName)!=0){
        int kk1 = nameTable.add(insideName);
        refb=((NameSet)nameTable.table[kk1]).bd.append(ref1.object);
        int kk2 = nameTable.add(outsideName);
        refb=((NameSet)nameTable.table[kk2]).bd.insertBefore(
                    ((NameSet)nameTable.table[kk2]).bd.finish,ref1.object);
        int kk3 = nameTable.add(newInside);
        refb=((NameSet)nameTable.table[kk3]).bd.append(ref1.object);
        int kk4 = nameTable.add(newOutside);
        refb=((NameSet)nameTable.table[kk4]).bd.append(ref1.object);
    }else{
        int kk5 = nameTable.add(newInside);
        refb=((NameSet)nameTable.table[kk5]).bd.append(ref1.object);
        int kk6 = nameTable.add(newOutside);
        refb=((NameSet)nameTable.table[kk6]).bd.append(ref1.object);
    }
}else{
    if(outsideName.compareTo(insideName)!=0){
        int kk7 = nameTable.add(insideName);
        refb=((NameSet)nameTable.table[kk7]).bd.append(ref1.object);
        int kk8 = nameTable.add(outsideName);
        refb=((NameSet)nameTable.table[kk8]).bd.insertBefore(
                    ((NameSet)nameTable.table[kk8]).bd.finish,ref1.object);
    }
}
```

```
            insideName = newInside;
            outsideName = newOutside;
            newOutside = outsideName.copy();
            newInside = insideName.copy();
            ref1=ref1.right;
      }while(ref1!= start);

      int kk1 = nameTable.add(newInside);
      refb=((NameSet)nameTable.table[kk1]).bd.append(ref2.object);
      int kk2 = nameTable.add(newOutside);
      refb=((NameSet)nameTable.table[kk2]).bd.insertBefore(
                          ((NameSet)nameTable.table[kk2]).bd.finish,ref1.object);
```

The names of the original areas are chained together to form the new area names using the *NameSet* class, which allows these name sequences to be created, stored and compared in various ways.

```
class NameSet implements Comparable{
    ListElement ref1,ref2;
    List ls=  new List();
    int test;
    NameSet(){}
    public ListElement add(String str,TextWindow IO){
        ref1 = ls.start;
        while(ref1 != null){
            if(str.compareTo((String)ref1.object)>0)break;
            ref1=ref1.right;
        }
        if(ref1!=null){ ls.insertBefore(ref1,str); return ref1.left;}
        else{ ls.append(str);return ls.finish;}
    }
    public void delete(String str){
        ref1 = ls.start;
        while (ref1!= null){
            if(str.equals((String)ref1.object)){ ls.delete(ref1);return;}
            ref1=ref1.right;
        }return;
    }
    public NameSet copy(NameSet names){
        NameSet newNames = new NameSet();
        ref1 = ls.start;
        while(ref1!= null){
            newNames.ls.append((String)ref1.object);
            ref1=ref1.right;
        }
        return newNames;
    }
}
```

```
public boolean equals(Object str){
    if((test = this.compareTo(str))==0)return true; else return false;
}
public int compareTo(Object names){
    ref1 = ls.start;
    ref2 = ((NameSet)names).ls.start;
    while((ref1!=null)&&(ref2!=null)){
        if((test=((String)ref1.object).compareTo((String)ref1.object))<0)return-1;
        if(test >0)return +1;
        ref1=ref1.right; ref2=ref2.right;
    }
    if(ref1==null){
        if(ref2==null) return 0;
        else return 1;
    }
    else if(ref2==null)return-1;
    return 0;
}
public void write(TextWindow IO){
    ref1=ls.start;
    while(ref1!=null){
        IO.writeString((String)ref1.object+" ");
        ref1=ref1.right;
        if(ref1!=null)IO.writeString("/ ");
    }
}
}
```

```
class Table{
    int count = 100;
    private Object[] table = new Object[count];
    private int size = 0;

    Table(){}
    public int locate(Object str){
        for(int i=0; i<size; i++){ if (str.equals(table[i]))return i;} return -1;
    }
    public int add(Object str){
        int index = locate(str);
        if(index== -1){ index = size; table[index]= str; size=size+1;}
        if(size== count){
            Object[] tbl = new Object[count + 100];
            for(int i=0;i<count;i++){ tbl[i]=table[i];}
            table = tbl;
            count=count+100;
        }return index;
    }
```

```
    public Object getElement(int i){
        if((i>=0)&&(i<size))return table[i];  else return "";
    }
    public boolean contain(String str){
        int k= locate(str);
        if (k<0)return false;else return true;
    }
    public int getSize(){   return size;}
    public void setTable(Object[] tbl){   table = tbl;  size= tbl.length;}
}
```

The output from this process will be a set of line segments for each name combination. These line segments can be processed to generate new polygon loops and new polygon sets if required. The line segments can be transferred directly to the hatching method for shading without this pre-structuring in the way shown in Figure 14.20.

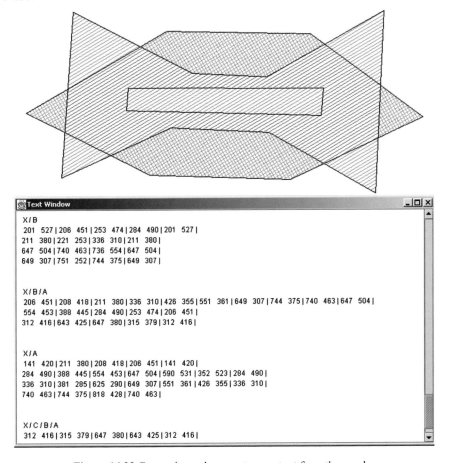

Figure 14.23 Generating polygon sets as output from the overlay

```
for(i=0; i<nameTable.size; i++){
    NameSet nm = (NameSet)nameTable.table[i];
    nm.ba = new ListOfTuples();
    ref= nm.bd.start;
    while(ref!=null){
        nm.ba.append((Point)ref.object); nm.ba.finish.link1 = ref;
        ref.link1 = nm.ba.finish; nm.ba.finish.tag = ref.tag;
        ref=ref.right;
    }
    if(nm.compareTo(exterior)!=0){
        IO.writeString("  "); ((NameSet)nameTable.table[i]).write(IO);
        nm.ba = nm.ba.sortn(IO);
        refa = nm.ba.start;   count = 0;
        while(refa!=null){
            if(refa.link1.tag > 0){
                if(refa.link1.tag==1)match=2;  if(refa.link1.tag==2)match=1;
                refb=refa.right;
                if(((Point)refa.link1.object).compareTo(
                                                (Point)refb.link1.object)==0){
                    while((refb!=null0&&(((Point)refa.link1.object).compareTo
                                                ((Point)refb.link1.object)==0)){
                        if(refb.link1.tag == match)break;
                        refb=refb.right;
                    }
                    if(refb!=null){
                        switch(match){
                        case 2: refb.link1.right= refa.link1; refa.link1.left = refb.link1;
                            break;
                        case 1: refa.link1.right= refb.link1; refb.link1.left = refa.link1;
                            break;
                        }
                        refb.link1.tag= -refb.link1.tag; refa.link1.tag= -refa.link1.tag;
                    }
                }
            }
        } refa=refa.right;
    } IO.writeString("\n  ");
```

Each name list contains a list of line segments *bd* making up the boundary of the area associated with the name, but, this area may well consist of multiple polygon loops. The first step is to generate a secondary list *ba* from the original list *bd*. This list *ba* can then passed to the *sortn()* procedure to sort *ba* into an ordered list, cross linked back to the original list *bd*. The original list contains point pairs defining line segments tagged as either 1 or 2 to indicate which is the first and which is the second in each segment pair. The ordered list is then processed to link together identical points within the *bd* list where the end of one line segment tagged 2 matches the beginning point of a second line segment tagged 1, illustrated in Figure 13.3. The second step is then to process the *bd* list to output continuous chains of points. In this case these chains will close to form polygon loops. Each loop can be added to a list associated with the area name within a polygon set object: *Polygons*.

```
...
ref= nm.ba.start;
Polygons p = nm.pg;
List pgl = new List();
p.pL= pgl;
do{
    refa= ref.link1;
    if((ref!=null)&&(ref.link1.tag == -1)){
        refa=ref.link1;
        refb=refa.right;
        ListElement st = refa;
        Polygon ppq =  new Polygon(IO,dW);
        pgl.append(ppq);
        List ply = new List();
        ppq.ply = ply;
        count = 0;
        do{
            count++;
            ply.append((Point)refa.object);
            refa.tag= - refa.tag;
             refb.tag = -refb.tag;
            do{
                refa = refb.right;  refb = refa.right;
            }while((refa!=null)&&(refa!=st)
                &&(((Point)refa.object).compareTo((Point)refb.object)==0));
        }while((refa!=null)&&(refa.tag <0)&&(count <64)&&(refa != st));
        ppq.length = count+1;
        ppq.p = new Point[count+1];
        ListElement refx = ply.start;
        ppq.p[count]=(Point)refx.object;
        for(int ii=0; ii < count; ii++){
            ppq.p[ii] = (Point)refx.object;
            refx=refx.right;
            IO.writeString(ppq.p[ii].xi()+"   "+ppq.p[ii].yi()+" | ");
            if((ii+1)%10==0)IO.writeString("\n    ");
        }
        IO.writeString(ppq.p[count].xi()+"   "+ppq.p[count].yi()+" | ");
        if(i<6)n=1;else n=2;
        IO.writeString("\n   ");
    }
    if(ref!=null)ref=ref.right;
}while(ref!=null);
IO.writeString("\n\n");
}
```

Figure 14.24 Shading polygon sets output from the overlay program

```
public NTuple mid(TextWindow tW){
    ListElement ref = this.start;  if (ref == null) return null;
    NTuple v = (NTuple)ref.object;
    NTuple min=new NTuple(v.dimension); NTuple max=new NTuple(v.dimension);
    NTuple midTuple = new NTuple(v.dimension);
    for(int i=0;i<v.dimension; i++)
        {min.n[i] = Double.MAX_VALUE;  max.n[i] = Double.MIN_VALUE; }
    while(ref != null){
        v = (NTuple)ref.object;
        if(v.dimension!= 0){
            for(int i=0;i<v.dimension; i++){
                if(v.n[i]<min.n[i])min.n[i]=v.n[i]; if(v.n[i]>max.n[i])max.n[i]=v.n[i];
            }ref = ref.right;
        }
    }boolean test = true;
    for(int i=0;i<v.dimension;i++){
        if(min.n[i]!=max.n[i]) test = false;
        midTuple.n[i]= (min.n[i]+max.n[i])/2.0;
    }if (test)midTuple=null;
    return (NTuple)midTuple;//(Comparable)midTuple;
}
public ListOfTuples joinTo(ListOfTuples b){
    if(this.start==null)return b;  if(b.start==null)return this;
    ListOfTuples a = new ListOfTuples(); a.start = this.start; a.finish = b.finish;
    this.finish.right= b.start;  b.start.left = this.finish; return a;
}
public ListOfTuples sortn(TextWindow tW){
    this.setLength();
    NTuple test = this.mid(tW);
    if(test== null)return this; if (this.start == this.finish) return this;
    ListElement ref = this.start;
    ListOfTuples leftList = new ListOfTuples();
    ListOfTuples rightList= new ListOfTuples();
    while(ref!=null){
        if(((NTuple)ref.object).compareTo(test)<0){
            leftList.push(ref.object);
            leftList.start.link1 = ref.link1; leftList.start.link2 = ref.link2;
        }else{
            rightList.push(ref.object);
            rightList.start.link1 = ref.link1; rightList.start.link2 = ref.link2;
        }ref = ref.right;
    }
    if (leftList.start==null) return  rightList = rightList.sortn(tW);
    if (rightList.start==null)return leftList   = leftList.sortn(tW);
    leftList = leftList.sortn(tW);  rightList = rightList.sortn(tW);
    return leftList.joinTo(rightList);
}
```

Although the plane sweep algorithm is potentially not as efficient as the recursive four-way subdivision it has some very convenient properties in the way it orders data and the demands it makes on the order of input data. Developed in the OBLIX program in 1969 to allow local processing to increase the complexity of drawings that could be managed in a computer with a relatively small fast memory space (compared with modern systems), it was also employed in 1970 in Utah to calculate the display data in the order needed to feed the raster sweep of TV monitors before frame store facilities existed. Currently it provides the same service for large electrostatic plotters that produce high-resolution technical drawings with sizes in excess of 1 metre wide. In practice both algorithms can be used for the first step in a polygon overlay operation to locate line intersection points.

Set Theoretic Data Structures and Land Parcel Systems

The second stage of the process is essentially a renaming exercise. This can be a system that creates the Boolean combination of the original names. This is useful for data base access languages used in geographic information systems. Storing the set of minimal area polygons, from the overlay operation allows the set operations supporting spatial data base queries to be executed as a series of merge sort operations, that are very fast once the initial geometrical overlay operation has been completed.

In geographic information system databases the basic geometric overlay operation does not have to be executed very often. However, where the same overlap testing is needed in computer aided design systems, changes in the geometry are often required repeatedly at great speed. This context poses problems where the data is merely stored as sets of polygons. The reason is that if a polygon is changed then all its neighbours in the overlay network also have to be changed.

Where the collection of polygons is large this will demand large repeated searches to find adjacent polygons. It is necessary to include the links from one polygon to its neighbours if these searches are to be avoided. In the overlay operation each edge is used to generate two new boundaries, one for each of the newly defined cells lying on each side of the line. This introduces an alternative mode of storage. Instead of storing polygons with duplicated edges and cross-linked references to adjacent polygons, the basic geometry can be stored as dual labelled edges. This approach is explored in a later chapter, as a way of representing networks along with the related topic of tessellating surfaces in different ways.

15
Spatial Relationships
Overlap &Adjacency
Rectangle-Rectangle
Window on Window

Introduction

In this chapter the general topic, highlighted yellow in Table 15.1, is again an exploration of the relationships of overlap and adjacency between areas.

Table 15.1 Basic Spatial Relationships

Objects	Point	Line	Area	Volume
Point	Point on Point	Point in Line	Point in Area	Point in Volume
Line		Line on Line	Line in Area	Line in Volume
Area			Area on Area	Surface in Volume
Volume				Volume on Volume

Rectangle Set on Rectangle Set

An overlay task that needs to be carried out very often in an interactive window system is establishing the interrelationship of window rectangles as they are moved relative to each other. In early systems with limited memory the way this was done

A. Thomas, *Integrated Graphic and Computer Modelling*,
DOI: 10.1007/978-1-84800-179-4_15, © Springer-Verlag London Limited 2008

was critical for keeping memory usage within bounds. The key operation is that shown in Figure 15.1. The process is one of partitioning the display space into rectangles. This allows fast block memory transfer operations to be used to update the display. Block memory transfer operations are primitive operations, which can be implemented at the hardware or firmware level in the display system.

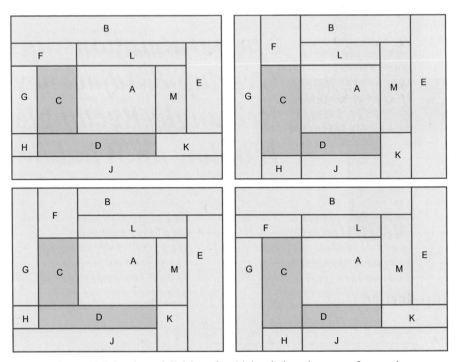

Figure 15.1 Overlay-subdivision of multiple windows into sets of rectangles

Figure 15.1 shows four possible subdivision patterns, each giving the same number of rectangles. Window rectangles are added serially on top of each other. Each window subdividing the existing lower level set of rectangles, if and where they overlap to give the configurations shown. Although this process can be implemented in a relatively direct recursive program, it requires fairly complex linked list structures to build a window manipulation system based on it.

An alternative approach that generates a larger number of subdivision rectangles can be set up in the way illustrated in Figure 15.2. As the labelling in Figure 15.2 shows this subdivision scheme creates a grid of variable sized rectangles that can be treated as an array, each cell being referenced by a pair of indexes. This makes it unnecessary to store each cell explicitly. Its dimensions can be calculated from the indexes used to access it. To simplify this task it is necessary to store the x and y values of the window edges in two arrays accessible by the cell indexes of the overall grid to give the corner values of the cells. The cells can then be systematically processed using two nested *for* loops. Each window can be represented by a set of rectangles, and each set can be reconstructed from the left and right index of the window boundaries and also the top and bottom boundary indexes. These can also be processed using two repeat loops. When windows are moved, resized, added or deleted from this system all that needs to be done is to rearrange the entries in these index based arrays, in practice this operation is more simply handled by a pair of ordered linked lists. Code to implement these ideas using keyword commands can be set up as follows.

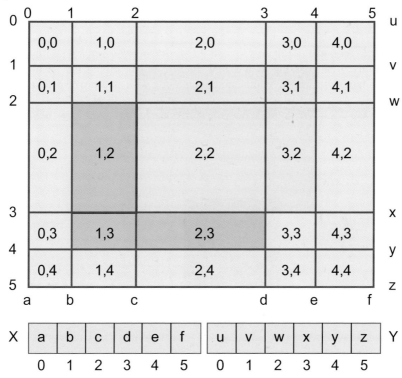

Figure 15.2 Overlay-subdivision of multiple windows into an array of rectangles

```
public class Windows{
    static TextWindow IO = new TextWindow(15,405,600,110);
    static DisplayWindow dW = new DisplayWindow(IO,15,5,600,400,Color.white);
    static int xmin = 0, xmax = dW.c.width, ymin = 0, ymax = dW.c.height;
    static List LofWs = new List();  static Win bw = null;
    static ListOfEdges lx= new ListOfEdges(),  ly= new ListOfEdges();
    public static void main(String[] args){
        Win win = null;  Color cc = Color.white; String colour=null; String str = null;
        int xl, yl, x1, y1, x2, y2, xr, yr,rx,ry;
        Point p1=new Point(2), p2=new Point(2);Point p = null;
        p1.x("=",xmin); p2.x("=",xmax); p1.y("=",ymin);p2.y("=",ymax);
        int width= xmax-xmin; int height = ymax-ymin;
        Picture pict = new Picture(null,cc,"rectangle",null,xmin,ymin,width,height);
        bw= makeWindow(pict,Color.white);  dW.repaint();
        ListElement winref=null, refwn=new ListElement();
```

```
do{
    IO.writeString("Please enter command code: ");
    str = IO.readString(); IO.readLine();
    if(str.equals("make")){
        cc=getColour();
        Picture pict = getWindowRectangle(cc);
        win = makeWindow(pict,null);
    }
```

The *make* command requires a call to *getWindowRectangle()* followed by a *makeWindow(..)* call passing to it a new *Picture* object giving the location of the window rectangle, its height, width and colour along with an array of pixel colours.

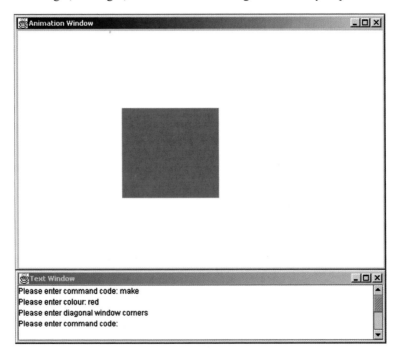

Figure 15.3 Make a new red window

```
if(str.equals("resize")){
    p=new Point(2);
    IO.writeString("Please identify the window \n");
    win = (Win)(refwn = identifyWindow(p)).object;
    if(win!=bw){
        pict = resizeWindow(win,bw);
        Color ccc= win.cc;
        removeWindow(refwn);
        win = makeWindow(pict,ccc);
    }
}
```

The *resize* command requires a window and a corner to be identified using the mouse. A second point is then requested to relocate the corner. The dimensions of the window rectangle then have to be updated, ensuring that the window is never reduced below a minimum size. This is followed by removing the image of the old version of the window from the screen, before replacing it with the new one.

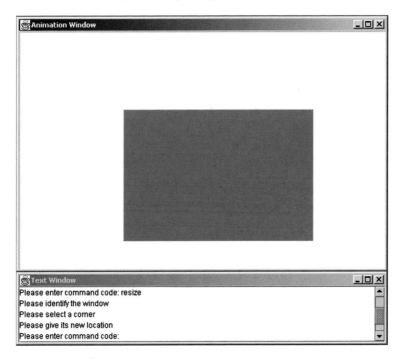

Figure 15.4 Resize the red window

```
if(str.equals("raise")){
    p= new Point(2);
    IO.writeString("Please identify a window\n");
    win = (Win)(refwn = identifyWindow(p)).object;
    if(win!=bw){
        pict = win.pic;
        removeWindow(refwn);
        win = makeWindow(pict,null);
    }
}
```

The *raise* command requires the target window to be identified, the visible portions of this window to be removed, followed by entering the window back on top of the other windows using the *makeWindow(..)* command. The edge index lists are changed but then changed back again. The stack of windows has the raised window removed from its previous location and reentered at the top of the stack.

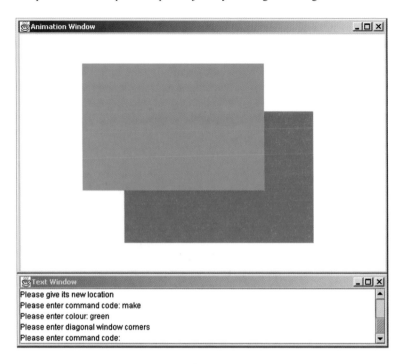

Figure 15.5 Overlay multiple windows

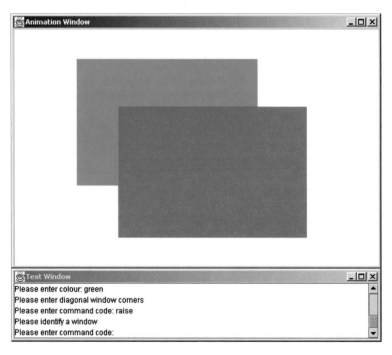

Figure 15.6 Raise the red window

```
if(str.equals("move")){
    p= new Point(2);
    IO.writeString("Please identify a window\n");
    win = (Win)(refwn = identifyWindow(p)).object;
    if(win!=bw){
        pict = moveWindow(win,p);
        removeWindow(refwn);
        win = makeWindow(pict,null);
    }
}
```

The *move* command requires the target window to be identified using a point entered by the mouse. A second point is then requested to define the movement relative to the first point. The coordinates for the rectangle at its new location are then calculated, followed, as before, by the old version being removed before the new rectangle is painted in. Again the index lists have to be updated for the new window location. The background window cannot be moved using this command.

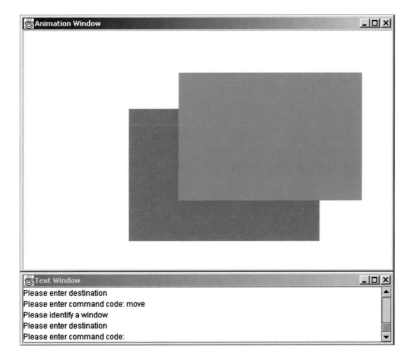

Figure 15.7 Move the green window

The *repaint* command requires the target window to be identified again by entering a point using the mouse, its new colour requested from the text window, followed by repainting the visible regions of the window in the new colour. This command does not raise the identified window to the top of the stack. This allows the command to be used to recolour the background window.

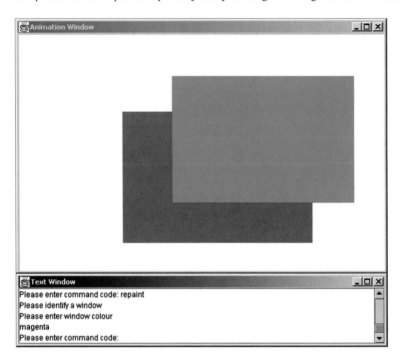

Figure 15.8 Repaint the red window magenta

```
if(str.equals("repaint")){
    p=new Point(2);  IO.writeString("Please identify a window\n");
    win = (Win)(refwn = identifyWindow(p)).object;
    Color col = getColour();  win.cc = col;
    repaint(p,col);
}
```

The *copy* command duplicates a window, while the remove command requires the targeted window, again located using the mouse, to be removed. The *end* command terminates the program.

```
if(str.equals("remove")){
    p=new Point(2);  IO.writeString("Please identify a window\n");
    win = (Win)(refwn = identifyWindow(p)).object;
    if(win!=bw)removeWindow(refwn);
}
```

```
if(str.equals("copy")){ cc=getColour();  copy(cc);}
```

```
} while(!str.equals("end"));
    IO.writeString("Program complete \n");
}
```

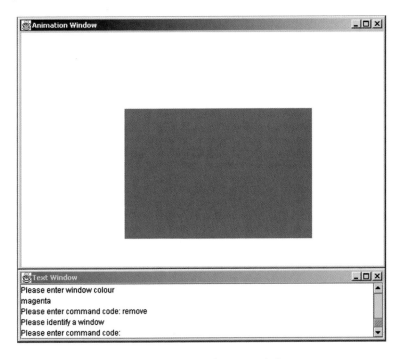

Figure 15.9 Remove the green window

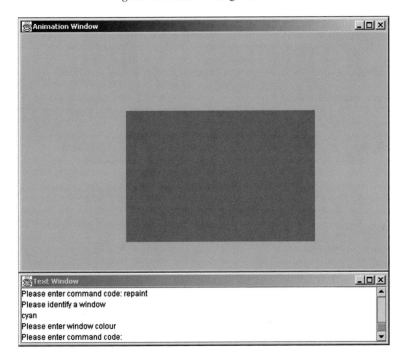

Figure 15.10 Repaint the background cyan

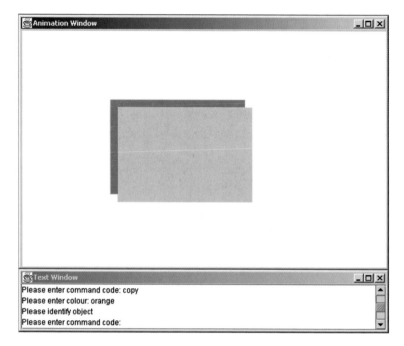

Figure 15.11 Copying an existing rectangle

Figure 15.12 Rarranging a set of rectangles using keyword commands

```
class ListOfEdges extends List{
    public ListOfEdges(){};
    public ListElement enter(Integer e){
        ListElement ref= this.start;
        if(ref==null){ref = push(e); ref.tag++; return ref;}
        Integer r = (Integer)ref.object;
        if(e.intValue() < r.intValue()){ ref= push((Integer)e); ref.tag++; return ref;}
        while((ref!=null)&&(r.intValue()<e.intValue())){
            ref=ref.right;   if(ref!=null)r = (Integer)ref.object;
        }
        if(ref==null){ref=append(e); ref.tag++; return ref;}
        if(r.intValue()==e.intValue()){ ref.tag++; return ref;}
        ref= insertBefore(ref,e); ref.tag++;
        return ref;
    }
    public void destroy(ListElement ref){
        ref.tag--;   if(ref.tag <=0) delete(ref);
    }
}
```

The *ListOfEdges* class manages the ordered lists of ordinate values: one for the left and right edges, and one for the top and bottom edges of the window rectangles. Keeping these ordinates in order allows two simple *for* loops to systematically access the rectangles that are contained between each pairs of edges defining the extent of a particular window. Duplicate elements in these lists are tagged with a count value so that window deletion does not lose a necessary entry in the list.

```
class Win{
    public int x1,x2,y1,y2,w,h;
    public Picture pic =null;
    public Color cc = null;
    public int tag=0;
    public ListElement rx1= null,rx2= null,ry1= null,ry2= null, rx=null, ry=null;
    public Win(int x,int y, int w, int h, Color cc, Picture pict){
        this.cc=cc;this.pic=pict;
        this.x1=x;this.y1=y; this.x2=x+w; this.y2=y+h; this.w=w; this.h=h;
        if(pict==null){
            pic = new Picture(new Color[w][h],cc,"rectangle",null, x1,y1, w ,h);
            for(int i=0;i<w;i++){ for(int j=0;j<h;j++){ pic.c[i][j]= cc; } }
        }else this.pic=pict;
    }
    public void setWin(ListElement rx1,ListElement rx2,
            ListElement ry1,ListElement ry2,ListElement rx,ListElement ry){
        this.rx1= rx1; this.rx2= rx2; this.rx=rx;
        this.ry1= ry1; this.ry2= ry2; this.ry=ry;
    }
}
```

Windows, Map Sheets and the Desktop Model

The *Win* class shown above provides the basic minimum definition of a window needed to run the examples shown in the Figures 15.3 to 15.12. By itself manipulating rectangles is not of great use. Two properties make it important. The first is, because a rectangle's sides are parallel to the coordinate axes, it can be used to manage min-max boxes used to enclose more complex shapes. This combined with divide and conquer spatial algorithms can give very efficient programs. The second property for raster based display systems is the way that rectangular areas can be accessed from memory using base displacement hardware addressing schemes. This can support a series of hardware and firmware primitive display operations that can also be used to support fast and efficient higher-level programs.

As graphics user interfaces developed with the introduction of raster displays and the way integrated circuit memory reduced the cost and raised the resolution of real time refreshed displays, the desk top model emerged as the higher-level concept that controlled their form. Nearly all text and graphic information was traditionally held in the form of documents of one kind or another, and these were usually used in combinations laid out on a work surface. As soon as virtual documents could be generated using computer graphic displays, the small restriction that aligning document boundaries with the pixel grid of the display system made, provided a system fast enough to be used interactively, and provide interfaces such as the frame or window based system supported, among others, by the Java language.

Geographic data held in map sheets were easy to transfer to this new environment. Also geographic databases set up as a collection of single subject spatial distributions could be presented in map-overlays by combining the contents of two or more map sheets. Architectural and engineering drawings exhibited a similar arrangement: separating out structures, services and detailed finishes into different cross-referenced drawings. Overlays of differently sourced image components make up many graphic art operations, for example making appliqués, montages and the compositing techniques used in printing to combine a mixture of text and graphics elements.

Clearly to make use of the rectangle overlay system it is also necessary to be able to draw or paint images onto the rectangles and to handle this "content" appropriately when the rectangles are moved, changed in size, or clipped by overlaid rectangles. There seem to be two general approaches to handling the content of images placed on these rectangular panels. The first where the image only exists as a pixel array is to process the source image to generate new images and place them into new display memory positions. The second where the image has been constructed by draw and fill commands is to redraw the image into its new display memory locations. Simple moves can be achieved for both sources by memory transfer operations where these are supported by the display system. Scaling and other changes to the content require more processing.

Changing the content of a window in interactive work will only occur in one window at once so rearranginging other windows can still be handled using pixel array based operations. Where more than one window is active, in the case of simulations for example, then each window may need its own program thread with redraw capabilities to keep a complex display correctly up to date.

In order to hold the content of each rectangle an array of pixel colour values needs to be generated and retained. In these examples this is done using a Picture class. Each window rectangle contains a picture object that has to be repainted whenever the window is moved.

```
class Picture{
    public Object obj =null;  public CoordinateFrame fm = null; public String name = "";
    public Color[][] c = null;  public Color cc = null;  public TextWindow IO = null;
    public Rectangle r= new Rectangle();  public int x=0, y =0, w=0, h =0;
    public Picture(Color[][] ic,Color col,String it,Object ob, int ix, int iy, int iw ,int ih){
        this.c = ic; this.cc=col; this.obj = ob;
        this.x=ix; this.y= iy; this.w= iw; this.h= ih; this.name=it;
        if((it!=null)&&(it.equals("rectangle"))){
            if (this.c == null){  this.c = new Color[w][h];
                for(int i=0;i<w;i++){for(int j=0;j<h;j++){this.c[i][j] = col;} }
            }
        }
    }
    public CoordinateFrame setScaling(int x,int y){
        Point p1=new Point(2); p1.n[1]= 0 ;p1.n[2] =  0;
        Point p2=new Point(2); p2.n[1]= x ;p2.n[2] =  y;
        return setScaling(p1,p2);
    }
    public CoordinateFrame setScaling(Point p1,Point p2){
        Point pa=new Point(2); pa.n[1]= 0 ;pa.n[2] = 0;
        Point pb=new Point(2); pb.n[1]= this.w ;pb.n[2] = this.h ;
        fm = new CoordinateFrame();  fm.setScales(p1,p2,pa,pb); return fm;
    }
    public Color getColour(int x,int y){
        if(name.equals("line")){
            LineSeg ls = (LineSeg)this.obj;
            if(ls.up){ if(y==ls.pnts[x])return ls.cc; else return null; }
            else{ if(ls.pnts[y]==x)return ls.cc;else return null; }
        }else if(name.equals("polygon")||name.equals("rectangle")) return this.c[x][y];
        return null;
    }
    public void setColour(Color cc){
        if(this.name.equals("line")){((LineSeg)this.obj).setColour(cc);}
        if(name.equals("polygon")){this.cc=cc;}
        if(name.equals("rectangle")){this.cc=cc;}
    }
    public void setColour(Color cc,int x,int y){
        if((this.name==null)||(name.equals("rectangle")))this.c[x][y]=cc;
        if(this.name.equals("polygon"))this.c[x][y]=cc;
        if(this.name.equals("line"))this.setColour(cc);
    }
}
```

Rewriting Rectangular Pixel Arrays

Where there is a simple memory transfer operation then there is no need to reconstruct the contents of a rectangle. Where in contrast a window rectangle needs to be up dated, for example to a different size, then it becomes necessary to calculate the scaling factors that match the change in the rectangle size, so, if required, they can be applied to create new contents for the modified window. Changing the contents of a Picture's array of pixel colors and also transferring this data to a *BufferedImage* is a relatively fast operation. The *BufferedImage* is the Java system's interface to its display facilities that can employ block memory transfers.

The Java system can, in one form or another, provide all the window facilities that have been discussed so far. However, because the Java system has to cope with a large number of display operations and the large number of possible interactions among them, it gets complex. It was decided for the purposes of this book that only the simplest primitive display facilities offered by the language would be employed. The remaining functionality would be provided by home-written basic Java programs, but, set up to illustrate key ideas, rather than to build a complete system. The choice to write programs rather than explain the existing Java libraries was made to allow alternative approaches to be explored and show the merits of some of the deceptively simple solutions that have emerged, that are easy to overlook unless they are compared with less elegant alternatives.

The first basic Java display facility that has been used as a building block is the use of the *Color* class. Representing colours is a complex topic that will be explored more fully in a later chapter. Currently the set of simple colours made available by the *Color* class as *Color* objects has been used, and it has been sufficient for the relatively diagrammatic images constructed so far. In the antialiasing exercises and the surface shading examples the Java facility to represent and render different shades of grey, has been used but with its full explanation deferred to a later chapter.

Upto this point the interface between these programs and the Java system and therefore the operating system and the display processor has been limited to the ability to change the colour of single pixels in a display raster, one by one, selected and modified by explicit program code. This approach is sufficient to allow the *Picture* object colour arrays to be displayed, but very slowly.

In order to explore more complex interactive graphic facilities it is necessary to take advantage of any block memory transfer operations provided that speed up interactive raster displays. This requires the extension to the *setPixel()* minimum interface used upto this point, to include the ability to transfer arrays of pixel values to the display screen as a fast primitive operation. These extensions are made in the *WorkPanel* class that provides the working display surface for the *DisplayWindow* class that has been used throughout this book. The extensions are based on the *BufferedImage* construct provided by the Java libraries to implement double buffered image-refresh and fast array operations. When the *repaint()* command is made for the *WorkPanel* the system calls its *paintComponent()* procedure which then transfers the data in the *BufferedImage* to the display screen in one step. The *WorkPanel* functions are listed below.

```java
class WorkPanel extends JPanel{
    private BufferedImage ds;  private WritableRaster raster = null;
    private ColorModel smp = null;  public int width, height;  public TextWindow IO;
    protected boolean newMouseEvent = false;  protected int mx=0, my=0;
    public WorkPanel(TextWindow tw,int width,int height){
        setSize(width,height); this.width = width; this.height = height; this.IO = tw;
        addMouseListener( new MouseAdapter() { public void mousePressed(MouseEvent e){
        mx = e.getX(); my = e.getY(); newMouseEvent = true;}} );
        ds = new BufferedImage(width,height,BufferedImage.TYPE_INT_ARGB);
        raster = ds.getRaster(); Graphics g = getGraphics(); smp = ds.getColorModel();
        generate(ds,Color.white,g);
    }
    public void paintComponent(Graphics g){
        super.setSize(width,height); super.paintComponent(g);
        g.drawImage(ds,0,0,width,height,null);
    }
    public void generate(BufferedImage ds,Color cc,Graphics g){
        int width = ds.getWidth(); int height = ds.getHeight();
        int argb = cc.getRGB(); smp = ds.getColorModel();
        Object colorData = smp.getDataElements(argb,null);
        for(int i=0;i<width;i++){ for(int j=0; j<height;j++){raster.setDataElements(i,j,colorData);}}
    }
    public int getPixel(int x,int y){ return ds.getRGB(x,y); }
    public Color getPixelColor(int x,int y){ return new Color(ds.getRGB(x,y),true); }
    public void setPixel(int x,int y ,Color cc){
        int argb = cc.getRGB(); Object colorData1 = smp.getDataElements(argb,null);
        raster.setDataElements(x,y,colorData1);
    }
    public void setPixels(int n,int m,int ii,int jj,int width,int height,Picture pic){
        int argb=0;
        for(int i=0;i<width;i++){
            for(int j=0;j<height;j++){
                Color cc = pic.getColour(n+i,m+j);
                if(cc!=null){argb = cc.getRGB();
                    Object colorData1 = smp.getDataElements(argb,null);
                    raster.setDataElements(ii+i,jj+j,colorData1);
        } } }
        this.repaint();
    }
    public void clearDisplayPanel(Color cc){ //background colour
        int width = ds.getWidth(); int height = ds.getHeight();
        int argb = cc.getRGB(); smp = ds.getColorModel();
        Object colorData = smp.getDataElements(argb,null);
        for(int i=0;i<width;i++){
            for(int j=0; j<height;j++){ raster.setDataElements(i,j,colorData); }
        }this.repaint();
    }
```

```java
public void setColor(ListElement[] lse,Win wn,int kk,
                                   int ii,int jj,int width,int height,Color c){
    int argb = c.getRGB();
    Object colorData1 = smp.getDataElements(argb,null);
    wn.cc=c;wn.pic.setColour(c);
    for(int i=0;i<width;i++){
        for(int j=0;j<height;j++){
            boolean done = false;int k=0;
            while(!done &&(k<kk)){
                Win w = (Win)lse[k].object;
                if(w==wn){                          //matching windows
                    wn.cc=c; wn.pic.setColour(c);
                    if(wn.pic.getColour(ii+i-wn.x1,jj+j-wn.y1)!=null){
                        raster.setDataElements(ii+i,jj+j,colorData1);  done = true;
                    }
                }else if(w.pic.getColour(ii+i-w.x1,jj+j-w.y1)!=null)done = true;
                k++;
            }
    } } } this.repaint();
}
public void line(int x1,int y1,int x2, int y2,Color color,Shadings s,boolean set){
    int kx,ky,dx,dy;
    dx = x2-x1; dy = y2-y1; kx = 1; ky = 1;
    if (dx < 0){ kx = -1; dx = -dx; }  if (dy < 0){ ky = -1; dy = -dy; }
    if (dx < dy) {this.octant(y1,x1,y2,x2,ky,kx,dy,dx,2,color,s);}
    else {this.octant(x1,y1,x2,y2,kx,ky,dx,dy,1,color,s);}
    if(set)this.repaint();
}
private void octant(int x,int y,int xend,int yend,int kx,int ky,
                                    int dx, int dy,int dir,Color cc,Shadings s){
    int argb = cc.getRGB();  Object colorData1 = smp.getDataElements(argb,null);
    int d, j ;   Point p = new Point(2);
    d = 2*dy-dx; dx = 2*dx; dy = 2*dy;
    if (ky < 0){ d = -d; dx = -dx; dy = -dy;}
    while(true){
        if(s==null){
            if (dir == 1) raster.setDataElements(x,y,colorData1);
            else raster.setDataElements(y,x,colorData1);
        }else{
            if (dir == 1){p.n[1] = x; p.n[2] = y;}else {p.n[2] = x; p.n[1] = y;}
            s.defineEdgePoint(p);
        }
        if (x == xend) return;
        if (d < 0)j = -ky; else j =ky;
        if (j > 0){d = d+dy-dx; y = y+ky;}else d= d+dy;
        x = x +kx;
    }
}
```

```
    public Color[][] setColor(int ii,int jj,int width,int height,Picture p,Color c){
        int argb = c.getRGB();Color[][] pict=null;
        if(p.c==null)p.c= new Color[width][height];
        Object colorData1 = smp.getDataElements(argb,null);
        for(int i=0;i<width;i++){
            for(int j=0;j<height;j++){
                p.c[i][j]=c;  raster.setDataElements(ii+i,jj+j,colorData1);
            }
        }
        if(p.c==null) p.c=pict;
        this.repaint();
        return pict;
    }
    public void setPixels(int ii,int jj,int width,int height,Picture pic){
        setPixels(0,0,ii,jj,width,height,pic);
    }
    public Picture polygonfill(Polygon p,Color cc,Color color){
        Picture pict=null;
        int ymin = Integer.MAX_VALUE;   int ymax = Integer.MIN_VALUE;
        int xmin = Integer.MAX_VALUE;   int xmax = Integer.MIN_VALUE;
        for (int i= 0; i< p.length-1; i++){
            if(ymin > p.p[i].yi())ymin = p.p[i].yi();
            if(ymax < p.p[i].yi())ymax = p.p[i].yi();
            if(xmin > p.p[i].xi())xmin = p.p[i].xi();
            if(xmax < p.p[i].xi())xmax = p.p[i].xi();
        }
        Color[][] pic = new Color[xmax-xmin+1][ymax-ymin+1];
        int len = ymax-ymin+1;
        Shadings S = new Shadings(null,len,ymin,false);
        for(int i= 0; i< p.length-1; i++){
            this.line(p.p[i].xi(),p.p[i].yi(),p.p[i+1].xi(),p.p[i+1].yi(),cc,S,false);}
            for(int j= 0;j < ymax-ymin+1;j++){
                for(int i= 0;i < xmax-xmin+1;i++){
                    if((i>=S.leftedge[j]-xmin)&&(i<=S.rightedge[j]-xmin)) pic[i][j]=cc;
            } }
        pict=new Picture(pic,null,"polygon",p,xmin,ymin,xmax-xmin+1,ymax-ymin+1);
        setPixels(0,0,xmin,ymin,xmax-xmin+1,ymax-ymin+1,pict);
        pict.cc= cc;
        return pict;
    }
  }
}
```

Based on the procedures in the *WorkPanel* and the *Picture* classes the following procedures were written to implement the command line instructions used to generate the displays in Figures 15.3 to 15.12, initially for simple rectangles but then extended to display polygons and overlaid sets of polygons and rectangles.

The *getWindowRectangle* procedure requests two points using the mouse to locate two diagonally opposite vertices of a new window rectangle. These are then used to create a new *Picture* object to return to the calling program.

```
static Picture getWindowRectangle(Color cc){
    IO.writeString("Please enter diagonal window corners\n");
    int minx,miny,maxx,maxy;
    Point p1 = dW.getCoord();Point p2 = dW.getCoord();
    if(p1.xi() < p2.xi()){minx=p1.xi(); maxx = p2.xi();} else{minx = p2.xi(); maxx=p1.xi();}
    if(p1.yi() < p2.yi()){miny=p1.yi();maxy=p2.yi();} else{miny = p2.yi(); maxy = p1.yi();}
    Picture pic =new Picture(null,cc,"rectangle",null,minx,miny, maxx-minx, maxy-miny);
    return pic;
}
```

The *makeWindow()* procedure, given the corner points of the new window in the *Picture* object obtained from the *getWindow()* procedure, enters them into the edge index lists, then clips them to the display frame boundary. It then places the new window on the top of the stack of windows and draws in its visible region with the requested colour by calling the *resetRectangle()* program. This in turn calls the *setPixels()* procedure from the *WorkPanel* object, which sets up the *Picture* object in the new *Window* object, and enters it into the *BufferedImage* for fast display.

```
static Win makeWindow(Picture pict,Color cc){
    int xl=pict.x,xr=pict.x+pict.w,yl=pict.y,yr=pict.y+pict.h;
    Win wint = new Win(xl,yl,(xr-xl),(yr-yl),cc,pict);
    ListElement ww= LofWs.push(wint);
    wint.rx1= lx.enter(new Integer(xl)); wint.rx2= lx.enter(new Integer(xr));
    wint.ry1= ly.enter(new Integer(yl)); wint.ry2= ly.enter(new Integer(yr));
    //drawRectangle(bw, wint, xl,yl,xr,yr,cc);   // only to process rectangles
    ListElement [] lse = new ListElement[1]; lse[0] = ww;
    resetRectangle(bw, lse, xl,yl,xr,yr,1);    //  processes rectangles,polygons and lines
    dW.repaint();
    return wint;
}
```

```
static boolean drawRectangle(Win bw, Win ws, int xl,int yl,int xr,int yr,Color cc){
    int xleft=xl;  int xright=xr;  int yleft=yl;  int yright=yr;    // only for rectangles
    if(bw!=null){                                          // replaced by resetRectangle()
        if(xr<=bw.x1)return false; if(xl>=bw.x2)return false;
        if(yr<=bw.y1)return false; if(yl>=bw.y2)return false;
        if(xl<bw.x1)xleft= bw.x1;  if(xr>bw.x2)xright=bw.x2;
        if(yl<bw.y1)yleft= bw.y1;  if(yr>bw.y2)yright=bw.y2;
    }
    int x1 = ws.x1; int y1= ws.y1;   int i = xleft-x1; int j = yleft-y1;
    dW.c.setPixels(i,j,xleft,yleft,xright-xleft,xright-xleft,ws.pic.c);
    return true;
}
```

The extension in this chapter is to employ the Java facilities that allow an array of pixels to be transferred as fast as the system will allow to the display screen. In the code given above this allows the fast rendering of rectangles, which links back to the tiling operations introduced in Chapter 3. In particular it raises the possibility of overlaying rectangles containing transparent, translucent and opaque areas as an image building operation. In its most general form this requires a further extension in the use of the colour coding facilities provided by the Java system, which will be explored in a later chapter. It also provides an alternative approach to the topics explored in chapter 14 where overlays of different polygons and polygon networks are combined together to give a composite display. The binary property of *transparent* or *opaque* is all that is needed to extend the current display facilites to handle polygons and triangles as well as the rectangles that are already catered for.

This on or off property can be provided using the existing *WorkPanel* interface to Java system facilities. If transparent areas are rendered, by leaving the color array entries in their *Picture* objects *null*, and only copying the opaque, colour values into the *BufferedImage*, the Java system treats the pixels with no value transferred to them as transparent. Using this facility requires several of the basic procedures set up to display overlaid rectangles to be modified. The first of these has to be the procedure used to identify an object in the display using the mouse. It is not simply the top window rectangle but the top window rectangle containing a visible pixel at the mouse pointer position that needs to be identified.

The procedure *identifyWindow()* initially used a point input from the mouse to identify a window by calling *locateWindow()* to find the top window rectangle containing the mouse pointer coordinate. The highlighted code shows the modification needed to extend the system to handle the transparent regions round more complex shapes such as polygons and triangles if they are to be displayed by the same process used for rectangles. Once the mouse identifies the top window rectangle containing its pointer, a futher pixel level test is required to see if this point is transparent in that window. If it is then the next window down has to be tested in the same way, until either the base window is located or a rectangle containing a visible pixel is found.

```
static ListElement identifyWindow(Point p){
    Point pp = dW.getCoord(); p.n[1]=pp.n[1]; p.n[2]=pp.n[2];
    ListElement refwn= LofWs.start;  return refwn=locateWindow(refwn,p);
}
static ListElement locateWindow(ListElement ref,Point pc){
    while(ref!= null){
        Win wn=((Win)ref.object);
        if((pc.xi()-wn.x2)*(pc.xi()-wn.x1)<=0)
            if((pc.yi()-wn.y2)*(pc.yi()-wn.y1)<=0)  //return ref to only identify the top rectangle
                if(wn.pic.getColour(pc.xi()-wn.x1,pc.yi()-wn.y1)!=null)return ref;
        ref=ref.right;
    }return ref;
}
```

A keyword command *"drawpolygon"* to display a polygon can be added to the main program in the following way.

```
if(str.equals("drawpolygon")){
    cc=getColour();
    pict = getPolygon(cc);
    win = makeWindow(pict,null);
}
```

As with the rectangle *"make"* command, the colour of the polygon is requested. The *getColour()* procedure asks for the colour to be entered by name, and then passes the *String* to the *setColour()* procedure, which returns the matching Java *Color* object for use in the display process. Then *getPolygon()* is called to obtain a polygon definition.

```
static Color getColour(){
    IO.writeString("Please enter colour: ");
    String colour = IO.readString(); IO.readLine();
    Color cc = setColour(colour);
    return cc;
}
```

The *getPolygon()* procedure enters the polygon as a series of boundary points defined by the mouse, which it displays as a boundary outline before generating a *Polygon* object. This is then passed to the *polygonfill()* procedure in the *WorkPanel* class, where a *Picture* object representing the shaded polygon is created inside the polygon's min-max box rectangle. This rectanglar area of pixels is then rendered by the *makeWindow()* procedure, extended to only paint the coloured pixels that lie inside the polygon's boundary leaving other pixels in the *BufferedImage* unchanged.

```
static Picture getPolygon(Color cc){
    IO.writeString("Please enter the number of vertices: ");
    int num=IO.readInteger(); IO.readLine();
    Point [] poly = new Point[num+1];
    IO.writeString("Please enter the vertices with the mouse: \n");
    poly[0]=dW.getCoord();
    for(int kk=1;kk<num;kk++){
        poly[kk]= dW.getCoord();  dW.plotLine(poly[kk-1],poly[kk],Color.blue);
    }
    poly[num]=new Point("=",poly[0]);  dW.plotLine(poly[num-1],poly[num],Color.blue);
    dW.getCoord();
    Polygon ppp= new Polygon(num+1);  ppp.p=poly;
    Rectangle r = new Rectangle();
    Color[][] c = dW.c.polygonfill(ppp,cc,null,r);
    Picture pict = new Picture(c,cc,"polygon",ppp,r.x,r.y,r.width,r.height);
    pict.cc= cc;
    return pict;
}
```

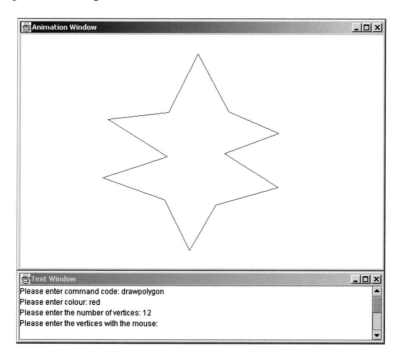

Figure 15.13 Draw a polygon boundary

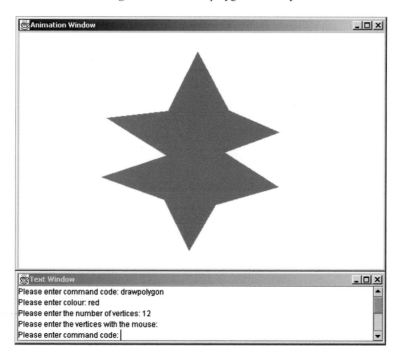

Figure 15.14 Colour-fill the polygon boundary

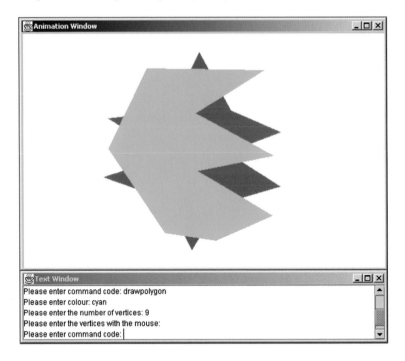

Figure 15.15 Overlay a second polygon

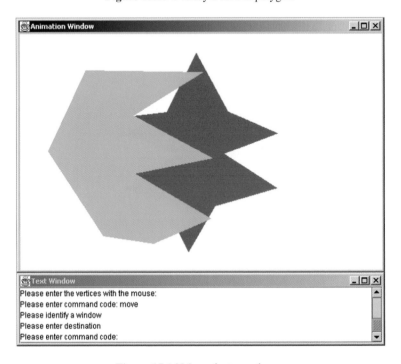

Figure 15.16 Move the top polygon

The *moveWindow()* procedure uses two points entered by the mouse. The first point is used to identify the window, the second to give its new location. The difference between these two points defines a movement vector, which is applied to each corner of the old window to give the coordinates of the new window. These are then used to generate a new *Picture* object which when passed back to the calling procedure allows the old version of the window to be deleted and a new version in the new location to be created.

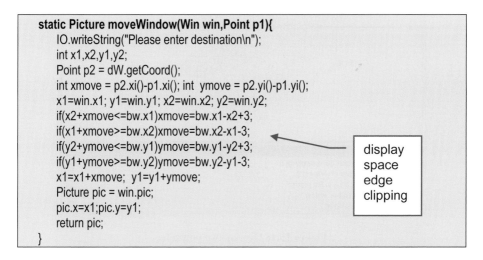

```
static Picture moveWindow(Win win,Point p1){
    IO.writeString("Please enter destination\n");
    int x1,x2,y1,y2;
    Point p2 = dW.getCoord();
    int xmove = p2.xi()-p1.xi(); int  ymove = p2.yi()-p1.yi();
    x1=win.x1; y1=win.y1; x2=win.x2; y2=win.y2;
    if(x2+xmove<=bw.x1)xmove=bw.x1-x2+3;
    if(x1+xmove>=bw.x2)xmove=bw.x2-x1-3;
    if(y2+ymove<=bw.y1)ymove=bw.y1-y2+3;
    if(y1+ymove>=bw.y2)ymove=bw.y2-y1-3;
    x1=x1+xmove;  y1=y1+ymove;
    Picture pic = win.pic;
    pic.x=x1;pic.y=y1;
    return pic;
}
```
display space edge clipping

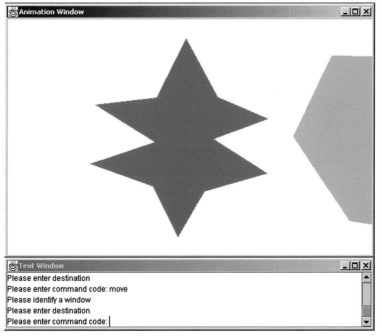

Figure 15.17 Move the cyan polygon showing edge clipping

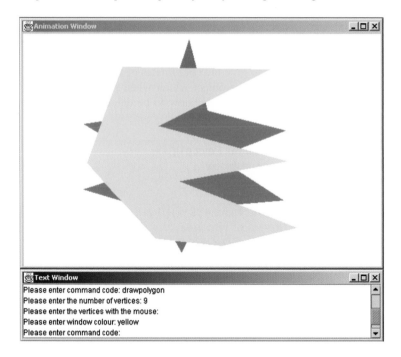

Figure 15.18 Overlay two polygons

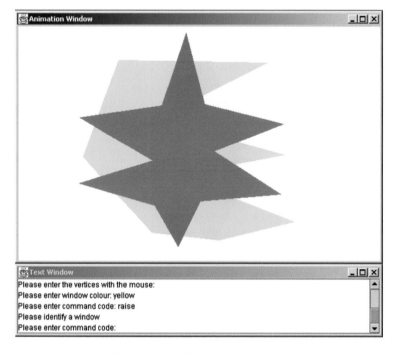

Figure 15.19 Raise the lower polygon

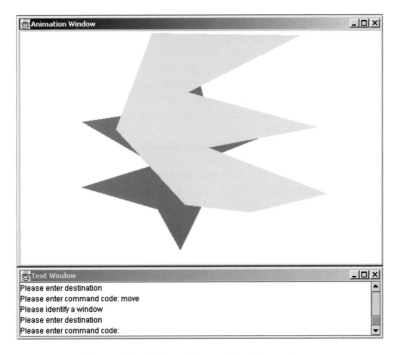

Figure 15.20 Raise and move the yellow polygon

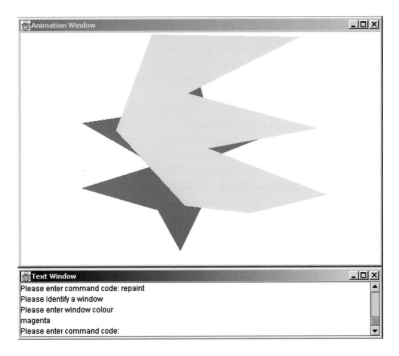

Figure 15.21 Repaint the lower polygon magenta

The *repaint()* procedure allows the displayed object selected by the mouse to be repainted: in the rectangle examples given above this merely means having its visible subrectangles repainted a new colour. The procedure identifies all the sub-rectangles within the target window and where they are visible it recolours them. For simple rectangles the list of windows is searched in stacking order from the top and the first window to contain a target sub-rectangle is the visible one, if this matches the selected window then this sub-rectangle can be modified. The code for this is highlighted light brown in the procedure list given below. Again for more complex shapes only the visible pixels in the target window's subrectangles are changed. In this case all the window rectangles above the target window also have to be tested at the pixel level to see if there is a hole revealing the target window. This is done by the *WorkPanel* procedure *setColor()*. The code for this step will handle both rectangles and polygons and is highlighted light green below.

```
static void repaint(Point p,Color cc){
    int llx,rrx,lly,rry;   ListElement refw= locateWindow(LofWs.start,p);
    Win w = (Win) refw.object;  w.pic.cc=cc;
    if(w.pic.name.equals("line")){w.pic.setColour(cc);}
    else{
        for(int i=0;i<w.w;i++){  for(int j=0;j<w.h;j++){
            if(w.pic.getColour(i,j)!=null)  w.pic.setColour(cc,i,j);
    }}}
    ListElement refx=w.rx1,strx= w.rx1,stry= w.ry1,endx= w.rx2,endy= w.ry2;
    while((llx=((Integer)refx.object).intValue())<xmin)refx=refx.right;
    while((lly=((Integer)stry.object).intValue())<ymin)stry=stry.right;
    while((rrx=((Integer)endx.object).intValue())>xmax)endx=endx.left;
    while((rry=((Integer)endy.object).intValue())>ymax)endy=endy.left;
    while(((Integer)refx.object).intValue()<((Integer)endx.object).intValue()){
        llx = ((Integer)refx.object).intValue();  rrx = ((Integer)refx.right.object).intValue();
        ListElement refy=stry;
        while(((Integer)refy.object).intValue()<((Integer)endy.object).intValue()){
            lly = ((Integer)refy.object).intValue();
            rry = ((Integer)refy.right.object).intValue();
            Point pp = new Point(2); pp.n[1]=(llx+rrx)/2;  pp.n[2]=(lly+rry)/2;
            // ListElement ref= locateWindow(LofWs.start,pp);  // repaint rectangles only
            // drawRectangle(bw,((Win)ref.object),llx,lly,rrx,rry,((Win)ref.object).cc);
            ListElement[] lse = new ListElement[LofWs.length+1];
            int kk= windowList(lse,LofWs.start,pp);
            dW.c.setColor(lse,w,kk,llx,lly,rrx-llx,rry-lly,cc);
            refy=refy.right;
        }refx=refx.right;
    } dW.c.repaint();
}
```

The *removeWindow()* procedure is used to remove the old version of a window that is being deleted or replaced by a new version, either because it is being moved or changed in some way.

The *remove* command requires the window to be identified by a point provided by the mouse. For a simple rectangle all the visible subrectangles for the target window need to be visited and the next window below in the window stack repainted. Again for non-rectangular shapes further pixel level tests need to be carried out to find the first visible pixel below each visible pixel of the targeted window that is being removed. The reference to the deleted window needs to be removed from the stack of windows list *LofWs*, and finally the edge index lists need to be updated by removing the entries for the target window.

```
static void removeWindow(ListElement refw){
    int llx,rrx,lly,rry;
    Win w = (Win) refw.object;
    ListElement refx=w.rx1, refy= w.ry1,  endx= w.rx2,  endy= w.ry2;
    while((llx=((Integer)refx.object).intValue())<xmin)refx=refx.right;
    while((lly=((Integer)refy.object).intValue())<ymin)refy=refy.right;
    while((rrx=((Integer)endx.object).intValue())>xmax)endx=endx.left;
    while((rry=((Integer)endy.object).intValue())>ymax)endy=endy.left;
    if(w!= bw){
        LofWs.delete(refw);
        ListElement stry=refy, strx=refx;
        while(((Integer)refx.object).intValue()<((Integer)endx.object).intValue()){
            llx = ((Integer)refx.object).intValue();
            rrx = ((Integer)refx.right.object).intValue();
            stry=refy;
            while(((Integer)stry.object).intValue()<((Integer)endy.object).intValue()){
                lly = ((Integer)stry.object).intValue();
                rry = ((Integer)stry.right.object).intValue();
                Point pp = new Point(2); pp.n[1]=(llx+rrx)/2; pp.n[2]=(lly+rry)/2;
                // ListElement ref= locateWindow(LofWs.start,pp); //to remove rectangles only
                // drawRectangle(bw,((Win)ref.object),llx,lly,rrx,rry,((Win)ref.object).cc);
                ListElement [] lse = new ListElement [LofWs.length+1];
                int kk = windowList(lse, LofWs.start,pp);
                resetRectangle(bw,lse,llx,lly,rrx,rry,kk);
                stry=stry.right;
            }refx=refx.right;
        }
        lx.destroy(w.rx1);lx.destroy(w.rx2);ly.destroy(w.ry1);ly.destroy(w.ry2);
    }dW.repaint();
}
```

In this case the code highlighted in light brown needs to be replaced by that highlighted in light green. After the reference to the target window has been removed from the stack of wndows: a list of sub rectangles containing the identifying point is passed to the *resetRectangle()* procedure for repainting, rather than the new top rectangle containing the point being repainted by the *drawRectangle()* procedure. To raise a polygon or rectangle to the top layer merely requires it to be identified, removed from its current level and repainted at the top level using *makeWindow()*.

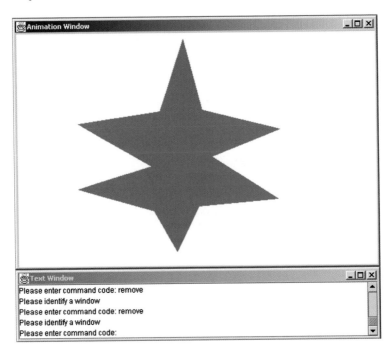

Figure 15.22 Remove the yellow polygon

```
static void copy(Color cc){
    Point p=new Point(2); Picture pict=null;
    IO.writeString("Please identify object \n");
    ListElement refwn=null; Win win = (Win)(refwn = identifyWindow(p)).object;
    if(win!=bw){
        int x= win.x1+12; int y= win.y1+12;
        if(win.pic.name.equals("line")){
            LineSeg newLs= ((LineSeg)win.pic.obj).copy(cc);
            pict = new Picture(null,cc,"line",newLs, x, y,win.w ,win.h);
            Win wint = makeWindow(pict,cc);  return;
        }else {
            Color[][] col= new Color[win.pic.w][win.pic.h];
            pict = new Picture(col,cc,win.pic.name,null, x, y, win.w , win.h);
            for(int i=0;i<win.pic.w;i++){  for(int j=0;j<win.pic.h;j++){
                if(win.pic.getColour(i,j)!=null)pict.setColour(cc,i,j);
            } }
            if(win.pic.name.equals("polygon")){
                pict.obj = ((Polygon)win.pic.obj).copyPolygon(dW,IO);}
            Win wint = makeWindow(pict,cc);
            wint.pic.cc=cc;  dW.c.repaint();
        }
    }
}
```

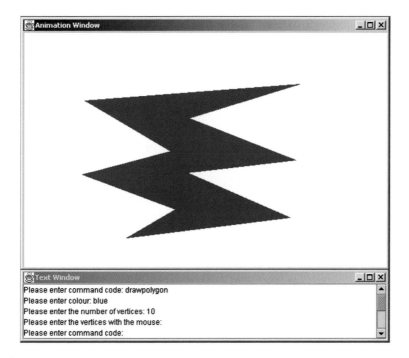

Figure 15.23 Enter a blue polygon

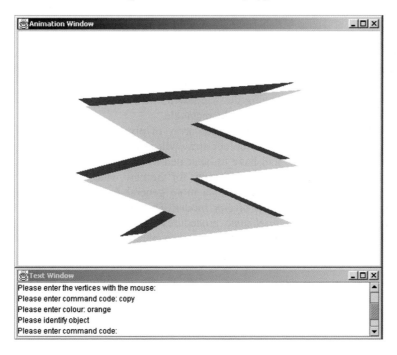

Figure 15.24 Create an orange polygon copy

Copying a rectangle is a relatively simple task. At the minimum a duplicate *Picture* element is created with a direct copy of its colour array, followed by changing its location coordinate and constructing and displaying the new Window. In the example in Figure 15.11 the extra facility to change the rectangle colour has been added to distinguish the source and copy rectangles. Copying a polygon can be started in the same way, however it is also necessary to duplicate the polygon definition in the *Picture* object. This is done by calling the *copyPolygon()* procedure from the *Polygon* class. The reason for doing this is that any subsequent use of a *resize* operation will require the polygon to be redrawn and the new boundary filled to give a new *Color* array in the copy's *Picture* object. The result of applying this operation to a polygon is shown in Figure 15.24. Again the new polygon has had its colour changed to distinguish it from its source.

```
class Polygon{ ....
    public Polygon copyPolygon(DisplayWindow dW,TextWindow IO){
        Polygon poly = new Polygon(this.length,dW);
        for(int k=0;k<this.length;k++){
            Point np = new Point(2);
            np.n[1]=this.p[k].n[1]; np.n[2]=this.p[k].n[2];
            poly.p[k] = np;
        }return poly;
}
```

The *resize* command is the simplest *transformation* command, but it still is more complicated to implement than the *move* commands. There are two ways of viewing the resize command when it is applied to a window rectangle. The first is to treat the rectangle as a "viewport" through which the contents of the window can be viewed. In this case the change in size of the rectangle will either reveal more or less of the contents. The second is as the min-max box defining the extent of the content *Picture* it contains. In this case changing the window frame will require the contents to be scaled up or down to continue to fill the rectangle.

In the first case either selecting a sub-array of pixel values or transferring the original *Color* array into a larger colour array will implement the task. In the second case a rectangular picture will have to be scaled by mapping the values held in the source *Color* array to a resized pixel array for the new window. This is a texture-mapping task that will be revisited in a more general form in a later chapter. A polygon however exists as a *Polygon* model. This can be scaled to fit the new window using the *CoordinateFrame* procedures introduced in Chapter 3. In both cases the original definition, either as a source picture-array or as a construction model, will have to be held separately from the current transformed model or *Color* array used for display.

The *resizeWindow()* procedure requires two points to define a movement vector. The first is used to select a window corner. The second is used to move this corner to a new location. Depending on which corner is moved, the rectangle vertex coordinates are adjusted accordingly. To apply this approach to a polygon requires the borders of the min-max box to be displayed in order for the user to identify a corner to move.

```
static Picture resizeWindow(Win win,Win wb){
    int x1=win.x1;int y1=win.y1;int x2=win.x2;int y2=win.y2;
    int xa=wb.x1;int ya=wb.y1;int xb=wb.x2;int yb=wb.y2;
    if((x1<xa)||(y1<ya)||(x2>xb)||(y2>yb)){ IO.writeString("cannot resize clipped object \n");
    }else{
        drawBoundingRectangle(win,wb);
        IO.writeString("Please select a corner \n");
        Point p1 = dW.getCoord(); int xx= p1.xi(), yy= p1.yi();
        int [] x= new int[4];  x[0]=win.x1; x[1]=win.x2; x[2]=win.x2; x[3]=win.x1;
        int [] y= new int[4];  y[0]=win.y1; y[1]=win.y1; y[2]=win.y2; y[3]=win.y2;
        int j=0;int d=Integer.MAX_VALUE;int dd=0;;
        for(int i=0;i<4;i++){
            dd = (xx-x[i])*(xx-x[i])+(yy-y[i])*(yy-y[i]);
            if(d>dd){ j=i; d=dd;}
        }
        IO.writeString("Please give its new location \n");
        Point p2 = dW.getCoord(); xa=p2.xi();  ya=p2.yi();
        int width=10,height=10;
        switch(j){
        case 0:if(xa>x[2]-10)xa =x[2]-10; if(ya>y[2]-10)ya =y[2]-10;
                xx=xa;  yy=ya;  width= x[2]-xa;  height = y[2]-ya;  break;
        case 1:if(xa<x[3]+10)xa=x[3]+10;  if(ya>y[3]-10)ya=y[3]-10;
                xx= x[3];  yy= ya;  width= xa-x[3];  height= y[3]-ya;  break;
        case 2:if(xa<x[0]+10)xa=x[0]+10;  if(ya<y[0]+10)ya=y[0]+10;
                xx=x[0];  yy=y[0];  width= xa-x[0];  height= ya-y[0];  break;
        case 3:if(xa>x[1]-10)xa=x[1]-10;  if(ya<y[1]+10)ya=y[1]+10;
                xx=xa;  yy=y[1];  width= x[1]-xa;  height= ya-y[1];  break;
        }
        Picture pict=null;
        if(win.pic.name.equals("rectangle")){
            Color[][] pic = new Color[width][height];  Color bb = Color.lightGray;
            if(win.pic.cc!=null)bb= win.pic.cc;
            pict = new Picture(null,bb,"rectangle",null,xx,yy,width,height);
        }
        else if(win.pic.name.equals("polygon")){
            CoordinateFrame fm= win.pic.setScaling(width+1,height+1);
            Polygon poly = ((Polygon)win.pic.obj).scalePolygon(fm);
            Rectangle r= new Rectangle();
            Color c[][] = dW.c.polygonfill(poly,win.pic.cc,Color.gray,r);
            pict = new Picture(c,win.pic.cc,"polygon",poly ,xx,yy,c.length,c[0].length);
        }
        else if(win.pic.name.equals("line")){
            LineSeg newLn = ((LineSeg)win.pic.obj).scaleLine(xx,yy,width,height);
            pict=new Picture(null,newLn.cc,"line",newLn, xx,yy,width,height);
        }return pict;
    } return win.pic;
}
```

Once the new rectangle frame has been defined it can be used to display a new window rectangle. In the example given above this merely paints a uniform colour into the new rectangle *Picture*, however if the rectangle contained an image this would have to be scaled. In the case of the polygon the *Polygon* model is scaled to the new window size and repainted a new colour. A new CoordinateFrame is generated relating the original polygon size to the new size using the original rectangle vertex coordinates and the new min-max box, vertex, coordinate-values. This is then passed to the polgon's *scalePolygon()* procedure to generate the new polygon model, which is then passed to the *polgonfill()* procedure of the *WorkPanel* class to get the new *Picture* object for the polygon. The *resizeWindow()* procedure generates a new *Picture* object for both polygons and rectangles, which is then passed back to the calling program to be rendered in a new window once the original one has been removed. The resize operation is illustrated for both polygons and rectangles in Figures 15.25 to 15.28.

```
class Polygon{ ....
    public Polygon scalePolygon(CoordinateFrame fm){
        Polygon poly = new Polygon(this.length,dW);
        for(int k=0;k<this.length;k++){
            Point np = fm.scaleWtoS(this.p[k]);
            poly.p[k] = np;
        }return poly;
    }
}
```

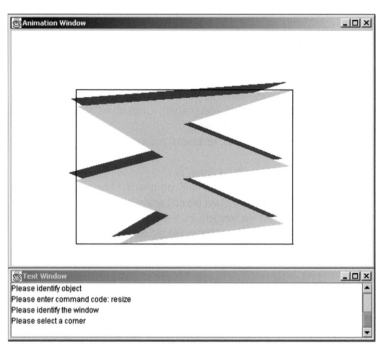

Figure 15.25 Select the orange polygon for resizing

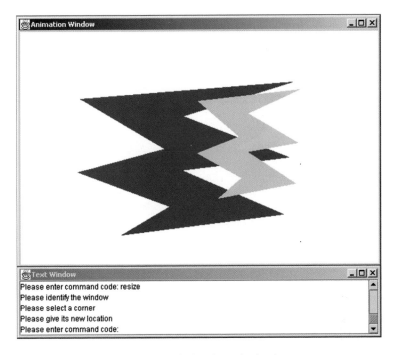

Figure 15.26 Display the resized polyon

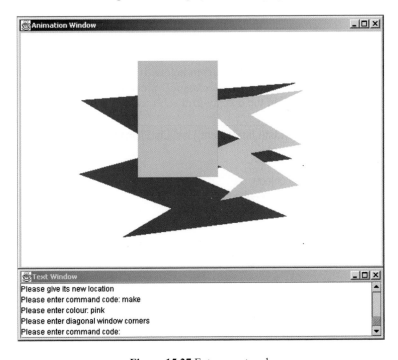

Figure 15.27 Enter a rectangle

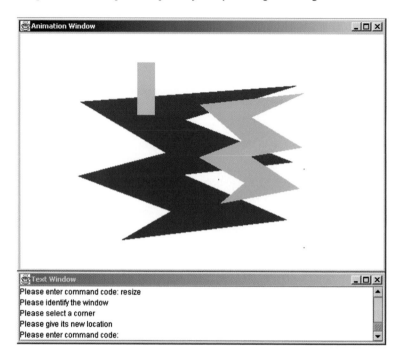

Figure 15.28 Resize the rectangle

When moving from simple rectangles to rectangles with transparent entries, it is useful to collect an ordered sub list of the window rectangles that contain a test point to reduce the number of pixel level tests on each of the overlaid window layers. This is done by the *windowList()* procedure given below. It generates an array of window references which it passes to the *resetRectangle()* procedure which searches it from the top for the visible pixels in the display-grid sub-rectangle containing the point.

```
static private int windowList(ListElement[] lse, ListElement ref,Point pc){
    int k=0;
    while(ref!= null){
        Win wn=((Win)ref.object);
        if((pc.xi()-wn.x2)*(pc.xi()-wn.x1)<=0){
            if((pc.yi()-wn.y2)*(pc.yi()-wn.y1)<=0){
                lse[k++]=ref;
                if(k>= lse.length){
                    IO.writeString("too many layers for array length "+k+"\n");
                    return k;
                }
                int wx= ((Win)ref.object).x1;  int wy=((Win)ref.object).y1;
        } }
        ref=ref.right;
    }return k;
}
```

The *resetRectangle()* code is given below. Given a list of windows potentially visible in a sub-rectangle of the screen it processes all the pixel locations within the rectangle identifying the non-transparent, in other words the first visible pixel encountered processing the list of windows in order from the top. This is used to generate a *Picture* object for the visible pixels within the rectangle, which is then rendered by the workPanel procedure *setPixels()*.

```
static boolean resetRectangle(Win bw, ListElement[] lse,
                              int xl,int yl,int xr,int yr,int kk){
    int xleft=xl;  int xright=xr;  int yleft=yl;  int yright=yr;
    if(bw!=null){
        if(xr<=bw.x1)return false; if(xl>=bw.x2)return false;
        if(yr<=bw.y1)return false; if(yl>=bw.y2)return false;
        if(xl<bw.x1)xleft= bw.x1;  if(xr>bw.x2)xright=bw.x2;
        if(yl<bw.y1)yleft= bw.y1;  if(yr>bw.y2)yright=bw.y2;
    }
    Win wn =null;  Color cc=null;  Color [][] pict = new Color[xright-xleft+1][yright-yleft+1];
    for(int i=0;i<xright-xleft;i++){
        for(int j=0;j<yright-yleft;j++){
            int k=0; boolean transparent = true;
            while(transparent && k<kk){
                wn = (Win) lse[k].object;    int wx= wn.x1; int wy = wn.y1;
                int xa = xleft-wx+i; int ya = yleft-wy+j;
                cc= wn.pic.getColour(xa,ya);
                if(cc!=null){ pict[i][j]= cc; transparent = false;}
                k++;
            }
        }
    }
    Picture pic = new Picture(pict,null,"rectangle",null,xleft, yleft,xright-xleft ,yright-yleft);
    dW.c.setPixels(xleft,yleft,xright-xleft,yright-yleft,pic);
    return true;
}
```

The *setColour()* procedure converts the name of a colour entered as a String to the matching *Color* object from the Java *Color* class. The *drawBoundingRectangle()* procedure draws a line round a window rectangle to support the *resize()* operation.

```
static void drawBoundingRectangle(Win ws,Win wb){
    int x1=ws.x1;int y1=ws.y1;int x2=ws.x2;int y2=ws.y2;
    int xa=wb.x1;int ya=wb.y1;int xb=wb.x2;int yb=wb.y2;
    if(x1<xa)x1=xa; if(y1<ya)y1=ya;if(x2>xb)x2=xb;if(y2>yb)y2=yb;
    dW.c.line(x1,y1,x1,y2-1,Color.black,null,true);
    dW.c.line(x1,y2-1,x2-1,y2-1,Color.black,null,true);
    dW.c.line(x2-1,y2-1,x2-1,y1,Color.black,null,true);
    dW.c.line(x2-1,y1,x1,y1,Color.black,null,true);
}
```

Drawing Lines

This system can be extended by a series of further command-line functions. The obvious next addition is that of drawing lines. It was possible to handle lines using the same approach used for polygons, however it required a large amount of memory setting up a min-max box of pixels merely to represent a line segment. Since most of such a pixel array ends up being empty or null, one possibility to save memory is to use the edge-line representation used in the polygon fill algorithm: a one-dimensional array of x values indexed by the lines y values, or an array of y values indexed by the associated x values, depending on the orientation of the line. The *Shadings* class is already able to generate this representation.

```
if(str.equals("drawline")){
    cc = getColour();
    pict = getLine(cc);
    win = makeWindow(pict,null);
}
```

The *drawline* command can be included in the same way that the *drawpolygon* command line was added. What this approach achieves is the ability to delete lines without leaving gaps, which provides the basis for an interactive drawing system. Previously the only option was to redraw a deleted line in the background colour, but this clearly generated holes in any other lines or polygons that the line crossed.

```
static Picture getLine(Color cc){
    IO.writeString("Please enter the line's end points\n");
    int minx,maxx,miny,maxy,w,h;
    Point p1= dW.getCoord();
    Point p2= dW.getCoord();
    LineSeg ln= new LineSeg(IO,dW,p1,p2,cc);
    Picture pict = new Picture(null,cc,"line",ln, ln.xmin, ln.ymin,ln.w ,ln.h);
    return pict;
}
```

There are still a variety of improvements and extensions. The test to remove the line in Figure 15.31 had to identify a pixel point belonging to the line. In this example this required two attempts to identify the line. Testing the pixel point provided by the mouse for its proximity to the line in the way outlined in Chapter 13 makes this interaction easier to execute. The *resize* procedure had to be extended to have the ability to redraw a scaled line. The memory space used to hold the min-max box for each line was reduced to the linear array set up in the *Shadings* objects, but this required the tests previously made on the objects *Picture Color* array to be changed to a function call: both to locate which objects have visible pixels in a search rectangle, and to allow a polygon or line object to be selected by the mouse pointer.

if(pict.cc[i][j]!=null){...}   changed to   if(pict.getColour (i,j)!=null){....}

The *LineSeg* class like the *Polygon* class provides line objects of the required type.

```
class LineSeg{
    public boolean up = false;  public Color cc = null;  public int pnts [] = null;
    public int length = 0,w=0,h=0;  public Point p1=null;Point p2=null;
    public int xmin=0,xmax=0,ymin=0,ymax=0;
    public TextWindow IO;  public DisplayWindow dW;
    public LineSeg(TextWindow txt,DisplayWindow dw,Point pa,Point pb,Color col){
        this.p1=pa;this.p2=pb;this.cc=col;this.IO=txt;this.dW=dw;
        if (p1.xi()<p2.xi()){xmin = p1.xi();xmax= p2.xi();}
        else {xmax = p1.xi();xmin= p2.xi();}
        if (p1.yi()<p2.yi()){ymin = p1.yi();ymax= p2.yi();}
        else {ymax = p1.yi();ymin= p2.yi();}
        p1.n[1]= p1.n[1]-xmin;  p1.n[2]= p1.n[2]-ymin;
        p2.n[1]= p2.n[1]-xmin;  p2.n[2]= p2.n[2]-ymin;
        w= xmax-xmin+1; h=ymax-ymin+1;
        this.up =true;
        int len = w;
        if(w<h){this.up=false;len= h;}
        pnts = new int[len];
        Shadings s= new Shadings(IO,len,0,up);
        dW.c.line(p1.xi(),p1.yi(),p2.xi(),p2.yi(),cc,s,true);
        for(int i=0;i<len;i++) this.pnts[i]= s.leftedge[i];
    }
    public LineSeg copy(Color col){
        Point pa= new Point(2); Point pb= new Point(2);
        pa.n[1]= this.p1.n[1]; pa.n[2]= this.p1.n[2];
        pb.n[1]= this.p2.n[1];pb.n[2]= this.p2.n[2];
        return new LineSeg(IO,dW,pa,pb,col);
    }

    public void setColour(Color cc){ this.cc = cc; }
    public Color getColour(int x,int y){
        if((x<0)||(x>=this.w)||(y<0)||(y>=this.h))return null;
        if(up){if(y==pnts[x])return cc;else return null;}
        else{ if(pnts[y]==x)return cc;else return null;}
    }

    public LineSeg scaleLine(int x,int y, int ww, int hh){
        Point pc=new Point(2);Point pd=new Point(2);
        if(this.p1.xi()== this.w-1){pc.n[1]=ww-1;}else{pc.n[1]=0;}
        if(this.p1.yi()== this.h-1){pc.n[2]=hh-1;}else{pc.n[2]=0;}
        if(this.p2.xi()== this.w-1){pd.n[1]=ww-1;}else{pd.n[1]=0;}
        if(this.p2.yi()== this.h-1){pd.n[2]=hh-1;}else{pd.n[2]=0;}
        LineSeg lineSg = new LineSeg(IO,dW,pc,pd,this.cc);
        return lineSg;
    }
}
```

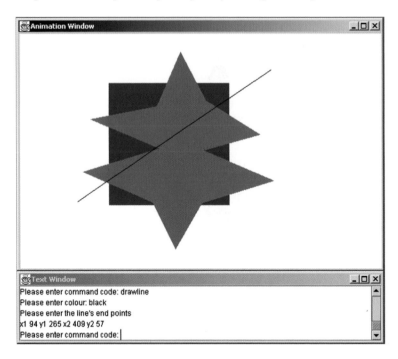

Figure 15.27 Drawing a line on top of a polygon and a rectangle

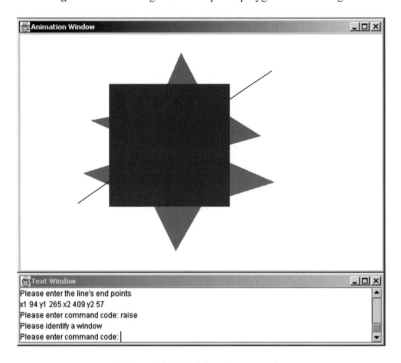

Figure 15.28 Raising the rectangle

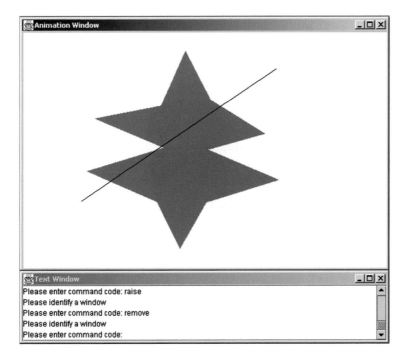

Figure 15.29 Delete the rectangle

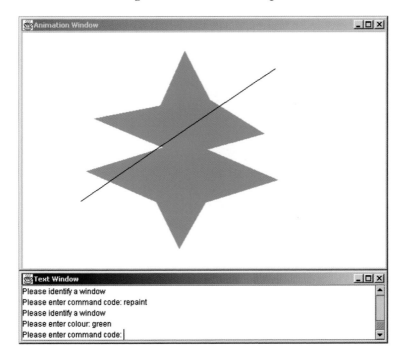

Figure 15.30 Repaint the polygon

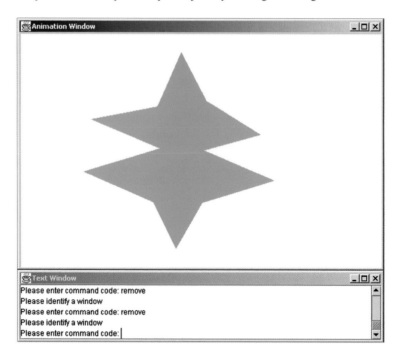

Figure 15.31 Deleting the line at the second attempt

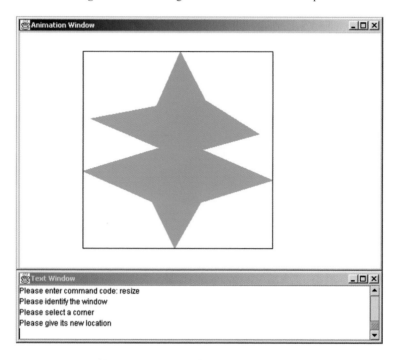

Figure 15.32 Resize the remaining polygon

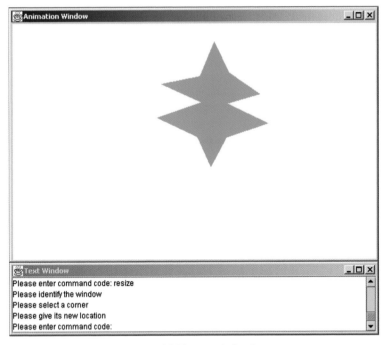

Figure 15.33 The rescaled polygon

The use of the drawing system outlined in this chapter is based on a simple command line language. This is cumbersome to use but has several merits. Each command employed in a drawing task identifies a step in producing the current display. If a mistake is made then if the sequence was a long one it would have to be repeated. The alternative is to make each command reversible and then keep the command and the data needed to undo it. This is a very necessary and powerful facility, for interactive work, that makes it possible to return step by step to an earlier state in preparing a display. This allows a trial and error approach to be adopted, which greatly enhances the flexibility of much interactive work.

DisplayWindow

The *DisplayWindow* class has provided the display facilities used to illustrate the graphic algorithms, up to this point with the minimum of interaction. The *getCoord()* procedure from the *DisplayWindow* has provided this interaction using the mouse to get point coordinates to build display objects. In the next chapter ways of improving interactive working are explored. This starts by extending the interactive operations that can be set into motion using the mouse. The procedures provided for the current graphics interface are given in the *DisplayWindow* class listed below.

```
class DisplayWindow extends JFrame {
    public Graphics g;  public int w,h;
    public TextWindow tW = null;
    protected WorkPanel c;
```

```
DisplayWindow(TextWindow tw,int x, int y, int width,int height,Color cc){
    w = width; h = height;this.tW = tw;
    Container pn= this.getContentPane();
    pn.setLayout(new FlowLayout(FlowLayout.LEFT,2,2));
    setSize(width,height);   Dimension d = this.getSize();
    addWindowListener(new WindowAdapter (){
        public void windowClosing(WindowEvent e){ System.exit(0); } });
    pn.add(c = new WorkPanel(tW,width,height));
    setTitle("Animation Window");   setLocation(x,y);
    pn.setBackground(Color.white);
    show();
}
public void setTextWindow(TextWindow tw){ this.tW = tw; }
public boolean clipBox(Box bx,Point p1,Point p2){
    int j=0, i=0;   i = classify(p1.xi(),p1.yi(),bx);   j = classify(p2.xi(),p2.yi(),bx);
    if((i==0)&&(j==0))return true;
    if((i&j&15)!=0)return false;
    if(edgeClip(p1,p2,bx.minP.xi(), 1,false))return false;
    if(edgeClip(p1,p2,bx.maxP.xi(),-1,false))return false;
    if(edgeClip(p1,p2,bx.minP.yi(), 1, true))return false;
    if(edgeClip(p1,p2,cbx.maxP.yi(),-1, true))return false;
    return true;
}
private int classify(int x, int y, Box bx){
    int j=0;
    if(x < bx.minP.xi())j=j+1;   if(x > bx.maxP.xi())j=j+2;
    if(y < bx.minP.yi())j=j+4;   if(y > bx.maxP.yi())j=j+8;
    return j;
}
private boolean edgeClip(Point p1,Point p2,double e,double d,boolean up){
    double x1,x2,y1,y2,k1,k2;
    if(up){x1= p1.yd();x2=p2.yd();y1=p1.xd();y2=p2.xd();}
    else{x1= p1.xd();x2=p2.xd();y1=p1.yd();y2=p2.yd();}
    k1= x1-e;k2=x2-e;
    if(k1*k2<=0){
        if(k1*d<0){ y1=(k2*y1-k1*y2)/(k2-k1);
            if(up){p1.y("=",e);p1.x("=",y1);}else {p1.x("=",e);p1.y("=",y1);}
        }if(k2*d<0){ y2=(k2*y1-k1*y2)/(k2-k1);
            if(up){p2.y("=",e);p2.x("=",y2);}else {p2.x("=",e);p2.y("=",y2);}
        }
    } else if(k1*d<0)return true;
    return false;
}
public void line(int x1,int y1,int x2, int y2,Color color){ c.line( x1, y1, x2, y2, color); }
public void plotPoint(Point p){ c.setPixel(p.xi(),p.yi(),Color.black);}
public void plotPoint(Point p,Color cc){ c.setPixel(p.xi(),p.yi(),cc); }
public void plotPoint(int x, int y,Color cc){ c.setPixel(x,y,cc); }
```

```
public void plotPoint(Point pp,CoordinateFrame b)
    {  Point p = b.scaleWtoS(pp); c.setPixel(p.xi(),p.yi(),Color.black); }
public void plotPoint(Point pp,CoordinateFrame b,Color cc)
    {  Point p = b.scaleWtoS(pp); c.setPixel(p.xi(),p.yi(),cc);}
public void plotLine(Point p1,Point p2)
    {  c.line(p1.xi(), p1.yi(), p2.xi(), p2.yi(),Color.black);}
public void plotLine(Point p1,Point p2,Shadings s)
    {  c.line(p1.xi(), p1.yi(), p2.xi(), p2.yi(),Color.black,s); }
public void plotLine(Point p1,Point p2,Color C)
    {  c.line(p1.xi(), p1.yi(), p2.xi(), p2.yi(),C); }
public void plotLine(Point p1,Point p2,Color C,Shadings s)
    {  c.line(p1.xi(), p1.yi(), p2.xi(), p2.yi(),C,s);  }
public void plotLine(int x1,int y1,int x2,int y2,Color C) {  c.line(x1, y1, x2, y2,C);}
public void plotLine(int x1,int y1,int x2,int y2,Color C,Shadings s)
    {  c.line(x1, y1, x2, y2,C,s); }
public void polygonFill(Polygon p,Color color){ c.polygonfill(p,color); }
public void polygonFill(Polygon p,Color color,CoordinateFrame b){
    Polygon np = new Polygon(p.length);
    for(int i =0;i< p.length;i++){  np.p[i] = b.scaleWtoS(p.p[i]); }
    c.polygonfill(np,color);
}
public void plotRectangle(int x, int y,int r,Color cc){
    Point p1 = new Point(2); Point p2 = new Point(2);
    p1.x("=",x-r);  p1.y("=",y-r);  p2.x("=",x+r+1);  p2.y("=",y+r+1);
    plotRectangle(p1,p2,cc);
}
public void plotRectangle(Point p1,Point p2,Color cc, CoordinateFrame b){
    Point pa = b.scaleWtoS(p1);   Point pb = b.scaleWtoS(p1);   plotRectangle(pa,pb,cc);
}
public void plotRectangle(Point p,int r,Color cc){
    Point p1 = new Point(2);  Point p2 = new Point(2);
    p1.x("=",p.xi()-r);  p1.y("=",p.yi()-r);  p2.x("=",p.xi()+r);  p2.y("=",p.yi()+r);
    plotRectangle(p1,p2,cc);
}
public void plotRectangle(int x,int y,int w,int h,Color cc){
    Color [][] col = new Color[w][h];
    for(int i = 0;i<w;i++){ for(int k = 0; k<h; k++){ col[i][k] = cc;}  }
    Picture pic= new Picture(col,null,null,null,x,y,w ,h);
    c.setPixels(0,0,x,y,w,h,pic);this.repaint();
}
public void plotRectangle(Point p1,Point p2,Color cc){
    int x1,x2,y1,y2;
    if (p1.xd() < p2.xd()) { x1 = p1.xi(); x2 = p2.xi(); } else {x1 = p2.xi(); x2 = p1.xi();}
    if (p1.yd() < p2.yd()){ y1 = p1.yi(); y2 = p2.yi(); } else {y1 = p2.yi(); y2 = p1.yi();}
    plotRectangle(x1,y1,x2-x1,y2-y1,cc);
}
public void clearDisplay(Color cc){     this.c.clearPanel(cc);}
```

```
public void plotTriangle(Point p1,Point p2,Point p3,Color color,Color cc){
    Polygon p = new Polygon(4);   p.p[0]=p1; p.p[1]=p2; p.p[2]=p3; p.p[3]=p1;
    this.c.polygonfill(p,color);     for(int i=0;i<3;i++) plotLine(p.p[i],p.p[i+1],cc);}
public void plotTriangle(Point p1,Point p2,Point p3,Color cc,Color c,
                                                            CoordinateFrame b){
    Point pa = b.scaleWtoS(p1);     Point pb = b.scaleWtoS(p1);
    Point pc = b.scaleWtoS(p1);  plotTriangle(pa,pb,pc,cc,c);
}
public Point getCoord(){
    this.c.newMouseEvent = false;   Point point = new Point(2);
    while (!this.c.newMouseEvent)Dummy.dummy();
    point.n[1] = this.c.mx;  point.n[2] = this.c.my;  return point;
}
public Point getCoord(CoordinateFrame b){
    this.c.newMouseEvent = false;     Point point = new Point(2);
    while (!this.c.newMouseEvent)Dummy.dummy();
    point.n[1] = this.c.mx;   point.n[2] = this.c.my;
    return b.scaleStoW(point);
}
public void quit(){System.exit(0);}
}
```

The overlay operation using matching pixel grids makes spatial operations for many applications simpler than the approach outlined in chapter 14. Values from matching grid cells can be combined together using what is in effect a non-spatial algorithm. This is the same approach employed in early cartography systems that generated graphic output for lineprinters. This was used in the Laboratory for Computer Graphics and Spatial Analysis in Harvard University, during the 1960's for both the SYMAP system and the Geographic Information System used by planners called GRID. The subsequent use of vector graphics in GIMMS and OBLIX in order to use line plotters was motivated by the desire to change scale automatically where grid based systems were memory intensive and the main computer system only had 32 K words of core memory.

The use of vector graphics in SKETCHPAD by Ivan Sutherland in 1963, had introduced the fast interactive editing facilities of which rubber banding is a notable example, to refresh cathode ray tube display systems. The return to raster displays in the 1970's in Utah, regained the ability to shade and colour images, but when transferred to TV display systems lost this new capability until fast block memory transfers and double-buffered display systems were introduced. This in turn depended on the development of cheap, mass integrated circuit memory. The two approaches offer different advantages. The ability to change size for polygons is best executed using the vector model, while simple movements and overlap operations are easiest to implement using raster systems. The interactive speed of editing has totally changed the task of drawing and combined with the way raster systems can also use camera captured images, provides one of the main advance that computer graphics provides the artist, designer and animator, over traditional manual techniques.

16

GUI: Graphic User Interfaces: Control Design Animation & Simulation Systems

Introduction

Having explored the initial task defined in chapter one: which was how graphic products can be generated using a computer language, the next task is to examine the role of graphics as a modelling medium following the explosive development of computer based systems. The idea that graphics could provide a view of complex computer based operations and their outcome, grew up fairly early on, programs were designed using flow charts, tables of data were turned into graphs and pie charts, and the desire to automate drawing for aircraft and automobile design work emerged fairly naturally from the need to be able to understand the results from mathematically modelling their surface shapes to improve their aerodynamic properties by exactly defining the surfaces in a repeatable way.

The concept that a graphics display could provide the communication interface between the different levels of processing information carried out by computer systems and their users, was the motivation behind developing the specialised hardware to create displays. Traditionally graphics had three roles: information storage, information communication and for many applications information analysis. The ability of the computer system to greatly extend the power and role of mathematical modelling, and the consequent ability to store these models numerically or symbolically, has left graphics the remaining role of communication between people and machines, but this has turned out to be critical to harnessing the developing potential of computer based systems.

A. Thomas, *Integrated Graphic and Computer Modelling*,
DOI: 10.1007/978-1-84800-179-4_16, © Springer-Verlag London Limited 2008

Generating Displays

The artist, sculptor, designer and draughtsman, traditionally, have built up their pictures, models, and drawings, interactively working directly with the display medium. The main technical problem addressed in this book, so far, has been how to create traditional graphic forms of output merely using language commands, without interaction. The exception, perhaps, has been capturing point coordinates using the mouse, but this is interactive in a limited sense that the position of the mouse pointer is adjusted on the screen until it gives the required location. The convenience this provides identifies an important requirement: in most design and construction work it is difficult to get what is wanted without a feedback and modification cycle. The keyword command language developed in the last chapter starts to introduce a level of interaction but still only provides the user with a clumsy alternative to working directly with a physical system or medium.

Windows, Icons, Mice and Pointers: WIMP Interfaces

This chapter explores some of the simple facilities that allow an interactive graphics user interface to be built to provide better access to visual feedback. They are implemented in a basic form to illustrate the way they work, and to examine the extra capabilities they provide the computer system.

The light pen used in the SKETCHPAD system allowed vector graphics displays to be interactively modified in a laboratory system in 1963. The digitising tablet followed a series of devices developed through the 60's from digitising tables used for large drawings and cartographic maps, to tracker balls and joysticks. Finally the mouse evolved as an essential component in the WIMP interfaces, composed of Windows, Icons, Mice and Pointers, that provide the current common form of the Graphics User Interface.

Essentially what these systems provide in one form or another is a refresh display that contains the target image, which can be divided into a series of overlaid virtual rectangular surfaces -- *Windows* -- on which text and graphics can be displayed. A position capturing device that feeds the computer system with coordinates that can be modified by moving the device: a pen or scribe on a sensitive panel, a tracker ball that senses the movement of a ball, or the inverted tracker ball that was the early *Mouse*, which again captures the movement of the unit from a ball that is rotated when it moves. The inputs from these devices are transferred to control a *Pointer* or cursor shown in the display. A function key either on the keyboard or on the tracking device can capture the position of the pointer, relative to the display and this can then be used by the computer system to control subsequent actions either in the display or on running programs.

It is the use of *Icons* that gives the extra fluency over the interactive command language explored in the last chapter and provides a less clumsy system to use. The icon is a small symbol or image displayed on the screen that, when it is selected by the pointer, can be used to activate a command. There is a variety of ways in which basic interactive mechanisms are implemented in current systems using "buttons" and "menus", and "sliders", and even haptic devices that allow three-dimensional sculpture to be implemented in a hands-on manner.

As was outlined in the chapter on hardware systems controlled by "language" inputs, the display system has its own machine code commands, and this output unit with input devices such as the mouse, digitising tablets or TV cameras, have to be controlled by programs in the operating system of the host computer. The programs managing these devices are generally run as independent processes so they can act autonomously in their own time frame. This means that providing the feedback loop between mouse inputs, for example, and the display system has to be provided through the operating system or indirectly through the language system.

The original basic WIMP structure has been extended and modified in a variety of ways but in principle remains the same and supports the wide range of applications that now depend on man–machine communication through a graphic user interface. The development of computer systems with interactive capabilities has extended in the many ways summarised by the original diagram in Figure 1 in the preface.

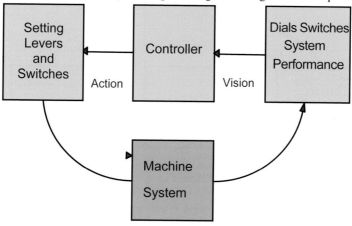

Figure 16.1 Manual control of a machine system

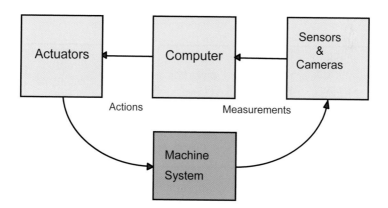

Figure 16.2 Computer control of a machine system

One of the early targets using computer "intelligence" was to automate menial repetitive tasks. The aim was the totally autonomous system in Figure 16.2. This was difficult to achieve in many cases. Human intervention was still necessary even where some tasks could be automated in this way. The result was a mixed system shown in Figure 16.3

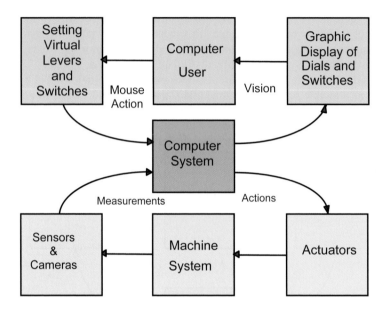

Figure 16.3 Interactive control using a computer system

In order for this kind of system to work it was usually necessary to have a model of the machine system within the computer system. Which comes first the machine or its model is becoming a chicken and egg question. The role of system simulation for estimating potential system behaviour and for design purposes is developing as an important separate line of development.

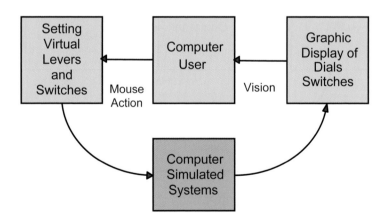

Figure 16.4 Interactive control of a computer simulation

The facilities needed to build the graphic interface for a simulated computer system have already been developed using the Java libraries. Figure 16.5 shows the use of various graphic feedback elements. Buttons, display windows for text giving input and output data flows, switches showing the system configuration to implement each instruction and a table giving the contents of memory seen through a view-port that can have its field of view adjusted with a slider.

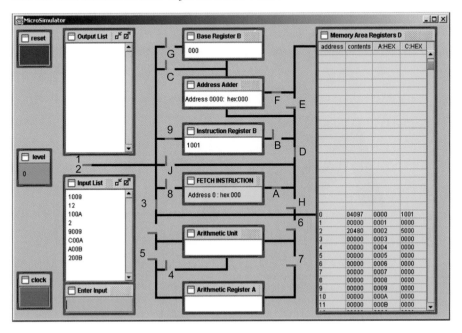

Figure 16.5 Computer based (computer) machinery simulation

Building Icons, Buttons, Menus and Sliders

Given the ability to obtain mouse inputs using the Java system, the keyword command language of chapter 15 can be implemented in a way that is intuitively easier to operate by using buttons, icons and menus.

For many operations, it is possible to extend the existing basic graphic algorithms to create the icons and buttons to simplify interactive work. The first task is to extend the polygon shading to include the procedures developed in chapter 11. This will allow a full range of single boundary polygons to be rendered, as well as giving the capability to draw areas with curved boundaries. The first will be input extending the *"drawpolygon"* command; the second will require a new command *"drawshape"*. Once this is done the next step is to implement a drawing grid to make it easier to get horizontal and vertical lines, in diagrams. Finally a *"make-icon"* command is needed to convert a polygon or curved shape into an icon. This command will scale the polygon or shape to a standard sized rectangle, allow it to be placed in its working position in the display space and then request an icon identifier for the new icon object. Once all the icons are in place then a command to activate the icons will make the act of clicking on the icon execute the command it represents. Before this is done it is necessary to implement a *deactivate-icons* icon, in order to return control to the command language if so required. A final system may be set up totally using icons however, while developing the system it proved very useful to be able to move backwards and forwards between the two approaches.

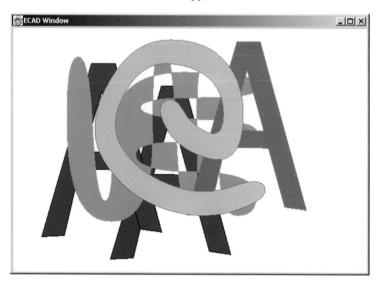

Figure 16.6 "Drawpolygon" and "drawshape" objects

Figure 16.7 Overlaying a grid to help drawing

Icons can be set up in the way shown in Figures 16.8 to 16.11. The *make-icon* command requests a shape or a polygon and then scales it to a standard size. It then places it where ever is convenient in the display space.

```
private static Win drawPolygon(){
    Color cb=getColour("please enter the boundary colour\n");
    Color cc=getColour("please enter the fill colour\n");
    IO.writeString("Please enter the number of vertices: ");
    int num=IO.readInteger(); IO.readLine();
    Point [] poly = new Point[num+1];
    IO.writeString("Please enter the vertices with the mouse: \n");
    poly[0]=dW.getCoord();
    for(int kk=1;kk<num;kk++){
        poly[kk]= dW.getCoord();  dW.plotLine(poly[kk-1],poly[kk],Color.blue);
    }
    poly[num]=new Point("=",poly[0]);
    dW.plotLine(poly[num-1],poly[num],Color.blue);
    dW.getCoord();
    Polygon ppp= new Polygon(num+1);  ppp.p=poly;
    int nn = ppp.length*2-1;
    Rectangle r = new Rectangle();
    Color [][]c= dW.c.polygonfill(ppp,cc,cb,r);
    Picture pict = new Picture(IO,c,cc,"polygon",ppp,r.x,r.y,r.width,r.height);
    pict.cc= cc;
    Win win = makeWindow(pict,cc);
    win.pic.cb= cb;
    return win;
}
private static Win drawshape(){
    Color cb=getColour("please enter the boundary colour\n");
    Color cc=getColour("please enter the fill colour\n");
    IO.writeString("Please enter the number of vertices: ");
    int num=IO.readInteger(); IO.readLine();
    Point [] poly = new Point[num+1];
    IO.writeString("Please enter the vertices with the mouse: \n");
    poly[0]=dW.getCoord();
    for(int kk=1;kk<num;kk++){
        poly[kk]= dW.getCoord(); dW.plotLine(poly[kk-1],poly[kk],Color.blue);
    }
    poly[num]=new Point("=",poly[0]);
    dW.plotLine(poly[num-1],poly[num],Color.blue);
    dW.getCoord();
    Polygon ppp= new Polygon(num+1); ppp.p=poly;
    Rectangle r = new Rectangle();
    Color [][]c= dW.c.shapefill(ppp,cc,cb,r);
    Picture pict = new Picture(IO,c,cc,"shape",ppp,r.x,r.y,r.width,r.height);
    pict.cc= cc;
    Win win = makeWindow(pict,cc);
    win.pic.cb=cb;
    return win;
}
```

```
private static void makeicon(){
    IO.writeString("is the icon curved y/n \n");
    Win w =null;
    String str=IO.readString(); IO.readLine();
    if(str.equals("n"))w= drawPolygon();
    else w= drawshape();
    ListElement refwn = LofWs.start;
    IO.writeString("please enter location \n");
    Point p = dW.getCoord();
    int xx= p.xi(); int yy= p.yi();
    int width = 25, height = 25;
    if((w.pic.name.equals("polygon"))||(w.pic.name.equals("shape"))){
        CoordinateFrame fm= w.pic.setScaling(width+1,height+1);
        Polygon poly = ((Polygon)w.pic.obj).scalePolygon(fm);
        Rectangle r= new Rectangle();Picture pict=null;
        if(w.pic.name.equals("polygon")){
            Color c[][] = dW.c.polygonfill(poly,w.pic.cc,w.pic.cb,r);
            pict = new Picture(IO,c,w.pic.cc,"polygon",poly ,xx,yy,c.length,c[0].length);
        }else if(w.pic.name.equals("shape")){
            Color c[][] = dW.c.shapefill(poly,w.pic.cc,w.pic.cb,r);
            pict = new Picture(IO,c,w.pic.cc,"shape",poly ,xx,yy,c.length,c[0].length);
        }Color ccc= w.cc;
        removeWindow(refwn);
        Win win = makeWindow(pict,ccc);
        win.pic.cb=w.pic.cb;
    }
}
```

Figure 16.8 Make-icon command: enter the control polygon for a curved shape

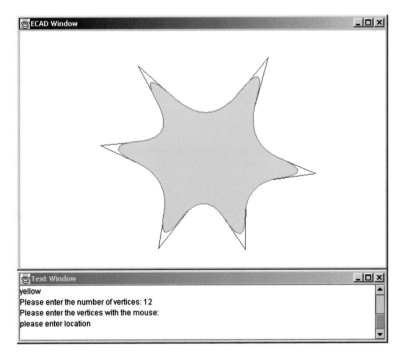

Figure 16.9 Make-icon command: infill the curved shape

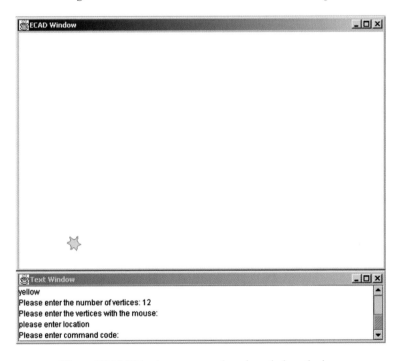

Figure 16.10 Make-icon command: scale and place the icon

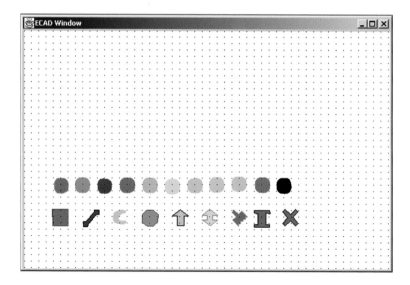

Figure 16.11 A test generating icons using a make-icon command

A directly related use for these operations is to generate text characters and fonts. Once a letter is in the form of an array of pixels it can be copied and placed in lines as text. Figure16.9 shows that a letter A with serifs is possible to construct with the *drawshape* command. Once the letter has been designed it can be converted to any size using the *resize* command. The implementation of the *repaint* command in this system illustration allows the self-overlaping area boundary shown in Figure 16.9 to be rendered removing the distinction between the boundary and its fill. The same result is given by initially making the boundary the same colour as the fill.

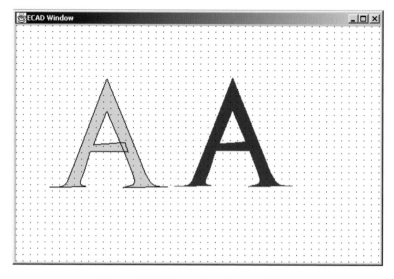

Figure 16.12 Generating character fonts

Constructing icons in the way illustrated in Figure 16.8 to 16.11 is time consuming and once a set of commands has been established is better executed automatically from internally stored data. However, as a developmental or experimental system this capability to extend an existing set of commands is very valuable. It allows alternative interface-system variations to be designed and tested. Figure 16.13 shows a built in set of command icons matching the text commands developed so far.

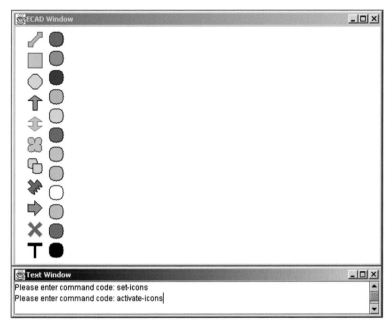

Figure 16.13 Initialising icons

```
private static void setIcons(){
    Color[] colours=new Color[] {
    Color.red,  Color.green,  Color.blue, Color.cyan, Color.yellow, Color.magenta,
    Color.pink, Color.orange, Color.white,Color.lightGray,Color.gray,Color.black,};
    String [] ics = new String[30];  Color [] cc = new Color[30]; Color [] cb = new Color[30];
    int [][] xlst = new int [30][];  int [][] ylst = new int [30][];
    Point [] plst = null;
    Point pa = new Point(1,1,1);  Point pb = new Point(1,14,14);
    Point p1 = new Point(1,1,1);  Point p2 = new Point(1,25,25);
    int num=23;
    ics[0]= "polygon";cc[0]=Color.cyan;cb[0]=Color.blue;                            //drawline
    xlst[0] = new int[]{10,14,14,12, 6, 6, 2, 2, 4,10,10};
    ylst[0] = new int[]{ 2, 2, 6, 6,12,14,14,10,10, 4, 2};
    ics[1]= "polygon";cc[1]=Color.pink;cb[1]=Color.blue;                            //drawlrectangle
    xlst[1] = new int[]{2,14,14, 2, 2};   ylst[1] = new int[]{2, 2,14,14, 2};
    ics[2]= "polygon";cc[2]=Color.yellow;cb[2]=Color.black;                         //drawpolygon
    xlst[2] = new int[]{ 6,10,14,14,10, 6, 2, 2, 6}; ylst[2] = new int[]{ 2, 2, 6,10,14,14,10, 6, 2};
```

```
ics[3]= "polygon";cc[3]=Color.green;cb[3]=Color.black;                    //raise
xlst[3] = new int[]{ 8,14,10,10, 6, 6, 2, 8};  ylst[3] = new int[]{ 2, 6, 6,14,14, 6, 6, 2};
ics[4]= "polygon";cc[4]=Color.orange;cb[4]=Color.red;                     //resize
xlst[4] = new int[]{ 8,14,10,10,14, 8, 2, 6, 6, 2, 8};
ylst[4] = new int[]{ 2, 6, 6,10,10,14,10,10, 6, 6, 2};
ics[5]= "shape";cc[5]=Color.pink;cb[5]=Color.blue;                        //drawshape
xlst[5] = new int[]{ 2, 7, 8, 8, 9,14,14,12,12,14,14, 9, 8, 8, 7, 2, 2, 6, 6, 2, 2};
ylst[5] = new int[]{ 2, 2, 6, 6, 2, 2, 7, 8, 8, 9,14,14,12,12,14,14, 9, 8, 8, 7, 2};
ics[6]= "shape";cc[6]=Color.yellow;cb[6]=Color.black;                     //copy
xlst[6] = new int[]{ 6, 6,10,10,10,14,14,14,14,10,10, 6, 6, 6, 6,10,10, 6, 6, 6, 2, 2, 2, 2, 6};
ylst[6] = new int[]{ 2, 2, 2, 6, 6, 6,10,10,14,14,14,14,10,10, 6, 6, 6, 6,10,10,10, 6, 6, 2, 2};
ics[7]= "polygon";cc[7]=Color.magenta;cb[7]=Color.black;                  //repaint
xlst[7] = new int[]{ 4, 7,10,13,14,11,12, 9,10, 7, 8, 5, 2, 5, 2, 4};
ylst[7] = new int[]{ 2, 5, 2, 5, 8, 7,10, 9,12,11,14,13,10, 7, 4, 2};
ics[8]= "polygon";cc[8]=Color.green;cb[8]=Color.black;                    //move
xlst[8] = new int[]{ 2, 8, 8,14, 8, 8, 2, 2};  ylst[8] = new int[]{ 5, 5, 2, 8,14,11,11, 4};
ics[9]= "polygon";cc[9]=Color.red;cb[9]=Color.red;                        //remove
xlst[9] = new int[]{ 4, 8,12,14,10,14,12, 8, 4, 2, 6, 2, 4};
ylst[9] = new int[]{ 2, 6, 2, 4, 8,12,14,10,14,12, 8, 4, 2};
ics[10]= "polygon";cc[10]=Color.black;cb[10]=Color.black;                 //text input
xlst[10] = new int[]{ 2,14,14, 9, 9, 7, 7, 2, 2}; ylst[10] = new int[]{ 2, 2, 4, 4,14,14, 4, 4, 2};
ics[11]= "shape";cc[11]=Color.green;cb[11]=Color.black;                   //colours
xlst[11] = new int[]{2,14,14, 2, 2};  ylst[11] = new int[]{2, 2,14,14, 2};
int xx= 20; int yy= 8; int width = 25, height = 25;
for(int ii=0;ii<num;ii++){
    int i=ii;  if(ii>=11){i=11; cc[11]=colours[ii-i]; cb[11]=Color.black;}
    plst = new Point[xlst[i].length];
    for(int j=0; j<xlst[i].length;j++){
        Point pp= new Point(2); pp.n[1]= xlst[i][j]; pp.n[2]= ylst[i][j]; plst[j]=pp;
    }
    Polygon pol= new Polygon(xlst[i].length);  pol.p=plst;
    CoordinateFrame fm= new CoordinateFrame();  fm.setScales(p1,p2,pa,pb);
    Polygon poly = pol.scalePolygon(fm);
    Rectangle r= new Rectangle();  Picture pict=null;
    if(ics[i].equals("polygon")){
        Color c[][] = dW.c.polygonfill(poly,cc[i],cb[i],r);
        pict = new Picture(IO,c,cc[i],"polygon",poly ,xx,yy,c.length,c[0].length);
    }else if(ics[i].equals("shape")){
        Color c[][] = dW.c.shapefill(poly,cc[i],cb[i],r);
        pict = new Picture(IO,c,cc[i],"shape",poly ,xx,yy,c.length,c[0].length);
    }
    Win win = makeWindow(pict,cc[i],true);
    win.pic.cb=cb[i];    win.tag=ii+1;
    if(ii>10)yy=yy+30;else yy = yy+33;
    if(ii==10){xx=55; yy= 8;}
}
}
```

Once control has been passed to the icons the basic interactive input is through the mouse. This means if a text input is required either the current command has to request it or an icon to receive text has to be activated. In the example given this is done using the T shaped icon. Using text commands, choices such as the colour for a boundary or a fill were made through the *TextWindow*. Changing the system to take advantage of the more direct interaction that the use of icons and the mouse permits, still appears to require the use of the *TextWindow*, but converting the function of the text window to prompting for the next command input. The current exceptions to this would be the requests for the number of vertices when entering polygons and for the colour to draw lines and infill areas. To remove these cases the input of polygons and colours is modified. Clicking colour buttons, coded as icons, is used to make colour choices, while numbers are entered using a slider.

In both the *drawpolygon* and *drawshape* procedures given above the polygon input uses the same code. This was highlighted and this section of the code is replaced by a procedure inputPolygon (..) that offers two input techniques: the existing one for the *TextWindow* and a modified one for icon inputs.

```
public static Point[] inputPolygon(int tag){
    int num=100; Point [] poly = new Point[num];
    if( ! iconSet){
        IO.writeString("Please enter the number of vertices: ");
        num=IO.readInteger(); IO.readLine();
        poly = new Point[num+1];
    }
    IO.writeString("Please enter the vertices with the mouse: \n");
    Point p = getCoordinate(tag);
    if(p==null) IO.writeString("point is null\n");
    poly[0]= p; int kk=1;
    while ((kk<num)&&(p!=null)){
        p = getCoordinate(tag);
        if(p!=null) {
            poly[kk]= p;
            dW.plotLine(poly[kk-1],poly[kk],Color.blue);
            kk++;
        }else num= kk;
    }
    poly[num]=new Point("=",poly[0]);
    dW.plotLine(poly[num-1],poly[num],Color.blue);
    Point []tp = new Point[num+1];
    for(int i=0;i<num+1;i++){tp[i]=poly[i];}
    dW.getCoord();
    return tp;
}
```

Using icons, the boundary sequence of vertex coordinates is completed by re-clicking on the *drawpolygon* or the *drawshape* icon to close the loop. This requires a new version of the get coordinate procedure to check whether a new input point is free or is on the terminating icon. The tag value is the number given to the icon that is calling this procedure to allow it to be identified to close the boundary loop.

```
private static Point getCoordinate(int tag){
    if(iconSet){
        Win win=null;ListElement refwn=null;
        Point p= new Point(2);
        win = (Win)(refwn = identifyWindow(p)).object;
        if(win.tag==tag)return null;  else return p;
    }return dW.getCoord();
}
```

```
do{                                                         // text input
    IO.writeString("Please enter command code: ");
    str = IO.readString(); IO.readLine();
    ...
    if(str.equals("activate-icons")){                       // transfer to icon input
        iconSet=true;  icons();  if (gridset) resetgrid();}
}while(!str.equals("end"));
```

```
private static void icons(){                                // icon input
    Win win=null;
    do{ListElement refwn=null;
        IO.writeString("please enter an icon command\n");
        Point p= new Point(2);  win = (Win)(refwn = identifyWindow(p)).object;
        if(win.tag!=0){
            switch(win.tag){
            case 1:  drawLine();if (gridset) resetgrid(); break;
            case 2:  drawRectangle();if (gridset) resetgrid(); break;
            case 3:  drawPolygon();if (gridset) resetgrid(); break;
            case 4:  raise();if (gridset) resetgrid(); break;
            case 5:  resize();if (gridset) resetgrid(); break;
            case 6:  drawshape();if (gridset) resetgrid(); break;
            case 7:  copy();if (gridset) resetgrid(); break;
            case 8:  repaint();if (gridset) resetgrid(); break;
            case 9:  move();if (gridset) resetgrid(); break;
            case 10: remove();if (gridset) resetgrid(); break;
            }
        }
    }while(win.tag!=11);
        iconSet=false;                                      // transfer back to text input
    IO.writeString("returning to text command\n");
}
```

A similar approach allows colours to be input using the mouse to identify a colour icon and then using its tag value to select the colour it represents.

```
static Color getColour(String str){
    if (str==null)IO.writeString("Please enter colour\n");
    else IO.writeString(str);
    if(iconSet){ return Colours();}
    String colour = IO.readString(); IO.readLine();
    Color cc = setColour(colour);
    return cc;
}
```

```
private static Color Colours(){
    Win win=null;ListElement refwn=null;
    Point p= new Point(2);
    win = (Win)(refwn = identifyWindow(p)).object;
    if(win.tag!=0){
        switch(win.tag){
            case 12:  return Color.red;
            case 13:  return Color.green;
            case 14:  return Color.blue;
            case 15:  return Color.cyan;
            case 16:  return Color.yellow;
            case 17:  return Color.magenta;
            case 18:  return Color.pink;
            case 19:  return Color.orange;
            case 20:  return Color.white;
            case 21:  return Color.lightGray;
            case 22:  return Color.gray;
            case 23:  return Color.black;
            default: return null;
        }
    }return null;
}
```

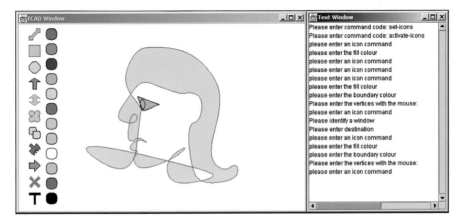

Figure 16.14 Overlaying two filled shapes using icon command-inputs

Computer Aided Design

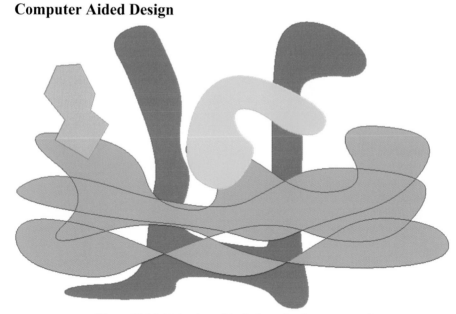

Figure 16.15 Abstract graphic design using icon commands

This scheme can clearly be extended with more detailed commands, however, the current set of commands allows a picture to be composed using the mouse-pointer very much in the way a pen, pencil and paintbrush might be used.

Figure 16.16 Simple picture composition using icon commands

Figure 16.17 Composing a face from separate shapes

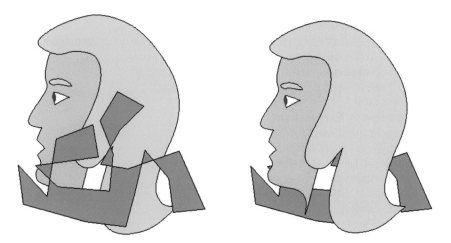

Figure 16.18 Raising the face elements grouped as a single layer

Group, Ungroup and Layers

An important command for composing more complex displays is the *group* command, with its inverse *ungroup*. Figures 16.17 and 18 illustrate the way a face can be composed from five simple elements. Once these are grouped together into a single layer then the whole face can be moved around raised and lowered and changed in size in a single step using a single command. In order to do this, however, it is necessary to introduce a further object type the *layer*. The *layer* object is merely a list of its component elements. However since this list may itself contain sub layers it creates a tree structure that has to be processed recursively by the existing commands. This requires further refactoring to include the necessary extensions.

```
private static void group(){
    IO.writeString("please identify first elements in the group\n");
    ListElement refwn=null;  Point p = new Point(2);
    Win win = (Win)(refwn = identifyWindow(p)).object;
    while(win!= bw){
        win.selected = true;
        IO.writeString("please identify next element in the group\n");
        win = (Win)(refwn = identifyWindow(p)).object;
    } makegroup();
}
private static Win makegroup(){
    int nx1= Integer.MAX_VALUE,ny1= Integer.MAX_VALUE;
    int nx2= Integer.MIN_VALUE,ny2= Integer.MIN_VALUE;
    ListElement ref = LofWs.start;  List tmplst = new List();
    Win win = (Win)ref.object;
    while ((win!=bw)&&(ref != null)){
        if(win.selected){
            win.selected = false;  tmplst.append(win);
            if(win.x1<nx1)nx1=win.x1;   if(win.y1<ny1)ny1=win.y1;
            if(win.x2>nx2)nx2=win.x2;   if(win.y2>ny2)ny2=win.y2;
            ListElement storeref=ref.right;
            lx.destroy(win.rx1);lx.destroy(win.rx2);ly.destroy(win.ry1);ly.destroy(win.ry2);
            LofWs.delete(ref);
            ref=storeref;
        } else ref=ref.right;  if(ref!=null)win = (Win)ref.object;
    }
    Rectangle r = new Rectangle();
    r.x=nx1;r.y=ny1;r.width=nx2-nx1;r.height=ny2-ny1;
    Color[][] col = new Color[nx2-nx1+1][ny2-ny1+1];
    ref= tmplst.finish;
    win = (Win)ref.object;
    while (ref != null){
        if(win!=bw){   //copy each object into the new array
            Picture pc= win.pic;
            int x=pc.x,y=pc.y,w=pc.w,h=pc.h;
            for(int i=0;i<w;i++) for(int j=0;j<h;j++){
                Color ct= pc.getColour(i,j);  if(ct!=null)col[x-nx1+i][y-ny1+j]= ct;
            }
        }ref=ref.left;
        if(ref!=null)win = (Win)ref.object;
    }
    Picture pict = new Picture(IO,col,null,"layer",tmplst,r.x,r.y,r.width,r.height);
    win = makeWindow(pict,null,true);
    win.x1=nx1;win.y1=ny1;win.x2=nx2;win.y2=ny2;
    IO.writeString("group complete\n");
    return win;
}
```

The simplest change is that required by the *move* command. All the elements grouped into a layer have to have their locations updated when the layer is moved in the way illustrated below.

```
private static void move(){
    ListElement refwn=null; Point p= new Point(2);
    IO.writeString("Please identify a window\n");
    refwn = identifyWindow(p);
    moveWindow(refwn,p);

}
private static void moveWindow(ListElement refwn,Point p1){
    Win win = (Win)refwn.object;
    if(win!=bw){IO.writeString("Please enter destination\n");
    int x1,x2,y1,y2;
    Point p2 = dW.getCoord();
    int xmove = p2.xi()-p1.xi(); int  ymove = p2.yi()-p1.yi();
    x1=win.x1; y1=win.y1; x2=win.x2; y2=win.y2;
    if(x2+xmove<=bw.x1)xmove=bw.x1-x2+3;
    if(x1+xmove>=bw.x2)xmove=bw.x2-x1-3;
    if(y2+ymove<=bw.y1)ymove=bw.y1-y2+3;
    if(y1+ymove>=bw.y2)ymove=bw.y2-y1-3;
    x1=x1+xmove;  y1=y1+ymove;
    x2=x2+xmove;  y2=y2+ymove;
    Picture pict = win.pic;
    pict.x = x1; pict.y = y1;
    if(pict.name.equals("layer"))moveLayer((List)pict.obj,xmove,ymove);
    removeWindow(refwn);
    win = makeWindow(pict,win.pic.cc,true);
    win.x1=x1; win.y1=y1; win.x2=x2; win.y2=y2;
    }

}
private static void moveLayer(List ls,int xmove,int ymove){
    ListElement ref= ls.start;
    while(ref !=null){
        Win wn = (Win)ref.object;
        wn.pic.x = wn.x1= wn.pic.x+xmove; wn.pic.y = wn.y1 = wn.pic.y+ymove;
        wn.x2= wn.x2+xmove; wn.y2=wn.y2+ymove;
        if(wn.pic.name.equals("layer"))
            moveLayer((List)wn.pic.obj,xmove,ymove);            //recursive call
        ref=ref.right;
    }

}
```

The same kind of change has to be carried out for the *copy* and the *resize* commands except in their cases the change is more complicated in that duplicate copies of all the elements need to be generated and kept in the case of *copy*, but the original deleted in the case of the *resize* command.

```
private static void copy(){
    ListElement refwn=null;  Point p= new Point(2);
    IO.writeString("Please identify a window\n");
    refwn = identifyWindow(p);
    Win wint = copyWindow(refwn,50 ,50);
    int xl= wint.x1,yl= wint.y1,xr= wint.x2 ,yr= wint.y2;
    ListElement ww= LofWs.push(wint);
    wint.rx1= lx.enter(new Integer(xl)); wint.rx2= lx.enter(new Integer(xr));
    wint.ry1= ly.enter(new Integer(yl)); wint.ry2= ly.enter(new Integer(yr));
    ListElement [] lse = new ListElement[1]; lse[0] = ww;
    resetRectangle(bw, lse, xl,yl,xr,yr,1);
    dW.c.repaint();
}
static Win copyWindow(ListElement refwn,int incx,int incy){
    Point p=new Point(2); Picture pict=null;
    Win win = (Win)refwn.object; Win wint=null;
    if(win!=bw){
        int x= win.x1+incx; int y= win.y1+incy;
        if(win.pic.name.equals("line")){
            LineSeg newLs= ((LineSeg)win.pic.obj).copy(win.cc);
            pict = new Picture(IO,null,win.cc,"line",newLs, x, y,win.w ,win.h);
            wint = new Win(IO,x,y, win.pic.w, win.pic.h, win.pic.cc, pict);
        }else {
            Color[][] col= new Color[win.pic.w][win.pic.h];
            pict = new Picture(IO,col,win.cc,win.pic.name,null, x, y, win.pic.w , win.pic.h);
            pict.cb=win.pic.cb;
            for(int i=0;i<win.pic.w;i++){
                for(int j=0;j<win.pic.h;j++)pict.setColour(win.pic.getColour(i,j),i,j);
            }
            if(win.pic.name.equals("layer")){
                List nls= new List();
                List ls = (List)win.pic.obj;  ListElement ref= ls.start;
                while(ref !=null){
                    wint = copyWindow(ref,incx,incy);          //recursive call
                    nls.append(wint);
                    ref=ref.right;
                }pict.obj = nls;
            }
            if(win.pic.name.equals("polygon"))
                {pict.obj = ((Polygon)win.pic.obj).copyPolygon(dW,IO);}
            if(win.pic.name.equals("shape"))
                {pict.obj = ((Polygon)win.pic.obj).copyPolygon(dW,IO);}
            wint = new Win(IO,x,y, win.pic.w, win.pic.h, win.pic.cc, pict);
            wint.pic.cb=win.pic.cb;
        }
    }return wint;
}
```

```
private static void resize(){
    ListElement refwn=null; Point p=new Point(2);
    IO.writeString("Please identify the window \n");
    Win win = (Win)(refwn = identifyWindow(p)).object;
    Picture   pict= win.pic;
    if(win!=bw){
        int x1=win.x1;int y1=win.y1;int x2=win.x2;int y2=win.y2;
        int xa=bw.x1;int ya=bw.y1;int xb=bw.x2;int yb=bw.y2;
        if((x1<xa)||(y1<ya)||(x2>xb)||(y2>yb)){
            IO.writeString("cannot resize clipped object \n");
        }else{
            drawBoundingRectangle(win,bw);
            IO.writeString("Please select a corner \n");
            Point p1 = dW.getCoord();  int xx= p1.xi(), yy= p1.yi();
            int [] x= new int[4];                      // identify the corner to be moved
            x[0]=win.x1;x[1]=win.x2;x[2]=win.x2;x[3]=win.x1;
            int [] y= new int[4];
            y[0]=win.y1;y[1]=win.y1;y[2]=win.y2;y[3]=win.y2;
            int j=0;int d=Integer.MAX_VALUE;int dd=0;;
            for(int i=0;i<4;i++){
                dd = (xx-x[i])*(xx-x[i])+(yy-y[i])*(yy-y[i]);
                if(d>dd){j=i;d=dd;}
            }
            IO.writeString("Please give its new location \n");
            Point p2 = dW.getCoord();  xa=p2.xi();  ya=p2.yi();
            int width=10, height=10;
            switch(j){                                 // define the new rectangle
                case 0:if(xa>x[2]-10)xa =x[2]-10;  if(ya>y[2]-10)ya =y[2]-10;
                        xx=xa;  yy=ya;  width = x[2]-xa;  height = y[2]-ya; break;
                case 1:if(xa<x[3]+10)xa=x[3]+10;  if(ya>y[3]-10)ya=y[3]-10;
                        xx= x[3];  yy= ya;  width = xa-x[3];  height = y[3]-ya; break;
                case 2:if(xa<x[0]+10)xa=x[0]+10; if(ya<y[0]+10)ya=y[0]+10;
                        xx=x[0];  yy=y[0];  width = xa-x[0];  height = ya-y[0]; break;
                case 3:if(xa>x[1]-10)xa=x[1]-10;  if(ya<y[1]+10)ya=y[1]+10;
                        xx=xa;  yy=y[1];  width = x[1]-xa;  height = ya-y[1]; break;
            }
            CoordinateFrame fm= win.pic.setScaling(width+1,height+1);
            win= scaling(win,xx,yy,win.x1,win.y1,fm);
            Color ccc= win.cc;
            removeWindow(refwn);
            win = makeWindow(win.pic,ccc,true);
        }
    }
}
```

It is convenient to divide the *resize* procedure into two the first identifying the change in size the second executing the change, in order to allow the recursive calls to work.

```
private static Win scaling(Win win,int x, int y,int bx, int by,CoordinateFrame fm){
    Picture pict=null;
    int xx= fm.scaleX_WtoS(win.x1-bx), yy= fm.scaleY_WtoS(win.y1-by);
    int x2= fm.scaleX_WtoS(win.x2-bx), y2= fm.scaleY_WtoS(win.y2-by);
    int width=x2-xx, height= y2-yy;
    if(win.pic.name.equals("rectangle")){
        Color[][] pic = new Color[width][height];
        Color cb = win.pic.cb;  Color bb = Color.lightGray;
        if(win.pic.cc!=null)bb= win.pic.cc;
        pict = new Picture(IO,null,bb,"rectangle",null,xx+x,yy+y,width,height);  pict.cb=cb;
    }else if(win.pic.name.equals("polygon")){
        Polygon poly = ((Polygon)win.pic.obj).scalePolygon(fm);
        Rectangle r= new Rectangle();
        Color c[][] = dW.c.polygonfill(poly,win.pic.cc, win.pic.cb,r);
        Color cb= win.pic.cb;
        pict = new Picture(IO,c,win.pic.cc,"polygon",poly ,xx+x,yy+y,c.length,c[0].length);
        pict.cb=cb;
    }else if(win.pic.name.equals("shape")){
        Polygon poly = ((Polygon)win.pic.obj).scalePolygon(fm);
        Rectangle r= new Rectangle();
        Color c[][] = dW.c.shapefill(poly,win.pic.cc,win.pic.cb,r);
        Color cb= win.pic.cb;
        pict = new Picture(IO,c,win.pic.cc,"shape",poly ,xx+x,yy+y,c.length,c[0].length);
        pict.cb=cb;
    }else if(win.pic.name.equals("line")){
        LineSeg newLn = ((LineSeg)win.pic.obj).scaleLine(xx,yy,width,height);
        pict=new Picture(IO,null,newLn.cc,"line",newLn, xx+x,yy+y,width,height);
    }else if(win.pic.name.equals("layer")){
        List nls= new List();
        List ls = (List)win.pic.obj;  ListElement ref= ls.finish;
        Color col[][] = new Color[width][height];
        while(ref !=null){
            Win wn= (Win)ref.object;
            Win wint =scaling(wn,x,y,bx,by,fm);                    //recursive call
            Picture pc= wint.pic;
            int ax=pc.x,ay=pc.y,aw=pc.w,ah=pc.h;
            for(int i=0;i<aw; i++)  for(int j=0;j<ah;j++){
                Color ct= pc.getColour(i,j);
                if(ct!=null)col[ax-x+i][ay-y+j]= ct;
            } nls.push(wint);  if(ref!=null)ref=ref.left;
        } pict = new Picture(IO,col,win.cc,win.pic.name,nls, x+xx, y+yy,width ,height);
        pict.obj = nls;
    }
    Win wint = new Win(IO,xx+x,yy+y,width,height, win.pic.cc, pict);
    wint.pic.cb=win.pic.cb;
    return wint;
}
```

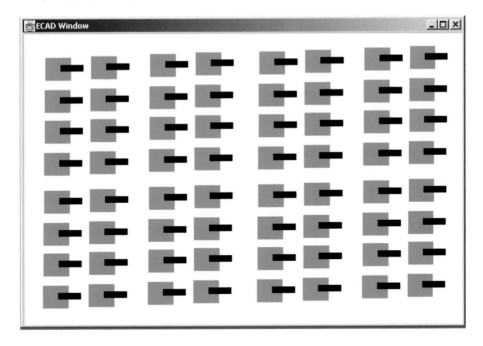

Figure 16.19 Duplicating elements

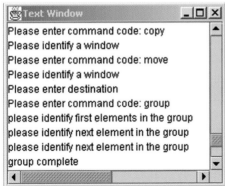

A powerful application of the group and copy commands is the ability to create a large number of duplicate objects. The first copy creates two. When these are grouped together and copied the number is four. This can then be increased to eight, sixteen and so on, in the way illustrated for 64 objects in Figure 16.19, executed by six group commands

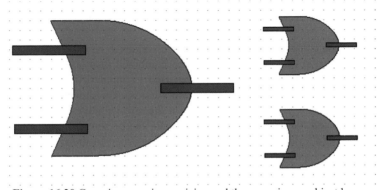

Figure 16.20 Grouping, copying, resizing and then copying an object layer

Active and Passive Graphic Elements

The ability to compose a drawing in a traditional way and output it as hardcopy merely creates a passive entity that still needs human interpretation. What is powerful about a computer graphics system is that the elements in a display are actively coupled to computer models representing their presence in the display. Not only this but they can also be coupled to a model that simulates the behaviour of the system they represent, in time.

Initially computer aided design systems were developed to generate the drawings needed to pass to manufacturers or builders to get a new product constructed. Carefully cross- referenced drawings ensured that the spatial relationships needed to construct the product were correct. However, the possibility for two extensions quickly became apparent: firstly the behaviour of the new product could be modelled and many further aspects of its structure and performance tested before committing to building a real system. Secondly once a design was checked out its computer model could be used to control the manufacturing process initially through the use of numerically controlled (NC) tools, and then more recently through more and more sophisticated robotic systems.

The use of icons for commands is the first example of this step to create active graphic elements. Other elements in a display can also be made active. Where a window is completely covered or is made too small to be practical, an icon can replace it. Clicking on this icon can recover the window to the top of the stack of windows when it is needed.

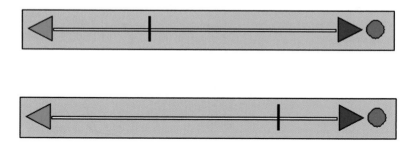

Figure 16.21 Active graphic elements: sliders to enter numerical values

Sliders can be constructed to allow numerical values to be entered into a program. In this case a group of elements need to be combined in a layer. Clicking on the various elements in the layer can then be used to input differing values. In the example shown in Figure 16.21 a simplified version of a slider is shown. When a numerical input value is required, clicking on the arrows moves the vertical marker along the scale and the current value it represents can be shown in the TextWindow. Once the required value has been set, it can be returned to the calling program by clicking on the red button. In order to implement this scheme it is necessary to have a *makeslider* command.

```
private static Win makeslider(int n,int nx1,int ny1){
    int nx2= nx1+n; int ny2= ny1+52;  int num=6;
    String [] ics = new String[num];
    Color [] cc = new Color[num];
    Color [] cb = new Color[num];
    int [][] xlst = new int [num][];
    int [][] ylst = new int [num][];
    int [] x= new int[num];  int [] y= new int[num];
    Picture pict =null; Point [] plst = null;
    ics[0]= "polygon";cc[0]=Color.lightGray; cb[0]=Color.black;  x[0]=0; y[0]=0;
    xlst[0] = new int[]{1, n, n,  1,  1};  ylst[0] = new int[]{1, 1, 52, 52, 1};
    ics[1]= "polygon";cc[1]=Color.white;cb[1]=Color.black;  x[1]= 42; y[1]=24;
    xlst[1] = new int[]{1, n-120, n-120, 1, 1};  ylst[1] = new int[]{1,   1,    4, 4, 1};
    ics[2]= "polygon";cc[2]=Color.green;cb[2]=Color.black;  x[2]= 8; y[2]=8;
    xlst[2] = new int[]{0,32,32, 0};  ylst[2] = new int[]{16,0,32, 16};
    ics[3]= "polygon";cc[3]=Color.blue;cb[3]=Color.black;  x[3]= n-76; y[3]= 8;
    xlst[3] = new int[]{0, 32, 0, 0 };  ylst[3] = new int[]{0, 16, 32,0 };
    ics[4]= "shape";cc[4]=Color.red;cb[4]=Color.black;  x[4]= n-42; y[4]= 8;
    xlst[4] = new int[]{16,32,16,0,16 };  ylst[4] = new int[]{ 0,16,32,16,0};
    ics[5]= "polygon";cc[5]=Color.black;cb[5]=Color.black;  x[5]= 42; y[5]=8;
    xlst[5] = new int[]{0,2, 2, 0,0};  ylst[5] = new int[]{0,0,32,32,0};
    int xx= nx1; int yy= ny1;  int width = n, height = 52;
    for(int ii=0;ii<num;ii++){
        xx=x[ii]+nx1;yy=y[ii]+ny1;
        int i=ii;
        plst = new Point[xlst[i].length];
        for(int j=0; j<xlst[i].length;j++){
            Point pp= new Point(2);  pp.n[1]= xlst[i][j];  pp.n[2]= ylst[i][j];
            plst[j]=pp;
        }
        Polygon poly= new Polygon(xlst[i].length);
        poly.p=plst;
        Rectangle r= new Rectangle();
        if(ics[i].equals("polygon")){
            Color c[][] = dW.c.polygonfill(poly,cc[i],cb[i],r);
            pict = new Picture(IO,c,cc[i],"polygon",poly ,xx,yy,c.length,c[0].length);
        }else if(ics[i].equals("shape")){
            Color c[][] = dW.c.shapefill(poly,cc[i],cb[i],r);
            pict = new Picture(IO,c,cc[i],"shape",poly ,xx,yy,c.length,c[0].length);
        }
        Win win = makeWindow(pict,cc[i],true);
        winb.selected=true;
        win.pic.cb=cb[i];
        win.tag=ii+1;
    }Win wint= makegroup(lst,nx1,ny1,nx2,ny2);
    return wint;
}
```

```
public static int readslider(){
    Win win = makeslider(500,50,50);
    Point p1 = new Point(2);
    IO.writeString("please set the slider\n");
    ListElement refwn= identifyWindow(p1);
    if(LofWs.start==refwn){
        List ls= (List)win.pic.obj;
        ListElement ref = ls.start;
        int count = 0;
        Win components[] = new Win[6];
        ListElement elements[] = new ListElement[6];
        while(ref!=null){
            Win wn = (Win)ref.object;
            LofWs.insertBefore(refwn,wn);
            components[count]= wn;elements[count++]= refwn.left;
            wn.rx1= lx.enter(new Integer(wn.x1));
            wn.rx2= lx.enter(new Integer(wn.x2));
            wn.ry1= ly.enter(new Integer(wn.y1));
            wn.ry2= ly.enter(new Integer(wn.y2));
            ref=ref.right;
        }
        lx.destroy(win.rx1);lx.destroy(win.rx2);
        ly.destroy(win.ry1);ly.destroy(win.ry2);
        LofWs.delete(refwn);
        int value= 0;
        while(true){
            ref= locateWindow(LofWs.start,p1);
            if(ref==elements[0]){
                Point p2 = dW.getCoord(); int xmove = p2.xi()-p1.xi(); int ymove = 0;
                value = movePointer(components,elements,xmove,ymove);
                IO.writeString("value = "+value+"\n");
            }else if(ref==elements[1]){
                for(int i=0;i<6;i++){ ref=elements[i]; removeWindow(ref);}
                return value;
            }else if(ref==elements[2]){
                int xmove = 1; int ymove = 0;
                value = movePointer(components,elements,xmove,ymove);
                IO.writeString("value = "+value+"\n");
            }else if(ref==elements[3]){
                int xmove = -1; int ymove = 0;
                value = movePointer(components,elements,xmove,ymove);
                IO.writeString("value = "+value+"\n");
            }
            p1= dW.getCoord();
        }
    }else return -1;
}
```

```
private static int movePointer(Win[] c,ListElement[] refs, int xmove,int ymove){
    Win sc= c[4];  Win win= c[0];
    ListElement ref= refs[0];
    int x1,x2,y1,y2;  x1=win.x1; y1=win.y1; x2=win.x2; y2=win.y2;
    if(x1+xmove<=sc.x1)xmove=sc.x1-x1;  if(x2+xmove>=sc.x2)xmove=sc.x2-x2;
    x1=x1+xmove; x2=x2+xmove;
    Picture pict = win.pic;  pict.x = x1; pict.y = y1;
    removeWindow(ref);
    win = makeWindow(pict,win.pic.cc,true);
    win.x1=x1; win.y1=y1;  win.x2=x2; win.y2=y2;
    c[0]=win;  refs[0]=LofWs.start;
    return win.pic.x-sc.pic.x;
}
private static void slider(){
    IO.writeString("please enter a points to locate the slider\n");
    Point p = dW.getCoord();
    IO.writeString("please enter a points to give the slider width\n");
    Point q = dW.getCoord();
    int nx1 = p.xi(),ny1= p.yi() ,nx2= q.xi(),ny2= q.yi();
    if(p.xi()>q.yi()){nx1=q.xi();nx2=p.xi();}  if(p.yi()>q.yi()){ny1=q.yi();ny2=p.yi();}
    int n = q.xi()-p.xi();
    if(n<126)n=126;
    makeslider(n,nx1,ny1);
}
```

```
int value = readslider();
IO.writeString(" value returned from the slider is "+value+"\n");
```

Figure 16.22 Reading values from a slider

Clicking the mouse on the green arrow reduces the value by one. Clicking the mouse on the blue arrow increases the value by one. Selecting the vertical marker with the mouse and relocating it with a second mouse click gives a new value, which can then be fine-tuned with the coloured arrows. Once the value is selected it can be returned by clicking on the red button.

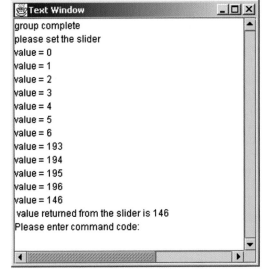

```
private static void ungroup(){
    IO.writeString("please identify the group\n");
    ListElement refwn=null;  Point p = new Point(2);
    Win win = (Win)(refwn = identifyWindow(p)).object;
    if(win.pic.name.equals("layer")){
    List ls = (List)win.pic.obj;  ListElement ref = ls.start;
    while(ref!=null){
        Win wn = (Win)ref.object;
        LofWs.insertBefore(refwn,wn);
        wn.rx1= lx.enter(new Integer(wn.x1)); wn.rx2= lx.enter(new Integer(wn.x2));
        wn.ry1= ly.enter(new Integer(wn.y1)); wn.ry2= ly.enter(new Integer(wn.y2));
        ref=ref.right;
    }
    lx.destroy(win.rx1);lx.destroy(win.rx2);ly.destroy(win.ry1);ly.destroy(win.ry2);
    LofWs.delete(refwn);
    }
}
```

The *ungroup* command separates out the components in a *layer* and, in the same order, places them back in the window stack in the position previously occupied by the *layer*.

Computer Aided Electronic Logic Circuit Design System ECAD

Graphic user interfaces allows graphic displays to be constructed interactively. They can also create a model of a system that is being designed to show how it will work as an active system once it is built. An example of a computer aided design system that depends on a graphics user interface to design electronic logic circuits, which illustrates the main point being made here, is presented in outline below.

Logic circuits can be designed using simple two-dimensional diagrams. So they can be constructed using the relatively simple facilities developed so far in this and earlier chapters. Linking together graphic components representing function blocks in a diagram to represent a circuit, can in parallel, link together the elements of the model needed to simulate the behaviour of the circuit. Icons can be set up to represent *and-gates, or-gates, not gates* and *latches* or memory register cells. They can also generate free *input* and *output* elements that can be used to feed the circuit new values, and present the results when the circuit is run. Circuits can be designed and then encapsulated as a new *function icon* in the display, which allows multiple copies of a sub-circuit to be used in a hierarchical system layout. An icon to generate wires between these components to link out ports to in-ports, and a similar icon to provide clock signals to latches completes a basic circuit design system. .

The icons for this system are set up by the command line *makecircuit*. This command also initialises the model of the circuit that will be used to simulate its behaviour. Each construction command using the icons to place components in a new circuit will add to this model. A wire must link each output from a function block, either to an input port for another function block or a free output port. Each input to a function block must be linked to an output from another block or to a free input port. When the free input units are clicked with the mouse they toggle the input values from true to false and back again, allowing different input configurations to be interactively entered into the circuit.

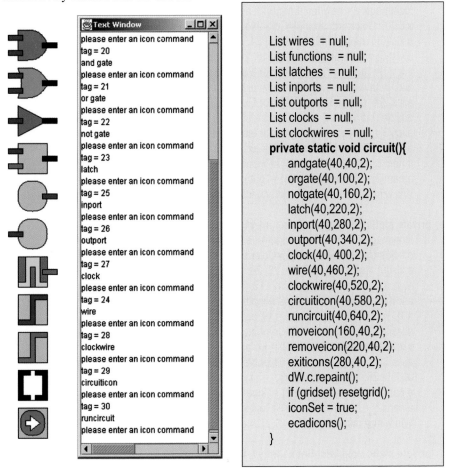

Figure 16.23 Setting up the circuit design icons

```
if(str.equals("makecircuit")){circuit( );if (gridset) resetgrid( );}
```

Figure 16.23 illustrates the first step: which is to set up the command icons needed to construct and run the circuit. The icons representing components will create a copy of themselves in the display space when clicked with a mouse. In effect they are an extension of the copy command with the extra code needed to create the function block that is associated with its graphic representation. The code to create the icons is fairly repetitive and is given below. When a component is copied and placed in the display workspace, it is given a different tag value so that when the mouse clicks on it a different action is carried out.

```
static  int [] xx=new int[]{0, 12,  12,  0,  0};
static  int [] yy=new int[]{0,  0,   4,  4,  0};
```

```
private static void andgate(int x1,int y1, int scale){
    IconImage a = new IconImage(4); a.ics[0]= "shape";
    a.cc[0]=Color.red; a.cb[0]=Color.black; a.x[0]=8; a.y[0]=0;
    a.xlst[0] = new int[]{0, 0, 16, 24, 24, 16,  0,  0,  0};
    a.ylst[0] = new int[]{0, 0,  0,  8, 16, 24, 24, 24,  0};
    a.ics[1]= "polygon";a.cc[1]=Color.blue;a.cb[1]=Color.black;
    a.x[1]= 0; a.y[1]=4; a.xlst[1] = xx; a.ylst[1] = yy;
    a.ics[2]= "polygon";a.cc[2]=Color.blue;a.cb[2]=Color.black;
    a.x[2]= 0; a.y[2]=16; a.xlst[2] = xx; a.ylst[2] = yy;
    a.ics[3]= "polygon";a.cc[3]=Color.black;a.cb[3]=Color.black;
    a.x[3]= 28; a.y[3]= 10; a.xlst[3] = xx; a.ylst[3] = yy;
    a.nam[0] = "body"; a.nam[1] = "inport"; a.d[0]=0; a.d[1]=0; a.d[2]=0;a.d[3]=0;
    a.nam[2] = "inport";a.nam[3] = "outport";
    Win wint = makeobject(4,a,x1,y1,scale); wint.name="andgate";  wint.tag=20;
}
private static void orgate(int x1,int y1, int scale){
    IconImage a = new IconImage(4); a.ics[0]= "shape";
    a.cc[0]=Color.green; a.cb[0]=Color.black;a.x[0]=8; a.y[0]=0;
    a.xlst[0] = new int[]{0, 0, 16, 26, 26, 16,  0,  0,  4, 0};
    a.ylst[0] = new int[]{0, 0,  0, 12, 12, 24, 24, 24, 12, 0};
    a.ics[1]= "polygon";a.cc[1]=Color.blue;a.cb[1]=Color.black;
    a.x[1]= 0; a.y[1]=4; a.xlst[1] = xx; a.ylst[1] = yy;
    a.ics[2]= "polygon";a.cc[2]=Color.blue;a.cb[2]=Color.black;
    a.x[2]= 0; a.y[2]=16; a.xlst[2] = xx; a.ylst[2] = yy;
    a.ics[3]= "polygon";a.cc[3]=Color.black;a.cb[3]=Color.black;
    a.x[3]= 28; a.y[3]= 10; a.xlst[3] = xx; a.ylst[3] = yy;
    a.nam[0] = "body"; a.nam[1] = "inport";a.d[0]=0;a.d[1]=0;a.d[2]=0;a.d[3]=0;
    a.nam[2] = "inport";a.nam[3] = "outport";
    Win wint = makeobject(4,a,x1,y1,scale); wint.name="orgate"; wint.tag=21;
}
private static void latch(int x1,int y1, int scale){
    IconImage a = new IconImage(4); a.ics[0]= "polygon";
    a.cc[0]=Color.orange; a.cb[0]=Color.black; a.x[0]=8; a.y[0]=0;
    a.xlst[0] = new int[]{0, 24,  24,  0,  0}; a.ylst[0] = new int[]{0,  0,  24, 24,  0};
    a.ics[1]= "polygon";a.cc[1]=Color.blue;a.cb[1]=Color.black;
    a.x[1]= 0; a.y[1]=4; a.xlst[1] = xx; a.ylst[1] = yy;
    a.ics[2]= "polygon";a.cc[2]=Color.magenta;a.cb[2]=Color.black;
    a.x[2]= 0; a.y[2]=16; a.xlst[2] = xx; a.ylst[2] = yy;
    a.ics[3]= "polygon";a.cc[3]=Color.black;a.cb[3]=Color.black;
    a.x[3]= 28; a.y[3]= 10; a.xlst[3] = xx; a.ylst[3] = yy;
    a.nam[0] = "body"; a.nam[1] = "inport";a.d[0]=0;a.d[1]=0;a.d[2]=0;a.d[3]=0;
    a.nam[2] = "clock";a.nam[3] = "outport";
    Win wint = makeobject(4,a,x1,y1,scale);wint.name="latch"; wint.tag=23;
}
```

```
private static void circuiticon(int x1,int y1, int scale){
    IconImage a = new IconImage(3);
    a.ics[0]= "polygon";a.cc[0]=Color.white; a.cb[0]=Color.black;
    a.x[0]=8; a.y[0]=0; a.x[1]= 8; a.y[1]=0; a.x[2]= 22; a.y[2]=0;
    a.xlst[0] = new int[]{0, 24,  24,  0,  0}; a.ylst[0] = new int[]{0,  0,  24, 24,  0};
    a.ics[1]= "polygon";a.cc[1]=Color.black;a.cb[1]=Color.black;
    a.xlst[1] = new int[]{0, 10, 10, 4,  4, 10, 10,  0, 0};
    a.ylst[1] = new int[]{0,  0,  4, 4, 20, 20, 24, 24, 0};
    a.ics[2]= "polygon";a.cc[2]=Color.black;a.cb[2]=Color.black;
    a.xlst[2] = new int[]{0, 10, 10,  0,  0,  6, 6, 0, 0};
    a.ylst[2] = new int[]{0,  0, 24, 24, 20, 20, 4, 4, 0};
    a.nam[0] ="body"; a.nam[1] ="body"; a.nam[2] = "body";a.d[0]=0;a.d[1]=0;a.d[2]=0;
    Win wint = makeobject(3,a,x1,y1,scale); wint.name="circuit"; wint.tag=29;
}
private static void notgate(int x1,int y1, int scale){
    IconImage a = new IconImage(3);a.ics[0]= "polygon";
    a.cc[0]=Color.magenta; a.cb[0]=Color.black;
    a.x[0]=8; a.y[0]=0; a.x[1]= 0; a.y[1]=10; a.x[2]= 28; a.y[2]=10;
    a.xlst[0] = new int[]{0, 24, 0, 0}; a.ylst[0] = new int[]{0, 12, 24, 0};
    a.ics[1]= "polygon";a.cc[1]=Color.blue;a.cb[1]=Color.black;
    a.xlst[1] = xx; a.ylst[1] = yy; a.xlst[2] = xx; a.ylst[2] = yy;
    a.ics[2]= "polygon";a.cc[2]=Color.black;a.cb[2]=Color.black;
    a.nam[0] = "body"; a.nam[1] = "inport"; a.nam[2] = "outport";a.d[0]=0;a.d[1]=0;a.d[2]=0;
    Win wint = makeobject(3,a,x1,y1,scale); wint.name="notgate"; wint.tag=22;
}
private static void inport(int x1,int y1, int scale){
    IconImage a = new IconImage(3); a.ics[0]= "shape";
    a.cc[0]=Color.magenta; a.cb[0]=Color.black; a.x[0]=8; a.y[0]=0; a.x[1]= 28; a.y[1]=10;
    a.xlst[0] = new int[]{0, 24,  24,  0,  0}; a.ylst[0] = new int[]{0,  0,  24, 24,  0};
    a.ics[1]= "polygon";a.cc[1]=Color.blue;a.cb[1]=Color.black; a.xlst[1] = xx; a.ylst[1] = yy;
    a.nam[0] = "body"; a.nam[1] = "outport";a.nam[2] = "button";
    a.ics[2]= "shape";a.d[0]=0;a.d[1]=0;a.d[2]=0;
    a.cc[2]=Color.red; a.cb[2]=Color.black;  a.x[2]=12; a.y[2]=4;
    a.xlst[2] = new int[]{4, 20,  20,  4,  4}; a.ylst[2] = new int[]{4,  4,  20, 20,  4};
    Win wint = makeobject(3,a,x1,y1,scale); wint.name="inport"; wint.tag=25;
}
private static void outport(int x1,int y1, int scale){
    IconImage a = new IconImage(3);  a.ics[0]= "shape";
    a.cc[0]=Color.pink; a.cb[0]=Color.black; a.x[0]=8; a.y[0]=0; a.x[1]= 0; a.y[1]=10;
    a.xlst[0] = new int[]{0, 24,  24,  0,  0}; a.ylst[0] = new int[]{0,  0,  24, 24,  0};
    a.ics[1]= "polygon";a.cc[1]=Color.blue;a.cb[1]=Color.black; a.xlst[1] = xx; a.ylst[1] = yy;
    a.nam[0] = "body"; a.nam[1] = "inport";a.nam[2] = "led";
    a.ics[2]= "shape";a.d[0]=0;a.d[1]=0;a.d[2]=0;
    a.cc[2]=Color.white; a.cb[2]=Color.black; a.x[2]=12; a.y[2]=4;
    a.xlst[2] = new int[]{4, 20,  20,  4,  4}; a.ylst[2] = new int[]{4,  4,  20, 20,  4};
    Win wint = makeobject(3,a,x1,y1,scale); wint.name="outport"; wint.tag=26;
}
```

```
private static void runcircuit(int x1,int y1, int scale){
    IconImage a = new IconImage(3);
    a.ics[0]= "polygon";a.cc[0]=Color.cyan; a.cb[0]=Color.black;
    a.x[0]=8; a.y[0]=0; a.x[1]= 10; a.y[1]=2; a.x[2]= 14; a.y[2]=6;
    a.xlst[0] = new int[]{0, 24, 24, 0, 0}; a.ylst[0] = new int[]{0, 0, 24, 24, 0};
    a.ics[1]= "shape";a.cc[1]=Color.red;a.cb[1]=Color.black;
    a.xlst[1] = new int[]{0, 20, 20, 0, 0}; a.ylst[1] = new int[]{0, 0, 20, 20, 0};
    a.ics[2]= "polygon";a.cc[2]=Color.white;a.cb[2]=Color.black;
    a.xlst[2] = new int[]{4, 12, 4, 4, 0, 0, 4, 4}; a.ylst[2] = new int[]{0, 6, 12, 8, 8, 4, 4, 0};
    a.nam[0] = "body"; a.nam[1] = "body"; a.nam[2] = "body";a.d[0]=0;a.d[1]=0;a.d[2]=0;
    Win wint = makeobject(3,a,x1,y1,scale); wint.tag=30;
}
private static void clock(int x1,int y1, int scale){
    IconImage a = new IconImage(3);
    a.ics[0]= "polygon";a.cc[0]=Color.gray; a.cb[0]=Color.black;
    a.x[0]=8; a.y[0]=0; a.x[1]= 8; a.y[1]=0; a.x[2]= 28; a.y[2]=10;
    a.xlst[0] = new int[]{0, 24, 24, 0, 0}; a.ylst[0] = new int[]{0, 0, 24, 24, 0};
    a.ics[1]= "polygon";a.cc[1]=Color.yellow;a.cb[1]=Color.black;
    a.xlst[1] = new int[]{6, 18, 18, 24, 24, 14, 14, 10, 10, 0, 0, 6, 6};
    a.ylst[1] = new int[]{0, 0, 20, 20, 24, 24, 8, 8, 24, 24, 20, 20, 0 };
    a.ics[2]= "polygon";a.cc[2]=Color.magenta;a.cb[2]=Color.black;
    a.xlst[2] = xx; a.ylst[2] = yy;  a.nam[0] ="body"; a.nam[1] ="body";
    a.nam[2] = "clockoutport";a.d[0]=0;a.d[1]=0;a.d[2]=0;
    Win wint = makeobject(3,a,x1,y1,scale); wint.name="clock"; wint.tag=27;
}
private static void wire(int x1,int y1, int scale){
    IconImage a = new IconImage(2);
    a.ics[0]= "polygon";a.cc[0]=Color.lightGray; a.cb[0]=Color.black;
    a.x[0]=8; a.y[0]=0;   a.x[1]= 8; a.y[1]=0;
    a.xlst[0] = new int[]{0, 24, 24, 0, 0}; a.ylst[0] = new int[]{0, 0, 24, 24, 0};
    a.ics[1]= "polygon";a.cc[1]=Color.blue;a.cb[1]=Color.black;
    a.xlst[1] = new int[]{10, 24, 24, 14, 14, 0, 0, 10, 10};
    a.ylst[1] = new int[]{ 0, 0, 4, 4, 24, 24, 20, 20, 0};
    a.nam[0] ="body"; a.nam[1] ="body";a.d[0]=0;a.d[1]=0;
    Win wint = makeobject(2,a,x1,y1,scale); wint.name="wire"; wint.tag=24;
}
private static void clockwire(int x1,int y1, int scale){
    IconImage a = new IconImage(2);
    a.ics[0]= "polygon";a.cc[0]=Color.lightGray; a.cb[0]=Color.black;
    a.x[0]=8; a.y[0]=0; a.x[1]= 8; a.y[1]=0;
    a.xlst[0] = new int[]{0, 24, 24, 0, 0}; a.ylst[0] = new int[]{0, 0, 24, 24, 0};
    a.ics[1]= "polygon";a.cc[1]=Color.magenta;a.cb[1]=Color.black;
    a.xlst[1] = new int[]{10, 24, 24, 14, 14, 0, 0, 10, 10};
    a.ylst[1] = new int[]{ 0, 0, 4, 4, 24, 24, 20, 20, 0};
    a.nam[0] ="body"; a.nam[1] ="body";a.d[0]=0;a.d[1]=0;
    Win wint = makeobject(2,a,x1,y1,scale); wint.name="clockwire"; wint.tag=28;
}
```

```
private static Win makeobject(int num,IconImage a,int nx1,int ny1,int sc){
    Point [] plst = null;  Picture pict =null;  int xx= nx1; int yy= ny1;
    for(int i=0;i<num;i++){
        xx=a.x[i]*sc+nx1; yy= a.y[i]*sc+ny1;
        plst = new Point[a.xlst[i].length];
        for(int j=0; j<a.xlst[i].length;j++){
            Point pp= new Point(2);
            pp.n[1]= a.xlst[i][j]*sc; pp.n[2]= a.ylst[i][j]*sc; plst[j]=pp;
        }
        Polygon poly= new Polygon(a.xlst[i].length);
        poly.p=plst;
        Rectangle r= new Rectangle();
        if(a.ics[i].equals("polygon")){
            Color c[][] = dW.c.polygonfill(poly,a.cc[i],a.cb[i],r);
            pict = new Picture(IO,c,a.cc[i],"polygon",poly ,xx,yy,c.length,c[0].length);
        }else if(a.ics[i].equals("shape")){
            Color c[][] = dW.c.shapefill(poly,a.cc[i],a.cb[i],r);
            pict = new Picture(IO,c,a.cc[i],"shape",poly ,xx,yy,c.length,c[0].length);
        }
        Win winb = makeWindow(pict,a.cc[i],false);
        winb.dir = a.d[i]; winb.name = a.nam[i];
        winb.selected=true; winb.pic.cb=a.cb[i]; winb.tag = i+1;
    }
    Win wint = makegroup(null);
    return wint;
}
```

This scheme sets up the icons needed to build a circuit. The *ecadicons()* procedure activates the icons so a circuit can be built in the display space. The tag values are the links to the procedures that execute the various commands. The tag value of 0 covers any click on the background window. The tags between 1 and 11 cover the drawing commands developed in the beginning of this chapter. The tag value of 12 returns the system to text window input commands. The new icons with tags in the range 20 to 30 create circuit-building icon-commands in the display. The tag values from 100 upwards, cover the actions of the circuit elements themselves. In the code given below this covers the in-port and out-port circuit elements which can have the logic value they represent toggled when clicked by the mouse to interactively provide different input signals to the circuit, and the clock input button.

The procedures used to generate the circuit elements are extensions of the copy command with the exception of the *wire* and *clockwire* commands and the *circuiticon* and *runcircuit* commands. The *wire* and *clockwire* commands however are more complicated and allow the wiring elements to be placed and linked together in the way illustrated in Figure 16.24. As the circuit elements are placed in the circuit layout these commands also have to set up the corresponding function objects needed to build the computer model that will be used to simulate the behaviour of the final circuit when the *runcircuit* command is given.

```
private static void ecadicons(){
    FunctionBlock circuit = new FunctionBlock( IO,null,null);  circuit.name="component";
    Win win=null;  int circuitNumber =1;
    do{ListElement refwn=null;
        IO.writeString("please enter an icon command\n");
        Point p= new Point(2); win = (Win)(refwn = identifyWindow(p)).object;
        switch(win.tag){
            case 0:  break;
            case 1:  drawLine(); break;
            case 2:  drawRectangle(); break;
            case 3:  drawPolygon(); break;
            case 4:  raise(); break;
            case 5:  resize(); break;
            case 6:  drawshape(); break;
            case 7:  copy(); break;
            case 8:  repaint(); break;
            case 9:  move(); break;
            case 10: remove(); break;
            case 11: raise(); break;
            case 12: iconSet=false; break;
            case 20: andgate(circuit,refwn,p); break;
            case 21: orgate(circuit,refwn,p); break;
            case 22: notgate(circuit,refwn,p); break;
            case 23: latch(circuit,refwn,p); break;
            case 24: wire(circuit,Color.black,"wire"); break;
            case 25: inport(circuit,refwn,p); break;
            case 26: outport(circuit,refwn,p); break;
            case 27: clock(circuit,refwn,p); break;
            case 28: clockwire(circuit,Color.magenta,"clockwire"); break;
            case 29: circuit=circuitIcon(circuit, circuitNumber);circuitNumber++; break;
            case 30: runcircuit(circuit); break;
            case 31: circuitBlock(circuit,refwn,p); break;
            case 104: inportExec(circuit,refwn); break;
            case 105: outportExec(circuit,refwn); break;
            case 106: clockExec(circuit,refwn); runcircuit(circuit); break;
        } if (gridset) resetgrid();
    }while(iconSet);
    IO.writeString("returning to text command\n");
}
private static void andgate(ListElement refwn,Point p1){
    setComponent(refwn, p1,100,"andgate") }
private static void orgate(ListElement refwn,Point p1){
    setComponent(refwn, p1,101,"orgate") }
private static void notgate(ListElement refwn,Point p1){
    setComponent(refwn, p1,102,"notgate") }
private static void latch(ListElement refwn,Point p1){
    setComponent(refwn, p1,103,"latch") }
```

```
private static void inport(ListElement refwn,Point p1){
    setComponent(refwn, p1,104,"inport") }
private static void outport(ListElement refwn,Point p1){
    setComponent(refwn, p1,105,"outport") }
private static void clock(ListElement refwn,Point p1){
    setComponent(refwn, p1,106,"clock") }
Private static void setComponent(ListElement refwn,Point p1,int tag,String str){
    IO.writeString("please locate the "+str+" with the mouse\n");
    Point p2 = dW.getCoord(); int dx = p2.xi()-p1.xi(); int dy = p2.yi()-p1.yi();
    Win win= copyWindow(refwn,dx,dy);
    FunctionBlock fn = new FunctionBlock();
    win.fref=fn; win.tag = tag; entercopy(win); //enters the window into the display space
    fn.refwn=LofWs.start; fn.wn=win;
}
private static void outportExec(ListElement refwn){
    Win wint= (Win)refwn.object;
    List ls = (List)wint.pic.obj;  ListElement rf=ls.start;
    while(!((Win)rf.object).name.equals("led")){rf=rf.right;}
    if(rf!=null){
        Win win =(Win)rf.object;
        int nx1= wint.x1;int ny1=wint.y1;
        if(win.name.equals("led")){
            Picture pc= win.pic;  Polygon poly = (Polygon)win.pic.obj;
            Rectangle r= new Rectangle();
            if(win.cc==Color.red)win.cc=Color.green;  else win.cc=Color.red;
            win.pic.c = dW.c.shapefill(poly,win.cc,win.pic.cb,r);
            Color cb= win.pic.cb;  int x=pc.x, y=pc.y, w=pc.w, h=pc.h;
            dW.c.setPixels(0,0,x,y,w,h,win.pic);
        }
    }
}
private static void inportExec(ListElement refwn){
    Win wint= (Win)refwn.object;
    List ls = (List)wint.pic.obj; ListElement rf=ls.start;
    while(!((Win)rf.object).name.equals("button")){rf=rf.right;}
    if(rf!=null){
        Win win =(Win)rf.object; int nx1= wint.x1;int ny1=wint.y1;
        if(win.name.equals("button")){
            Picture pc= win.pic;  Polygon poly = (Polygon)win.pic.obj;
            Rectangle r= new Rectangle();
            if(win.cc==Color.red)win.cc=Color.green; else win.cc=Color.red;
            win.pic.c = dW.c.shapefill(poly,win.cc,win.pic.cb,r);
            Color cb= win.pic.cb; int x=pc.x, y=pc.y, w=pc.w, h=pc.h;
            dW.c.setPixels(0,0,x,y,w,h,win.pic);
        }
    }
}
```

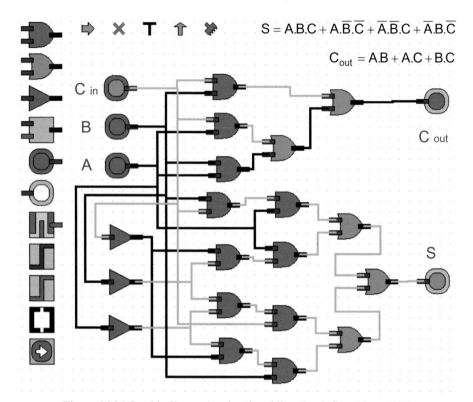

$$S = A.B.C + A.\overline{B}.\overline{C} + \overline{A}.\overline{B}.C + \overline{A}.B.\overline{C}$$

$$C_{out} = A.B + A.C + B.C$$

Figure 16.24 Graphically constructing the adding circuit from Figure 5.37

Function Blocks

As a circuit is being set up graphically, in the display space, a corresponding *FunctionBlock* is set up to capture its structure as a computer model for the circuit. This is a similar step to that taken when setting up the maze problems in Chapter 4, where the array of colours representing the maze was matched with an array of integers used to automate the maze solving process. In this case as each active element is placed in the display, a matching *FunctionBlock* to model its behaviour, is collected in lists in the overall circuit *FunctionBlock*.

```
class FunctionBlock{
    public Win wn = null; public ListElement refwn = null; // reference to associated window
    public int[] in= null;                              //list of input wire references
    public int[] out =null;                             // list of output wire references
    public List functions=null; public FunctionBlock[] fncts= null;   // list of function blocks
    public List wires=null; public Wire[] wrs= null;              // list of wires
    public List inPorts=null; public FunctionBlock  input[]=null;   // list of free inPorts
    public List outPorts=null; public FunctionBlock output[]=null;   // list of free outPorts
    public List clock=null;  public FunctionBlock[] clck=null;      // list of clocks
    public List clckwrs=null; public Wire[] clwrs= null;          // list of clock wires
    public String name ="functionblock";  public int tag=0;
```

```
public FunctionBlock(){ }
public FunctionBlock(TextWindow IO, Win w, ListElement rfn){ //constructor for circuits
    this.wn = w; this.refwn= rfn; this.IO=IO;
    functions=new List();          // list of function blocks in this function block
    wires=new List();              // list of wires in this layer of function blocks
    inPorts=new List();            // list of free inports
    outPorts=new List();           // list of free outPorts
    clock=new List();              // list of clock buttons
    name ="component";             // item name.
}
public FunctionBlock(TextWindow IO,String nam, Win w, ListElement rfn){
    this.wn = w; this.refwn= rfn; this.IO=IO;
    List ls = (List)wn.pic.obj;    // list of elements in the component's window icon.
    ListElement ref = ls.start;
    int i=0,j=0,k=0;
    while(ref!=null){        //give the function icon tabs identifiers for cross linking to wires
        Win tab = (Win)ref.object;
        if (tab.name.equals("inport")) {tab.tag=i; i++;}
        if (tab.name.equals("outport")){tab.tag=j; j++;}
        if (tab.name.equals("clockinport")){tab.tag=k; k++;}
        ref=ref.right;
    }
    in = new int[i]; out = new int[j]; // index links to circuit or component wire arrays
    this.name =nam; // item name.
}
```

Wire Links

Setting up the *make-wire* process has many possibilities. However, the complexity grows very fast as more options are added. In line with the policy adopted in earlier chapters the simplest schemes that illustrate the key ideas have been adopted. This involves starting a wire link at an *out-port* for a function block, a free in-port or on a an existing wire. Similarly the end of a wire is restricted to be linked, to either an *in-port* to a function block or a free out-port. The direction of a link to an *in-port* or an *out-port* has to be the same as that of the port tab in question. Wire links are made up from horizontal or vertical rectangles. The tricky part is to make sure that the links to in-ports and from out-ports are correctly aligned. These rules allow a wire to be constructed as a tree structure with one input and multiple outputs. This is in contrast to the simple function blocks, which can have multiple inputs but only single outputs.

The finite state diagram in Figure 16.25 defines the *make-wire* function. The main steps are relatively simple and the task of laying out the circuit in a reasonable way is left to the designer. However where links that are impossible to make are attempted a basic level of error checking is provided. The action that cannot be implemented is replaced by a "try again" message put out on the text window. The clockwires are placed using the same construction and display procedures the only difference being that the type of function blocks, they can link to, is restricted to latches, and the colour of the wires is magenta instead of black.

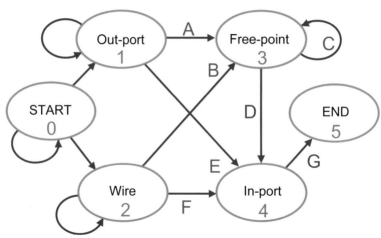

Figure 16.25 Wiring state transition diagram

```
private static int transitionA(Win wn, Point p2,Color cc){
    Point p1=new Point(2);
    if(p2.xi()>wn.x1) {
        p1.n[1]= (wn.x1+wn.x2)/2; p1.n[2]=(wn.y1+wn.y2)/2 ; p2.n[2]= p1.n[2];
        Win wnt= drawwire(p1,p2,cc); wnt.selected=true; wnt.name="wire";
        return 3;
    } IO.writeString("try again \n"); return -1;
}
private static int transitionB(Win wt,Point p1,Point p2,Color cc){
    if((wt.dir == 1)||(wt.dir == 2)){ p1.n[1]=(wt.x1+wt.x2)/2; }
    if((wt.dir == 0)||(wt.dir == 3)){ p1.n[2]=(wt.y1+wt.y2)/2; }
    int dir=direction(p1,p2);
    if((dir == 1)||(dir == 2)){ p2.n[1]= p1.n[1];  }else{ p2.n[2]= p1.n[2]; }
    Win wnt=drawwire(p1,p2,cc); wnt.selected=true;wnt.name="wire";  return 3;
}
private static int transitionC(Point p1,Point p2,Color cc ){
    Point pa=new Point(2); Point pb= new Point(2);
    int dir = direction(p1,p2);
    if((dir==1)||(dir==2)){p2.n[1]=p1.n[1];}else{p2.n[2]=p1.n[2];}
    Win wnt= drawwire(p1,p2,cc);wnt.name="wire";
    wnt.selected=true;return 3;
}
private static int transitionE(Win wt,Win wn,Color cc){
    Point p1= new Point(2); Point p2= new Point(2);
    if(wt.y1 == wn.y1){
        p1.n[1]= (wt.x1+wt.x2)/2; p1.n[2]=(wn.y1+wn.y2)/2 ;
        p2.n[1]=(wn.x1+wn.x2)/2;; p2.n[2]= p1.n[2];
        Win wnt= drawwire(p1,p2,cc); wnt.selected=true;wnt.name="wire";return 4;
    }IO.writeString("try again \n"); return -1;
}
```

```
private static int transitionD(Point p1, Win wn,Color cc ){
    Point pa= new Point(2);
    if((p1.yi()!=(wn.y1+wn.y2)/2)&&(p1.xi()<wn.x1-2)){
        pa.n[2]= (wn.y1+wn.y2)/2; pa.n[1]=p1.n[1];
        Win wnt= drawwire(p1,pa,cc);wnt.selected=true;wnt.name="wire";
        p1.n[1]=(wn.x1+wn.x2)/2;p1.n[2]=(wn.y1+wn.y2)/2;
        wnt= drawwire(pa,p1,cc);wnt.selected=true;wnt.name="wire";
        return 4;
    }else if(p1.xi()<wn.x1-2){
        pa.n[1]=(wn.x1+wn.x2)/2;pa.n[2]=(wn.y1+wn.y2)/2;
        Win wnt= drawwire(p1,pa,cc);wnt.selected=true;wnt.name="wire";
        return 4;
    } IO.writeString("try again \n"); return -3;
}
private static int transitionF(Point p,Point p2,Win wt, Win wn,Color cc){
    Point p1=new Point(2);
    if((wt.dir==1)||(wt.dir==2)){                                    //vertical
        p1.n[1]= (wt.x1+wt.x2)/2;  p1.n[2]= p2.n[2]; ;
        int k=transitionD(p1,wn,cc);if(k<0)return -2; else return k;
    }else{
        p1.n[1]= p2.n[1]; p1.n[2]= (wt.y1+wt.y2)/2;
        int k=transitionD(p1,wn,cc );if(k<0)return -2; else return k;
    }
}
private static Win drawwire(Point pa,Point pb,Color c){
    Point p1=new Point(2);  Point p2=new Point(2);  p1.c("<-",pa); p2.c("<-",pb);
    Color cc= c;  Color cb= c;    int minx,miny,maxx,maxy;
    int dir=direction(p1,p2);
    int w= p1.xi() - p2.xi(); int h= p1.yi() - p2.yi();
    if(w<0)w= -w;  if(h<0)h= -h;
    if(w<h){
        p2.n[1] = p1.xi();
        if(p1.yi()<p2.yi()) miny = p1.yi()-2; else miny = p2.yi()-2;
        minx= p1.xi()-2; w= 4; h=h+4;
    }else{
        p2.n[2] = p1.yi();
        if(p1.xi()<p2.xi()) minx= p1.xi()-2; else minx = p2.xi()-2;
        miny= p1.yi()-2; h= 4; w=w+4;
    }
    Picture pict =new Picture(IO,null,cc,"rectangle",null,minx,miny,w,h);
    if(cb!=null){
        for(int i=0;i<w;i++){pict.c[i][0] = cb; pict.c[i][h-1] = cb;}
        for(int j=0;j<h;j++){ pict.c[0][j] = cb; pict.c[w-1][j] = cb;}
    }
    Win win = makeWindow(pict,cc,true);   win.name="wire"; win.dir= dir; pict.cb=cb;
    return win;
}
```

```
private static Win wire(FunctionBlock circuit,Color cc,String type){
    String id="";  Point p1 =new Point(2);  Point p2 =new Point(2);
    ListElement ref1=null,saveref=null;
    int x1=0,y1=0,x2=0,y2=0,xa=0,ya=0,xb=0,yb=0;Wire wr=null;
    Win win=null,wnt=null,wint=null, wn=null,lstwn=null;  int state =0;
    do{
        Point p =new Point(2);
        ref1=identifyWindow(p);  win = (Win)ref1.object; if(win.tag == 10 )state=5;
        wnt = findElement(win,p);  if(wnt!=null)id=wnt.name;
        switch(state){
            case 0:  if(wnt!=null){
                        if(id.equals("output")||id.equals("outport")){
                            wn=wnt;                              // save for later
                            state = 1;
                            wr = new Wire(wnt.tag);          //outports for simple functions
                            FunctionBlock fb= (FunctionBlock)win.fref;
                            wr.tag=circuit.wires.length;
                            if(win.name.equals("component")) fb.output[wnt.tag].out[0]=wr.tag;
                            else fb.out[wnt.tag]=wr.tag;
                            circuit.wires.append(wr);
                        } else if(id.equals("wire")){ wr = win.wref;   saveref= ref1;  state = 2; }
                    }else state = 0; break;
            case 1:  if(wnt!=null){
                        if(id.equals("input")||id.equals("inport")){
                            state=transitionE(lstwn ,wnt,cc);
                        } else{IO.writeString("try again \n");state=-1;}break;
                    }else {state =transitionA(wn, p,cc);} break;
            case 2:  if(wnt!=null){
                        if(id.equals("input")||id.equals("inport")){
                            state = transitionF(p,p1,lstwn ,wnt,cc );
                        } else {state = -2; IO.writeString("try again \n");}break;
                    }else {state=transitionB(lstwn,p1,p,cc);} break;
            case 3:  if(wnt!=null){
                        if(id.equals("input")||id.equals("inport")){state= transitionD(p1,wnt,cc);
                        } else {state = -3; IO.writeString("try again \n");}
                    }else {state =transitionC(p1,p,cc );} break;
        }
        if(state<0) state= -state; else {p1=p; lstwn =wnt;}
    }while (state<4);
    if(state==4 ){
        wint = transitionG(saveref,type); wint.wref=wr;  wr.wn=wint;
        ListElement rf= LofWs.start; wr.winref=rf;
        if(win.name.equals("component"))win.fref.input[wnt.tag].in[0]=wr.tag;
        else((FunctionBlock)((Win)ref1.object).fref).in[wnt.tag]= wr.tag;
    }
    wint.wref = wr;   return wint;
}
```

```
private static Win transitionG(ListElement ref,String type){
    Win wint=null;
    if (ref!=null){
        Win win = (Win)ref.object;
        if(win.name.equals(type))wint = makegroup(ref);
        else wint = makegroup(null);
    }else wint = makegroup(null);
    wint.name= type;  wint.tag=2;  return wint;
}
public static int direction(Point p1,Point p2){
    int w =p2.xi()-p1.xi();  int h =p2.yi()-p1.yi();  int j=0; if(w<h)j=j+1; if(w< -h)j=j+2;  return j;
}
```

When the mouse is clicked on an *outport* tab for a function icon the tab is identified by searching the elements grouped in the single window-layer that is the function icon. The tab has had its tag set to identify which output it represents in the *out[]* array when its function icon was created and placed in the display. Each new wire is pushed down into the circuit's list of wires and its position in the list is placed in the *out[]* array of the function it is being linked to, ready for when the circuit is run as a simulation. In a similar way when the wire is terminated on an *inport* its location in the list of circuit wires is placed in the function's *in[]* array in the position that corresponds to the input tab the wire is linked to. If a wire is started, linked to an existing wire, then it adopts the identity of the existing wire. This arrangement ensures that each function block obtains references to the wires linking its *inports* and its *outports* as the circuit is constructed in the display space.

Circuit Simulation

Given the capability to represent the circuit graphically and structurally, the next task is to explore the way the same circuit can be modelled as a computer model that simulates its behaviour. There are again several levels at which this task can be done. The simplest is to treat the display model as an example of a more general diagram called a flow chart. This is possible because though the wires have no implicit direction to the way signals or electricity flows through them the gates or "function blocks" in this case do. If the internal elements in the gates the transistors and resistors were to be modelled then this directionality could not be assumed and a more complex modelling scheme would have to be adopted.

Each function block as the name suggests can be modelled as a function that takes the input values presented to it on its in-port tags and calculates the value to be placed on its out-port tag. This makes it possible to represent a whole circuit as a list of in-port values and a list of out-port values. One step in running the circuit will clearly be to take all the input values for a list of function blocks and calculate their output values that need to be placed in their out-port list of values. A second step is then to transfer the out-port values to the in-port cells that they are linked to by wires. If this is repeated until inputs at the front of the system have had a sufficient number of cycles to reach the backend of the system then the behaviour of the whole system with the appropriate values on all the wires will have been evaluated.

One way of achieving this in a general way is to keep cycling the transfers until there has been no change in any of the input values or output value. This will accommodate any circuit layout but clearly handles time in a rather unrealistic way. The order in which things occur will be correct but how long it takes will vary. One restriction has to be placed on this scheme. Feedback routes through the wires and function blocks have to be avoided or handled with care. It opens up the kind of problems that real circuits can exhibit producing *races, hazards* and *oscillations* that will never finish if the only terminating test is "*when nothing changes*". These circuit design problems can be found treated in more specialised texts.

One of the solutions to create correct behaviour with feedback loops is to introduce latches controlled by clock signals and to only allow feedback through loops containing a clocked latch. The circuits can still create oscillating behaviour but the results will be correct, and not depend on the arbitrary timing this simplified scheme can create when feedback is not controlled by the timing framework imposed by a clock signal.

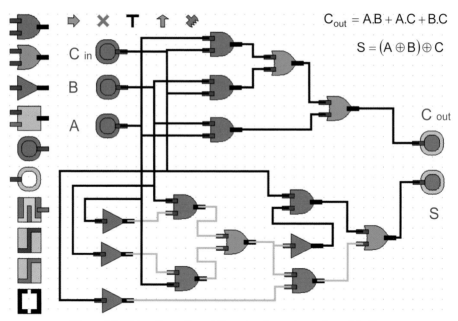

$$C_{out} = A.B + A.C + B.C$$

$$S = (A \oplus B) \oplus C$$

Figure 16.26 An alternative adding-circuit constructed graphically

For the circuit shown in Figure 16.26 setting up the associated computer model is relatively simple. When components that contain internal circuitry are introduced the task becomes more interesting! The nature of the computer model in this case appears to offer a relatively simple approach. The behavioural model of the overall circuit can be constructed from the graphic model, but it does not have to reflect the hierachical arrangments of components in the graphic display. This can be done because each wire can only hold a single value at any moment in time. If a connection is made up from a series of separate wire objects, then as long as they all refer to the same value-variable then, on one hand, there is no need to modify the graphic model, and on the other hand, to match the graphic model with the computer model element by element. Consider the circuit shown in Figure 16.27.

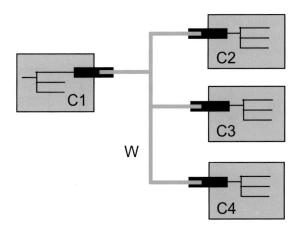

Figure 16.27 Hierarchical wiring problem

In this diagram there are five distinct wiring elements in the graphic model. The internal wires in circuit C1 linked to the outport shown; the three internal wires linked to the input ports in C2, C3 and C4; and the external wire labelled W linking these elements together. When the wire W is set up it is started from the output port for C1. Because this port is linked to the internal wires C1w the reference this wire has to its value variable can be transferred to the new wire W and then from W to all the internal wires C2w C3w and C4w linked to the input ports that W is subsequently attached to, giving the arrangement shown diagrammatically in Figure 16.28. This way of handling the wires should makes it easy to compose the computer model made up from collections of basic gates and subcircuit components. It merely consists of concatenating the function-block lists for all the components into a single list and doing the same with the lists of wires.

The difficulty with this approach arises if specific references are made from the function blocks to the wires that link them together. If circuit components need to be duplicated it makes it necessary to duplicate these references. This could be done using the object-oriented facilities of the Java language setting up a new class for each new circuit, but this requires a complex setting up process. A related scheme using the facilities of Java that have already been introduced can be set up based on separating the data that identifies a particular circuit from the functions that are needed to simulate its actions. Using arrays rather than linked lists to hold the references to wires, used by the functions to access the values held on particular wires, does this. A logic function *and*() can then be represented by the combination of the input values specified by array location indexes for the array of wires, rather than direct references to the wires in question.

```
and(){  wire.value[0][out[0]]= wire.value[1][in[0]]&&wire.value[1][in[1]];}
```

Where *out[0]* and *in[0]* and *in[1]* hold indexes to the array of wires that represent this particular instance of the circuit in question. This makes it possible to duplicate function blocks of the same type as the same object but with different value pairs for each of its wires. In other words there is a class of *FunctionBlocks* where new *functionblocks* are objects with a different internal arrangement of wires and basic functions, but duplicates of the same type of *functionblock* are the same class object but distinguished by being associated with different *wire data-values*. Essentially it is the wire values that define the duplicated circuit components.

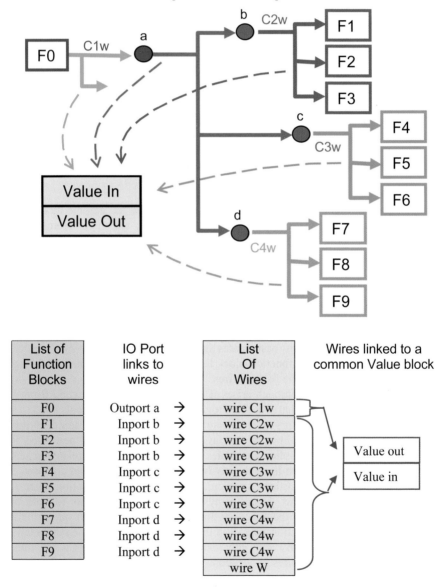

List of Function Blocks	IO Port links to wires		List Of Wires	Wires linked to a common Value block	
F0	Outport a	→	wire C1w		
F1	Inport b	→	wire C2w		
F2	Inport b	→	wire C2w		
F3	Inport b	→	wire C2w	Value out	
F4	Inport c	→	wire C3w	Value in	
F5	Inport c	→	wire C3w		
F6	Inport c	→	wire C3w		
F7	Inport d	→	wire C4w		
F8	Inport d	→	wire C4w		
F9	Inport d	→	wire C4w		
			wire W		

Figure 16.28 Labelling wires

```
private static void runcircuit(FunctionBlock cb){
    FunctionBlock c= makecircuit(cb,false);
    boolean exit = false;
    while(!exit){
        for(int i=0;i<c.fncts.length;i++){ c.fncts[i].execute(dW,c.wrs); }
        exit = true;
        for(int i=0;i<c.wrs.length;i++){
            Wire wr= c.wrs[i];
            if(wr.transfer())exit=false;
            if(wr.v[1])repaintWindow(wr.wn,Color.red);
            else repaintWindow(wr.wn,Color.green);
        }
    }for(int i=0;i<c.output.length;i++){
        boolean a = c.wrs[c.output[i].in[0]].v[1];
        outportExec(dW,c.output[i].wn,a);
    }
}
```

A circuit set up in this way can then be simulated by two loops the first evaluating the functions and the second transferring values from one end of a wire to the other, in the way shown for the adding circuits in Figures 16.29 to 16.32. A simple solution to handling hierarchically structured circuits is to convert the free- imports and free outports into function elements that transfer inputs to outputs.

```
public void execute(DisplayWindow dW,Wire[] wr){
    boolean exit = false;
    if(this.name.equals("component")){
        for(int i=0;i< this.input.length;i++){ this.input[i].linkIn(this.wrs,wr); }
        while(!exit){
            for(int i=0;i<this.fncts.length;i++){ fncts[i].execute(dW,this.wrs); }
            exit = true;
            for(int i=0;i<wrs.length;i++){
                Wire wrr= this.wrs[i];
                if(wrr.transfer())exit=false;
            }
        }
        for(int i=0;i<this.output.length;i++){ this.output[i].linkOut(this.wrs,wr); }
        return;
    } if(this.name.equals("andgate")){
        boolean a=false,b=false; a=wr[in[0]].v[1]; b=wr[in[1]].v[1]; wr[out[0]].v[0]= a&&b;
    } if(this.name.equals("orgate")){
        boolean a=false,b=false; a=wr[in[0]].v[1]; b=wr[in[1]].v[1]; wr[out[0]].v[0]= a||b;
    } if(this.name.equals("notgate")){ boolean a=false; a=wr[in[0]].v[1]; wr[out[0]].v[0]= !a; }
    return;
}
public void linkIn(Wire[] wrs, Wire[] wr){ wrs[out[0]].v[0] = wr[in[0]].v[1]; }
public void linkOut(Wire[] wrs, Wire[] wr){ wr[out[0]].v[0] = wrs[in[0]].v[1]; }
```

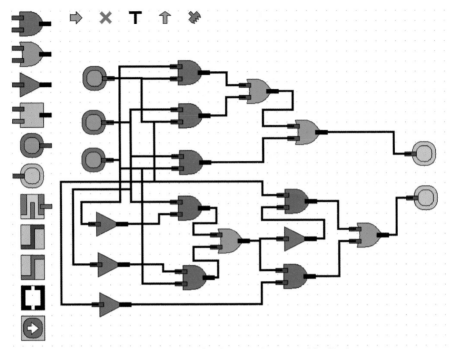

Figure 16.29 Adding circuit entered as a graphic display

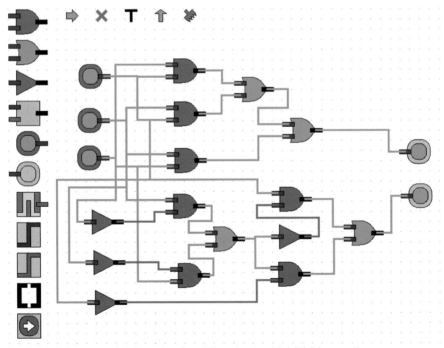

Figure 16.30 Running the adding circuit with initial input values

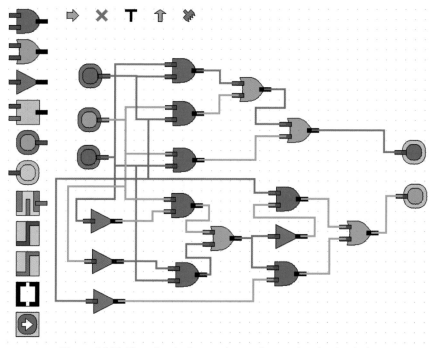

Figure 16.31 Changing the input values and rerunning the simulation

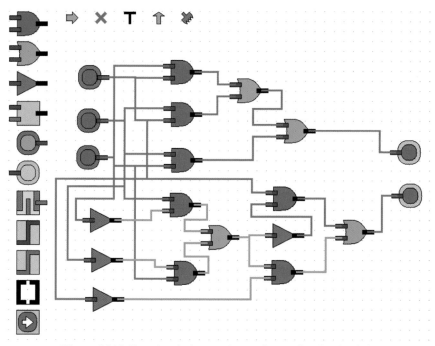

Figure 16.32 Changing the input values and rerunning the simulation

Encapsulating a Circuit as a Component-generating Icon

Once a simple circuit can be modelled in this way the next necessary step in developing a design system is to introduce hierarchy, in this case implementing the *circuit-icon* command. The system discussed so far is capable of setting up a circuit design of the form shown in Figure 16.26 for a one-bit adder. The annotation was added afterwards to explain the circuit. Although characters and therefore text can be generated in this system as Figure 16.9 demonstrated, only key operations are illustrated in these examples to keep the tasks as simple as possible. What is clear from the image in Figure 16.26 is that a relatively simple circuit has filled the display space. The next necessary step is to encapsulate circuit units such as this as circuit-icons so that more complex circuits such that shown in Figure 16.33 can be built up in the same display space. A circuit can be reduced to a simple box-icon with the appropriate number of in-port tags and out-port tags in the way shown in Figure 16.33.

```
private static void makeCircuit(int In, int Out,int circuitNumber,Point p ){
    int dim = In +Out + 1 +circuitNumber;
    int [] xx=new int[]{0, 12,  12,  0,  0};  int [] yy=new int[]{0,  0,  4,  4,  0};
    int [] xxx=new int[]{0, 4,  4,  0,  0};  int [] yyy=new int[]{0,  0,  4,  4,  0};
    int span =In; if(Out>span)span = Out;  int height = 12* span; int width = 24;
    IconImage a = new IconImage(dim);
    a.ics[0]= "polygon";a.cc[0]=Color.pink; a.cb[0]=Color.black;
    a.x[0]=8; a.y[0]=0; a.nam[0] ="body"; a.d[0]=0;
    a.xlst[0] = new int[]{0, width,  width,  0,  0}; a.ylst[0] = new int[]{0, 0,  height, height,  0};
    for(int i= 1;i<=In;i++){
        a.ics[i]= "polygon";a.cc[i]=Color.blue; a.cb[i]=Color.black;
        a.x[i]= 0; a.y[i]=12*i-8; a.nam[i] ="inport"; a.d[1]=0; a.xlst[i] = xx; a.ylst[i] = yy;
    } for(int i= 1+In;i<=In+Out;i++){
        a.ics[i]= "polygon";a.cc[i]=Color.blue; a.cb[i]=Color.black;
        a.x[i]= 28; a.y[i]=12*(i-In)-8; a.nam[i] = "outport"; a.d[i]=3; a.xlst[i] = xx; a.ylst[i] = yy;
    } for(int i=1+In+Out;i<=In+Out+circuitNumber;i++){
        a.ics[i]= "polygon";a.cc[i]=Color.red; a.cb[i]=Color.black;
        a.x[i]= 14; a.y[i]=8*(i-In-Out)-4; a.nam[i] ="id"+circuitNumber; a.d[1]=0;
        a.xlst[i] = xxx; a.ylst[i] = yyy;
    } Win wint = makeobject(dim,a,p.xi(),p.yi(),2);
    wint.name="circuit";  wint.tag=31;
}
private static void circuitIcon(){
    IO.writeString("circuiticon\n");
    IO.writeString("please enter the number of inports\n");
    int numIn = IO.readInteger(); IO.readLine();
    IO.writeString("please enter the number of outports\n");
    int numOut = IO.readInteger(); IO.readLine();
    IO.writeString("please use the mouse to locate the circuit\n");
    Point p = dW.getCoord();
    makeCircuit(numIn,numOut,2,p);
}
```

Executing the "*makeCircuit*" into an icon command is done using the "*encapsulate*" icon **[ ]**. For the example in Figure 16.33 this was initially setup using the *circuitIcon* procedure given above. This requests the number of in-ports and the number of out-ports from the text window in order to calculate the correct size for the icon. It also gives it an arbitrary circuit number: 2 in this case. This is shown by the two red squares on the icon, which allows the circuit component to be identified in a larger layout. The next stage is to replace a circuit constructed in the display space by such a circuit icon, in which case these parameters, obtained for test purpose from the text window, must be obtained from the displayed circuit *FunctionBlock*.

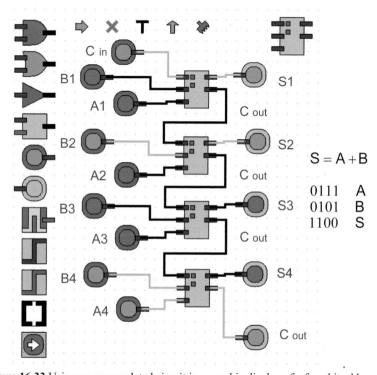

Figure 16.33 Using an encapsulated circuit in a graphic display of a four-bit adder

The ability to encapsulate a circuit makes it necessary to match the in-port and out-port tabs to the free in-ports and free out-ports in the original circuit before it is replaced by the new component, This is shown in progress in Figures 16.34 to 16.36. The reason this has to be done is that the order in which the original circuit is constructed will be reflected by the order in which components are placed in lists. This can be arbitrary and need not reflect the spatial order of the final layout that makes the unit easy to use. All circuit icons are given the tag value of 31. When any *circuit* icon is clicked by the mouse it calls the procedure *circuitBlock()* to generate a working copy of the circuit component it represents, with the new tag value of 110. This identifies the copy as a circuit element and not a component-generating icon, requiring a new, copied *FunctionBlock*. This unit can then be added to the list of functions in the current circuit as a new active component.

```
private static FunctionBlock circuitIcon(FunctionBlock fn, int circuitNumber){
    FunctionBlock fb= makecircuit(fn,false);  // convert lists to arrays leave the existing wires
    int numIn = fb.input.length;   int numOut= fb.output.length;
    IO.writeString("please use the mouse to locate the component\n");
    Point p = dW.getCoord(); Win wnn = makeCircuit(numIn, numOut, circuitNumber, p);
    ListElement refwn= LofWs.start;                          //window placed ontop of the pile
    fb.wn = wnn;  fb.refwn = refwn;  wnn.fref= fb;  fb.name="component";
    fb.input = new FunctionBlock[numIn];              //new arrays for input tabs
    fb.output = new FunctionBlock[numOut];  //new arrays for output tabs
    IO.writeString("please match new inport tabs with circuit free inports\n");
    ListElement ref = fn.inPorts.start;
    while(ref!=null){
        IO.writeString("click on the tab matching the highlighted inport\n");
        repaintWindow(((FunctionBlock)((ListElement)ref).object).wn,Color.cyan);
        Point p1 = dW.getCoord();  Win wnt = findElement(wnn,p1);        // new icon tab
        while((wnt==null)||!(wnt.name.equals("inport"))){
            IO.writeString("try again\n"); p1 = dW.getCoord(); wnt = findElement(wnn,p1); }
        fb.input[wnt.tag]=new FunctionBlock(IO,1,1,((FunctionBlock)ref.object).wn,ref);
        fb.input[wnt.tag].out[0] = ((FunctionBlock)ref.object).out[0];
        fb.input[wnt.tag].in[0] = 0;   fb.input[wnt.tag].name = "input";
        repaintWindow(((FunctionBlock)((ListElement)ref).object).wn,Color.yellow);
        ref=ref.right;
    }IO.writeString("please match new outport tabs with circuit free outports\n");
    ref = fn.outPorts.start;
    while(ref!=null){
        IO.writeString("click on the tab matching the highlighted outport\n");
        repaintWindow(((FunctionBlock)((ListElement)ref).object).wn,Color.orange);
        Point p1 = dW.getCoord();  Win wnt = findElement(wnn,p1);
        while((wnt==null)||!(wnt.name.equals("outport"))){
            IO.writeString("try again\n"); p1 = dW.getCoord(); wnt = findElement(wnn,p1); }
        fb.output[wnt.tag]=new FunctionBlock(IO,1,1,((FunctionBlock)ref.object).wn,ref);
        fb.output[wnt.tag].in[0] = ((FunctionBlock)ref.object).in[0];
        fb.output[wnt.tag].out[0] = 0;                   // to be defined externally
        fb.output[wnt.tag] = ((FunctionBlock)ref.object); fb.output[wnt.tag].name = "output";
        repaintWindow(((FunctionBlock)((ListElement)ref).object).wn,Color.yellow);
        ref=ref.right;
    } ref=fn.wires.start;
    while(ref!=null){ removeWindow(((Wire)((ListElement)ref).object).winref);ref=ref.right;}
    removeComponent(fn.functions.start);
    removeComponent(fn.inPorts.start);
    removeComponent(fn.outPorts.start);
    return new FunctionBlock(IO,null,null);
}
private static void removeComponent(ListElement ref){
    while(ref!=null){
        removeWindow(((FunctionBlock)((ListElement)ref).object).refwn); ref=ref.right; }
}
```

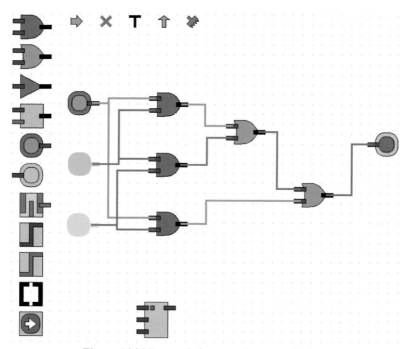

Figure 16.34 Encapsulating a one bit carry circuit

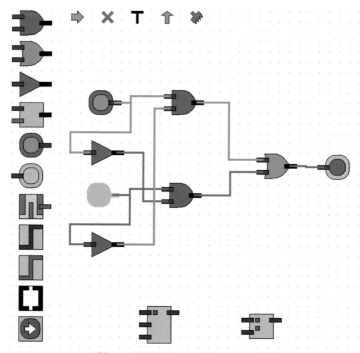

Figure 16.35 Encapsulating an exclusive or

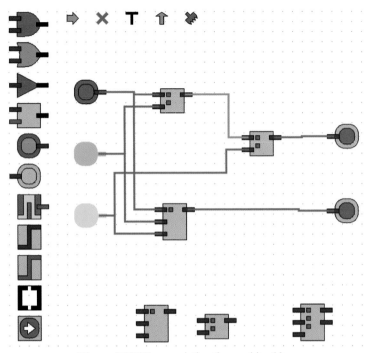

Figure 16.36 Encapsulating the one bit adder

Duplicating Circuit Components

```
private static void circuitBlock(FunctionBlock circuit,ListElement refwn,Point p1){
    IO.writeString("please locate the circuit with the mouse\n");
    Point p2 = dW.getCoord();
    int dx = p2.xi()-p1.xi(); int dy = p2.yi()-p1.yi();
    Win win= copyWindow(refwn,dx,dy);
    FunctionBlock oldfb = ((Win)refwn.object).fref;
    FunctionBlock fn = oldfb.copy(IO);
    fn.name = "component";  win.fref=fn;  win.tag = 110;
    entercopy(win);  fn.refwn = LofWs.start; fn.wn=win;
    circuit.functions.push(fn);
}
public FunctionBlock copyfunct(TextWindow IO, String nam, int[] ins, int[] outs){
    FunctionBlock fn = new FunctionBlock();
    fn.in = new int[ins.length];  fn.out = new int[outs.length];
    for(int i=0;i<ins.length;i++){ fn.in[i] = ins[i]; }
    for(int i=0;i<outs.length;i++){ fn.out[i] = outs[i]; }
    fn.name= nam;
    return fn;
}
```

Once a circuit icon has been set up it can be used to create working units in the way illustrated in Figures 16.34-36 sequentially developing the one bit adding circuit.

```
public FunctionBlock  copy(TextWindow IO){
    if(!this.name.equals("component")){
        FunctionBlock fn= copyfunct(IO,this.name,this.in, this.out); return fn; }
    int ff= this.fncts.length, ww= this.wrs.length;
    int ii= this.input.length, oo= this.output.length,  cc=0;
    if(this.clock!=null) cc= this.clock.length;
    FunctionBlock fn = new FunctionBlock(IO,ff,ww,ii,oo,cc,wn,refwn);
    for(int i=0;i<ff;i++){
        fn.fncts[i]= this.fncts[i].copy(IO); }  // needs new wires
    for(int i=0;i<ww;i++){ fn.wrs[i]= new Wire(0,"wire"); }
    for(int i=0;i<ii;i++){
        fn.input[i]=  this.input[i].copy(IO);}
    for(int i=0;i<oo;i++){
        fn.output[i]= this.output[i].copy(IO);}
    name ="component"; // item name.
    return fn;
}
```

These copy operations employ alternative *FunctionBlock* constructors:

```
public FunctionBlock(TextWindow IO,int f,int w,int i, int o,int c,Win wn, ListElement rfn){
                                            //constructor for components
    this.wn = wn; this.refwn= rfn; this.IO=IO;
    fncts=new FunctionBlock[f];             // list of function blocks in this function block
    wrs =new Wire[w];                       // list of wires in this layer of function blocks
    input=new FunctionBlock[i];             // list of free inports
    output=new FunctionBlock[o];            // list of free outPorts
    clck = new FunctionBlock[c];            // list of clock sources
    name ="component";                      // item name.
}
```

```
public FunctionBlock(TextWindow IO, int ins, int outs, Win w, ListElement rfn){
                                            //constructor for functions
    this.wn = w; this.refwn= rfn; this.IO=IO;
    in = new int[ins];
    out = new int[outs];
    List ls = (List)wn.pic.obj;             // list of elements in the component's window icon.
    ListElement ref = ls.start;
    int i=0,j=0;
    while(ref!=null){                       //setting up the IO tabs linked to wires
        Win tab = (Win)ref.object;
        if (tab.name.equals("input")) { in[i] = i;  tab.tag=i; i++;}
        if (tab.name.equals("output")){ out[j] = j; tab.tag=j; j++;}
        ref=ref.right;
    }
    this.name= w.name;
}
```

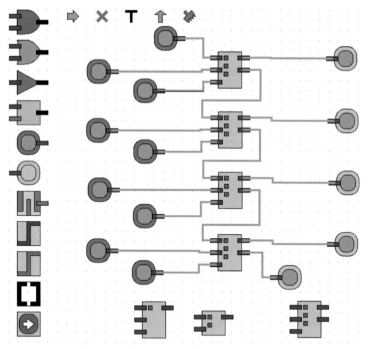

Figure 16.37 Running a four-bit adder circuit

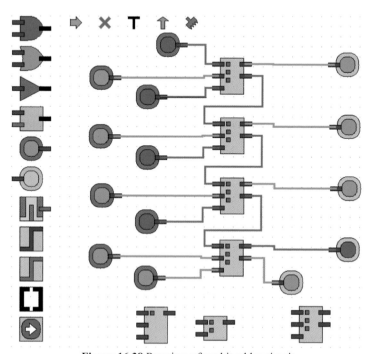

Figure 16.38 Running a four-bit adder circuit

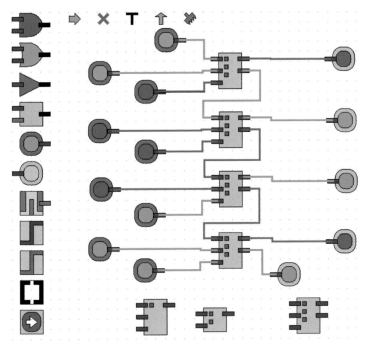

Figure 16.39 Running a four-bit adder circuit

Figures 16.37 to 39 illustrate the use of the new components to build a four-bit adder and then run it in simulation.

The final step in this sequence of examples is to implement latches and clock circuits to give a simple but complete logic design and simulation system. The clock-generating icon can be created and used in the same way as the in-port icon. It acts as a button, when it is clicked with the mouse it changes colour from yellow to green and back again. The difference being that the clock signal's effect, in this scheme is modelled directly. Rather than setting values on the clock wires used to link it to the latches it controls, it acts directly on the latches. When the clockwires are placed in the display linking the latches to the clock-generator, a list of the latches is created in the clock's function-block. When the clock button is clicked this list is used to activate all the latches it contains. When a latch is activated it transfers the value from the standard wire linked to its input tab to the wire linked to its output tab.

When the circuit is run the latches are processed as *FunctionBlocks* between the *Wire* value transfer-cycles, but they do nothing. It is only when they are activated by the clock circuit that data is transferred from their inputs to their outputs. The clock button, after reseting the latches, has to rerun the circuit to transfer the results through the rest of the circuit. The Figures 16.40 to 16.49 show a simulation of a four bit counting circuit using these facilities. The first step is to create the one bit adding and carry circuit for two inputs, shown in Figure 16.40 being encapsulated. The following sequence shows a count from zero to eight.

```
private static void clockExec(FunctionBlock cf,ListElement refwn){
    FunctionBlock c= makecircuit(cf,false);  IO.writeString("toggle input\n");
    Win wint= (Win)refwn.object;
    ListElement latchref = ((FunctionBlock)wint.fref).clock.start;
    if(latchref==null) return;
    List ls = (List)wint.pic.obj;  ListElement rf=ls.start;
    while(!((Win)rf.object).name.equals("button")){rf=rf.right;}
    if(rf!=null){
        while(latchref!=null){
            ListElement lstch = ((Wire)latchref.object).clk.start;
            while(lstch!=null){
                FunctionBlock latch= ((FunctionBlock)lstch.object);
                c.wrs[latch.out[0]].v[0] = c.wrs[latch.in[0]].v[1];
                lstch=lstch.right;
            }latchref=latchref.right;
        }
        Win win =(Win)rf.object;
        int nx1= wint.x1;int ny1=wint.y1;
        if(win.name.equals("button")){
            Picture pc= win.pic;  Polygon poly = (Polygon)win.pic.obj;
            Rectangle r= new Rectangle();
            if(win.cc==Color.yellow){win.cc=Color.green;} else {win.cc=Color.yellow; }
            win.pic.c = dW.c.polygonfill(poly,win.cc,win.pic.cb,r);
            Color cb= win.pic.cb;  int x=pc.x, y=pc.y, w=pc.w, h=pc.h;
            dW.c.setPixels(0,0,x,y,w,h,win.pic);
        }
    }
}
private static Win clockwire(FunctionBlock circuit,Color cc,String type){
    String id="";  Point p1 =new Point(2);  Point p2 =new Point(2);
    ListElement ref1=null,saveref=null;
    int x1=0,y1=0,x2=0,y2=0,xa=0,ya=0,xb=0,yb=0;Wire wr=null;
    Win win=null,wnt=null,wint=null, wn=null,lstwn=null;  int state =0;
    do{
        Point p =new Point(2);
        ref1=identifyWindow(p);  win = (Win)ref1.object; if(win.tag == 10 )state=5;
        wnt = findElement(win,p);  if(wnt!=null)id=wnt.name;
        switch(state){
            case 0:
                if(wnt!=null){
                    if(id.equals("clockoutport")){
                        wn=wnt;  state = 1;  wr = new Wire(wnt.tag,"clockwire");
                        FunctionBlock fb= (FunctionBlock)win.fref;
                        fb.clock.push(wr);
                    } else if(id.equals("clockwire")){
                        wr = win.wref; saveref= ref1;  state = 2; }
                }else state = 0; break;
```

```
            case 1:
                if(wnt!=null){
                    if(id.equals("clockinport")){state=transitionE(lstwn ,wnt,cc,type);}
                    else{IO.writeString("try again \n");state=-1;}break;
                }else {state =transitionA(wn, p,cc,type);} break;
            case 2:
                if(wnt!=null){
                    if(id.equals("clockinport")){state = transitionF(p,p1,lstwn ,wnt,cc,type );}
                    else {state = -2; IO.writeString("try again \n");}break;
                }else {state=transitionB(lstwn,p1,p,cc,type);} break;
            case 3:
                if(wnt!=null){
                    if(id.equals("clockinport")){ state= transitionD(p1,wnt,cc,type);
                    } else {state = -3; IO.writeString("try again \n");}
                }else {state =transitionC(p1,p,cc,type );}  break;
        } if(state<0) state= -state; else {p1=p; lstwn =wnt;}
    }while (state<4);
    if(state==4 ){
        wint =  transitionG(saveref,type);  wint.wref=wr;  wr.wn=wint;
        ListElement rf= LofWs.start; wr.winref=rf;
        wr.clk.push((FunctionBlock)((Win)ref1.object).fref);
    } wint.wref = wr;  return wint;
}
```

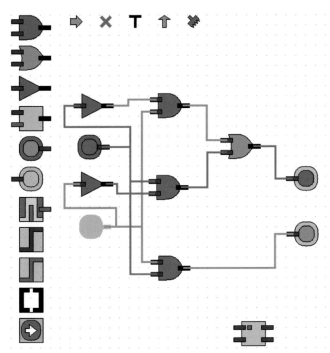

Figure 16.40 Creating a one-bit counting circuit component

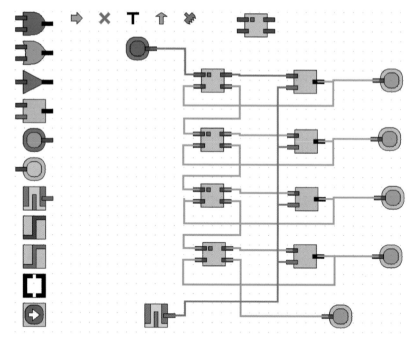

Figure 16.41 Setting up the four-bit counter

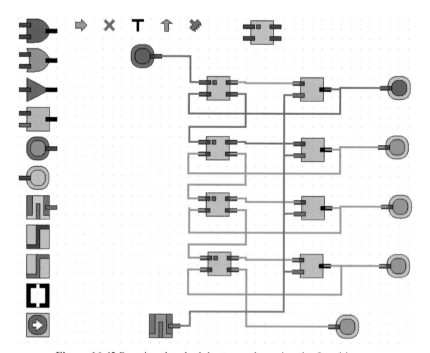

Figure 16.42 Pressing the clock button and running the four-bit counter

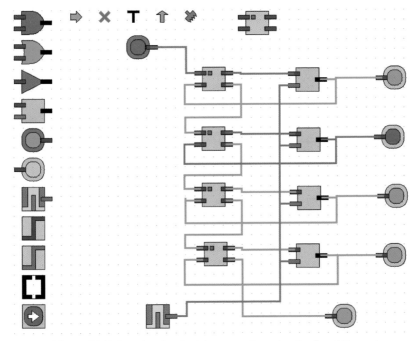

Figure 16.43 Pressing the clock button and running the four-bit counter

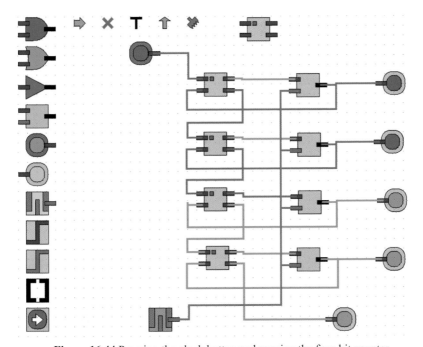

Figure 16.44 Pressing the clock button and running the four-bit counter

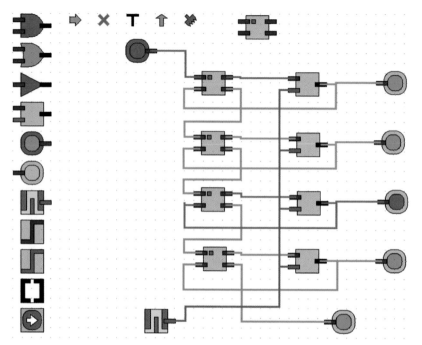

Figure 16.45 Pressing the clock button and running the four-bit counter

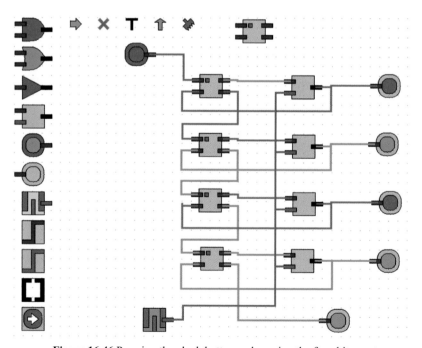

Figure 16.46 Pressing the clock button and running the four-bit counter

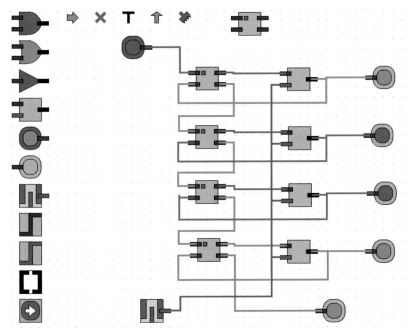

Figure 16.47 Pressing the clock button and running the four-bit counter

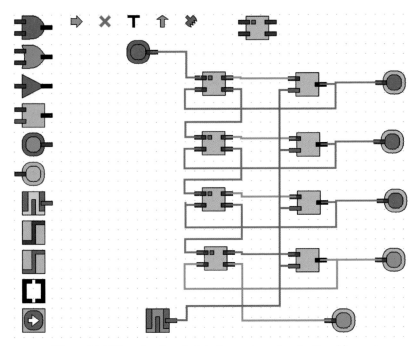

Figure 16.48 Pressing the clock button and running the four-bit counter

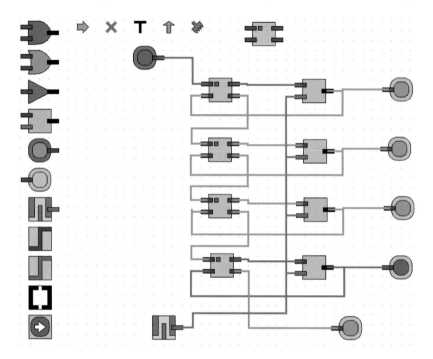

Figure 16.49 Pressing the clock button and running the four-bit counter

This presentation of the simulation employs conventional graphic images and to show the action of the circuit has to give a time series of nine separate Figures. If this system is run as a live simulation then the benefit of using computer graphics is clear. Though the graphics are the same in both cases, the use of the computer to model the display makes it possible to have one active image as opposed to many passive images to work with.

To make this scheme into a robust system will require more work. In its current state of development if a component is misplaced and deleted as an image the corresponding element in the computer model is not removed, and the whole circuit construction has to be started over again. What is presented demonstrates the kernel of an operational system and shows the relationship between the graphic model and the functional model and the way the two together give a more powerful modelling scheme.

The graphic model in this case is a variation on a flow chart, and consequently this form of graphics user interface can be used for a number of related applications, where a similar diagram can be used to represent a system. Computer programs themselves can be constructed within the same kind of framework, as can control system simulation models. Even mathematical equations and formulae can be presented in a related way where the structure of the display is directly coupled to the computer model of the equations and can consequently be used to manipulate or evaluate them.

Conclusions

In this book what has been covered is the way standard graphic production can be modelled by the computer system. Two approaches have been outlined the first based on line drawing techniques the second on infill and area painting effects. Both these approaches depend on primitive hardware operations that affect the way they are implemented. It is possible to scale and transform the line based models much more easily than the pixel array schemes. However the pixel array schemes, now that there is sufficient memory to support them, provide a powerful and fast range of image manipulation effects. The use of overlay and transparency, by matching standard manual picture building techniques, provides an intuitive and natural way of working with a picture building system.

Three-dimensional Models

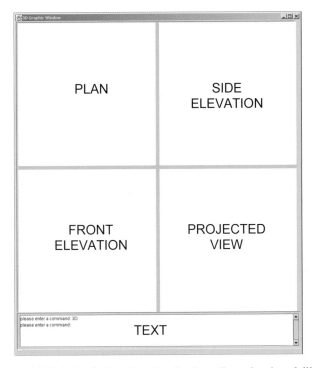

Figure 16.50 A Grapic User Interface for three dimensional modelling

The design of real world objects traditionally used sets of cross-referenced drawings as the most manageable way of handling the information. Three-dimensional physical models often had to be built to resolve complex spatial problems but they were cumbersome to store and still often required geometry worked out on paper to determine how they could be built. The automation of drawing was a great advantage when drawings needed to be edited and changed, but computer models of these drawings did not allow the properties of the final product to be modelled by the computer. Drawing sets still needed human interpretation to understand the three dimensional interrelationships they contained. It was only by changing the modelling process to build a three dimensional model and then working out how to generate the two dimensional images from it, for communication purposes, that the advantages of the new technology could provide the three dimensional designer the same advantages that have been demonstrated for the circuit simulation using a two dimensional diagram. The interface to generate these models still employs the two dimensional projections discussed in chapter 1 as a standard graphic model containing a plan, two elevations and a projected drawing in four connected windows shown in Figure 16.50. The development and use of three-dimensional models and their graphic display is explored in the next book.

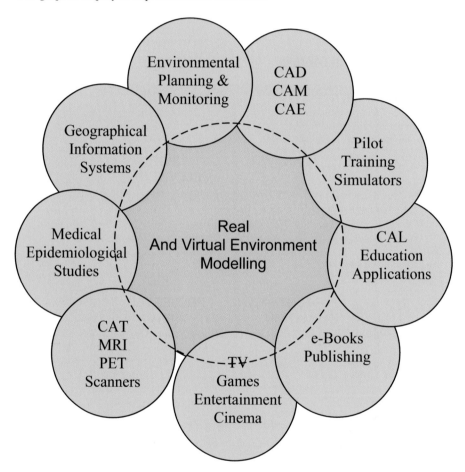

Figure 16.51 Application areas for real and virtual environmental modelling

The integration of graphic and computer modelling is now supporting many application areas. Figure 16.51 gives a selection of related topics that can be grouped round the subject emerging from these studies that has been given the title of "Real and Virtual Environment" modelling. The framework is provided by the computer system the content by computer models suitable for the application areas in question.

Index